BIOLOGY AND SYSTEMATICS
OF
COLONIAL ORGANISMS

THE SYSTEMATICS ASSOCIATION
SPECIAL VOLUME NO. 11

BIOLOGY AND SYSTEMATICS OF COLONIAL ORGANISMS

Proceedings of an International Symposium
held at the University of Durham

Edited by

G. LARWOOD

Department of Geology, University of Durham

and

B. R. ROSEN

Department of Palaeontology, British Museum (Natural History)

1979

Published for the
SYSTEMATICS ASSOCIATION

by
ACADEMIC PRESS · LONDON · NEW YORK · SAN FRANCISCO

ACADEMIC PRESS INC. (LONDON) LTD.
24–28 Oval Road
London NW1 7DX

U.S. Edition published by
ACADEMIC PRESS INC.
11 Fifth Avenue
New York, New York 10003

Copyright © 1979 by
THE SYSTEMATICS ASSOCIATION

All Rights Reserved

No part of this book may be reproduced in any form by photostat, microfilm, or any other means, without written permission from the publishers

Library of Congress Catalog Card Number: 78-18029
ISBN: 0 12 436960X

Printed in Great Britain at the
University Press, Cambridge

Contributors

BOSENCE, Mr D. W. J., *University of London, Goldsmiths' College, New Cross, London SE14 6NW, England.*

BUSS, Mr L. W., *Department of Earth and Planetary Sciences, The Johns Hopkins University, Baltimore, Maryland 21218, U.S.A., and Discovery Bay Marine Laboratory, Box 35, Discovery Bay, St. Ann, Jamaica, West Indies.*

CARLILE, Dr M. J., *Department of Biochemistry, Imperial College of Science and Technology, London SW7 2AZ, England.*

COOK, Ms P. L., *Department of Zoology, British Museum (Natural History), Cromwell Road, London SW7 5BD, England.*

CORNELIUS, Dr P. F. S., *Department of Zoology, British Museum (Natural History), Cromwell Road, London SW7 5BD, England.*

COULSON, Dr J. C., *Department of Zoology, University of Durham, Science Laboratories, South Road, Durham City, DH1 3LE, England.*

CRISP, Professor D. J., *N.E.R.C. Unit, Marine Science Laboratories, University College of North Wales, Menai Bridge, Gwynedd LL59 5EH, Wales.*

CROWTHER, Mr P. R., *Department of Geology, University of Cambridge, Sedgwick Museum, Downing Street, Cambridge CB2 3EQ, England.*

CURDS, Dr C. R., *Department of Zoology, British Museum (Natural History), Cromwell Road, London SW7 5BD, England.*

CURTIS, Professor A. S. G., *Department of Cell Biology, University of Glasgow, Glasgow G11 6NU, Scotland.*

DIXON, Ms F., *Department of Zoology, University of Durham, Science Laboratories, South Road, Durham City DH1 3LE, England.*

FEDOROWSKI, Dr J., *Polska Akademia Nauk, Zakład Paleozoologii Pracownia, 61–725 Poznań, ul. Mielżyńskiego, 27/29, Poland.*

FRY, Mr W. G., *Faculty of Sciences, Luton College of Higher Education, Park Square, Luton, Bedfordshire LU1 3JU, England.*

TEN HOVE, Dr H. A., *Rijksuniversiteit Utrecht, Laboratorium voor Zoölogische Oecologie en Taxonomie, Plompetorengracht 9–11, Utretcht 2501, Holland.*

HOWSE, Dr P. E., *Chemical Entomology Unit, Departments of Biology and Chemistry, Building 3, The University, Southampton SO9 3TU, England.*

HUGHES, Dr R. N., *Department of Zoology, The Brambell Laboratories, University College of North Wales, Bangor, Gwynedd LL57 2UW, Wales.*
JACKSON, Dr J. B. C., *Department of Earth and Planetary Sciences, The Johns Hopkins University, Baltimore, Maryland 21218, U.S.A.*
KAŹMIERCZAK, Dr J., *Zakład Paleobiologii, Polska Akademia Nauk, Al. Żwirki i Wigury 93, 02–089 Warszawa, Poland.*
KIRK, Dr N. H., *Department of Geology, Llandinam Building, University College of Wales, Aberystwyth, Dyfed SY23 3DB, Wales.*
LARWOOD, Dr G. P., *Department of Geological Sciences, Science Laboratories, University of Durham, South Road, Durham City DH1 3LE, England.*
NUDDS, Dr J. R., *Department of Geological Sciences, Science Laboratories, University of Durham, South Road, Durham City, DH1 3LR, England. But now at: Department of Geology, Trinity College, Dublin 2, Republic of Ireland.*
ORLOVE, Mr M. J., *School of Biological Sciences, University of Sussex, Falmer, Brighton, Sussex BN1 9QG, England.*
PETERSEN, Dr K. W., *Universitatets Zoologiske Museum Universitetsparken 15, 2100 København, Denmark.*
RICKARDS, Dr R. B., *Department of Geology, University of Cambridge, Sedgwick Museum, Downing Street, Cambridge CB2 3EQ, England.*
ROBERTS, Mr C. D., *20 High Street, Winfrith, Dorchester, Dorset, England.*
ROSEN, Dr B. R., *Department of Palaeontology, British Museum (Natural History), Cromwell Road, London SW7 5BD, England.*
RYLAND, Professor J. S., *Department of Zoology, University College of Swansea, Singleton Park, Swansea SA2 8PP, Wales.*
SABBADIN, Professor A., *Istituto de Biologia Animale, Università de Padova, Via Loredan 10, 35100 Padova, Italy.*
SHELTON, Dr G. A. B., *Department of Zoology, University of Oxford, South Parks Road, Oxford OX1 3PS, England.*
SKELTON, Dr P. W., *Jane Herdman Laboratories of Geology, University of Liverpool, P.O. Box 147, Liverpool L68 3BX, England. But now at: Department of Earth Sciences, The Open University, Walton Hall, Milton Keynes, UK7 6AA, England.*
WARNER, Dr G. F., *Department of Zoology, University of Reading, Whiteknights, Reading RG6 2AJ, England.*
WERNER, Dr B., *Biologischer Anstalt Helgoland, Zentrale 2000 Hamburg 50, Palmaille 9, Federal Republic of Germany.*

Preface

The Symposium, organized on behalf of the Systematics Association, on the Biology and Systematics of Colonial Organisms, was held in the University of Durham in April 1976. The help of the University and of the staff of Collingwood College and members of the academic, technical and secretarial staff of the Science Laboratories is gratefully acknowledged.

At the meeting specialists concerned with a wide range of living and fossil organisms exchanged ideas and opinions on the problems and concepts of coloniality. Treated in the broadest terms "coloniality" was intended also to cover a variety of associations, including gregariousness and aggregation, and thus a very wide range of life forms, relationships and conditions was compared. Similarities and differences in coloniality in many groups were considered and the implications for taxonomy examined in a great variety of fields including palaeontology, ecology, genetics, cell biology, physiology and behaviour and in growth, function and development.

The organizers are most grateful to the following chairmen of technical sessions: Dr A. G. Coates (Department of Geology, George Washington University), Professor D. J. Crisp (Marine Science Laboratories, University College of North Wales), Dr R. H. Hedley (British Museum (Natural History)), Professor J. S. Ryland (Department of Zoology, University College of Swansea). Thanks are due also to all speakers most of whom have contributed papers to the present volume.

During the meeting the following demonstration exhibits were presented: "Gregariousness and 'Reef' Formation in *Serpula vermicularis* L." by Mr D. W. J. Bosence (Geology Department, Goldsmith's College, London), "A Gregarious Tournaisian tube-builder, Gastropod or Polychaete?" by Mr T. P. Burchette (Department of Geology, The University, Newcastle upon Tyne), "Bacterial, Fungal and Slime Mould Colonies" by Dr M. J. Carlile (Department of Biochemistry, Imperial College, London), "Brood Chambers in Bryozoa" by Miss P. L. Cook (Department of Zoology, British Museum, Natural History), "Population structure in Lunulitiform Bryozoa" by Dr E. Håkansson (Institut for Historisk Geologi, Copenhagen), "Graptolites" by Dr R. B. Rickards (Department of Geology, Sedgwick Museum, Cambridge) and "Lamellar Growth in some Jurassic Bryozoans" by Dr P. D. Taylor (Department of Geological Sciences, University of Durham). The organizers extend their thanks to these exhibitors and to other contributors among the 87 participants at the meeting. The resulting present publication reflects their expertise and enthusiasm and the wide biological interests of coloniality particularly for invertebrate specialists, for marine biologists, ecologists and for palaeontologists.

One of the editors (Dr B. R. Rosen) has set out, in an extended introductory paper, the background, origin and aims of the Symposium and discusses the relationship of the papers presented here to the broad field of concepts generated in studies of the population biology of plants and concludes with a suggested terminological scheme for social organisms.

The editors wish particularly to extend their thanks to authors both for their written papers and for their forebearance in the long preparation period necessary for the volume. The valuable services of colleagues acting as referees are also gratefully acknowledged. Finally, thanks are due to the staff of Academic Press (London) for their contribution in realizing the publication.

G. P. LARWOOD

September 1978

Contents

LIST OF CONTRIBUTORS		v
PREFACE		vii
INTRODUCTION: Modules, Members and Communes		
B. R. ROSEN		xiii

1 The Affinity and Palaeobiology of Stromatoporoids: A Critical Review (Abstract)
 J. KAŹMIERCZAK 1
2 Bacterial, Fungal and Slime Mould Colonies
 M. J. CARLILE 3
3 Group Phenomena in the Phylum Protozoa
 C. R. CURDS 29
4 Individuality and Graft Rejection in Sponges, or a Cellular Basis for Individuality in Sponges
 A. S. G. CURTIS 39
 Coloniality (Addendum)
 A. S. G. CURTIS 48
5 Taxonomy, the Individual and the Sponge
 W. G. FRY 49
6 Coloniality in the Scyphozoa: Cnidaria
 B. WERNER 81
7 Development of Coloniality in Hydrozoa
 K. W. PETERSEN 105
 Colonies of Colonies in *Physalia* (Addendum to paper by Petersen)
 P. F. S. CORNELIUS 140
8 Co-ordination of Behaviour in Cnidarian Colonies
 G. A. B. SHELTON 141
9 On Some Aspects of Coloniality in Permian Corals
 J. FEDOROWSKI 155
10 Coloniality in the Lithostrotionidae (Rugosa)
 J. R. NUDDS 173

11	Some problems in Interpretation of Heteromorphy and Colony Integration in Bryozoa	
	P. L. COOK	193
12	Structural and Physiological Aspects of Coloniality in Bryozoa	
	J. S. RYLAND	211
13	Coloniality in Vermetidae (Gastropoda)	
	R. N. HUGHES	243
14	The Colonial Behaviour of *Modiolus modiolus* (L.) (Bivalvia), and its Ecological Significance (Abstract)	
	C. D. ROBERTS	255
15	Gregariousness and Proto-cooperation in Rudists (Bivalvia)	
	P. W. SKELTON	257
16	Different Causes of Mass Occurrence in Serpulids	
	H. A. TEN HOVE	281
17	The Factors Leading to Aggregation and Reef Formation in *Serpula vermicularis* L.	
	D. W. J. BOSENCE	299
18	Dispersal and Re-aggregation in Sessile Marine Invertebrates, Particularly Barnacles	
	D. J. CRISP	319
19	The Social Insects as a Special Case of Coloniality	
	M. J. ORLOVE	329
20	The Uniqueness of Insect Societies: Aspects of Defence and Integration	
	P. E. HOWSE	345
21	Aggregation in Echinoderms	
	G. F. WARNER	375
22	New Observations on the Mode of Life, Evolution and Ultrastructure of Graptolites	
	R. B. RICKARDS and P. R. CROWTHER	397
23	Thoughts on Coloniality in the Graptolithina	
	N. H. KIRK	411
24	Colonial structure and Genetic Patterns in Ascidians	
	A. SABBADIN	433
25	Colonial Breeding in Sea-birds	
	J. C. COULSON and F. DIXON	445
26	Habitat Selection, Directional Growth, and Spatial Refuges; Why Colonial Animals Have More Hiding Places	
	L. W. BUSS	459

27 Morphological Strategies of Sessile Animals
   J. B. C. JACKSON 499
AUTHOR INDEX 557
INDEX OF GENERA AND SPECIES 569
SUBJECT INDEX 579
SYSTEMATICS ASSOCIATION PUBLICATIONS 591

Modules Members and Communes: A Postscript Introduction to Social Organisms

B. R. ROSEN

Department of Palaeontology, British Museum (Natural History) London, England

ORIGIN AND SCOPE OF THE SYMPOSIUM

A specialist working on a colonial group of organisms is continuously impressed by the difficulties of applying concepts and methods used by colleagues who work with non-colonial organisms. This statement of course turns partly on definitions and they will be discussed later. For the moment, I am referring to coloniality as understood, for example, by coelenterate or bryozoan specialists (2.2b in Table I).

The superficial origin of these difficulties lies in the sheer dissimilarity between the organic groups concerned, but these difficulties may also have a more fundamental origin in human perception and cognition. It may be easier for us to envisage life processes in other beings which, like ourselves, are free to move in their environment in physical independence from each other, than it is to envisage life processes in an organism which is seemingly both one and many "individuals" at the same time, and which may even be attached to one spot. Perhaps the closest human experience is that of pregnancy, if aberrant occurrences such as siamese twins are ignored. Aspects of our social organization may provide certain parallels (a rowing team?), but in general, coloniality in the present sense is a very alien notation to us. As further, if extreme, support for this view, Bertalanffy (1952, in Mackie, 1963), in considering the converse concept of individuality, has stated that it

> originates in a sphere quite different from that of science and objective observation. Only in the consciousness of ourselves as beings different from others are we immediately aware of individuality that we cannot define rigidly in the living organisms around us.

Systematics Association Special Volume No. 11, "Biology and Systematics of Colonial Organisms", edited by G. Larwood and B. R. Rosen, 1979, pp. xiii–xxxv. Academic Press, London and New York.

There are therefore basic conceptual problems here which underlie any semantic difficulties. The origin of these problems can be linked to the view that we most readily develop theories and explanations in terms of what is already most familiar to us (Smith, 1975), often in a system of extended anthropocentrisms (Young, 1975). (It is notable that colonial organisms rarely if ever appear as characters in folklore and children's literature.) Our perception of life processes in, say, a slug or grasshopper, may even suffer from this without our always being aware of it. The advantages in day-to-day research terms however outweight the disadvantages in that a wider range of familiar experience provides a more accessible source of ideas (mistaken or otherwise) about the organisms concerned.

This familiarity provides a basis for inductive reasoning when a specialist working with a group of colonial organisms is at a severe disadvantage, since all generalizations "such as expectations of events and laws of nature" are inductively derived (Gregory, 1974). This does not mean that serious scientific ideas about non-colonial organisms are simply based on crude holistic anthropocentrisms and analogies with human existence, but rather that initial premises are constructed inductively from familiar elements which are then put to work in the deductive framework succinctly summarized by Gregory (1974). In non-colonial organisms (present provisional sense of 2.2a in Table I) this mental process is demonstrated in the papers in this volume by Orlove and by Coulson and Dixon, where human metaphors predominate in their premises. In contrast, understanding the life processes of integrated colonial organisms demands the kind of deductive approach demonstrated in the paper by Jackson. The conversely limited scope for inductive reasoning probably accounts for the essential difficulties of working with such organisms. Perhaps the most successful way of bringing colonial organisms within a familiar area is to regard them in some respects as higher plants. I shall return to this later.

The problems raised by colonial organisms are real enough whatever their philosophical or practical origins. In the first place one wonders about the functional significance of coloniality. This is especially so since the solitary habit is evidently successful (i.e. common and widespread), and in some groups like coelenterates it occurs side by side with the colonial habit even within single families or single environments (Fedorowski, this volume). Is coloniality indeed a single phenomenon, or is it really different things in different major organic groups? And if it is indeed several phenomena, not one, then is it necessary to look at the possibility that phenomena like gregariousness (2.2a in Table I) may intergrade with coloniality and need to be distinguished in their own right? Clearly there exists a case for specialists in different groups to talk to each other about these problems, without organizers prejudging the issues and thereby limiting the range of possible contributions at such a meeting. Apart from anything else, the word "coloniality" and its conjugates are used very broadly and any attempt to be more rigorous should emerge from as wide a range of users as possible.

It was thoughts like these that suggested this symposium and the idea was presented to the Systematics Association in October 1973. At about this time a volume on the subject of coloniality in animals (Boardman et al., 1973a) was published in the U.S.A. The present organizers did not learn of this until after the suggestion had been formally adopted by the Systematics Association, though some of us were certainly aware of the extended seminar on the subject which later led to the American symposium and its volume (Boardman al., et 1973b). This volume is a stimulating collection of papers by authors who are mostly based in North America, and when their volume appeared, the Systematics Association felt nonetheless that European specialists might also like to have the opportunity to exchange ideas around the same broad subject, and perhaps even to reply on particular points. The scope of the present symposium as a whole differs in being generalist (13 major organic groups and 30 contributions) rather than specialist (4 major organic groups and 24 contributions).

Our basic aim was to explore the nature of coloniality in as many different groups of organisms as possible, with contributors using the term in whatever way they understood it. At the inevitable risk of producing too generalized or superficial views of any single major group of organisms, the symposium was therefore conceived as a group-by-group survey, presenting a combination of current research findings and reviews which go beyond routine literature surveys. It is hoped that readers who consult this volume with interest in a particular group bear in mind that the primary aim was to review different aspects o coloniality, not to review the groups except as a means to this end. Thus not every major group is covered by a review (e.g. Anthozoa, and molluscan groups). It would take several volumes to serve both purposes thoroughly. Reviews, some of which are linked to a particular emphasis, are given by Carlile (bacteria, fungi, and slime moulds), Curds (protozoans), Crisp (marine invertebrates particularly barnacles), Fry (sponges), ten Hove (serpulids), Howse (social insects), Kirk (graptolites), Petersen (hydrozoans), Ryland (bryozoans), Warner (echinoderms). Two further papers (Buss, Jackson) commence with a marine ecological viewpoint and consider the implications for a whole range of marine organisms. In the remaining papers, the emphasis is on lower level taxonomic groups, or on particular investigations or approaches as outlined below.

One of the problems posed by symposia of this kind is that while they serve to bring specialists together and to gather diverse contributions on a special theme into a single book, genuine interactive discussion of the implications of everyone's papers is rarely achieved. There is simply not time for participants to absorb and then consider the salient points of all the contributions. Many important thoughts and ideas occur to people only after the meeting is over. This is undoubtedly good for their individual research, but it leaves only the editors (rather arbitrarily) to attempt to find the patterns and themes of a meeting. A communal viewpoint, or communally expressed range of viewpoints remains out of range. A possible solution would appear to be the preparation of a set of contributions to precede a meeting of its contributors and other participants. This would be made available about six months before the symposium. A second set of papers could then consist of contributions made by the same authors and others in response to the first set of papers.

It is customary for an introduction to a symposium volume to provide a guide to the contributions which follow. The themes of coloniality are so broad however, and the previous literature now so extensive, that I have chosen instead to give a personal view of just two key areas of interest:

1. Basic terminology and concepts of coloniality and individuality.
2. Function.

I have referred throughout to present authors' papers as well as to other work, and in this way I have tried to give readers a synthesis of the volume, within these particular topics. We have previously published a brief summary of the symposium (Larwood and Rosen, 1976).

TOWARDS SOME DEFINITIONS

1. Colonies

This symposium was convened to consider, above all else, the range and characteristics of coloniality in the widest sense of its usage. In the absence of any final discussion which attempted to offer definitions based on present contributions (or on the literature) readers will notice immediately, if they do not already know, that coloniality means different things to different authors (2.2a, 2.2b in Table I). Before attempting to resolve the different usages however, an adjacent term must first be distinguished, that of *colonialism*.

One of the few (we hope) examples where the editors of this volume disputed authors' terminology was in the use of the word *colonialism* (2.1 in Table I). All authors agreed to substitute *coloniality*, *gregariousness* (etc.) though the word may be found elsewhere in the coloniality literature. *Colonialism* is

Introduction

an altogether different phenomenon from anything considered here, not because it is largely used in a human context, but because the human social phenomenon to which it usually refers is almost opposite to the collaborative, integrated, communal or co-operative implications of *coloniality* in its general biological usage (2.2, 2.3 in Table I). *Colonialism* has recently (Andreski and Bullock, 1977) been described as the

> extension of the power of a state through the acquisition usually by conquest, of other territories; the subjugation of their inhabitants to an alien rule imposed on them by force, and their economic and financial exploitation [by that state]. A sharp and fundamental distinction [is maintained] between the ruling nation and the subordinate (colonial) populations.

To highlight the need for adequate terminological distinction from biological usage, there is the added difficulty that the subjugated population and territory are of course referred to as a *colony*. Phenomena parallel to the human one do appear to exist in species other than man, and Wilson (1975) includes a discussion of them in his Chapter 17, "Social symbioses". It might be argued that, if Wilson regards such symbioses as falling within his sociobiological heading (3 in Table I), then they, together with human colonialism, legitimately fall within the broadest meaning of *coloniality* (2.2 in Table I). The important difference is that in the human phenomenon, the emphasis is on intraspecific subjugation, whereas Wilson includes his examples of social symbioses because they are symbioses involving organisms which also have a social organization. His social symbioses moreover, are interspecific. *Colonialism* (2.1 in Table I) therefore remains a separate concept from *coloniality* (2.2 in Table I) and from Wilson's *sociality* (3 in Table I). Colonialism does have a place in discussions of coloniality but the two groups of ideas should not be confused or used interchangeably. They have unfortunately acquired etymologically similar names in English (a predicament comparable to "inflammable" and "non-flammable").

Someone determined to demonstrate that *coloniality* and *colonialism* were essentially the same, might look for intraspecific subjugation in the social insects, but the division of labour and origin of the social castes in these insects has a genetic, nutritional or hormonal basis (Wilson, 1975, Howse, this volume). This again distinguishes them from human colonialism.

The easiest way out of these terminological problems is to adopt Wilson's use of *society* for the general case (3 in Table I), and, notwithstanding the title of the present Symposium and volume, the embracing subject of the papers which follow is that of *sociality*. *Colony* can then be used as a particular case of *society* (3a.2, 3b.2 in Table I), as Wilson defines it (his p. 8). But if the conflict of usage of *colony*, as just discussed, is felt to be a major consideration, the solution would be to adopt *commune* and its conjugates for Wilson's *colony*. It is difficult to imagine biologists willingly adopting "coral communes" and "ant communes", but such terms would at least represent an approach to anthropocentric consistency in our terminology.

A second use of *colony* that also has to be separated from present usage is in its ecological meaning of an abstractly conceived space or a real habitat occupied by a population or a taxon (2.3 in Table I). This is directly derived from one of the Latin meanings of *colonia*, a "settlement of Roman citizens in a new or hostile country" (OED; 1 in Table I). The process of colonization is thus an act of occupation without necessarily implying any subjugation of other species or populations on the one hand, nor on the other hand, necessarily implying a communal or interactive process. *Colonization* covers a range of related phenomena from broad aspects of adaptive radiation to physical settlement on a substrate. Although it need have nothing whatever to do with social organization, and is altogether too broad in meaning for the present title, its use impinges considerably on present subject matter. Jackson considers, in effect, how "colonies" (2.2, especially 2.2b, in Table I) exhibit different adaptive strategies in "colonization" (in this broader sense of 2.3 in Table I). This theme is therefore also to be found in the papers by Bosence, Buss, Carlile, Coulson and Dixon, Crisp, ten Hove, Hughes, Skelton and Warner.

2. *Individuals*

Further thoughts on coloniality cannot progress without customary recourse to discussion of the converse idea of *individuality*. A colony can presumably be recognized if it can be seen to consist of some kind of repeated component unit, which is then regarded as an *individual*. The idea is simple enough, and raises no obvious problems with organisms which, like the social insects or gregarious and communal vertebrates, are not permanently united physically. Nor is there a problem with those organisms which live close together, perhaps in contact with each other, as do many barnacle species (Crisp), vermetids (Hughes), serpulids (ten Hove, Bosence) and the extinct rudistid bivalves (Skelton). The repeated unit in all these groups is still easily recognizable, and in the case of the first three, can be verified by observations of larval settlement and subsequent growth.

Difficulties arise in organisms which consist of organically united units, or if, as in the sponges (Curtis, Fry), one is trying to establish if such units are present at all. In some groups, units can be defined in terms of a solitary counterpart, if there is one. This is a unit which, if separated from the colony, could survive and regenerate another colony (cf. *ramet*, Harper, 1977. See p. xxx). Thus the polyps of a colonial coral are readily recognizable as a multiple version of a solitary coral. The apparently distinctive nature of the coelenterate polyp body-plan however is not an infallible indication of the coelenterate individual, as Shelton's remarks on chondrophores show (a comparable dilemma exists in the meandroid corals (Wells, 1973)). In the Bryozoa, there are no solitary counterparts (see Ryland's opening discussion), so unless there is ready comparison with the founder zooid (ancestrula), other criteria have to be found. The difficulties are reflected by the differing interpretations of many bryozoan structures, as reviewed by Ryland. Moreover, as with the Hydrozoa, the occurrence of polymorphism and repeated groupings of heterogeneous units (cormidia) present a choice of several kinds of hierarchically related individual. Most elusive of all, it seems, are the units of sponges (Fry, Curtis).

In order to tackle these difficulties, specialists in these particular groups have re-examined older ideas (for a review, see Mackie, 1963), in which a repeated structure can be found in almost any organism if one looks hard enough. Annelids might be thought of as colonies of segments, and metazoans as colonies of cells. As with bryozoan cormidia, hierarchies can also be recognized, but there is a risk that the search for repeated units might become an exercise in geometry removed from biological context.

Biological significance must be found, or at least postulated, for the repeated unit. To this end, many authors have tried to give a behavioural definition of individuality, in preference to or in addition to a morphological one. Three difficulties have arisen from this however. Firstly the word *individual* has become ambiguous, because in the absence of explanation, it may not be clear whether an author is referring to morphological or behavioural individuality. Different words are needed, as will be discussed later. Secondly, the essential character of individuality as a behaviour is that of autonomy, and it is clear from the literature, that, as with purely morphological units, it is possible to observe many levels of individuality, in this sense, within a single organism. Indeed many organisms can be viewed both as individuals and colonies at the same time. Beklemishev (1969) and many subsequent authors (see for example Boardman *et al.*, 1973b, Carlile, Nudds, Ryland, in this volume) in fact recognize a cyclical shift in autonomy from the solitary organisms to the individuals (so defined) within more primitive colonies, to advanced colonies with highly integrated and co-ordinated polymorphic individuals, which also possess autonomy as a complete colony. Mackie (1963) has called organisms which show this last development of coloniality, "superorganisms". Following Mackie, siphonophores are invariably cited as examples of this advanced state, though the bryozoans approach a similar level, held back it seems by their sessile habit (Ryland). Super-organisms are an illuminating idea but the paradox that defines them does not help in the search for a stable terminology for *individual* and *colony*.

The third difficulty picks up the anthropocentric thread of the opening section. *Individuality* eludes accurate behavioural definition probably because we can only really conceive it in terms of human experience, and we therefore have no real way of knowing how it applies to other organisms. (If the anthropocentric argument seems too removed from the main consideration, one might simply say that the behavioural approach is too subjective). As Mackie has pointed out in this connection, there is a danger that too great an emphasis on this approach leads to mystical rather than biological explanations, and I follow Mackie in leaving Bertalanffy (quoted earlier) with the final word on this problem.

As an alternative means of finding a biological basis for individuality, Mackie argues that the units should have evolutionary significance. Similarly, Dr R. P. S. Jefferies (personal communication) believes that the identical individual units of a colonial organism are those which "were once capable of independent existence, whether or not they still are. We deduce this by being able to homologize them with individuals that still exist separately, or, in the case of fossils, used to do so." Conversely a morphological feature like a vertebra or a worm segment is not an individual, even though it is repeated "because it never lives as a separate entity and never did." An evolutionarily defined individual is undoubtedly a valid objective, but many supposed evolutionary relationships are debatable especially in the lower invertebrates (e.g. Cook states that bryozoan brood chambers of similar appearance sometimes prove to be composed of non-homologous structures). It would still be useful moreover to have an unambiguous concept of *individual* whether or not one has established an evolutionary significance for the chosen unit.

I feel that one can only resolve the foregoing problems in defining *individual* by going back to the older literature and starting with morphologically defined repeated units. Initially, one can choose any morphological feature that seems potentially significant or interesting. The next step takes the common factor in both the foregoing behavioural and evolutionary approaches by using functional analysis of the chosen units to test, confirm and define their biological significance. A functional approach is in any case likely to find repeated units in many organisms, even without a rigidly morphological starting point (e.g. Cook). The scope for functional analysis is greatly increased moreover by population analysis of the repeated units (see Section on modules, taxonomy and function, p. xxx). The risk that the chosen morphological units might be superficial or trivial is justified so long as the units are being used for functional analysis of the organism concerned. In this respect units are really being used as a matrix for analysing function, or as part of an "operational model" (Fry, this volume). It follows that any organism that can be successfully viewed in this way is a colony (or society in Wilson's sense, above and 3 in Table I), if only provisionally for the duration of an investigation. Unsuccessfully defined units can be rejected, and organisms without any satisfactorily defined units should not be regarded as social.

The first advantage of this approach is that it does not depend on the extensive prior accumulation of knowledge implicit in designating units of evolutionary significance (though functional units are presumably of evolutionary significance, albeit unspecified in many cases). The second advantage is that it shifts emphasis away from the autonomy problems inherent in many behavioural arguments. As with evolutionarily defined units, autonomously defined units and functionally defined units may in fact correspond, but if Bertalanffy is right it is philosophically easier to specify different levels of function, than different levels of autonomously defined individuality.

The argument then is that individuality in the autonomous sense might be pursued in its own right without causing terminological confusion and paradoxes in defining the units within a colony. Autonomous corporate individuality in a particular organism does not prevent it being seen also as a social organization of units, if desired. All that is necessary is that units within organisms should not be called *individuals* unless one is simultaneously making a statement about individuality in this deeper sense ("individual" can of course be used in its usual adjectival way to mean one amongst several similar things: "individual zooid", "individual tentacle", etc.). The perennial conundrum of whether (for example) sponges, groups of siphonophore zooids or bryozoan cormidia are individuals or colonies, can therefore be regarded as a separate issue. Individuality and coloniality (sociality) are converse ideas but they should not be regarded as perfectly complementary concepts. The logical complement to a colony (society) is a repeated morphological unit with a known or supposed functional significance. It remains to find a term to distinguish it from *individual*.

When I was inviting contributors to the present symposium, I thought that it would be useful to find someone who would be prepared to discuss higher plants in the context of coloniality, but eventually decided that this might seem too removed from the central subject matter of the meeting. This was unfortunate, because it is apparent that the current approach to the population biology of plants (Harper 1977) answers many of the problems that have just been discussed:

> There are two levels of population structure in plant communities. One level is described by the number of individuals present that are represented by original zygotes....Such units will be called *genets*....An individual genet may be a tiny seedling or it may be a clone....Each genet is composed of modular units of construction—the convenient unit may be the shoot on a tree, the ramet of a clone, the tiller of a grass or the leaf with its bud in an annual.

Harper finds that,

> ...the populational structure of a genet is common in some animal forms, most notably the hydroids, corals and their allies....In the corals and colonial hydroids the population dynamics have, like higher plants, two distinct aspects of population growth: multiplication of the number of genets (or zygotes) and multiplication of the number of modules that compose a genet.

Harper also argues that this parallel can be extended to the social insects. (See also Lüscher (1955) in Mackie (1963)).

Harper does not state that higher plants are social organisms, but his unifying approach supports the present argument that a social organism might be defined most simply as one which consists of repeated units, rather than on considerations of individuality. By combining Harper's and Wilson's views, many higher plants become, in effect, social organisms, at least in their population biology.

It is worth noting that Fry (1970) has arrived at a similar conclusion from the viewpoint of a sponge zoologist:

> ...there can be no doubt that it is certain recent studies of sponges with the electron microscope which should compel us to pay more attention to the concept of sponges as populations of cells and their products....[This] has confirmed in the most striking way ...that it is the large numbers of interactions between individual cells which control the form and successful functioning of the sponge.
> Such a picture is highly reminiscent of mixed populations of whole organisms and as rigorous statistical procedures have proved so fruitful in ecology in the comparison of ecosystems, there seems to be no reason why some of the ecologists' statistical techniques should not prove fruitful in comparing populations of sponge cells and their products.

Consistent with this earlier paper Fry has concentrated in the present volume on populations of aquiferous units in sponges.

The idea that colonies (i.e. social organisms) are definable in terms of repeated morphological units has also been expressed by Finks (1973) in discussion of sphinctozoan sponges. Here, in a proof footnote to his abstract on the penultimate text page of the North American symposium volume, Finks arrives at the same terminology as Harper, for the repeated units: *modular units*. As it happens, his formal definition restricts *modular unit* to an aggregated structural unit having the form of an individual organism, but his text indicates that his approach is the same as that of Harper and Fry.

Harper, Finks and Fry (this volume) have thus (coincidentally?) all referred to repeated morphological units as modules or modular units. This is equivalent to bryozoologists' *members* (Cook) or *morphs* (Ryland). *Members* maintains anthropocentric consistency, but *modular unit* is best suited to structures and organisms, such as sponges, where it is difficult to find a feature as distinctive as a zooid or polyp. It is also better for referring to the general case within the whole phenomenon of sociality. *Members* on the other hand is the term which fits more easily into discussion of vertebrate groups, including ourselves.

Social organisms can therefore be defined as those which show modular structure or organization, or, using the phrase given by Harper (1979) in a recent lecture to the British Ecological Society, "iterative unitary construction" (3 in Table I). It seems that the problem of finding the "individual" in a colony can be resolved in an unexpected consensus of sponge and bryozoan zoologists and plant ecologists.

3. Genetic and Reproductive Aspects

Considerable attention has been paid to genetic definitions of *colony* and *individual*, especially in the context of the colony–individual paradox. Curtis provides a succinct summary of the genetic approach in his addendum. It will be argued that this approach throws more light on the concept of *individual*, as restricted here, than on *colony*. The principal difficulty is that of finding a genetic criterion for coloniality that does not exclude numerous modular organisms, and which avoids major inconsistencies. It is desirable nevertheless to find a link between modular and genetic approaches, which should give positive insights into some of the taxonomic problems that confront specialists in colonial groups.

A colonial organism defined genetically apparently must have cells of the same genotype (Curtis). Many authors have argued that this same criterion should be applied to the *individual*. Urbanek (1973) has pointed out that rigid adherence to this argument would have to include as "individuals" pairs or groups of completely separate and autonomous organisms like identical human twins and clonally produced but dispersed offspring. By the same reasoning, the only organisms that may be considered colonial using this genotypic criterion, are those whose units are produced in exactly this way. All non-clonal aggregations would be excluded, together with certain kinds of colony in which members are very closely related but not actually identical. The first difficulty could easily be overcome by terminology. The second is more serious, because there appear to be no universal thresholds of genetic relatedness by which one might classify degrees of sociality.

Sabbadin has for instance described colonies of the ascidian *Botryllus* in which units may be genotypically dissimilar. These have grown by fusion of two distinct growths, an event which occurs if their genotypes share particular alleles. It would be clearly fallacious to regard colonies showing evidence of heterogeneous fusion as belonging to a different social category from those which do not. It would also be impracticable to have to investigate every growth before assigning it to its category. This kind of growth appears to be relatively common in ascidians (see also Carlile's discussion of heterokaryons in Fungi). This contrasts with the general case of histoincompatibility reactions recorded by Curtis in sponges. In the absence of comparable work on many other colonial organisms, one therefore has to assume the possibility of colonies of species in other groups consisting of units or portions which are genetically heterogeneous (but see Thorpe's bryozoan work in Ryland).

An additional inconsistency is raised by the adjacent occurrence of compatible and incompatible strains within a cluster of sponges (Curtis). A genetic criterion would make it necessary to distinguish two orders of aggregation within such a cluster: those which are genetically similar and can fuse, and those which are dissimilar and cannot fuse. A parallel of different origin is found in the social insects. Colony members in the social Hymenoptera show a range of genetic relatedness ratios within a single colony (e.g. workers to each other: 3/4; queen to worker: 1/2) (Hamilton, Trivers, in Wilson 1975). Authors of very different disciplinary backgrounds nevertheless concur that the social insects are colonial (Harper, 1977; Schopf, 1973; Wilson, 1975), and there would seem to be no neat way of genetically distinguishing such colonies from other kinds of organic colonies by using a qualifying terminology based on degrees of relatedness.

In attempting to bring together the modular and genetic approaches to social organisms, one can commence with the restriction that social organisms be monospecific, in order to distinguish societies from interspecific associations, however intimate (e.g. lichens, but see Carlile). Such associations fall more easily into the conceptual framework of community ecology or parasitology.

The next step follows Beklemishev (1969) in distinguishing societies (present terminology) whose units are organically united from those whose units are not. I propose to refer to the first as *continuous modular societies*, and the second as *discontinuous modular societies* to replace the frequent use of *colony* and *aggregation* for these two phenomena (2.2a, 3a, 2.2b, 3b in Table I). The new terms can be used in conjunction with the older terms, redefined as explained below. It is not necessary at this level to specify degrees of co-ordination, nor to exclude from continuous societies those which do not show evidence of co-ordination as Curtis suggests. This second criterion would make life difficult for palaeontologists (and we are trying to be as unified as possible here) by excluding many extinct organisms by default. It also excludes continuous societies whose living units detach from each other, but remain in skeletal contact. This occurs in some corals (Nudds refers to various extinct examples). The polyps detach from each other in the course of budding but continue to live in a single communal skeletal structure. Polyps would be co-ordinated in only small numbers for a finite period following budding, but the skeleton has permanent colony-wide functions in spatial competition (Jackson), support and protection. It should also be possible to analyse growth of such corals in terms of the modular population dynamics of higher plants (Harper, 1977, see p. xxx).

In both continuous and discontinuous societies, genotypic variation can range from complete identity to the relatively large order of dissimilarity found in any population whose members derive from different zygotes, as already discussed. In general, cells within a particular continuous society are more likely to be genotypically homogeneous or very closely related, and cells within a particular discontinuous society are more likely to show lower relatedness.

Parallel to the genotypic spectra in both kinds of societies, there is a spectrum of co-ordination now well established in the approach to continuous societies following Beklemishev (1969). This author is a major influence on papers in the symposium edited by Boardman *et al.* (1973a), and also on parts of the present volume (e.g. Nudds, Ryland). Curds and Carlile (independently of Beklemishev) have been able to trace a co-ordination series with protozoan, microbial and microfungal groups.

A co-ordination spectrum however is also to be found in discontinuous modular organisms, as can be inferred from many of the present papers. Examples of discontinuous societies which are least socially organized and co-ordinated would include concentrations originating from large scale or frequent larval settlements of marine organisms in one place. Although this may be inherent in larval behaviour and therefore represent a low level of co-ordination, environmental factors are often important or more important than any inherent tendencies (e.g. some serpulids: Bosence, ten Hove). In groups like barnacles (Crisp) however, concentrations are a fundamental part of the reproductive cycle, and this may also apply to some vermetid molluscs (Hughes). Skelton implies a more advanced state of definite "proto-co-operation" in units which grow close together and provide mechanical support for each other (rudistid bivalves). More highly developed co-ordination can be found in the feeding behaviour of many echinoderms (Warner). Beyond this is the whole realm of organization, co-operation and co-ordination found in many vertebrate groups (e.g. Coulson and Dixon, on birds) and in the social insects (e.g. Howse). The highest social insects, like the more advanced continuous societies, also exhibit polymorphism of their modular units (i.e. castes). Wilson's (1975) synthesis of this whole subject offers a good introduction to the range of higher level organizations found in discontinuous social organisms including Man.

The essential nature of the co-ordination series in both continuous and discontinuous modular organisms is that both kinds of societies acquire an increasing amount of corporate identity and autonomy along the series, until their behaviour approaches that of a single non-modular organism (Beklemishev), i.e. with a corporate *individuality* in the sense restricted here in an earlier section. The propensity of members or modular units to work for the benefit of the whole society to a greater or lesser extent, implies their altruism and the evolution of this somewhat surprising quality has led to a fascination with the genetics of the altruistic tendency in kin selection (see Howse, Orlove).

I have already argued, from the ideas of Harper and Wilson in particular, that a *colony* need not be continuously modular (e.g. social insects). The most important character of colonies is that they exhibit co-operation between their members or modular units. This has a higher level of biological significance than whether units are actually attached to each other or separate (the criterion stressed by Beklemishev, 1969; Boardman et al., 1973b; Curtis, this volume). On the other hand the physical relationship of units to each other is directly observable, whereas co-ordination often has to be discovered by investigation. Moreover, as with the genetic spectrum it is not possible (at present?) to define thresholds of degrees of co-ordination that might be used to classify all organisms. A terminology is needed therefore which clearly expresses what is readily determinable (i.e. modular structure—because this can be recognized by the observer), but which is flexible with regard to what is completely intergradational. Thus a *society* is something which consists of modules (as before). Societies with great genetic similarity and/or high intermodule co-ordination or integration can be regarded as *colonies* (or preferably as *communes*), and those which exhibit low levels of either, or both, can be regarded as *aggregations* (or *populations*, see Carlile) (3a.1, 3b.1, 3a.2, 3b.2 in Table 1). These terms can then be qualified by *continuous* or *discontinuous* as necessary.

If a scale of co-ordination is required, one can apply the scheme given by Beklemishev (e.g. Coates and Oliver, 1973) for continuous modular organisms,

TABLE I. Current and suggested usage of coloniality and related terms. Further division of 3a.2/3b.2 is shown in Table II.

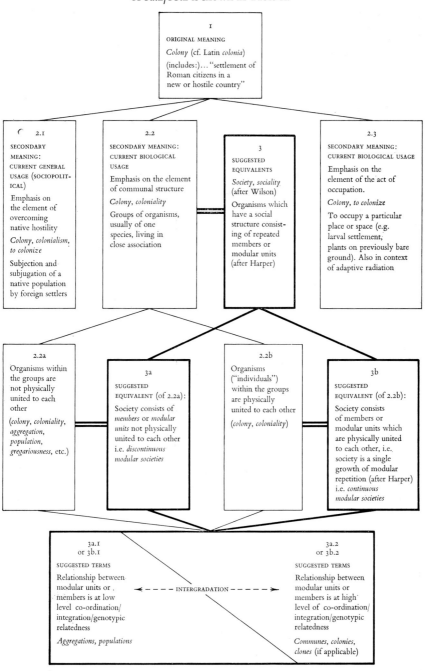

or Wilson's (1975) concept of the four peaks of social evolution which embraces both continuous and discontinuous modular organisms.

There are however two reasons why a single gradational scheme of co-

TABLE II. Suggested subdivision of coloniality (communes, colonies and clones) as shown in 3a.2/3b.2 of Table I.

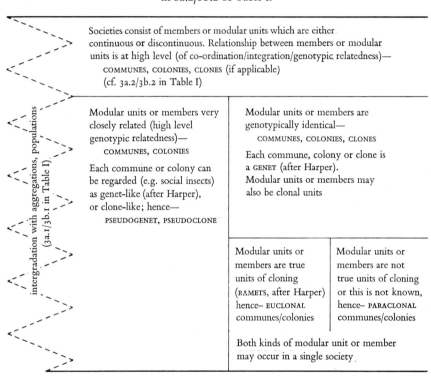

ordination is difficult to apply. Firstly, there are many levels of co-ordination within a single society, and these levels may even vary in time within the society (e.g. sponges; Fry); and human societies are certainly not universally and unvaryingly communal). Secondly, and this underlies the first point, assessment of degree of co-ordination between modules must obviously be related to choice of modules, especially in those species with several levels of modularity. This is most obvious in those polymorphic organisms like many bryozoans and some hydrozoans, which also have cormidial structures (p. xix). Many organisms moreover may even be continuously and discontinuously modular at the same time. That is, continuously social organisms may themselves occur in discontinuous aggregations (*sensu* 3a.1 in Table I). This is seen for example in stands of ecologically dominant vegetative organisms like higher plants and reef corals. Cornelius gives an example of what is perhaps the most extended range of modularity and co-ordination seen in any single organism, with his observation that the siphonophore, *Physalia*, habitually occurs in swarms (fleets?) of Portuguese Men o'War. Parallels occur in graptolites (synrhabdosomes; see Kirk, Rickards and Crowther), and in ant "supercolonies" (Howse).

To summarize so far therefore, one can recognize continuous and discontinuous societies which can each be loosely subdivided into the intergradational concepts of communes and aggregations on the amount of co-ordination and/or genetic overlap shown by their members or modular units (Table I). On the discontinuous side there is already an extensive vocabulary for such systems, especially for human societies and for the gregarious behaviour found in other vertebrates. In seeking a consistent terminology for all social organisms, I prefer *commune* to *colony* as a term signifying a high level of co-ordination between modular units. This is not only for anthropocentric consistency as discussed earlier, but because human societies should be accommodated in the present scheme, and human colonies (2.1 in Table I) are not to be confused with degrees of communality in human societies. It is doubtful however if *colony* would ever be ousted for most invertebrate groups, and there are many human communal structures which are not communes in the accepted sociopolitical sense (Labedz, 1977). But there is a problem of consistency here which deserves recognition.

Is it possible to offer further subdivision of sociality? Although I have stressed earlier that different modular organisms exhibit a range of genetic relatedness between their modules, a great number of them do fall into the special category of having units which are genotypically identical. This makes it possible to subdivide communes and colonies (3a.2 and 3b.2 in Table I) as shown in Table II. In discontinuous societies, this state is generally clonal in origin, and the term *clone* can obviously be retained without difficulty. Following Harper (1977) continuous modular units can also be thought of as clonal, but significant qualifications can be made here, one of which provides a direct link between *individual, modular unit* and genetic concepts of coloniality.

Curtis makes the point that a colony defined as a genotypically homogeneous (continuously modular) organism "differs in no respect from ... a unitary organism such as a vertebrate". He therefore makes the suggestion that the test for coloniality ought to be whether a separated unitary portion of the colony will regenerate a new colony. This test has apparently been rarely used (and he makes the stimulating prediction that many siphonophores may not show such regenerative properties). It is however fairly well documented in bryozoans

(Ryland gives examples). Beklemishev (1969) regards regenerating units as an indication of a stage in the evolutionary development of coloniality.

The idea of regenerating units is equivalent to the concept of the *ramet* in higher plants. This "is the unit of clonal growth, the module that may often follow an independent existence if severed from the parent plant" (Harper 1977). The ramet of a continuous modular organism, once identified, is a powerful candidate for the *individual*, and as such, should be much easier to identify and test than many other notions of *individual*. The ramet being the true unit of cloning, should also be used to distinguish *euclonal societies* from *paraclonal societies* and *pseudoclonal societies*, all these categories being either continuous or discontinuous (Table II). This system allows for the earlier discussed need for recognition that modular units may be freely chosen by the observer.

A society is *euclonal* if the chosen units are ramets. *Paraclonal* units share genotypic identity as do euclonal units, but they cannot regenerate (or are not yet tested for this). A single social growth can be referred to as both *paraclonal* and *euclonal* according to the modules being considered. *Pseudoclonal* can be used to refer to units which do not share genotypic identity, but are so closely related and co-ordinated that they can be treated as clone-like for certain ecological or functional purposes. This remains consistent with Harper's suggestion about the social insects, and also allows for heterogeneous coalescence in apparently continuous modular growths. The most likely source of complication is the possibility that more than one kind of unit within a clonal society may be able to regenerate, a suspicion aroused by horticultural propagation and by polymorphism.

It is significant with regard to clonal growth in continuous modular organisms that Harper (1977) has argued that asexual or vegetative reproduction is an unsatisfactory term. Their modular units are clonal and "formed by growth—not reproduction". They do not stem from a single initial cell. This argument is also supported by Young (1964) who has stressed that (sexual) reproduction and growth are very different orders of homeostatic response.

MODULES, TAXONOMY AND FUNCTION

Many of the taxonomic difficulties in continuous modular organisms arise from the enormous plasticity of the outward form of whole growths. This is an acute problem in sessile and vegetative groups like reef corals (Rosen, 1970; Veron and Pichon, 1976). No broad generalization can provide a solution to a particular taxonomic problem, but the modular approach opens up a whole range of established methodology, which should have application to taxonomy.

In continuous clonal organisms, the connection between function and form, is that the plasticity of the whole is achieved through the birth and death pattern of the modules (Harper, 1977). It should therefore be possible to extract generalized patterns from studies of phenotypic growths. Such patterns should be used as taxonomic characters additional to the established one of description of static morphological features, especially when the modules too may be very plastic as in reef corals (Wijsman-Best, 1972). The actual manner of budding of new modules is in itself genetically controlled (Harper, 1979), and provides a further range of characters. Use of budding pattern is already established in some groups (e.g. bryozoans) but in corals studies such as those by Fedorowski are still relatively rare. It is also worth remembering that the modules considered need not be restricted to the usual zooids or polyps. It should undoubtedly become regular procedure to include amongst a continuous modular taxon's characters, diagnostic aspects of its "structural demography" (Harper, 1979).

Function in modular organisms can be considered at two interrelated levels, that of the society as a whole, and that of the separate members or modules within the society. It would require undue space to draw together all the views expressed in this volume and in previous work on the subject of function in social organisms, but readers are invited to consider the present papers bearing these two levels in mind.

In general the functional significance of different kinds of module in certain groups (e.g. Bryozoa) is well understood, but there is still scope for the population approach. On the level of the function of the whole, one comes back to the question, "why be a modular organism?". For vegetative continuous modular organisms, the best summary answer is found once again in Harper (1977): "The population-like structure of an individual plant fits it admirably to respond to stresses by varying the birth rate and the death rate of its parts, leaves, branches, flowers, fruits, rootlets, etc." Response to "stress" can be extended to include the three major ecological strategies in plants defined by Grime (1977): resistance to stress, competitiveness and resistance to disturbance. Harper's statement and Grime's strategies should be equally applicable to sessile modular organisms other than higher plants. Thus one arrives at Jackson's paper, with ideas also derived from botany, but from a slightly different direction. His consideration of growth geometry, and growth rates of both the geometric whole and the individual parameters of the geometry, is an equivalent to Harper's structural demography. In practice, one would choose whichever methodology was most appropriate in the circumstances.

In an earlier section, I suggested that social organisms were those which were of modular construction, whose modules had functional significance. This by itself does not exclude worms (for example) as colonies or societies of segments. It is now possible to remove this difficulty by using the above approach of viewing modular units as a population contributing to the growth and reproduction of the whole organism or society in the manner stated by Harper (above). As before, observers can choose modular units how they wish, but a social organism should be one whose units can be shown to have this particular role. It remains to be shown how far this model of function in continuous modular organisms might apply to pelagic marine organisms (siphonophores, graptolites, salps, diatoms, etc.).

Reasons for sociality in discontinuous modular organisms are easier to construct, probably for the anthropocentric reasons already discussed. The literature is extensive. Warner's observation that most marine discontinuous modular organisms are filter feeders is of especial interest in understanding benthic marine sociality. Discontinuous sociality in the pelagic environment merits future attention as there is much information waiting to be interpreted in present terms.

It is commonly observed that amongst animals, continuous modular structure is typical of lower phyletic groups, and discontinuous modular structure is proportionately more typical of higher groups. This appears (intuitively) to be connected with a combination of size limitations and restricted functional range of the module body plan in lower organisms. Modular continuity is very often accompanied by polymorphism and is therefore thought to be equivalent to the development of organs in higher organisms (Mackie, 1963; Wilson, 1975). Greater overall size can also be achieved by modular repetition, and this would be critical in the development of strategies discussed by Jackson and Buss. In this way such organisms transcend the metabolic and mechanical limitations of the single module.

The tendency to become morphologically more integrated (e.g. corals: Coates and Oliver, 1973) may represent economy of tissue, especially where there is extensive biomineralization, and the energetics of coloniality therefore merit attention. In corals, the well known symbiosis with zooxanthellae would have assisted these developments, because of the effect of the zooxanthellae on growth rates and colony size. Werner believes however that in certain scyphomedusae, continuous modular growth developed in response to radiation into cooler waters, where temperatures are not high enough to permit continuous production of new medusae (strobilation) throughout the year, i.e. evolution of modularity by reproductive default.

Introduction xxxiii

In general higher organisms are much less frequently sessile, and can escape, or try to escape from adverse circumstances. Although they are often discontinuously modular, they are only rarely continuously modular, unless one includes their modular continuity at the cellular level. This view does at least contrive a continuation of the sociality series passing on from Wilson's highest level of "super-organisms" with their cormidial-based organs, into organic groups with cellular-based organs. Both share mobility and are considerably less plastic in their overall growth than continuous modular development in lower organisms.

SUMMARY

1. The general term for monospecific associations, often referred to as *colonies*, *aggregations*, etc., should be *societies*. *Colonialism* is rejected as a term having completely inappropriate implications.

2. Many of the difficulties encountered in the understanding of social organisms arise from a lack of a usable anthropocentric starting point. In particular, effort has been devoted in previous work to the idea of *individuality*, and this has led to a certain amount of confusion and terminological paradox.

3. A different approach to social organisms can be derived from the population biology of plants. The concept of the *modular unit* is an especially powerful one, and should be distinguished from the customary use of *individual*. *Modular unit or member* should replace *individual* as the working repeat unit of social organisms.

4. Social organisms can be separated structurally into those whose modular units are united to each other (*continuous modular* societies), and those whose units are separated (*discontinuous modular* societies). Both kinds of society show a complete intergradation of genotypic relatedness between modules, and of co-ordination and integration between modules. *Colony*, or preferably *commune*, should be used to indicate a high level in either or both series, and *aggregation* or *population* to indicate a low level in either or both series. Many organisms occur as discontinuous societies of continuous modular growths ("colonies of colonies"). These relationships, and their equivalents in existing terminology, are summarized in Table I.

5. A stable notion of *individual* is suggested by one particular kind of modular unit, the *ramet*, which is the unit of discontinuous cloning or of continuous modular regeneration.

6. Continuous modular organisms, by extension from plant ecology, are highly plastic in their overall growth habit. Their growths are a dynamic resultant of the life and death patterns of their member modules. Observers are free to define modular units as they wish, but truly social organisms consist of modular units which function as members of a population contributing to the growth and reproduction of the whole. At the moment this appears to apply more to those continuous societies which are sessile and vegetative than to those which are mobile. It appears to apply to discontinuous societies, especially to the social insects.

7. Modular units of functional significance in the population dynamics of continuous

modular growths, can provide a key to the understanding of the taxonomy of continuously modular invertebrate groups. Diagnostic features of their "structural demography" should be used to supplement more established characters.

Suggested terminological schemes are given in Tables I and II.

It is usual for symposium introductions and summaries to concentrate on the ideas expressed by its contributors. Here, I have found that many of the contributions can be linked by drawing on a number of stimulating ideas recently published elsewhere, in a field not conventionally regarded as "coloniality", that of the population biology of plants. In devoting considerable space to these ideas, I have tried, in effect, to write the paper which is "missing" from the volume. I hope contributors and readers will gain from seeing how the papers in this volume can thereby be drawn together notwithstanding their very broad subject spread.

ACKNOWLEDGEMENTS

I should like to thank the following friends and colleagues for helpful discussion, manuscript criticism and encouragement: Dr Dick Jefferies, Theya Molleson and Dr Noel Morris (Department of Palaeontology, British Museum (Natural History); Pat Cook, Dr Paul Cornelius, and John Peake (Department of Zoology, British Museum (Natural History); also my co-editor, Dr Gilbert Larwood, and my father, Professor Harold Rosen (Institute of Education, University of London).

REFERENCES

For the sake of completeness of citations, it should be noted that in addition to the works listed below, all the papers in this volume (see contents list) are also referred to in the text.

ANDRESKI, S. and BULLOCK, A. L. C. (1977). Imperialism. *In* "The Fontana dictionary of modern thought" (A. L. C. Bullock and O. Stallybrass, eds). pp. 301–303. Fontana Books, London.

BOARDMAN, R. S., CHEETHAM, A. H. and OLIVER, W. A., Jr. (eds), (1973a). "Animal colonies. Development and function through time". Dowden, Hutchinson and Ross, Inc., Stroudsburg, Pennsylvania.

BOARDMAN, R. S., CHEETHAM, A. H., OLIVER, W. A., Jr., COATES, A. G. and BAYER, F. M. (1973b). Introducing coloniality. *In* "Animal colonies. Development and function through time". (R. S. Boardman, A. H. Cheetham, W. A. Oliver, eds), pp. v–ix. Dowden, Hutchinson and Ross, Inc., Stroudsburg, Pennsylvania.

BEKLEMISHEV, W. N. (1969). "Principles of comparative anatomy of invertebrates. Volume 1. Promorphology". Oliver and Boyd, Edinburgh [translation of 3rd Edition in Russian, 1964, Nauka, Moscow].

COATES, A. G. and OLIVER, W. A., Jr. (1973). Coloniality in zoantharian corals. *In* "Animal colonies. Development and function through time". (R. S. Boardman, A. H. Cheetham and W. A. Oliver, eds), pp. 3–27. Dowden, Hutchinson and Ross, Inc., Stroudsburg, Pennsylvania.

FINKS, R. M. (1973). Modular structure and organic integration in Sphinctozoa. [Abstract]. *In* "Animal colonies, Development and function through time" (R. S.

Boardman, A. H. Cheetham and W. A. Oliver, eds), pp. 585-586. Dowden, Hutchinson and Ross, Inc., Stroudsburg, Pennsylvania.

FRY, W. G. (1970). The sponge as a population: a biometric approach. *In* "The biology of the Porifera." Symp. Zool. Soc. Lond. **25**, 135-162 W. G. Fry (ed.).

GREGORY, R. L. (1974). "Concepts and mechanisms of perception". Gerald Duckworth and Co. Ltd., London.

GRIME, J. P. (1977). Evidence for the existence of three primary strategies in plants and its relevance to ecological and evolutionary theory. *Am. Nat.* **111**, 1169-1194.

HARPER, J. L. (1977). "Population biology of plants". Academic Press, London.

HARPER, J. L. (1979). The significance of growth form and life-cycle in the population dynamics of higher plants [Lecture given to British Ecological Society]. *Symp. Br. ecol. Soc.* **19** (in press).

LABEDZ, L. (1977). Commune. *In* "The Fontana dictionary of modern thought" (A. L. C. Bullock and O. Stallybrass, eds). pp. 115-116. Fontana Books, London.

LARWOOD, G. P. and ROSEN, B. R. (1976). Facts of colonial life. *Nature, Lond.* **261**, 369-370.

MACKIE, G. O. (1963). Siphonophores, bud colonies, and superorganisms. *In* "The Lower Metazoa. Comparative biology and phylogeny". (E. C. Dougherty, ed.), pp. 329-336. University of California Press, Berkeley.

ROSEN, B. R. (1970). "A review of the species problem in Recent Scleractinia". Unpublished MS.

SCHOPF, T. J. M. (1973). Ergonomics of polymorphism: its relation to the colony as the unit of natural selection in species of the phylum Ectoprocta. *In* "Animal Colonies. Development and function through time". (R. S. Boardman, A. H. Cheetham and W. A. Oliver, eds), pp. 247-294. Dowden, Hutchinson and Ross, Inc., Stroudsburg, Pennsylvania.

SMITH, F. (1975). "Comprehension and learning. A conceptual framework for teachers." Holt, Rinehart and Winston, New York.

URBANEK, A. (1973). Organization and evolution of graptolite colonies. *In* "Animal colonies. Development and function through time". (R. S. Boardman, A. H. Cheetham and W. A. Oliver, eds). pp. 441-514. Dowden, Hutchinson and Ross, Inc., Stroudsburg, Pennsylvania.

VERNON, J. E. N. and PICHON, M. (1976). Scleractinia of eastern Australia. Part I. Families Thamnasteriidae, Astrocoeniidae, Pocilloporidae. *Aust. Inst. mar. Sci. Monogr. Ser.* **1**, 1-86.

WELLS, J. W. (1973). What is a colony in anthozoan corals? [Abstract]. *In* "Animal colonies. Development and function through time". (R. S. Boardman, A. H. Cheetham and W. A. Oliver, eds), p. 29. Dowden, Hutchinson and Ross, Inc., Stroudsburg, Pennsylvania.

WIJSMAN-BEST, M. B. (1972). Systematics and ecology of New Caledonian Faviinae (Coelenterata-Scleractinia). *Bijdr. Dierk.* **42** (1), 1-90.

WILSON, E. O. (1975). "Sociobiology. The new synthesis". The Belknap Press of Harvard University Press, Cambridge, Massachusetts.

YOUNG, J. Z. (1964). "A model of the brain". Clarendon Press, Oxford.

YOUNG, J. Z. (1975). "An introduction to the study of man." Oxford University Press, Oxford.

1 | The Affinity and Palaeobiology of Stromatoporoids: A Critical Review (Abstract)

J. KAŹMIERCZAK

Institute of Paleobiology of the Polish Academy of Sciences, Warszawa, Poland

Abstract: Stromatoporoids are important fossil rock-building organisms characteristic of many Lower Palaeozoic shallow-water carbonates. Forms ascribed to stromatoporoids are also known from some Mesozoic epicontinental and tethyan carbonate sediments. Even after 150 years of investigation the systematic position of stromatoporoids is still controversial. A critical review shows that both coelenterate (hydrozoan) and poriferan (actually mainly sclerosponge) models for stromatoporoid affinities are not satisfactory. The author's detailed studies on growth patterns, evolutionary trends at structural and substructural levels, reactions to the settlement of foreign organisms, and the character of occupied biotopes testify against the animal nature of stromatoporoids. Algal affinity is postulated for most stromatoporoids excluding from them some marginal forms of clearly poriferan (pharetronid and sclerosponge) nature. Stromatoporoids originated as a result of *in situ* calcification of variously organized colonies or aggregations of coccoid cyanophytes whose growth and permineralization was controlled by the environment to a considerable extent. Stromatoporoids, therefore, can be treated as a special kind of calcareous stromatolite with a highly ordered internal structure.

EDITORS' NOTE

The full text of this paper appeared in *Nature, Lond.* **264** (5581), 49-51 (Nov. 4th, 1976).

Systematics Association Special Volume No. 11, "Biology and Systematics of Colonial Organisms", edited by G. Larwood and B. R. Rosen, 1979, p. 1. Academic Press, London and New York.

2 | Bacterial, Fungal and Slime Mould Colonies

M. J. CARLILE

Department of Biochemistry, Imperial College of Science and Technology, London, England

Abstract: The growth, form and behaviour of representative bacterial, fungal and slime mould colonies are described. Microbial colonies are sometimes only accumulations of individuals, resulting from multiplication without subsequent dispersal, but often, especially in eukaryotic micro-organisms, individuals within a colony display cell-to-cell interactions and co-ordinated behaviour and seem to derive advantages from the colonial condition. A few micro-organisms, at some stages within the life-cycle, show a supression of the autonomy of individual cells to the point where the degree of organization verges on that of a multicellular individual. The role of taxes and tropisms, cell fusion and genetic incompatibility in the development of colonies and the maintenance of their integrity is considered, and the evolution of the colonial condition discussed.

INTRODUCTION

Commonly among micro-organisms the individual organism is a single cell. In some instances the individual is less readily defined—in the fungi, for example, it could be argued that a single hypha, or compartments within a hypha, or a branched system of hyphae, should be regarded as the individual. However, almost always the individual is so small that nearly all microbiological work is carried out on populations or colonies of micro-organisms.

The microbiologist usually obtains *populations* by growing the micro-organisms in a liquid nutrient medium. If the culture vessel is shaken, or the medium is otherwise agitated, large and homogeneous populations may be obtained very rapidly. For example, under ideal conditions, the bacterium *Escherichia coli* will undergo binary fission every 20 minutes. This means that a single cell inoculated into 100 ml of medium can give rise to a population of 10^9 cells at a density of 10^7 cells ml^{-1} in nine hours. The rapid growth rates and

---

Systematics Association Special Volume No. 11, "Biology and Systematics of Colonial Organisms", edited by G. Larwood and B. R. Rosen, 1979, pp. 3–27. Academic Press, London and New York.

homogeneous conditions obtainable in shaken liquid cultures lead to populations of cells produced in this way being the usual material for physiological and biochemical work.

The alternative method of culturing micro-organisms is to grow them on the surface of a medium gelled with agar. A single cell placed on an agar medium will normally yield a discrete *colony* of approximately circular outline. Such colonies lack the homogeneity of the populations produced in liquid culture—cells at the margin of the colony, for example, will be young and well-nourished, and those at the centre old and starved. However, such colonies often develop morphological features lacking in a dispersed population in liquid culture, so agar media are widely employed in taxonomic and genetical work. Microbiologists call all discrete aggregates of cells on agar media "colonies", although as will be indicated below, some would be regarded by other biologists as being a population not deserving the status of a colony.

Members of a *population* of unicellular micro-organisms in shaken liquid culture will influence each other indirectly since each cell will deplete the environment of nutrients and contaminate it with waste products. The individuals will compete with each other and the population as a whole will lack any characteristic form. Some microbial "colonies" on agar media may be little more than rather dense populations, which however will lack the homogeneity of a shaken liquid culture. A true *colony* can be envisaged as a population of unicellular micro-organisms which has a fairly characteristic but perhaps variable form, in which direct interactions between cells occur through chemical signals or cell surface properties, and in which the individuals derive benefits from the association, although some competition may still occur. In a *multicellular organism* there will be a highly characteristic form, all cells will be under central control, competition between cells will be eliminated and cells even sacrificed for the benefit of the organism. A few micro-organisms, at some stages in their life-cycle, approach the degree of organization expected in a multicellular organism. A formal, perhaps an evolutionary scheme, for the status of the microbial colony can hence be proposed in Table I.

It is possible to envisage a complete spectrum of forms between what is indisputably a population of unicellular individuals, through the colonial condition, to the undoubtedly multicellular organism. Hence there will be considerable scope for personal opinion, influenced by a judgement of which criteria are most significant in deciding whether a particular aggregation of cells is to be regarded as a population or a colony, or alternatively, a colony or a multicellular organism.

The simple structure of micro-organisms and the ease with which vast populations and numerous generations can be handled in the laboratory has led to a knowledge of their growth, physiology, biochemistry and genetics which is in many respects unequalled for any other forms. On the other hand, their small size and rapid growth (and disappearance) has resulted in microbial ecology being a subject of great difficulty, leading to a very poor knowledge of the form, size or even occurrence of the colonies of most micro-organisms in nature. It is therefore suggested that while the microbial colony on agar media may be a good model for the elucidation of the type of morphogenetic and

TABLE I. *Scheme for the status of the microbial colony*

Population of unicellular individuals	⟶	Colony of unicellular individuals	⟶	Multicellular organism
Spatial distribution of cells random; cell-to-cell interactions fortuitous; competition between cells total.		Spatial distribution of cells fairly well defined; significant cell-to-cell interactions and co-operation; some competition.		Well-defined form; suppression of cell autonomy and of competition between cells virtually complete.

biochemical interactions which mould a population of individuals into a colony, the microbiologist has much to learn from the botanist and zoologist on the role of the colony in the survival of the species in nature.

The present chapter covers such a wide field with so voluminous a literature, that in order to keep the reference section to a reasonable length, reviews that contain extensive bibliographies rather than research papers will be cited where possible, resulting in some injustice to individuals that have made major contributions. Prokaryotes will also receive, in relation to the amount of work done on their colonies, undue emphasis, because of the interest that lies in examining the evidence for coloniality at the most primitive level.

THE COLONY IN PROKARYOTES

The Prokaryotes—for a general textbook see Stanier, Adelberg and Ingraham (1977)—are those micro-organisms that lack a true nucleus with nuclear envelope and mitosis, their genetic apparatus consisting of a single circular DNA molecule packed into the centre of the cell to form an ill-defined nucleoid.

Reproduction is by binary fission, or in a few forms by budding. Mating is rare and, when it does occur, consists of the unilateral transfer of DNA from a donor to a recipient cell. These and other features make it clear that the Prokaryotes are structurally the most primitive of existing organisms and came into existence before all other surviving forms. It is hence of interest to see what degree of colonial organization can be achieved in the Prokaryotes. Taxonomic categories in the Prokaryotes above the generic level are highly controversial but for considering colony organization it will be convenient to discuss under the following groupings: Eubacteria, Actinomycetes, Myxobacteria, blue–green algae.

1. Eubacteria

The Eubacteria (Eubacteriales and Pseudomonadales) or "true bacteria" are those Prokaryotes which form the main subject of *bacteriology* and would be accepted by all bacteriologists as undoubtedly being bacteria. The commonest forms are cocci (spherical cells usually of diam. c. 1 μm) and bacilli (cylindrical rods commonly c. 1×2 μm). There are also intermediate forms (coccobacilli) and other shapes, such as curved (vibrios) or spiral (spirilla) rods. Some species are non-motile, although in liquid media these will be dispersed by Brownian movement, whereas other species can, under suitable conditions, develop one or more flagella, permitting motility in liquids, including thin films. If cells multiply but dispersal does not occur, colonies visible with the microscope (*microcolonies*) or even with the naked eye develop. The latter could be termed *macrocolonies* by contrast with microcolonies, although bacteriologists usually refer to them simply as "colonies".

(a) *Microcolonies*. Studies on the development of bacterial microcolonies have been reviewed by Hoffman (1964). The form of individual cells, the location of cell wall growth, the way in which cell division occurs, the rate and extent of cell separation, the occurrence of post-fissional movements and the nature of the cell surface and presence of slime or capsular material will all affect microcolony form.

Repeated binary fission of a coccus and its progeny along a single division plane followed by failure of the cells to separate will give a chain of cells, as is seen in *Streptococcus* (Fig. 1a). If the divisions alternate regularly between two planes at right angles to each other a plate of cells will arise (Fig. 1b). Plates will be small if separation occurs soon (e.g. *Tetracoccus*) but large if it is long delayed—plates of 64 cells occur in *Lampropedia* for example.

In *Sarcina* successive synchronous divisions occur in a sequence of three planes, leading to a cubic arrangement of cells (Fig. 1c). Less synchronous and regular cell divisions will result in a less regular colony, as occurs in most coccoid genera.

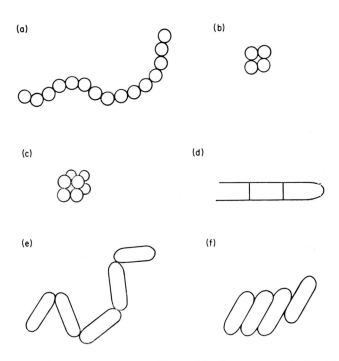

FIG. 1. Diagrams showing arrangement of bacterial cells in micro-colonies. (a) A chain, e.g. *Streptococcus*. (b) A small plate, e.g. *Tetracoccus*. (c) A cube, e.g. *Sarcina*. (d) A filament, e.g. *Bacillus mycoides*. Filaments can also be formed which lack cross-walls. (e) A zig-zag resulting from snapping, e.g. *Arthrobacter*. (f) A palisade resulting from slipping, e.g. *Escherichia coli*.

Rod-shaped bacteria multiply by elongation followed by cell division. Microcolony form will be influenced by the extent of subsequent cell separation and post-fission movements. In the absence of either, filaments will be produced, as in *Bacillus mycoides* (Fig. 1d). "Snapping"—common in coryneform bacteria, such as *Arthrobacter*—will produce a zig-zag chain of cells (Fig. 1e) and "slipping", often seen in *Escherichia coli*, a palisade (Fig. 1f). Under some circumstances, long range forces can orient rod-shaped bacteria in arrays that simulate the crystalline state (Goldacre, 1954). As a microcolony develops, cells become too numerous in the centre of the colony to be accommodated as a monolayer and are pushed up to form initially a second and then further layers, giving an appearance under the microscope resembling that of a contour map of an island. A further factor in microcolony form is whether and to what extent cells produce mucilaginous capsules.

In nature bacterial cells will occur as dispersed populations or as microcolonies. The latter have been demonstrated in a variety of natural habitats, such as soil, surfaces of animals and plants (Sieburth, 1975), dental plaque (Newman and Poole, 1974) and the rumen of sheep (Hungate, 1966).

(b) *Macrocolonies.* On the surface of agar media a microcolony will develop into a macrocolony visible to the naked eye. The form of such colonies is of taxonomic utility and there are a range of terms for describing their elevation and outline (Wilson and Miles, 1975). Macrocolonies probably do not occur in nature, but the growth and form of such colonies are readily studied and may illuminate other morphogenetic processes, including those of colonies of higher forms.

The growth dynamics of microbial colonies on the surface of solid media has been reviewed by Pirt (1975). At first there is an exponential increase in cell number, i.e. the rate of growth (dN/dt) is proportional to the number of cells present (N) and to the specific growth rate (α) of the organism under the prevailing conditions.

Hence
$$\frac{dN}{dt} = \alpha N$$

Integrating
$$N = N_0 e^{\alpha t}$$

N_0 being the number of cells present at the beginning of exponential growth, e the base for natural logarithms and t the elapsed time when N is measured.

Growth rate ceases to be exponential as local depletion of nutrients occurs, and a phase follows in which the diameter of the colony increases linearly with time (Pirt, 1967; Cooper et al., 1968; Palumbo et al., 1971). This linear phase is the result of exponential growth confined to a peripheral zone at the margin of the circular colony (Fig. 2a). Pirt (1975) calculated that the zone width was 46 μm in his *Escherichia coli* colonies and 25 μm in *Streptococcus faecalis*. Cooper et al. (1968) with *Klebsiella aerogenes* demonstrated by sprinkling colonies with carborundum particles that the centre of the colony did not contribute to radial spread. Cells at the centre of the colony will lack nutrients, will have ceased growth and may be dying while cells at the margin are still actively growing. Finally, radial spread of the colony, as a result of the effect of nutrient depletion and perhaps metabolic wastes on growth, will no longer be linear and may even cease. Cooper et al. (1968) demonstrated a phase in which the *area* increased

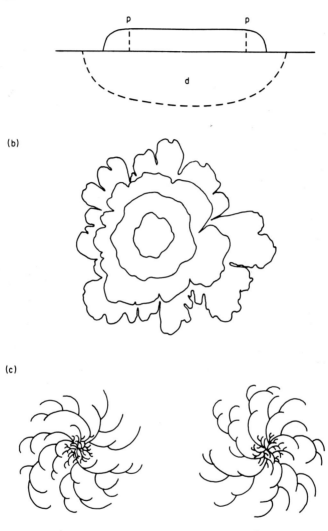

Fig. 2. Bacterial macrocolonies. (a) Side view of a sectioned bacterial colony indicating the peripheral growth zone (p) and the region of agar depleted of nutrients by diffusion (d). After Pirt (1975). (b) A colony of *Klebsiella* (*Aerobacter*) *aerogenes* showing successive colony boundaries—at 3, 7, 12 and 21 days. After Cooper *et al.* (1970). (c) Diagram of left-handed (counter-clockwise) and right-handed strains of *Bacillus mycoides*.

linearly with time and pointed out that this could correspond to the width of the radial growth zone becoming inversely proportional to the radius of the colony. The morphology of ageing colonies was considered by Cooper et al. (1970) who showed that the initially circular outline of the colonies of K. aerogenes became "floral" (Fig. 2b). They pointed out that when nutrients became severely depleted, any slight protrusion in the colony outline would have an advantage in nutrient availability and through further growth would increase its advantage. The initial irregularities presumably result from minor inhomogeneities in the medium.

An aspect of colony morphogenesis that has at times received considerable attention, especially from Russian microbiologists (e.g. Gause, 1940), is the spiral growth (Fig. 2c) of the filamentous bacterium *Bacillus mycoides* (*Bacillus cereus* var. *mycoides*). Most strains of this species appear to show a counterclockwise spiralling of their feathery, filamentous growth. Murray and Elder (1949) showed that the appearance was due to about 75% of the filaments of such strains curving to the left and 25% to the right. In the clockwise strains he studied, the proportions were approximately reversed. The explanation of the curvatures appears to be in the interaction of a slight tendency to axial rotation by the growing filament with the agar surface (Roberts, 1938).

Under appropriate conditions many bacteria can develop flagella and move. This offers further possibilities for colony spread and movement on a moist agar surface. Perhaps the best known of such effects is the swarming of colonies of *Proteus* (Smith, 1972) which will cover the entire agar surface of a Petri dish as a result of a succession of rapid radial advances with intervening pauses, to give a colony with a pattern of concentric zones (Fig. 3). Adler (1966), by using very soft agar has demonstrated swarming in *Escherichia coli*, accompanied by the production of concentric zones, and has established that these effects are due to positive chemotaxis to nutrients. The zones with the *E. coli* colonies, however, move outwards, whereas those of *Proteus* are stationary, so further study of the effect in *Proteus* is needed to establish its basis. Both positive and negative chemotaxis have received detailed study in *E. coli* (Adler, in Carlile, 1975).

When two swarming colonies of the same strain of *Proteus* meet they merge. However, when two swarming colonies of different strains meet a narrow demarcation zone (Fig. 3) develops between them a few hours after merging has begun, and within this zone many cells are found to be abnormal or moribund. The basis of this effect, the Dienes phenomenon (Smith, 1972) is not understood, but it is clear that incompatibility phenomena that prevent genetically different colonies from fusing exists even in organisms as primitive as the Prokaryotes.

In some *Bacillus* spp. the entire colony is able to migrate across a quite well dried agar surface—one on which single cells would be incapable of movement. The paths taken by such colonies have been studied in detail by Murray and

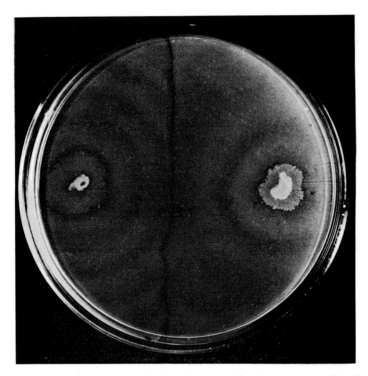

FIG. 3. Two strains of *Proteus mirabilis* have been inoculated at opposite sides of a 9 cm petri dish of nutrient agar. As the colonies have spread, radial zoning has occurred and after the colonies have met, a demarcation line due to incompatibility has formed between them—the Dienes phenomenon.

Elder (1949) in *Bacillus circulans* and two other species. A bullet-shaped colony commonly follows a curved path leaving cells which develop into subsidiary colonies in its wake. The path followed is often a spiral of decreasing radius so that finally the "head" of the colony catches up with the "tail" and a rotating colony is produced. About two-thirds of such colonies rotate in a counter-clockwise direction, about one-third clockwise. Little is known about the mechanism of migration of such colonies; presumably the colony can maintain moist conditions beneath itself and possibly alignment of cells and entrainment of flagellum activity occurs.

2. Actinomycetes

These organisms—"hyphal bacteria"—were at one time thought to be diminutive fungi (Actinomycete = ray-fungus), and indeed the colonies of the best-known genus, *Streptomyces*, with hyphae (filaments with apical growth)

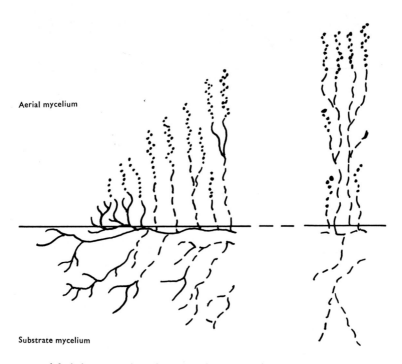

FIG. 4. Simplified diagram of a side view of a sectioned *Streptomyces* colony, showing living hyphae (continuous lines), dead hyphae (broken lines) and spores (dots). Margin of colony on left, centre at right. After Chater and Hopwood, in Ashworth and Smith (1973).

penetrating the substratum and spores borne on aerial mycelium (a collective term for a mass of hyphae), have a close superficial resemblance to a fungus colony. They are, however, Prokaryotes, and are so closely linked to the Eubacteria by a series of intermediate forms that drawing a boundary between the Eubacteria and the Actinomycetes is very much a matter of personal judgement. These transitional forms include *Arthrobacter* in which long and branched cells can develop, *Mycobacterium* which can be mycelial, and *Actinomyces* and *Nocardia* which, although they form mycelial colonies, reproduce by fragmentation and not by spores. It is in *Streptomyces* that the greatest exploitation of hyphal growth at the Prokaryote level occurs. Apical growth and branching permit efficient spreading on a surface and penetration into the substratum (Fig. 4). Aerial hyphae produce powdery spores above the surface thus providing an effective means of dispersal. Thus in Actinomycetes we see the evolution of mycelial colonies exhibiting differentiation, some cells being specialized for dispersal, whereas other parts of the colony exploit the substratum. Colony structure and differentiation in *Streptomyces coelicolor* is discussed by Chater and Hopwood (see Ashworth and Smith, 1973).

3. *Myxobacteria*

The Myxobacteria ("slime-bacteria") are the best known representatives of the gliding bacteria, in which cells, although lacking flagella, can glide over moist surfaces by a process which is still not understood. Cell-to-cell interactions in the fruiting Myxobacteria are reviewed by Dworkin (see Ashworth and Smith, 1973) and morphogenetic events discussed and illustrated by Quinlan and Raper (1965).

Fruiting bodies bear numerous resting spores, in some species in cysts. When a mass of spores, or a cyst, is deposited on a suitable substrate the spores germinate and the resulting vegetative cells emerge from the cyst as a swarm. A swarm migrates, or spreads, depending on the species, and cell numbers increase by binary fission. Cells at the edge of the swarm may move a short distance away from the swarm, but soon return. If two swarms come close together, they migrate towards each other and fuse. Finally, when the nutrients in the substratum are exhausted, the cells in a swarm converge to a centre and a fruiting-body bearing cysts develops. The stalk of the fruiting body is composed of slime secreted by the cells, which soon dries and hardens. The varied forms of the fruiting bodies are a basis for classification.

There are many forms of cell-to-cell interaction displayed in the life-cycle, as discussed by Dworkin in Ashworth and Smith (1973). Solitary spores of *Myxococcus xanthus* will not usually germinate, but spores at high density readily do so. Cells in a swarm tend to migrate together, and not disperse. In *Nannocystis* a swarm migrates away from high light intensities but individual cells are not negatively phototactic. In *Podangium lichenorum* patterns of oscillating waves, presumably caused by synchronous cell movement, move across a swarm. A mutant of *M. xanthus* is motile if in a swarm, but not when alone. On nutrient exhaustion, aggregation, shown to be chemotactic in some species, occurs. The development of a complex fruiting body bearing numerous cysts must involve some form of co-operation. Good progress has been made in elucidating the molecular basis of a few of the above interactions, such as the mutual stimulation of germination.

The advantages to Myxobacteria in maintaining themselves as a swarm of cells may be related to the food sources that they exploit. The Myxobacteria include both cellulolytic and bacteriolytic strains. It seems probable that some of the latter secrete antibiotics that kill the bacteria. In both instances the food source will be relatively large (cellulose fibres and bacterial cells) and highly refractory, particularly the bacterial cell wall. A swarm will be far more capable than will a single cell of producing the quantity and concentration of antibiotics and extracellular enzymes needed for rapid exploitation of such food sources. Other features of the life-cycle, such as the production of spore masses in cysts, and spore germination only in the presence of numerous other spores are probably related to the generation of optimal populations for nutrient exploitation.

4. Blue–Green Algae

The blue–green algae (sometimes called Cyanobacteria) are Prokaryotes that perform a photosynthetic process similar to that in plants, in which water is the ultimate hydrogen donor and oxygen is generated. Those species that are motile carry out gliding movements similar to the Myxobacteria, and it is clear that some of the gliding bacteria and blue–green algae are closely related. Recently two works have been published on the biology of the blue–green algae (Carr and Whitton, 1973; Fogg *et al.*, 1973).

The blue–green algae, like the Eubacteria, have a range of cell forms, including cocci and rods. They generate a range of microcolonies, including plates and cubes of cocci, and filaments of cocci and rods. In some species differentiation occurs within the filaments, a colourless cell, the heterocyst, which is capable of nitrogen fixation, occurring at intervals along the chain of green photosynthetic cells. Other specialized cell types are known in blue–green algal filaments and experimental work on life-cycles and differentiation is beginning (Carr and Bradley, in Ashworth and Smith, 1973; Lazaroff, in Carr and Whitton, 1973). The spacing of specialized cells, such as heterocysts, within a filament is of special interest, as it permits investigation of a uniquely simple morphogenetic situation, the differentiation of a one-dimensional pattern (Wilcox, 1970). Blue–green algal colonies are associated with the deposition and erosion of massive deposits of calcium carbonate (Golubíc, in Carr and Whitton, 1973) and are thought to be responsible for the reef-like rocks known as stromatolites, dating back to Pre-Cambrian times.

THE COLONY IN EUKARYOTIC MICRO-ORGANISMS

The Eukaryotes represent a structural advance on the Prokaryotes in the possession of a true nucleus bounded by a nuclear envelope, several or many chromosomes, mitosis, membrane-bound organelles such as mitochondria and sometimes chloroplasts, and in various other ways (Stanier, Adelberg and Ingraham, 1977). Features of interest in the context of colony development lacking in Prokaryotes but present in Eukaryotes are cell fusion, protoplasmic streaming and cell diameters commonly of the order of 10 μm rather than 1 μm. The occurrence of cell fusion (and phagocytosis and pinocytosis) is related to the possession of a more plastic cell membrane, in turn probably due to high steroid content of the membrane—Eukaryotes commonly contain about one hundred times the levels of steroids detectable in Prokaryotes. Protoplasmic streaming is possible in Eukaryotes because of a less viscous protoplasm and greater cell diameters—the resistance offered by the small prokaryote hyphae would render protoplasmic streaming virtually impossible, as the velocity of flow expected in a tube, other factors being equal, is proportional to the square of the radius of the tube (Hagen–Poiseuille Law). The larger size of eukaryote cells also permits the occurrence of chemotropism (growth oriented in response to a chemical gradient) and true chemotaxis (movement oriented in response to a chemical gradient) as cells are large enough for the ready detection of different concentrations of a chemical signal at different points on the cell (Carlile, 1975). The "chemotaxis" of Eubacteria, the only Prokaryotes in which chemotaxis has been thoroughly studied, is strictly speaking a klinokinesis, the cells tending to continue moving in the same direction if a concentration level becomes more favourable with time, and to make random turns if it becomes less favourable. The role of cell fusion, protoplasmic streaming and chemotaxis in colony formation in eukaryotic micro-organisms will be illustrated below. A feature widespread among Eukaryotes but apparently absent in Prokaryotes, is the occurrence of circadian and other endogenous temporal rhythms (Sweeney, 1969). These may influence the appearance of fungal colonies by producing during growth a daily alternation of sparse and densely branched hyphae, or sterile and fertile mycelium (Carlile, 1970). The occurrence of such rhythms in eukaryotic but not prokaryotic organisms may reflect more effective cell-to-cell communication in the former—the existence of detectable endogenous rhythms in colonies or multicellular organisms would seem to imply the entrainment of spatially separated biochemical oscillations (Winfree, 1973).

1. The Fungi

The fungi are hyphal heterotrophic Eukaryotes and some non-hyphal groups (of which the yeasts are the most important) which are closely related to hyphal forms. The essential feature of fungi, the hypha, will first be discussed, then fungal colonies and fruiting bodies and finally yeast colonies. Recent textbooks on the fungi, from physiological and morphological viewpoints respectively, are those of Burnett (1976) and Webster (1970) and the growth of fungal hyphae and colonies is discussed by Bull and Trinci (1977). The fungi can be divided into the lower fungi (Phycomycetes) in which the formation of cross-walls (septa) in the hyphae is very rare, and the higher fungi (Ascomycetes, Basidiomycetes and Fungi Imperfecti) in which cross-walls are common. The discussion of fungal hyphae and fungal colonies that follows applies to what occurs on the surface of agar media. A great deal of work has been done on pellet and filament formation in shaken liquid culture, but these studies, although of great significance in relation to fundamental microbiology and the fermentation industry, are of little direct relevance to the consideration of naturally occurring colonies.

(a) *Fungal hyphae.* The fungal hypha (Bartnicki-Garcia, in Ashworth and Smith, 1973) is a cylindrical tube which towards the tip tapers gently to end in an approximately hemispherical apex (Fig. 5a). Extension is confined to the tapering and apical region. The rate of growth of the hypha is, however, usually far too great to be accounted for solely by metabolic processes occurring at the apex, and it is clear that metabolites are carried towards the apex from the rear. Prominent protoplasmic streaming towards the apex can often be seen in the centre of a hypha, and sometimes very careful observation will reveal reverse streaming at the periphery. However, it would seem that there must be considerable water loss at the hyphal apex in order to account for the virtual absence of reverse streaming in most species. This streaming towards the apex is presumably the means by which metabolites are carried to the apex to permit rapid extension.

Fungal hyphae are coenocytic and have nuclei distributed through the cytoplasm. In the higher fungi cross-walls divide the hyphae into compartments which may contain one, two or several nuclei, depending on species. In actively growing hyphae the compartments communicate by means of septal pores, which permit protoplasmic streaming and even nuclear migration between compartments. Damaged compartments, or compartments in the older part of a colony, can be isolated by pore plugs.

Hyphae commonly produce branches some distance behind the apex,

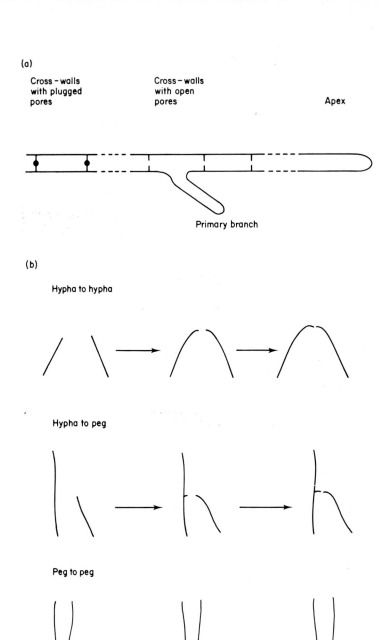

FIG. 5. Fungus hyphae. (a) Fungus hypha showing apex, cross walls with pores, a primary branch, and cross-walls with plugged pores. (b) Diagrams showing autotropism of fungal hyphae about to give rise to various patterns of hyphal anastomosis—hypha-to-hypha, hypha-to-peg, and peg-to-peg. After Buller (1933).

alternately on the two sides. These primary branches in turn produce secondary branches in a similar manner. The branch hyphae are smaller than the main hypha and have lower extension rates (Butler, 1961). The above monopodial type of branching is the commonest in the fungi and is the optimal system for "draining" an area of nutrients (Leopold, 1971). Dichotomous and sympodially branching hyphal systems also occur in fungi.

The efficient spacing of fungal hyphae would seem to imply some form of negative autotropism—a chemotropic response in which hyphae grow away from each other. Negative autotropism has been demonstrated with the germ-tubes (the hyphae emerging from germinating spores) of several fungi. In the higher fungi positive autotropism often occurs, with hyphae growing towards each other and fusing—hyphal anastomosis (Fig. 5b). These tropisms and anastomoses were studied in detail by Buller (1933) and recently reviewed by Gooday (see Carlile, 1975)

(b) *Fungal colonies*. A fungal colony can be initiated by the germination of a single spore. In the higher fungi, however, germ-tubes from many spores may fuse to give a single colony. A colony is at first a mass of undifferentiated hyphae, but soon a sharply defined outline is established. Bull and Trinci (1977) list the various features in which the hyphae of a colony margin differ from those of undifferentiated mycelium. An established colony increases in diameter linearly with time and, as with bacteria, there is a peripheral zone in which exponential growth is occurring. The rate of extension of the hyphae at the margin of the colony is a product of the width of the peripheral growth zone and the exponential growth rate in that zone (Trinci, 1971). Since the peripheral growth zone is of vastly greater width than in bacteria (e.g. 8500 μm of *Rhizopus stolonifer* instead of the 90 μm of *E. coli*), fungal colonies can spread far faster than bacterial colonies.

Yanagita and Kogane (1962) distinguish the following zones in the fungal colony:

1. Extending Zone.
2. Productive Zone.
3. Fruiting Zone.
4. Aged Zone.

The *extending zone*, which corresponds approximately with the peripheral growth zone of Trinci (1971), consists largely of rapidly advancing well-spaced hyphae. The *productive zone* is the site of a major increase in biomass. In the *fruiting zone* spore production, sometimes in or on fruit bodies of substantial size, occurs. The hyphae of the *aged zone* are commonly vacuolated and often become emptied of protoplasm and scaled off. Fungi may also show radial zoning patterns of sparse or dense hyphae, or sporing and non-sporing mycelium, as a direct result of diurnal fluctuations of light and temperature, or as a consequence of circadian or other endogenous rhythms (see page 15).

The occurrence of hyphal anastomosis in higher fungi raises the question of how often colonies in nature are heterokaryotic (i.e. have present in their cytoplasm two different nuclear types). Extensive use has been made of heterokaryons in fungal genetics (Fincham and Day, 1971) but such heterokaryons are usually made by bringing together strains produced by mutagenic treatment of a common ancestor, which hence differ little genetically except for specific markers. However, if two strains of a fungus are isolated from nature, they will probably be unable to form a heterokaryon with each other and it may be necessary to isolate many strains before two are found which can fuse with each other and establish a heterokaryon (Caten, 1971). Such somatic heterogenic incompatibility (Esser and Blaich, 1973) is widespread in fungi. On the other hand, mycologists who isolate from nature fungi which can exist either as parasites or saprophytes, such as *Fusarium oxysporum* or *Botrytis cinerea*, have much experience of the "dual phenomenon"—soon after isolation sectors appear with different characteristics from the rest of the colony, such as being sterile instead of sporing abundantly, or being flat instead of having aerial mycelium. Sub-cultures from such sectors will breed true, retaining the characteristics of the sectors. It would seem probable that the production of such sectors is due to the formation of homokaryotic areas (i.e. the breakdown of a naturally occuring heterokaryon) or alternatively the loss of a cytoplasmic factor (i.e. the breakdown of a heteroplasmon). Unfortunately relatively little genetical work has been done on those species which show the dual phenomenon best, so the possibility that natural colonies may often consist of balanced heterokaryons, with nuclear ratios changing to adapt to changed conditions, remains open. A classic discussion of heterokaryosis and sectoring is that of Pontecorvo (1946), and the role of heterokaryosis in natural populations of fungi is considered by Burnett (1975).

Although incompatibility barriers may prevent vegetative fusions between the hyphae of genetically different fungus strains, within a strain fusions readily occur between hyphae in most higher fungi. In consequence, the radiating and branching systems of hyphae in a fungal colony can, in the higher fungi, become converted into a network, permitting movement of protoplasm and nutrients in any direction.

(c) *Fungal Fruiting Bodies.* The above mentioned hyphal network of higher fungi has important consequences for fruiting-body formation. In the Basidiomycetes (mushrooms, toadstools and related forms) it can permit nuclei of another mating-type, which have gained access to the hyphae at one point, to multiply and migrate throughout the colony and produce a dikaryon of two mating types (Raper, 1966). Fruiting-body formation can be initiated at any point in such a colony. Once fruiting bodies have started to form, the network of hyphae permits protoplasm to be mobilized from a very extensive area and deployed in the construction of large fruiting bodies. Perhaps the climax in the development of the fungal colony is seen in the "fairy-ring", a colony many metres in diameter, bearing fruiting bodies of a construction sufficiently co-ordinated and precise as to perhaps qualify them as multicellular individuals.

(d) *Yeast colonies.* Yeasts are fungi that have reverted to the unicellular condition, and reproduce by budding or fission. In consequence, their colonies are very similar to those of Eubacteria, except that they are composed of much larger cells.

2. *The Cellular Slime Moulds*

The cellular slime moulds (Acrasiales) are the subject of a book by Bonner (1967) and of a major part of books of Ashworth and Dee (1975) and Olive (1975). A book has recently been devoted to the most intensively studied species, *Dictyostelium discoideum* (Loomis, 1975).

The organisms are usually cultured on a lawn of a suitable bacterium, such as *Escherichia coli*, on agar media. A spore placed upon the medium will germinate to give an amoeba, which reproduces by binary fission to give a population of amoebae. The amoebae, feeding on the bacteria, produce a clear circular plaque in the lawn through destruction of the bacteria. Amoebae tend to abandon the centre of the plaque, now almost devoid of bacteria, and move to the edge to consume further bacteria. The plaque—or colony—increases in radius and will merge with other expanding plaques. The outward movement of the amoebae towards bacteria is presumably chemotactic. A number of substances are known to be chemotactic for the amoebae of *D. discoideum*, and at least two of them, 3′,5′ cyclic adenosine monophosphate (cyclic AMP) and folic acid, are produced by bacteria.

When the supply of bacteria is exhausted, the amoebae aggregate, streaming towards centres, which may consist of one or many "founder cells". Here the role of a chemotactic hormone, termed acrasin, is firmly established. In *D. discoideum* it is clear that acrasin is cyclic AMP, however, some species do not respond to cyclic AMP and other acrasins must exist. A form of contact guidance is also involved as amoebae stream towards a centre, the "head" of an amoeba maintaining contact with the "tail" of the one in front. Amoebae release, along with cyclic AMP, a phosphodiesterase which destroys it—this perhaps helps to prevent premature aggregation and later on "noise" due to accumulation of the hormone. Cyclic AMP may be released in pulses, resulting in waves of movement towards the centre spreading over the aggregating amoebae. A relay system may also operate, an amoeba receiving the cyclic AMP signal, making a move towards the centre, and then itself releasing a pulse of cyclic AMP which will influence amoebae further out. It will be appreciated that, depending on circumstances, a wide variety of aggregation patterns can be generated; these have been analysed by Durston (1973). Many mutants affecting the aggregation process in different ways are known (Riedel et al., 1973). Chemotaxis in the cellular slime moulds has recently been reviewed (Konijn, in Carlile, 1975).

The aggregated amoebae give rise to a slug-like object, the grex or pseudoplasmodium. The grex contains up to 100000 amoebae and can be several millimetres in length. The grex is capable of much faster migration than are individual amoebae, and unlike amoebae, is phototactic. It migrates to a suitable site and there differentiates into a fruiting body, which consists of a stalk, composed of dead cells, bearing a spherical sporangium containing spores. The degree of co-ordination shown in grex migration and fruiting body development approaches that expected in a multicellular organism rather than a colony of amoeboid cells.

3. Plasmodial Slime Moulds

The plasmodial slime moulds (Myxomycetes) are the subject of a book by Gray and Alexopoulos (1968) and a substantial part of the books by Ashworth and Dee (1975) and Olive (1975).

Myxomycete spores germinate to yield amoebae, which like those of the cellular slime moulds, can be cultured on a lawn of bacteria on agar. Within a spore population, plasmodia can be produced, in many species as a result of the mating of amoebae that differ in mating type. The plasmodium grows, and nuclear division but not cell division occurs. The result is a coenocytic mass of protoplasm containing many millions of nuclei which divide synchronously approximately every twelve hours and which may cover many cm^2. Genetically identical plasmodia often fuse with each other to form larger plasmodia. Plasmodia may ingest and digest amoebae of the population from which the plasmodia have arisen. Within the plasmodium spectacular protoplasmic streaming occurs and the plasmodium is capable of migrating at speeds far greater than is possible for amoebae, and of engulfing and digesting relatively large objects, such as fungal colonies.

The plasmodium is a state of living matter that is rather difficult to classify. It is certainly not a population of individual cells, although it could perhaps be regarded as a population of nuclei in a common cytoplasm. The degree of coordination shown by a plasmodium in protoplasmic streaming and motility might justify it being regarded as a multinucleate individual, yet plasmodia are liable to fragment and (if of identical genotype) to fuse with each other very readily. A plasmodium has, as Buller (1931) recognized, much in common with a fungal colony; one might almost regard the latter, in the case of the lower fungi, as a plasmodium confined to a system of tubes, the hyphae.

Myxomycete plasmodia that differ genetically usually are unable to merge with each other, there being incompatibility genes that prevent fusion, and others that cause a post-fusion reaction, which takes the form of the destruction of nuclei or protoplasm of one strain or sometimes both (Carlile, 1973).

Nutrient exhaustion leads to the formation of resting structures (sclerotia) or, if light is present, fruiting bodies which in many species are complex. These contain spores which can be dispersed to initiate the amoeboid phase.

4. Lichens

Some of the most remarkable colonies in the microbial world are those formed by lichens. Lichens (Ahmadjian and Hale, 1973; Hale, 1974) are symbiotic associations of a fungus and an alga. The fungus—almost always an Ascomycete —is incapable of survival in nature without its algal partner. The algal partner—usually a member of the eukaryotic green algae or prokaryotic blue-green algae—is a species which can thrive in the free-living state. The algal partner will contribute to the lichen colony by fixing carbon dioxide, and in the case of the blue-green algae, nitrogen. The contribution of the fungal component is more difficult to define, but it would seem to be the major factor in determining the form of the colony and perhaps protects the algae, embedded in a matrix of fungal hyphae, from dessication. Certainly lichens can grow under conditions of low nutrition and dessication too severe for the growth of the algal partner. There are a great variety of lichens with a wide diversity of forms, divisible into crustose (encrusting rocks or trees), foliose (leaf-like) and fruticose (tree-like, either erect or pendulous). Colonies grow very slowly—the increase in radius ranges from less than 1 mm to about 30 mm year^{-1}. Dispersal is by means of soredia—fragments with both fungal hyphae and algal cells—or by fungal spores, which may chance to fall near cells of an appropriate alga.

THE EVOLUTION AND ADAPTIVE SIGNIFICANCE OF COLONIALITY

The above account indicates that coloniality has arisen many times and to differing extents in the microbial world. However, a general sequence of steps in the evolution of coloniality may be suggested as follows:

1. Multiplication of cells accompanied by failure to disperse, results in an aggregate, which, if it has a sufficiently well defined form, is likely to be termed a colony. The microcolonies of Eubacteria, for example, result from failure of cells to separate. Macrocolonies of Eubacteria on a surface result from their inability to move (by flagella action or Brownian movement) in the absence of an aqueous phase. Such colonies may, through interaction with the environment, grow uniformly and have a characteristic form. Failure to disperse will fortuitously bring some advantages. The individual cell will, from the presence of others, receive some protection against dessication, visible and ultraviolet radiation, and against any toxic factor in the environment, the absorption of which will be shared between many cells and thus have less effect. The advantages are accompanied by the disadvantages of intense competition between cells for nutrients and oxygen and of adversely affecting each other through metabolic wastes. If these disadvantages outweigh the advantages, further evolution towards a more highly organized form of colony will not occur. It would seem likely that most eubacterial colonies are of this type, where the colonial condition is accidental and of little or no benefit to the cell. However the eubacterial colony provides a simple model which illustrates many of the features of more complex colonies.

2. It can be anticipated that sometimes the advantages of cells being together can outweigh the advantages of dispersal. Differentiation of cells for specialized functions can occur (e.g. the heterocysts of blue–green algae, the dispersable spores of Actinomycetes): a cell mass may be able to migrate in conditions where single cells could not move, or may be able to create an enzyme concentration which can attack a substrate where solitary cells would be ineffective, as seems to occur with Myxobacteria. When the aggregated condition thus proves more efficient than the solitary state, we may expect the evolution of cell-to-cell interactions tending to keep cells together and to co-ordinate their activities. At the Prokaryote level this seems to have happened only in the Myxobacteria. In the Eukaryotes positive and negative taxes and tropisms, and cell and hyphal fusion, have been exploited repeatedly in the evolution of coloniality.

3. Colonies will commonly be clones—the nuclear genotype and the cytoplasmic genotype (mitochondrial and chloroplast DNA, plasmids, endosymbionts, viruses) being uniform throughout the colony. If the genotype is a successful one, then mechanisms may evolve to prevent merging with other colonies and contamination with alien nucleic acids—especially viruses. Mechanisms for achieving this exist even at the prokaryote level—for example, the restriction phenomenon through which alien DNA is destroyed (Boyer, 1971) and some such mechanism can be seen preventing the merging of colonies of *Proteus* (Fig. 3). In eukaryotic micro-organisms, such mechanisms are highly developed, as for example in Myxomycetes (Carlile, 1973) and constitute the phenomenon of heterogenic incompatibility (Esser and Blaich, 1973). Hence colonies will usually remain clones. The question remains open, however, as to whether in some groups adaptability may be gained by having colonies composed of two nuclear types, as in the balanced heterokaryons which occur in some fungi in the laboratory (Jinks, 1952). In a sense such a system does exist in the lichens, although here the two nuclear types come from wholly different sources—alga and fungus. The climax of the process of having many genotypes in a single organism—which, as proposed below, is a very highly integrated colony—is the immune system of mammal or bird, where numerous genetically different clones of cells await an appropriate immunological challenge as a signal for multiplication (Burnet, 1969).

4. Finally, the extent of co-ordination may reach the point where it may seem more appropriate to speak of a multicellular organism than of a colony, as with the grex and fruiting bodies of cellular slime moulds and the massive fruiting bodies of mushrooms and toadstools.

In the introduction the sequence "Population of unicellular individuals → Colony of unicellular individuals → Multicellular organism" was proposed as a formal scheme. In the cellular slime moulds it would make a valid ontogenetic sequence, since scattered amoebae (population) give rise to aggregation patterns (colony) and finally the grex and fruiting body (multicellular organism). There is also, as indicated above, a case for regarding the scheme as an evolutionary sequence which has been followed, in part at least, many times. Finally, it can be suggested that the process has been partially repeated in the many developments of coloniality in multicellular organisms, thus "Population of multicellular individuals → Colony of multicellular individuals → Colony integrated to the point where it verges on being a higher order individual."

ACKNOWLEGMENTS

I wish to thank Dr Gillian Butler, University of Birmingham, and Dr Tony Trinci, Queen Elizabeth College, for access to unpublished data and manuscripts in press, and for stimulating discussion of microbial growth and colony structure.

REFERENCES

ADLER, J. (1966). Chemotaxis in bacteria. *Science* **153**, 708–716.
AHMADJIAN, V. and HALE, M. E., eds (1973). "The Lichens" Academic Press, New York.
ASHWORTH, J. M. and DEE, J. (1975). "The Biology of Slime Moulds" Arnold, London.
ASHWORTH, J. M. and SMITH, J. E., eds (1973). "Microbial Differentiation." Symposium of the Society for General Microbiology, Vol. 23. Cambridge University Press, Cambridge.
BONNER, J. T. (1967). "The Cellular Slime Molds" 2nd edition. Princeton University Press, Princeton.
BOYER, H. W. (1971). DNA restriction and modification mechanisms in bacteria. *A. Rev. Microbiol.* **25**, 153–176.
BULLER, A. H. R. (1931). "Researches on Fungi" Vol. 4. Longmans, Green, London.
BULLER, A. H. R. (1933). "Researches on Fungi" Vol. 5. Longmans, Green, London.
BULL, A. T. and TRINCI, A. P. J. (1977). The physiology and metabolic control of fungal growth. *Adv. Microbial Physiol.* **15**, 1–84.
BURNET, M. (1969). "Cellular Immunology" Cambridge University Press, Cambridge.
BURNETT, J. H. (1975). "Mycogenetics" Wiley, London.
BURNETT, J. H. (1976). "Fundamentals of Mycology" 2nd edition. Arnold, London.
BUTLER, G. M. (1961). Growth of hyphal branching systems in *Coprinus disseminatus*. *Ann. Bot.* **25**, 341–352.
CARLILE, M. J. (1970). The photoresponses of fungi. *In* "The Photobiology of Microorganisms" (P. Halldal, ed.) ,Wiley, London.
CARLILE, M. J. (1973). Cell fusion and somatic incompatibility in Myxomycetes. *Ber. Deutsch. Bot. Ges.* **86**, 123–139.
CARLILE, M. J., ed. (1975). "Primitive Sensory and Communication Systems: the taxes and tropisms of micro-organisms and cells" Academic Press, London.
CARR, N. G. and WHITTON, B. A., eds (1973). "The Biology of Blue-Green Algae" Blackwell Scientific Publications, Oxford.
CATEN, C. E. (1971). Heterokaryon incompatibility in imperfect species of *Aspergillus*. *Heredity* **26**, 299–312.
COOPER, A. L., DEAN, A. C. R. and HINSHELWOOD, C. (1968). Factors affecting the growth of bacterial colonies on agar plates. *Proc. R. Soc.* B **171**, 175–199.
COOPER, A. L., DEAN, A. C. R. and HINSHELWOOD, C. (1970). Morphological changes in growing cultures of *Aerobacter* (*Klebsiella*) *aerogenes*. *Proc. R. Soc.* B **175**, 95–105.
DURSTON, A. J. (1973). *Dictyostelium discoideum* aggregation fields as excitable media. *J. theor. Biol.* **42**, 483–504.
ESSER, K. and BLAICH, R. (1973). Heterogenic incompatibility in plants and animals. *Adv. Genet.* **17**, 107–152.
FINCHAM, J. R. S. and DAY, P. R. (1971). "Fungal Genetics" 3rd edition. Blackwell Scientific Publications, Oxford.

Fogg, G. E., Stewart, W. D. P., Fay, P. and Walsby, A. E. (1973). "The Blue-Green Algae" Academic Press, London.
Gause, G. F. (1940). On the relation between the inversion of spirally twisted organisms and the molecular inversion of their constituents. *Biodynamica* **3**, 125–143.
Goldacre, R. J. (1954). Crystalline bacterial arrays and specific long range forces. *Nature* **174**, 732–734.
Gray, W. D. and Alexopoulos, C. J. (1968). "Biology of the Myxomycetes" Ronald Press, New York.
Hale, M. E. (1974). "The Biology of Lichens" 2nd edition. Arnold, London.
Hoffmann, H. (1964). Morphogenesis of bacterial aggregations. *A. Rev. Microbiol.*, **18**, 111–130.
Hungate, R. E. (1966). "The Rumen and its Microbes" Academic Press, London.
Jinks, J. L. (1952). Heterokaryosis: a system of adaption in wild fungi. *Proc. R. Soc.* B, **140**, 83–99.
Leopold, L. B. (1971). Trees and streams: the efficiencies of branching patterns. *J. theor. Biol.* **31**, 339–354.
Loomis, W. F. (1975). "*Dictyostelium discoideum*: a developmental system" Academic Press, New York.
Murray, R. G. E. and Elder, R. H. (1949). The predominance of counterclockwise rotation during swarming of *Bacillus* species. *J. Bact.* **58**, 351–359.
Newman, H. N. and Poole, D. F. G. (1974). Structural and ecological aspects of dental plaque. *In* "The Normal Microbial Flora of Man." Society for Applied Bacteriology Symposium Series (F. A. Skinner and J. G. Carr, eds), No. 3, pp. 111–134.
Olive, L. S. (1975). "The Mycetozoans" Academic Press, New York.
Palumbo, S. A., Johnson, M. G., Rieck, V. T. and Witter, L. D. (1971). Growth measurements on surface colonies of bacteria. *J. gen. Microbiol* **66**, 137–143.
Pirt, S. J. (1967). A kinetic study of the mode of growth of surface colonies of bacteria and fungi. *J. gen. Microbiol.* **47**, 181–197.
Pirt, S. J. (1975). "Principles of Microbe and Cell Cultivation" Blackwell Scientific Publications, Oxford.
Pontecorvo, G. C. (1946). Genetic systems based on heterokaryosis. *Cold Spring Harbor Symp. Quant. Biol.* **11**, 193–201.
Quinlan, M. S. and Raper, K. B. (1965). Development of the myxobacteria. "Encyclopedia of Plant Physiology" Vol. 15, pt. 1, pp. 596–611.
Raper, J. (1966). "Genetics of Sexuality in Higher Fungi" Ronald Press, New York.
Riedel, V., Gerisch, G., Muller, E. and Beug, H. (1973). Defective cyclic adenosine—3',5'-phosphate-diesterase regulation in morphogenetic mutants of *Dictyostelium discoideum*. *J. mol. Biol.* **74**, 573–585.
Roberts, J. L. (1938). Evidence of a rotational growth factor in *Bacillus mycoides*. *Science* **87**, 260–261.
Sieburth, J. M. (1975). "Microbial Seascapes" University Park Press, Baltimore.
Smith, D. G. (1972). The *Proteus* swarming phenomenon. *Sci. Prog.* **60**, 487–506.
Stanier, R. Y., Adelberg, E. A. and Ingraham, J. L. (1977). "General Microbiology" 4th edition., Macmillan, London.
Sweeney, B. M. (1969). "Rhythmic Phenomena in Plants" Academic Press, London.

TRINCI, A. P. J. (1971). Influence of the width of the peripheral growth zone on the radial growth rate of fungal colonies on solid media. *J. gen. Microbiol.* **67**, 325–334.

WEBSTER, J. (1970). "Introduction to Fungi" Cambridge University Press, Cambridge.

WILCOX, M. (1970). One-dimensional pattern found in blue-green algae. *Nature* **228**, 686–687.

WILSON, G. S. and MILES, A. (1975). "Topley and Wilson's Principles of Bacteriology, Virology and Immunity" 6th edition. Edward Arnold, London.

WINFREE, A. T. (1973). Polymorphic pattern formation in the fungus *Nectria*. *J. theor. Biol.* **38**, 363–382.

YANAGITA, R. and KOGANE, F. (1962). Growth and cytochemical differentiation of mould colonies. *J. gen. appl. Microbiol.* **8**, 201–213.

3 | Group Phenomena in the Phylum Protozoa

C. R. CURDS

British Museum (Natural History), London, England

Abstract: Groups and colonies of protozoan cells are commonly encountered. Colonies are of two fundamentally different types depending on the manner in which they have developed. The majority are formed as the result of a failure to separate following asexual binary fission. The other method of colony formation, almost exclusively restricted to the Sarcodina, involves the aggregation of a population of individual cells after binary fission and separation have been completed. The extent to which the collections of protozoa have become adapted to a true integrated colonial existence varies from simple gregarious behaviour to the formation of colonies where the resultant association of cells behaves more as a single entity than as a collection of individuals. Between these outer limits lies a whole range of organisms whose collective associations may be either temporary as a response to some environmental factor or of a more permanent nature.

INTRODUCTION

The tendency towards the formation of temporary or permanent groups of protozoan cells is widespread throughout the phylum and appears to have arisen independently on several different occasions during evolutionary times. Group phenomena are known in all the major taxa of the phylum although they are more common in some than in others. In the Sporozoa there are no truly colonial forms but even here we can find examples of organisms, such as the trophozoites of some monocystid gregarines (*Pleurocystis* and *Zygocystis*), living together in pairs for part of their life-cycle. The purpose of this paper is to give some of the many examples of collective associations known in the Protozoa in order to illustrate the wide variety and to introduce the possible ways in which coloniality may have developed. Some of the examples given here are well

---

Systematics Association Special Volume No. 11, "Biology and Systematics of Colonial Organisms", edited by G. Larwood and B. R. Rosen, 1979, pp. 29–37. Academic Press, London and New York.

known but the term "colony" has been deliberately interpreted in as wide a manner as possible so that less well known examples of cell aggregation can be included.

METHODS OF COLONY FORMATION

Protozoan colonies appear to be of two fundamentally different types depending upon the way in which the groups are formed. The great majority of colonies develop as a result of the failure of the cells to separate after nuclear fission and colonies of this type are widely distributed throughout several classes of Protozoa.

A series of colonial phytoflagellates illustrated in Fig. 1a, b, c and d shows the exponential series of cell numbers in the colony (2, 4, 8, 16, 32 . . .) that one would expect if colony formation was the result of a failure to separate after binary fission. In all these cases the individual flagellate cells lie together in a common matrix (Fig. 1e) which collectively may take the overall form of a flat plate or complete sphere. Other examples are given in Fig. 1 to illustrate groups of cells that appear to have been formed by incomplete division. *Polykrikos* (Fig. 1f) is a marine dinoflagellate which is composed of four cells joined end to end, though each cell retains its typical dinoflagellate morphology. In the case of the parasitic astomatous ciliate *Radiophrya* (Fig. 1g) the group is temporary, because chains of daughter cells, formed by transverse binary fission, eventually do break free from the mother cell. There are many other examples of incomplete division, and the formation of "monsters" by incomplete division is also well documented. For example, chains of incompletely separated cells of the ciliate *Paramecium caudatum* each with its own mouth, have been reported to occur when fed on specific bacteria that did not completely satisfy the dietary requirements of the ciliate (Curds and Vandyke, 1966).

The second method of colony formation is less widely distributed and is almost exclusively restricted to the Sarcodina. In this method, the colony is derived from a population of individual cells which aggregate after binary fission and complete separation have taken place. The best known examples of organisms which behave in this way are the slime mould amoebae such as *Dictyostelium discoideum* (Loomis, 1975) where the small completely separate amoebae aggregate to form a pseudoplasmodium as a response to food limitation. Slime moulds have already been discussed in detail by Carlile in the previous paper. Recently Olive (1975) noted the discovery of a ciliate which was observed to form an amoeboid colony by aggregation in a way similar to the slime moulds. There are no other records of ciliates or indeed protozoa outside the Sarcodina behaving in this manner and perhaps we should regard this preliminary report with caution until more evidence becomes available.

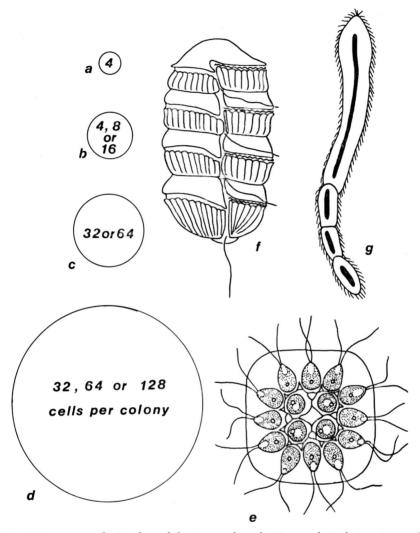

Fig. 1. Protozoan colonies formed by incomplete division. a–d. Relative sizes and numbers of cells in colonies of *Gonium sociale*, *G. pectorale*, *Eudorina elegans* and *Pleodorina californica* respectively, e. *Gonium pectorale*, f. *Polykrikos kofoidi*, g. *Radiophrya hoplites*.

TYPES OF PROTOZOAN COLONIES

The extent to which collections of protozoa have become adapted to colonial existence varies enormously. Perhaps the most simple form of group behaviour is that shown by certain solitary protozoa that are gregarious by nature. For example, the peritrich *Vorticella convallaria* is a solitary but gregarious ciliate that tends to settle in clumps even when there are plenty of other unoccupied sites available. However this feature is characteristic of only certain species, as there are many other species of *Vorticella* that are not gregarious.

Some groups of protozoa are of a temporary nature only and two examples are given in Fig. 2. The parasitic zooflagellate *Leptomonas* (Fig. 2a) is found in the

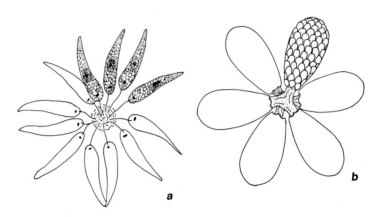

Fig. 2. Rosette formation in a. *Leptomonas* and b. *Euglypha rotunda* (after a scanning electron micrograph by Hedley and Ogden, 1973).

gut of certain insects, particularly in some Diptera and Hemiptera. These protozoa tend to attach themselves to the epithelial cells of the host to form rosettes of numerous, attached, swaying and rotating organisms. Rosette formation is well known in other flagellates such as *Trypanosoma* but the signicance is unknown. Certain amoebae are also reported to form rosettes. The testate amoeba *Euglypha rotunda* (Fig. 2b) was shown by Hedley and Ogden (1973) to produce pseudopodia which fused with those of other cells so that the cytoplasm was shared between several individuals. It would appear that temporary aggregations of this type in amoebae are responses to food limitation, an explanation which agrees with the data concerning the slime mould amoebae (Loomis, 1975). Some naked soil amoebae such as *Mayorella palestinensis* are known to clump together without cytoplasmic connections both with members of their own species and in some cases with other amoebae species which do not normally clump (Band and Mohrlok, 1969). Various environmental factors such as food limitation, pH and temperature have been suggested as the stimuli to clump.

The formation of temporary groups is also seen in some Heliozoa and in

Actinophrys sol there is a remarkable example of co-operation between individuals of this otherwise solitary organism. Looper (1928) found that when *A. sol* was fed upon small desmids the animals engulfed the prey with their pseudopodia in the normal way. However, if the population was fed upon large desmids an individual *Actinophrys* was not large enough to engulf them alone. In these circumstances several heliozoan individuals clumped together around a desmid, and fused their pseudopodia so that a sufficiently large jointly owned pseudopodium was formed with which to engulf the desmid. This therefore is an example of a protozoon taking advantage of a temporary collective association to meet the demands of the environment.

Many protozoa form groups of a more permanent nature by living clustered together either within a common organic secretion, joined by a common stalk or by the fusion of several shells. Certain species of the ciliate *Vaginicola* live in pairs within a common theca. Other peritrichs such as *Epistylis* have a common branching stalk, while several individuals of the flagellate *Bicosoeca* live within tests that are permanently cemented together. In all of these examples there are no cytoplasmic connections between the cells and the colonies behave more as a collection of separate individuals without co-ordination than as an integrated community.

CO-ORDINATION IN COLONIES

Although it is true to say that groups of cells do not begin to behave as a single entity until they are joined by some cytoplasmic connection there is some evidence to suggest that sometimes a limited amount of co-ordination can occur even when the cytoplasm of the cells is not continuous. In the simpler peritrichous ciliate colonies such as *Carchesium* where the "muscle strand" or myonemes of the stalks of the zooids are not vitally connected there is some co-ordination in the growth of the colony. In *Carchesium limneticum* the zooids are of the same size and morphology yet they do differ in reproductive potential and in the length of the stalks that they are able to produce (Fauré-Fremiet, 1948). An apical cell of *Carchesium* has the ability to reproduce indefinitely and at each division daughter branch cells are produced, each of which is capable of producing only seven other branch cells. During these six divisions there is a progressive reduction in the number of further possible divisions and in the length of stalk produced. Thus the zooid produced by the final division no longer reproduces and it produces the shortest stalk. In this way the colony increases in size presumably indefinitely and probably spreads from place to place by mechanical breakage of the stalk. Fauré-Fremiet (1948) explained this apparent differentiation of zooids on the basis of the segregation of some determining factor which is in continuous supply to the apical cells but limited in the branch cells so that it is reduced more and more with each successive division.

In the more advanced colonial peritrichs such as *Zoothamnium* there is complete cytoplasmic continuity through the continuous myonemes and in this genus there is certainly more co-ordination and integration. Furssenko (1929) investigated the growth and morphology of the freshwater species *Zoothamnium arbuscula* while Fauré-Fremiet (1930) and Summers (1938a, b) have studied the marine species *Z. alternans*. In *Zoothamnium* the zooids are morphologically dissimilar, serve different functions and the colony grows in a more integrated fashion so that it does not proliferate indefinitely. The life-cycle of *Zoothamnium* is illustrated in Fig. 3. A colony of this animal begins as a large free-swimming individual known as the migratory macrozooid or "ciliospore" which settles on a suitable substratum and secretes a stalk containing a central myoneme. At this stage the organism may encyst, but if environmental conditions are favourable it will feed and ultimately undergo an unequal division to produce a large terminal zooid and a smaller branch zooid. During all unequal divisions both macronucleus and cytoplasm divide unequally so that it is possible to obtain segregation of nuclear and/or cytoplasmic controlling factors. The terminal cell then divides unequally again to produce a second branch zooid on the opposite side to the first. After several branches have been formed in this way the branch zooids are then able to divide unequally to produce axial macrozooids and a second branch cell or microzooid. Only the terminal cell in the branches normally has the ability to divide and this soon ceases to do so. This results in the branch attaining a definite length, giving a particular form to the colony, appropriate to the species. Axial macrozooids do not appear until several branches have been produced and they are never formed on adjacent branches. Free-swimming ciliospores develop from the axial macrozooids and these serve to disperse the colony.

In common with other related peritrichs, *Zoothamnium* is able to reproduce sexually as well as by the asexual production of ciliospores. In sexual reproduction some of the subterminal branch zooids, apparently either at random or as a response to a local injury to one of the main branches, produce free-swimming micro-conjugants. These swarmers break free from the colony and swim to and fuse with either an apical or axial macrozooid. If the swarmer fuses with an axial macrozooid a fertilized ciliospore is produced which behaves as an asexual ciliospore in that it initiates a new colony. However, should an apical macrozooid be fertilized, the resultant zygote remains *in situ* and the branches below begin to behave like separate colonies with their terminal and subterminal

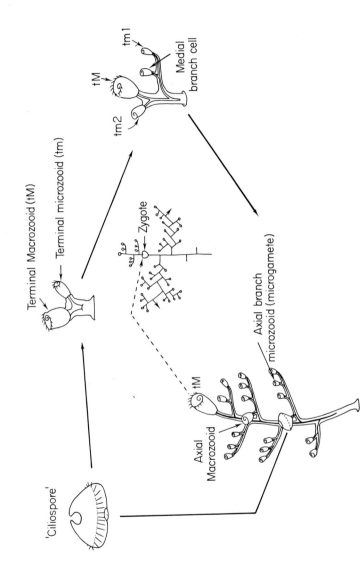

Fig. 3. Life-cycle and growth of a *Zoothamnium* colony (modified after Furssenko, 1929 and Summers, 1938a, b).

zooids multiplying to produce sub-branches with the original branches acting as new colonial axes. It would appear in these situations that either the apical macrozooid loses its dominance after fusion, or the nuclear reorganization has a rejuvenating effect which spreads to adjacent cells.

In the flagellated protozoa, there is a similar gradation of colonial entity leading to true colonial integration. Even in the simpler colonial flagellates as in *Pleodorina* there is a definite polarity to the colony in the way that it swims. Here, the anterior pole is denoted by the position of smaller cells that have larger eyespots which suggests that these cells are more sensitive to light.

In *Volvox* (Janet, 1912, 1922, 1923; Pocock, 1933a, b; Smith, 1944) there is also a definite anterior and posterior hemisphere. In this case the cells in the anterior hemisphere do not reproduce at all and daughter colonies are restricted to the posterior. The cells adjacent to each other are joined by narrow cytoplasmic connections and it is tempting to regard the continuity of behaviour and colonial organization as a whole, as a consequence of this protoplasmic continuity. This is particularly suggested by the evidence that cells detached from the colony in some way quickly lose their ability to continue the basic life processes in isolation and soon die. However not all species of *Volvox* have vital connections between the constituent cells yet they still retain colonial integrity.

Volvox has the ability to reproduce asexually and sexually and in each case specialized cells carry out these functions. In asexual reproduction the specialized cells are known as gonidia, but there may only be eight such cells out of a total of several thousands in a colony. Gonidia divide to form small daughter colonies in the posterior hemisphere and apparently these divide, at a progressively slower rate, to produce a more or less specific number of cells in the daughter whose actual number varies with the species. Cytoplasmic continuity with the mother colony is not broken until birth when the daughter colony has the correct number of cells. After birth, the daughter grows by the increase in the size of the constituent cells not by further division.

In sexual reproduction of *Volvox* we find for the first time that there can be colonies of different sexes, although some species are hermaphrodite. In some species the male gametes are produced by certain specialized cells, while in others all of the cells are capable of producing gametes. Female cells in these colonies are fewer in number and after fertilization develop into daughter colonies similar to those produced asexually.

Finally it is in the slime moulds that we find the most complex colonial behaviour which must presumably require the most precise co-ordination mechanisms. Here we really see behaviour that at least superficially appears to be more related to that of a multicellular organism than to that of a colony of single-celled organisms.

REFERENCES

BAND, R. N. and MOHRLOK, S. H. (1969). An analysis of clumping in the soil amoeba *Mayorella palestinensis*, *J. Protozool.* **16**, 35–44.
CURDS, C. R. and VANDYKE, J. M. (1966). The feeding habits and growth rates of some fresh-water ciliates found in activated-sludge plants. *J. appl. Ecol.* **3**, 127–137.
FAURÉ-FREMIET, E. (1930). Growth and differentiation of the colonies of *Zoothamnium alternans* (Clap. and Lachm.). *Biol. Bull. mar. biol. Lab. Woods Hole* **58**, 28–51.
FAURÉ-FREMIET, E. (1948). Croissance et morphogénèse des colonies de *Carchesium limneticum* Svec. *Anais Acad. bras. Cienc.* **20**, 103–115.
FURSSENKO, A. (1929). Lebenszyklus und Morphologie von *Zoothamnium arbuscula* Ehrenberg. (Infusoria Peritricha). *Arch. Protistenk.* **67**, 376–500.
HEDLEY, R. H. and OGDEN, C. G. (1973). Biology and Fine Structure of *Euglypha rotunda* (Testacea:Protozoa). *Bull. Br. Mus. nat. Hist.* (Zool.) **25**, 119–137.
JANET, C. (1912). "Le Volvox" Mém. 1. Limoges.
JANET, C. (1922). "Le Volvox" Mém. 2. Paris.
JANET, C. (1923). "Le Volvox" Mém. 3. Paris.
LOOMIS, W. F. (1975). "*Dictyostelium discoideum*. A developmental system". Academic Press, New York.
LOOPER, J. B. (1928). Observations on the food reactions of *Actinophrys sol*. *Biol. Bull. mar. biol. Lab. Woods Hole* **54**, 485–502.
OLIVE, L. S. (1975). "The Mycetozoans" Academic Press, New York.
POCOCK, M. A. (1933a). *Volvox* and associated algae from Kimberley. *Ann. S. Afr. Mus.* **16**, 473–521.
POCOCK, M. A. (1933b). *Volvox* in South Africa. *Ann. S. Afr. Mus.* **16**, 523–545.
SMITH, G. M. (1944). A comparative study of the species of *Volvox*. *Trans. Am. microsc. Soc.* **63**, 265–310.
SUMMERS, F. M. (1938a). Some aspects of normal development in the colonial ciliate *Zoothamnium alternans*. *Biol. Bull. mar. biol. Lab. Woods Hole* **74**, 41–55.
SUMMERS, F. M. (1938b). Form regulation in *Zoothamnium alternans*. *Biol. Bull. mar. biol. Lab. Woods Hole* **74**, 130–154.

4 | Individuality and Graft Rejection in Sponges
or
A Cellular Basis for Individuality in Sponges

A. S. G. CURTIS

Department of Cell Biology, University of Glasgow, Glasgow, Scotland

Abstract: Boundaries of non-coalescence which mark the individuality of particular sponge individuals of the same species are often seen in communities of sponges. The biological basis of these non-coalescence sponges has been investigated by testing for the acceptance or rejection of grafts. The results of these experiments suggest ways in which coloniality can be limited and controlled.

INTRODUCTION

One of the clearest tests of individuality in higher organisms is to graft between individuals. Grafts are rejected by immune mechanisms based on the fact that there is very little likelihood of two individuals possessing the same sets of histocompatibility antigens. However we might, at this stage, wonder whether these antigenic differences are only incidental consequences of the individuality of organisms. An apparently much simpler feature of individuality is that an organism will maintain its separateness from another individual of the same species. For instance van de Vyver (1971) observed that many individual masses of the freshwater sponge, *Ephydatia fluviatilis*, would fail to fuse with other sponges of this species when placed side by side. By studying the fusibility of sponges of this species from the area around Brussels, she discovered that there are a number of strain types in this species, each strain type being defined by the fact that it will not fuse with another strain. When two unlike strains are placed side by side a temporary fusion forms which is replaced by the appearance of

---

Systematics Association Special Volume No. 11, "Biology and Systematics of Colonial Organisms", edited by G. Larwood and B. R. Rosen, 1979, pp. 39–48. Academic Press. London and New York.

a zone of *non-coalescence* (also known as non-confluence). A space appears between the sponge bodies. These experiments of placing sponges side by side are, of course, very akin to grafting conditions and the first question investigated in this paper is whether non-coalescence correlates with graft rejection.

Curtis and van de Vyver (1971) showed that non-coalescence was probably due to the production of factors by the cells of each strain type that diminish the adhesiveness of the other strain type. Thus when two unlike strains are placed side by side the concentrations of these factors derived from the other strain rise in the contact zone of a sponge leading to a diminution in cell adhesion so that the cells in the contact zone fall out. These factors, which may have counterparts in vertebrates (Curtis, 1974 and Curtis and de Sousa, 1975) can be termed morphogens or interaction modulation factors. The second question examined in this paper is whether such factors operate in graft rejection in sponges.

MATERIALS AND METHODS

The majority of the experiments were carried out on specimens of *Hymeniacidon* sp. (probably *perleve*) though some preliminary development work was done by Mr. R. Harvey on *Halichondria panicea* and *Pachymatisma johnstoni*. Considerable attention was given to the identification of the sponge species used, but the difficulty with this group and the fact that the large scale morphology of these sponges does not correspond with any described species has made identification difficult.

Grafting was carried out underwater, using SCUBA equipment, leaving the sponges *in situ*. All the grafts were carried out on sponges in the Taynish channel off Loch Sween (Nat. Grid ref. NR 724827).

Grafts were removed from donors by cutting with a scalpel to remove a tissue mass approximately 1 cm by 1 cm by 2–4 cm deep. The grafts were immediately transferred to similar size holes cut in the hosts. Normally grafts were carried out in such a way that the graft reached down to the subjacent rock. Usually a graft was placed in the hole left after another graft had been derived. Sketches and photographs were made of the grafts and hosts and the set of sponges amongst which grafts had been made were labelled by fixing a marker to the rock. In general between four and fourteen non-coalescent individuals might be found on one rock. Each set or group of sponge individuals is referred to in the Tables by a number, while individuals are given a letter, so that 14A and 14E refer to two sponges in group 14. Individuals might have up to six non-coalescent contacts (see Fig. 1 and Plate I) though four contacts was the average. Grafts were examined usually at monthly intervals after grafting. A graft was scored as "accepted" if after one or two months:

PLATE I. Three individuals of *Hymeniacidon* sp. showing zones of non-coalescence. Note that the individual sponges have slight morphological differences. Picture height 10 cm.

(a) It could not be separated from the host on gentle palpation.
(b) No pinacoderms had grown down the sides of the graft and host.
(c) It was joined to the host by a continuous pinacoderm.
(d) It was in good health, as judged by colour, presence of a pinacoderm and active water transport.

A graft was scored as "rejected" if:
(a) There were pinacoderms lining the sides of graft and host.
(b) There was a clearly evident gap between host and graft.

A small number of grafts could not be found one month after grafting—these were ignored.

Sponge factors, interaction modulation factors, were prepared from sponges transferred to the Dunstaffnage Marine Laboratories, Oban and kept at 6°C for not more than 20 hours in a circulating sea water system. Sponges were washed in CMF seawater (composition NaCl 35 g litre^{-1}, KCl 0·7 g litre^{-1}, 10^{-2} M Hepes buffer) and then disaggregated in CMF seawater containing 1×10^{-2} M EDTA for six minutes, using mechanical dispersal. The cells were pelleted by centrifugation and the supernatants were filtered through Millipore 0·22 μm pore size filters. Each ml of supernatant was the product of between 1×10^7 and 3×10^7 cells.

The effect of these factors on cell adhesion was measured using my (Curtis, 1969) technique. Cells were prepared from the disaggregates described above by resuspending the pellets in CMF seawater, repelleting and repeating these stages again. The suspensions were finally made up in artificial seawater. 20% by volume of a homologous or heterologous factor preparation was then added and the adhesiveness of the cells measured taking four or five measurements over a 30 minute period. Shear rates of $c.$ 10 sec^{-1} were used. The addition of the factor preparations means that a certain amount of EDTA was added to the system but since the control sets (cells plus their homologous factor) also contained EDTA an adequate control was provided.

RESULTS

1. Non-coalescence

It was established by observation that non-coalescence occurred between the sponges listed in Table I. See also Fig. 1.

2. Graft Rejection

A number of autografts were made as well as allografts between non-coalescent individuals and remote sponges. Results are set out in Table II.

TABLE I. Correlation between non-coalescence and graft rejection

Non-coalescence between strain types		Graft rejection (tested in both directions)	Non-coalescence between strain types		Graft rejection (tested in both directions)
14 A	14 G	Not tested	14 J	14 K	Rejection
14 B	14 C	Rejection	14 L$_1$	14 M	Rejection
14 B	14 L$_2$	Rejection	14 M	14 N	Rejection
14 B	14 E	Rejection	13 A	13 B	Rejection
14 C	14 D	Rejection	13 A	13 C	Rejection
14 D	14 G	Rejection	13 B	13 C	Rejection
14 D	14 L$_1$	Rejection	13 B	13 E	Rejection
14 E	14 F	Rejection	13 C	13 D	Not tested
14 G	14 H	Rejection	13 C	13 E	Rejection
14 G	14 J	Rejection	13 D	13 A	Not tested
14 H	14 I	Not tested			

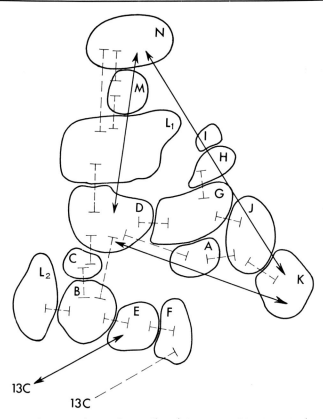

FIG. 1. Diagram showing the grafting effected in Group 14 *Hymeniacidon* sp. Heavy pointed arrowed lines indicate reciprocal accepted grafts, light broken barred lines indicate those reciprocal grafts that were rejected. Non-coalescences occurred wherever two individuals abut. The dimension B to N is about 0·8 m.

I observed that some of the grafts that were rejected appeared to become necrotic with the host slowly growing into the space previously wholly occupied by the graft: in other cases the grafts appeared to enlarge at the expense of the host. Nevertheless in many cases the grafts remain alive for long periods but do not appreciably enlarge or diminish in size.

TABLE II. Graft Rejection

Graft type	Fusion (Acceptance)	Rejection With necrosis	Graft survives
Autografts	24	0	0
Allografts			
(a) between non-coalescent neighbours	0	10	21
(b) between remote sponges	16	7	27

In particular it should be noted that all autografts are accepted and all allografts between non-coalescent individuals are rejected, while grafts between remote sponges are usually rejected. The incidences of acceptance and rejection can be used to calculate the probable number of strains present. In order to do this it has to be assumed that the strains are present in equal proportions in the population. If there are five strains, the chance of selecting any specified pair of sponges of the same strain type is 1 : 25, but because there are five types, the chance of selecting any like pair is 1 : 5. The data on allografts in Table II has to be adjusted against selection of either non-coalescent or remote graft, by calculating the number of neighbours and remote pairs to be expected in this population of 20 sponges. There are 190 possible pairs (380 if reciprocal grafts are considered) of which 80 (160) will be neighbour pairs since the average number of neighbours is four. Thus 50 remote grafts will be matched by a population of 37 neighbour grafts, 16 of the 87 grafts are accepted, i.e. 1 : 5 corresponding to five strains.

3. Interaction Modulation Factors

Measurements of collision efficiency, a parameter of adhesion, were made on all the sponges amongst which grafting had been done. The adhesiveness of those cells treated with a factor preparation made from the same sponge was always appreciable (see Table III). If the factor was taken from another strain which would accept grafts from and with the strain type of the cells there was little

TABLE III. Effect of interaction modulation factors on adhesion of various strains.

Cell type	Factor type	Adhesion (collision efficiency %)
(a) Autologous combinations		
14 B	14 B	18·6
14 D	14 D	12·2
14 E	14 E	15·6
14 F	14 F	22·5
14 K	14 K	12·1
14 L_2	14 L_2	9·6
14 N	14 N	21·7
13 A	13 A	13·7
13 B	13 B	14·2
13 C	13 C	12·8
13 E	13 E	12·4
(b) Accepted allograft combinations		
13 C	14 E	15·8
14 E	13 C	21·8
14 D	14 K	11·0
14 D	14 N	7·4
14 K	14 N	18·4
14 K	14 D	16·0
14 N	14 D	21·7
14 N	14 K	13·4
(c) Rejected allograft combinations or non-coalescent pairs		
13 A	13 E	0·0
13 A	13 B	0·0
13 B	13 E	0·0
13 C	14 F	0·0
13 C	13 E	0·0
13 E	13 C	0·0
14 B	14 E	1·4
14 B	14 L_2	0·06
14 D	14 L_1	1·2
14 E	14 F	0·03
14 E	14 B	0·0
14 F	14 E	1·8
14 L_1	14 D	0·0
14 L_1	14 N	0·0
14 L_2	14 B	0·9

effect on cell adhesion. However factor from those strain types that are not graft compatible with the strain type of the cells leads to a very considerable decrease in cell adhesion.

DISCUSSION

The results show that graft rejection may occur between individual sponge bodies of the same species. This does not appear to have been reported previously, though Paris' (1960) failure to obtain graft acceptance within *Tethya lyncurium* may well be an example of this phenomenon. Very few reports on graft rejection in sponges appear to have been made, though the interspecific grafting by Moscona (1968) should be noted. The incidence of graft rejection which suggests that there may be five strain types in the area studied is compatible with the concept that there are a number of histocompatibility alleles in *Hymeniacidon*.

The results also show that there is a very close correlation between graft rejection and non-coalescence and that interaction modulation factors diminish adhesion in all those combinations where rejection or non-coalescence would take place. Thus there is strong reason for suggesting that non-coalescence and graft rejection are due to the action of interaction modulation factors. Such factors would reduce the adhesiveness of the cells of both host and graft close to the contact zone. As a consequence a gap would form between graft and host.

It is remarkable how close a parallel there appears to be in graft rejection reaction throughout the animal kingdom for histocompatibility reactions involve two or more loci in sponges, ascidians (Oka, 1970), coelenterates (Hauenschild, 1956) as well in vertebrates. In all such reactions there is some evidence that diminution of adhesion may be an important feature. Recently Curtis and de Sousa (1975) have obtained evidence for the action of interaction modulation factors in vertebrates. It is possible to speculate that the basic graft rejection reaction is one of the production of and reaction to interaction modulation factors and that the involvement of antibody formation in graft rejection in higher animals is only a matter of secondary importance.

The existence of such a system is presumably important in producing the non-coalescence phenomenon in the field. It is clear that such a phenomenon will prevent the formation of animals which are genetic chimerae. Such a system will also prevent one sponge overgrowing another.

The observation that some of the grafts necrose, others grow at the expense of their host and many survive unchanged in size, suggests that there may be a hierarchy in the system such that some grafts are producers of so active an interaction factor that they slowly erode the host away, while at the other end of the hierarchy the graft is so damaged by factors from the host that it dies. Such a hierarchy might be of considerable importance in determining the structure of sponge communities in an area.

The close similarities between this graft rejection system and those seen in animals where we would have no difficulty in saying that a single body is a single individual suggests that the true definition of an individual is that it is a single assembly of cells of the same genotype. The coloniality clearly seen in coelenterates for example and rather less clearly in sponges is nothing more than the coloniality which is displayed by cells of, say, a vertebrate grown in culture. Thus these results lead us to the statement which Borojevic (1970) proposed, namely that the sponge individual is that which is enclosed by a single cellular pinacoderm. (Those sponges with a collagenous pinacoderm must be excluded from this definition.) It can now be added that individuality in sponges may be basically expressed and maintained by interaction modulation systems.

ACKNOWLEDGEMENTS

I should like to thank the following persons: Mr J. Pease and the officers of the Nature Conservancy who made access to the site possible; Mr R. Harvey for carrying out useful preliminary work on the site; Dr J. Blaxter and the staff of the Marine Biological Station, Dunstaffnage, for providing laboratory facilities; and Dr C. Evans for advice and encouragement.

REFERENCES

BOROJEVIČ, R. (1970). Le comportement des cellules d'éponge lors de processus morphogénétiques. *Ann. Biol.* **10**, 533–545.

CURTIS, A. S. G. (1969). The measurement of cell adhesiveness by an absolute method. *J. Embryol. exp. Morph.* **22**, 305–325.

CURTIS, A. S. G. (1974). The specific control of cell positioning. *Archs. Biol., Paris* **85**, 105–121.

CURTIS, A. S. G. and DE SOUSA, MARIA, A. B. (1975). Lymphocyte interactions and positioning. *Cellular Immunol.* **19**, 282–297.

CURTIS, A. S. G. and VAN DE VYVER, G. (1971). The control of cell adhesion in a morphogenetic system. *J. Embryol. exp. Morph.* **26**, 295–312.

HAUENSCHILD, C. (1956). Ueber die Vererbung einer Gewebertiaglichkeits-Eigenschaft bei dem Hydroidpolypen *Hydractinia echinata*. *Z. Naturforsch.* **11**, 132–183.

MOSCONA, A. A. (1968). Cell aggregation properties of specific cell ligands and their role in the formation of multicellular systems. *Dev Biol.* **18**, 250–277.

OKA, H. (1970). Colony specificity in compound ascidians. In "Profiles of Japanese Science and Scientists" (H. Yukawa, ed.), pp. 196–205. Kodansha Ltd., Tokyo.

PARIS, J. (1960). Greffes homoplastiques et bourgeonnement expérimental chez *Tethya lyncurium* Lamarck. *C.r. hebd. Séanc. Acad. Sci. Paris* **245**, 578–580.

VAN DE VYVER, G. (1970). La non confluence intraspécifique chez les spongiaires et la notion d'individu. *Ann. Embryol. Morph.* **3**, 251–262.

ADDENDUM

COLONIALITY

The definition of a colonial organism needs re-examination in the light of recent developments in cell and developmental biology. Obviously a colonial organism must be composed of cells of one genotype. Those assemblages formed by the co-settlement of several different organisms may at times superficially resemble a colonial organism. One of the tests of common genotypy is to carry out grafts between the two or more sections of the colony. Graft rejection will reveal that the grafts come from different organisms.

The next question is whether the colony behaves as an individual. This implies that there is some form of cell communication throughout the colony. Pavans de Ceccatty (1974) has shown that sponges show co-ordinated behaviour throughout an individual. If separate parts of the colony are not connected behaviourally they will never pursue a co-ordinated behaviour. Thus they cannot be regarded as being part of a colony.

Up to this point the definition of a colonial organism differs in no respect from that accorded to a unitary organism such as a vertebrate. The early definition of coloniality would emphasize that the colony consisted of repeated units each of which contained sufficient organs to operate independently of the others. This leads to the concept of regenerability in colonies. Thus the real test, very rarely used, is whether one of these morphological units can regenerate into a number of units. The non-colonial organisms cannot regenerate into further units. I suspect that some of the colonial organisms, which possess regenerating units have evolved into close relatives, e.g. the siphonophores, where the units have become morphogenetically interrelated in such a way that they have lost the ability to regenerate from apparent morphological units.

REFERENCE

PAVANS DE CECCATTY, M. (1974). Coordination in sponges. The foundations of integration. *Am. Zool.* **14**, 895–903.

5 | Taxonomy, the Individual and the Sponge

W. G. FRY

Department of Science, Luton College, Luton, England

Abstract: Since the acceptance of the animality of sponges in the first half of the nineteenth century, whole sponges, their component amoeboid cells, flagellate cells, lobes, branches, choanocyte chambers and oscula have been credited with individuality. To no one of these units can a definition of "individual" be fitted well. In such default the term colony can have no precise or useful meaning.

Sponge taxonomists rely upon three systems: cellular, skeletal and aquiferous, for the majority of their data. Cellular and skeletal units or modules are rarely discernible. However, aquiferous units (modules) can be recognized, both by their functions and by the structure of some of their parts. Although in life they are fugacious and overlapping in function, the forms and patterns of grouping of individual aquiferous modules have great ecological and functional significance, while in preserved collections they determine completely the practice of sound taxonomy.

Having a severely compromised and perhaps only mystical individuality in life, aquiferous modules have a real operational individuality in taxonomy.

INTRODUCTION

In both the earliest and the most recent writings whole sponges are usually considered as individuals, but the notion of sponges as colonies has a long and respectable history. Since the first half of the nineteenth century, when the animality of sponges became acceptable, branches and lobes, choanocyte chambers, amoeboid and flagellate cells have all been proposed as the individuals of the sponge colony.

Because the word "colony" has its origins in concepts of the State and in politics, its definition and its employment have been susceptible *ab initio* to metaphysical bias. Similarly—but here because of political and religious associations—the term "individual" must be used circumspectly in biology. The

---

Systematics Association Special Volume No. 11, "Biology and Systematics of Colonial Organisms", edited by G. Larwood and B. R. Rosen, 1979, pp. 49–80. Academic Press, London and New York.

satisfactory definition of "colony" in biology depends upon the satisfactory definition of "individual", but because of the implications accreted to the two words it does not seem likely that universally acceptable definitions of both terms can be sought with hope. Rather, it is to be feared that an endless semantic would have begun (Mackie, 1963). This situation may be all to the good, for biology progresses by case law rather than by statute law. Furthermore, within the science the concepts of colony and individual are most useful as operational models whose manipulation produces information outside the concepts but within other fields, such as genetics, evolution, immunology and morphogenesis.

The majority of spongologists of this and previous times have speculated on the colonial or individual nature of sponges and it is impossible to précis here all their arguments. The major features of the arguments have been discussed most ably by Hartman and Reiswig (1975), to whose paper and to the bibliography of this article the reader is referred.

It is, perhaps, not surprising to find that some of the earlier writings on individuality and coloniality of sponges have a polemical flavour. We must not forget that their authors lived and wrote in times when scientific concepts mattered on the grand scale. Undoubtedly, the writings reflect, with varying strengths, the achieved and developing philosophies of their authors. On that count the writings are historically very interesting, but we should be especially grateful that they engendered and maintained the intellectual ferment out of which came so many useful facts.

It seems that—in biology at least—we do not live at a time when we are prepared to stake our reputations upon concepts in science which have deeper philosophical and metaphysical implications. We seek data, we manipulate data, we are often swamped by data, and we tend to be pragmatic. Yet it seems to me also that in his practice the sponge taxonomist avoids consideration of the individual or colonial nature of his specimens at grave risk to the universality of his conclusions and to the predictive powers of his classifications. A sponge taxonomist would be wise not to accept passively the neutralist view of the sponge as an "individualized complexus (cormus)" (Hadži, 1963).

Sponge taxonomists concern themselves principally with three systems—aquiferous, cellular and skeletal. They take samples of a sponge and determine the attributes of the skeletal, the aquiferous and (sometimes) the cellular systems represented in the samples. Consequently, it is prudent to ensure that the data are adequate representations of the systems.

In the following sections I shall explore some of the problems of adequate sampling for taxonomic purposes by considering the repeating patterns of the subunits of the cellular, aquiferous and skeletal systems. The repeating patterns

5. Taxonomy, the Individual and the Sponge

are best understood in relation to the functions of the systems, which, in turn, are comprehensible only in relation to the ontogenies and the internal and external hydrodynamic regimes of sponges.

CELLS, MORPHOGENESIS AND CO-ORDINATION

In 1967 a group of spongologists met at Roscoff and considered, *inter alia*, the definition of the sponge individual. That their published definition of the individual as "A mass of sponge substance bounded by a continuous pinacoderm" (Borojevič *et al.*, 1968) suggests a resigned compromise is not surprising. Although Mergner (1964, 1970) has shown, in a fresh water sponge, induction processes which imply that different portions of the aquiferous system have some biochemical tissue integrity, a majority of spongologists would not accept a description of sponges or sponge components in terms of tissues. Rather, it is now accepted practice to refer to cell lineages (i.e. differentiation pathways, Fig. 1) and to consider the behaviour and the fates of individual cells within various assemblages of the lineages (Borojevič, 1966; Efremova, 1967; Fry, 1970a; Lévi, 1969, 1971; Sarà, 1974; cf. Kilian, 1964). Representatives of all lineages occur throughout the sponge, their particular states of differentiation being dependent upon their immediate neighbours and upon the local physico-chemical environment. Individual cells may also move long distances within the sponge, differentiating as they move.

In adopting this approach we are regarding afresh facts fastened upon some 100 years ago for quite different reasons. Thus, Carter (1848) and Perty (1852) considered sponges to be aggregations of amoeboid cells. Later, James Clark (1866, 1868), Carter (1872) and Saville Kent (1878a, b) came to regard sponges as aggregations of flagellate cells—the "spongozoa". The polarization of views was not, of course, complete, and Carter (1857, 1869, 1871) regarded the true sponge cells as amoeboid-flagellate (cf. Willmer, 1960). Nevertheless there was for a time a strong body of opinion that sponges are aggregations of Protozoa. Many new forms of choanoflagellate Protozoa were described during those years and the fact of their sometimes occurring in groups (e.g. *Codosiga botrytis*, *Desmarella moniliformis*, *Proterospongia haeckeli*), lent particular force to this opinion.

In even the smaller sponges there are many millions of cells, and at any moment the form and functions of a particular cell depend upon the lineages and the state of differentiation of that cell's immediate neighbours. Recent work on protozoan choanoflagellate cells (e.g. Leadbeater and Morton, 1974; Hibberd, 1975) has tended to emphasize their essential solitariness and self-

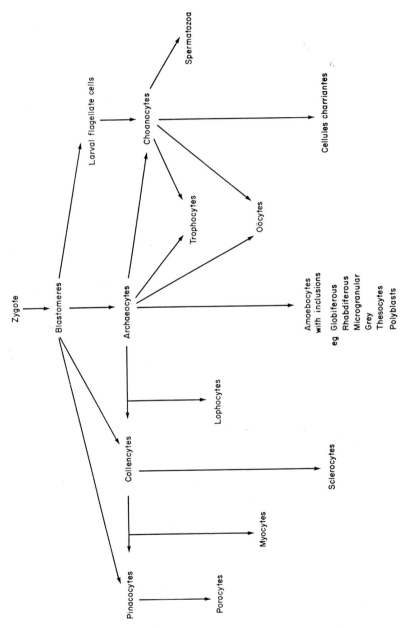

FIG. 1. Pathways of differentiation of cells in Demospongia and Calcarea during embryogenesis, metamorphosis, growth, regeneration and reproduction.

dependence, which contrasts strongly with the continuing complex cellular dynamics of the very smallest sponge.

In the late nineteenth century Ranvier and Pekelharing (see Vosmaer, 1908) likened the organization of sponges to that of the reticulo-endothelial system of vertebrates. The analogy has striking force today when our very incomplete picture of sponge cell dynamics nonetheless indicates a complexity equal to the subtleties of the migrations and interactions of the members of the cell lineages of the reticulo-endothelial system (Burnett, 1969; Roitt, 1975; Good and Fisher, 1972; Vitetta and Uhr, 1975).

Neuroid net conduction and transmission undoubtedly occur in at least some sponges, as does whole-sponge integration of various irregular or rhythmic activities (Parker, 1910, 1914; Pavans de Ceccatty, 1958, 1969, 1970, 1971; Prosser et al., 1962; Lentz, 1966; Reiswig, 1971, 1974). Pavans de Ceccatty's later conclusions imply the presence, throughout some sponges at least, of a neuromyoid network, one of whose functions is a pseudohormonal integration of morphogenesis. Such a structure, with this function, militates strongly against coloniality. Consequently, it cannot be systemically advisable to regard sponges as colonies of unusual protozoans. To do so would be conceptually as fruitful as deciding that we exclude the Choanoflagellida from the Porifera solely because they are incapable of organizing aquiferous systems.

For those nineteenth century workers who were predisposed to consider sponges as colonial Metazoa, groups of flagellate cells offered themselves as candidates for individuality. Their constancy of form and position as the walls of choanocyte chambers encouraged the notion that the choanocyte chambers were the individuals of the sponge colony. The notion was elaborately and confusingly explored by Haeckel (1889, 1896). It has now all but disappeared, leaving only some nomenclatorial problems among the Calcarea. Any scheme purported to encompass the variations in number, size and connections of choanocyte chambers solely on the basis of preserved anatomy and phylogenetic speculation was doomed to collapse in confusion. We shall see below that the variations of the choanocyte chambers are explicable only when the hydrodynamics of the whole sponge are considered.

The sponge larva is an obvious candidate for individuality. A single larva can give rise to an enormous sponge. However, there are many records of larval fusion within the parent and when free and during and after metamorphosis and some evidence of morphogenetic advantage in this phenomenon (Fry, 1971 and Van de Vyver, 1970, *ubi litt.*). Van de Vyver (1970, see also 1971a) has shown that in the fresh water *Ephydatia fluviatilis* and the marine *Crambe crambe* there are barriers to fusion of several different intraspecific strains. At this time these and other data reveal our almost total ignorance of the genetics and biochemistry of intra- and interspecific variation in sponges (for other investigations see also Bergquist *et al.*, 1969; Curtis, 1970, this volume; Humphreys, 1970; Moscona, 1968; MacLennan, 1970; Van de Vyver, 1970–71). The possibility of some sponges being genetic mosaics (Fry, 1973) adds another dimension to this problem (see Sabbadin, this volume).

The phenomenon of fusion of large sponges in the field has interested many workers. There are several records of failure to fuse upon contact of two or more sponge masses which have been identified as conspecific (e.g. Brien, 1967; Van de Vyver, 1970, 1971a; Bromley and Tendal, 1973; Simpson, 1973). However, it is unwise to extrapolate from such observations because, while inability to fuse may be detectable, ability to fuse will more often than not pass unrecognized in the field. Where ability to fuse has been recorded, the suspicion remains that recombination of previously regressed fragments of a single sponge was involved (Burton, 1949).

If we cannot recognize unequivocally sponge cells, spatially discrete groups of cells with common functions, larvae, or even all mature sponges as individuals, we are clearly in no little intellectual difficulty. However, a way out of this impasse, and therefore out of many taxonomic difficulties, may appear through further exploitation of immunological techniques (see MacLennan, 1970; Van de Vyver, 1970, 1971, and Curtis, this volume).

1. *The Aquiferous System*

All too often the external symmetry of sponge specimens appears indeterminate (cf. Burton, 1932). However, all sponges have at least one, and often several, obvious large openings on their surfaces. These are nowadays usually called oscula, but we shall see (p. 68) that it is not necessarily advantageous to use the term indiscriminately. They are the exits for water moving through the aquiferous system of the sponge. In the history of spongology, after the whole sponge, the oscula were the first candidates for individuality (Schmidt, 1864), largely because of the previous inclusion of sponges with other zoöphytes and lithophytes and because of preoccupations with gastral cavities. The homologization of polyp enterons and the larger sponge cavities has been abandoned by the majority of workers (but cf. Brien, 1967, 1968; Beklemishev, 1964), although a number of otherwise pleasing basic text-books still greet us with the labels "gastral cavity" or "paragaster".

It is essential, for any understanding of sponges, to appreciate that the movement of water through the aquiferous system influences immediately the

5. Taxonomy, the Individual and the Sponge

entirety of the sponge and not just some special part of the animal. We can contrast this situation with that in some other animal groups, such as bryozoans, molluscs, crustaceans, brachiopods and tunicates, in which coelomic cavities, guts, hearts and kidneys possess morphological and functional integrity quite divorced from the wholesale movement of water by the animal.

The oscula have very clear physiological functions. They are the exits for water which has entered the sponge through the more numerous smaller ostia and which has had removed from it oxygen and food material but has had added to it carbon dioxide, faecal matter and nitrogenous excretory products. The oscula, therefore, pass out water which the sponge would do well to avoid taking in again. In his classic paper, Bidder (1923) explored this idea and developed the concept of the Diameter of Supply (Fig. 2).

Figure 3 is based partly upon Bidder's (1923, 1937) concepts and some more recent conclusions of Reiswig (1971–1975). The figure relates the number, size, output, and degree of grouping of oscula to a sponge's mass and the environmental hydrodynamics. O_N is the number of oscula, O_D the mean oscular diameter, E_L the mean length of the exhalant aquiferous ducting, and E_D the mean diameter of that ducting. C_V is the mean volume of the pumping chambers (choanocyte chambers) and C_N their number. P is the pumping pressure in the chambers and G the amount and pattern of grouping of the oscula. H is the external hydrodynamic regime. The water output per unit time and the velocity of ejection of water from individual oscula are determined by the internal laminar flow rates, the numbers of choanocyte chambers, the working pressure within them, the dimensions of the various exhalant ducts and the oscular diameter. In turn the exhalant velocity and output determine how far stale water is ejected from each osculum, and thus the diameter of supply and the extent to which recently exhaled water is re-cycled. The external hydrodynamic regime may reduce the effective diameter of supply to zero in shallow, perfectly still water, and extend it to infinity in strong currents.

Because of the mode of operation of the choanocyte flagella, the hydrostatic pressure generated in a choanocyte chamber is inversely proportional to the volume of the chambers. Sponges with a low tissue mass can remain in optimum physiological state with a single osculum and a single, large, low pressure pump chamber (e.g. *Leucosolenia complicata*). Sponges with a larger cell mass require a larger throughput of water and therefore a larger diameter of supply. They can achieve this up to a certain point by developing more and smaller, higher pressure pumps, increasing E_D to compensate for increased E_L, and increasing the diameter of the single osculum (O_D). Here, as in other cases, the oscular diameter is critical because, when slightly less than the diameter of the supplying

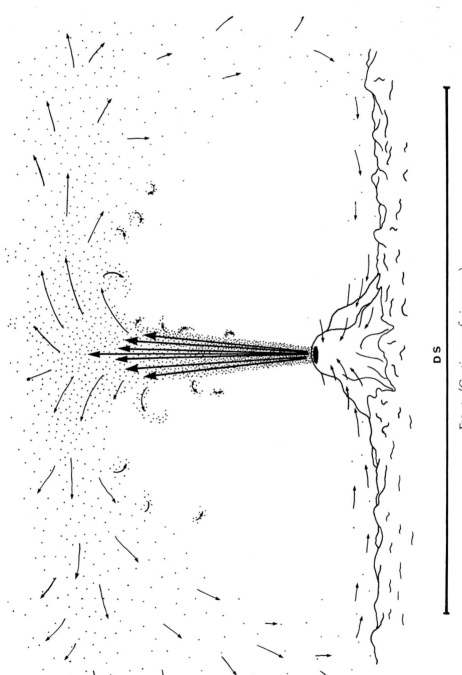

Fig 2. (Caption on facing page).

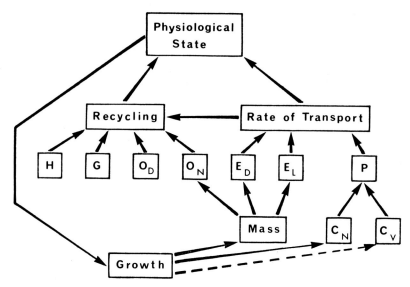

FIG. 3. Some factors determining the architecture of the aquiferous system of a sponge. The broken line indicates a relationship probably peculiar to Calcarea. O_D influences Rate of Transport only as $O_D \rightarrow$ Zero. See text for meanings of letters.

atrium, the oscular rim acts as a "stop", converting potential energy to kinetic energy and regulating the exit velocity of the water. All sponges so far studied have the ability to close their oscula partially or completely by expansion of sheets of the exopinacoderm and, in some cases, by contraction of the whole sponge. As sponges have no effectors of distortion organized on a scale above the cellular (see Bagby, 1966), there is a fairly low upper limit to the oscular diameter which can be occluded efficiently in this manner—apparently at about 1 cm.

We may regard the presence of an osculum as a response to the need for an

---

FIG. 2. The concept of Diameter of Supply (DS) exemplified by a sessile sponge, with single osculum or vent, pumping in still water. The density of the stippling is proportional to the concentration of recently exhaled water (i.e. with reduced O_2 and increased CO_2 concentrations).

Here the sponge has achieved a physiologically adequate Diameter of Supply, in that the force of the oscular/vent jet transports exhaled water sufficiently far for inhaled water to contain very little of the recently exhaled water. In still conditions this will be achieved only in a very large water volume.

DS is the diameter of the "shell" of water exhaled by the sponge. Based on Bidder (1923, 1937).

exit for water supplying a particular volume of sponge cells. Thus, associated with each osculum is a certain volume of sponge cells together with the aquiferous ducting which supplies and drains them, which we may call an aquiferous module. No such module (Fry, 1970b) can come into being and persist in a healthy state without an adequate diameter of supply.

Fig. 3 also indicates the factors which determine the volume of any aquiferous module. Different sponges have aquiferous modules of different volumes. Apparently, amongst only the Calcarea occur some few species which have high C_V and low C_N and P and which cannot decrease the former and increase the latter during growth. In these forms (see Fig. 8), when cellular volume increases beyond a certain value, a complete new aquiferous module develops as an outgrowth from the base or side of the original (e.g. species of *Leucosolenia*, *Clathrina* and *Dendya s.* Burton, 1963). Occasionally forms are found in which the original aquiferous module splits into roughly equal portions, e.g. *Clathrina coriacea* (Montagu) *s.* Burton (1963). Such new and older modules retain, at least initially, and sometimes apparently throughout life, wide connections of their aquiferous systems.

In some other calcareous forms, e.g. *Leuconia*, *Aphroceras*, *Scypha*, the sponges can undergo some increase in C_N and decrease in C_V while retaining only one osculum. When doing this they develop pouched outgrowths of the wall of the single choanocyte chamber and, by cellular bridges between these pouches, they thicken the sponge wall and increase E_L. We often meet the three separate sponge types of "Ascon", "Sycon" and "Leucon" in textbooks. In fact we can find all three types in one specimen of *Scypha ciliata* (Burton, 1963).

It appears that all of the demosponges and the sclerosponges (Hartman and Goreau, 1970, 1972, 1975) are committed from the earliest stages with a functional aquiferous system (i.e. rhagon) to low C_V and high C_N and they frequently achieve greater total volumes than can the calcareous sponges. However, since E_D, E_L, C_V, O_D and P have optimum values for maximal efficiency in any sponge, there must be an upper limit for transport through a single osculum without immediate recycling of exhaled water on a large scale. Therefore, in continued growth new aquiferous modules must be formed together with their subserving oscula. Suitably sited, new oscula can extend the effective diameter of supply of the sponge.

Among encrusting demosponges new modules arise at the periphery and often result in the formation of very large sheets bearing more than 50 oscula. In many such forms the oscula are clearly overdispersed (Fry 1970b, Brien, 1967). Often we may observe that such encrusting forms bear their oscula at the ends of chimneys, which increases the diameter of supply. Parker (1910,

1914) showed that in the digitate form *Stylotella heliophila* the oscula close in still water, presumably because the effective diameter of supply is too small to prevent recycling of water. In *Ophlitaspongia seriata* Grant, 1865, on the other

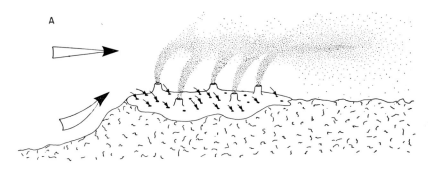

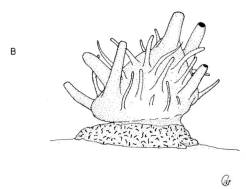

FIG. 4. A. Diagrammatic representation of the infinitely large effective diameter of supply of an encrusting sponge in a moderately strong current; B. A specimen of *Polymastia bursa* S. Koltun 1966 (8 cm h) with papillae extended. The narrower papillae are blind-ended and bear ostia only. The stouter papillae bear ostia and terminal oscula or vents.

hand, the oscular papillae elongate in still water (Fry, 1971), presumably to increase further the diameter of supply. Large specimens of such sponges as *O. seriata*, *S. heliophila* and encrusting *Halichondria panicea* are not normally found in continuously still water, and we may deduce that their diameters of supply are not in themselves sufficient to prevent recycling, but are normally extended by local water movements (Fig. 4A).

Hymeniacidon perleve (Fig. 5) which often grows as thin film encrustations with scattered, low oscular papillae, may also be found in more massive form when it bears, apparently, fewer oscula. In such cases the walls of the oscular

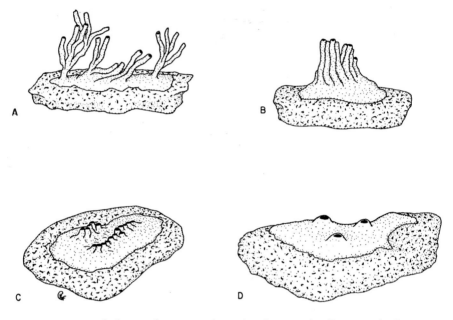

FIG. 5. Four growth forms of *Hymeniacidon perleve* (Montagu), all *c*. 3 cm h, showing various degrees of grouping of exhalant aquiferous apertures. A, B, and C with separate oscula; D, with vents only.

chimneys of neighbouring modules have fused and grown upwards together to a greater height than that achieved by solitary chimneys, thereby increasing the diameter of supply. Some of the more fantastic forms of *H. panicea* are perhaps due to similar methods of oscular grouping.

Many of the smaller globular demosponges and sclerosponges have extensible oscular chimneys which originate from low mounds e.g. *Aplysilla rosacea*, *Ceratoporella nicolsoni*. In yet other forms, notably the several species of *Polymastia* (Fig. 4b) the diameters of supply are greatly increased by all the oscula being borne at the tips of stout-walled papillae pointing away from the substratum (Boury-Esnault, 1974).

Many sponges are pedunculate, this condition being particularly common in deeper water (e.g. *Leucosolenia blanca*, *Stylocordyla borealis*, *Hyalonema hozawai*, *Chrondocladia gigantea*, *Rhizaxinella burtoni*; see Figs 6 and 8). The possession

5. Taxonomy, the Individual and the Sponge

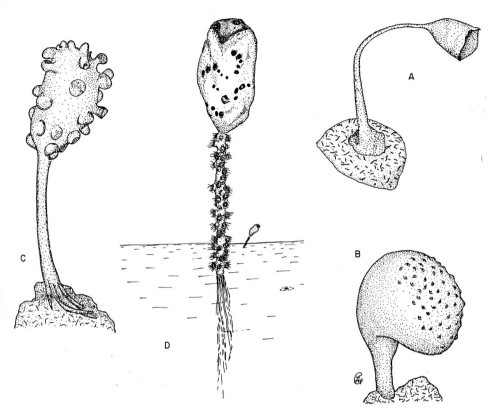

FIG. 6. Examples of stalked silicious sponges. A, *Stylocordyla borealis* (Demospongia) (13 cm h); B, *Rhizaxinella burtoni* (Demospongia) (8 cm h); C, *Chondrocladia gigantea* (Demospongia) (20 cm h); D, *Hyalonema hozawai* (Hexactinellida) (15 cm h). A and B after Koltun, 1966; C after Koltun, 1959.

of a stalk does not increase the diameter of supply but allows that diameter to be realized almost completely by the sponge (Bidder, 1923).

The development of papillae is particularly pronounced in the boring and other infaunal sponges, such as *Siphonodictyon* and *Cliona* species. In some of the latter species the papillae protruding from the substratum are, separately, oscular and ostial (Fig. 7). In the peculiar (infaunal?) form *Disyringa* the papillate condition has been refined further to widely opposite single ostial and oscular papillae.

Given that there is an upper limit to O_D for structural reasons, that growth can occur only in repeated aquiferous modules, and that for a maximum value of P and O_D there is a maximum diameter of supply, it is not immediately clear how sponges can achieve the regular massive volumes which are observed (up to 100 litres) without recycling exhaled water. A massive globular sponge bearing many regularly spaced oscula or a sponge consisting of many closely set,

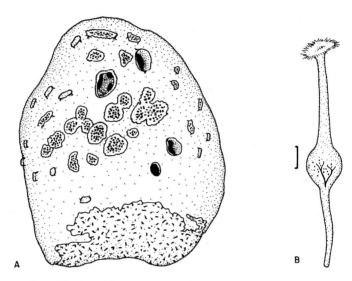

FIG. 7. Examples of "infaunal" sponges. A, *Cliona celata* boring limestone and showing vents and "frilled" ostial papillae. After Vosmaer (1932); B, Sollas' (1888) reconstruction of *Disyringa dissimilis* (scale–1 cm).

short tubes, would continuously inhale water from beneath a shallow canopy of recently exhaled water. Because their inhalant requirements would be large, in a still regime such sponges would be in continuous danger of recycling recently used water.

A stratagem, suggested by the more massive growths of *Hymeniacidon*, is commonly employed by many sponges to obviate such danger. If several oscular jets are grouped into a single stream, then the frictional resistance to the movement of the single large stream is less than the sum of the resistances of the individual jets. Such reduction in energy loss in the exhaled water will cause the exhalant stream to travel further away from the sponge before all its energy is dissipated, thereby increasing the diameter of supply. Many sponge forms amongst the Demospongia, Hexactinellida and Calcarea are the results of various methods and degrees of coupling of oscular jets.

Numerous specimens of *Leucosolenia* (Calcarea) have been described (Fig. 8) in which some or all of the asconoid branches reunite to form a single (*L. cordata*

5. Taxonomy, the Individual and the Sponge

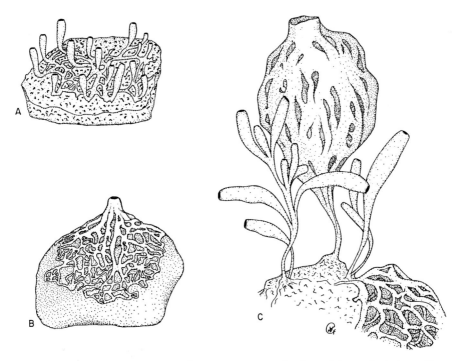

FIG. 8. Growth forms of some Calcarea. A, Asconoid tubes of *Leucosolenia botryoides* (*c*. 4 mm h); B, *Clathrina coriacea* (2 cm h). A clathriid form with asconoid tubes fused to form a single osculum (after Minchin, 1896); C, *L. blanca* (? = *C. coriacea*) (*c*. 1 cm h). Three growth forms of an apparently continuous mass (after Michlucho-Maclay, 1868).

(Haeckel)) or several oscula (*L. osculum* (Carter), *L. cavata* (Carter)). This form is referred to as clathriid.

Pachymatisma johnstonia (Demospongia) grows as thick masses in crevices on sublittoral rock faces (Fig. 9). In the larger specimens all of the oscula lie in one or a few small fields on swellings which protrude beyond the rim of the rock crevice. Although oscular coupling is here relatively low, it is presumably sufficient to ensure the removal of exhaled water far from the ostia on the surface of the sponge mass sunk below the crevice rims. A similar condition has been remarked by Hartman (1967) in the subtropical demosponge *Neofibularia mordens*, which does not inhabit crevices and grows to be more massive than *P. johnstonia*, but which has clusters of oscula on low mounds or ridges. A similar form is adopted by the hexactinellid *Semperella schulzei* (Schulze, 1887). Tighter grouping of oscula can be seen in the demosponges *Synops anceps* and *Tedania actiniformis* (Fig. 9), in which larger specimens

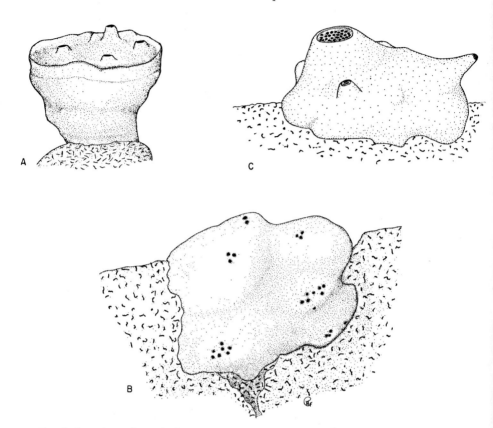

Fig. 9. Grouping of oscula in some Demospongia. A, *Tedania actiniformis* (3 cm h); B, *Pachymatisma johnstonia* (4 cm thick); C, *Synops anceps* (after Vosmaer, 1932).

bear groups of oscula, surrounded by a low rim, on the ends of cylindrical or conical projections (Vosmaer, 1932; Ridley and Dendy, 1887).

Koltun (1966) has illustrated a series of specimens of the sessile *Geodia phlegraei* (Demospongia) in which the oscula are progressively more grouped with increasing size and eventually come to lie in the floor of a cup-shaped depression (Fig. 10).

Many demosponges and hexactinellids proceed much further than the above mentioned species in grouping their oscula, and we can observe a wide range of tubular, vase- and cup-shaped sponges (Fig. 11) in which the tubes and calyces are depressions into which empty all of the oscula of the sponge, e.g. *Poterion neptuni, Callyspongia vaginalis, Haliclona permollis, Xestospongia muta, Phakellia arctica, Esperiopsis digitata*. In addition, some stalked forms become calicular from ovate as in the larger specimens of *Stylocordyla* and *Hyalonema*.

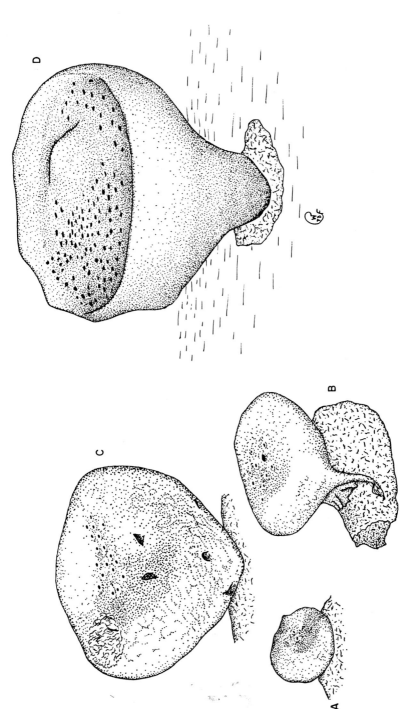

Fig. 10. Four specimens of *Geodia phlegraei* showing grouping of oscula. A, 2·5 cm h; B, 4 cm h; C, 12 cm h; D, 16 cm h. After Koltun (1966).

An alternative adaptation is the separation of oscula and ostia onto two surfaces of a flattened fan or sheet, e.g. *Haliclona ventilabra* s. Koltun, 1959, *Melonanchora kobjakovae*. Here the oscular jets are not coupled (Fig. 12) but the relatively

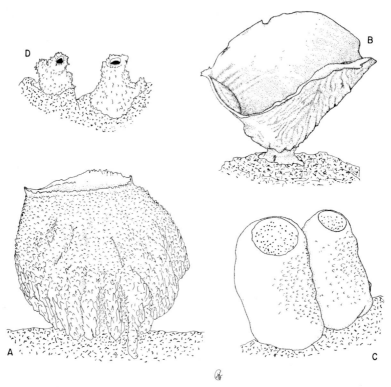

FIG. 11. Examples of tubular, vase- and cup-shaped demosponges; A, *Xestospongia muta* (1·5 m h); B, *Esperiopsis digitata* (30 cm h); C, *Verongia* sp. (25 cm h); D, *Ianthella* sp. (12 cm h).

weakly exhaled water is segregated from the ostia. Yet another adaptation is to carry the aquiferous modules up from the substratum as a solid digitate form. Here there is little or no oscular grouping and the underdispersed oscula come to occupy positions at which nearby branches interfere least with their individual diameters of supply (e.g. *Haliclona oculata* Pallas, Fig. 12c).

Where all oscula are grouped closely it is possible to distinguish between exhalant and inhalant surfaces, since oscula will effectively exclude ostia. Where the sponge is flat and encrusting, flat and free, tube-shaped or solid and erect it is possible to detect the positions of the aquiferous modules. This is not so

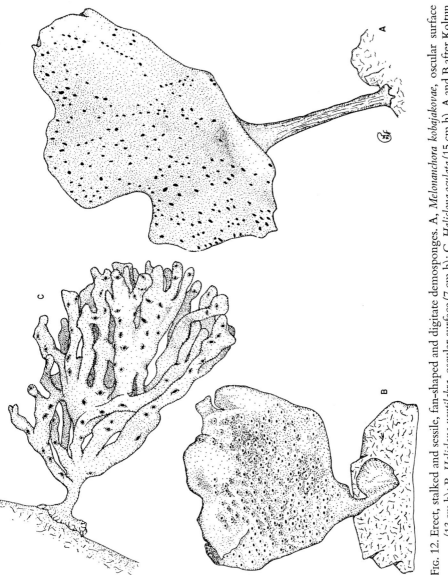

FIG. 12. Erect, stalked and sessile, fan-shaped and digitate demosponges. A, *Melonanchora kobajakovae*, oscular surface (13 cm h); B, *Haliclona ventilabra*, oscular surface (7 cm h); C, *Haliclona oculata* (15 cm h). A and B after Koltun (1959).

with vase- or cup-shaped sponges, for in them the inhalant and exhalant canals interweave irregularly over much of their distances from ostia to oscula, and a section of the sponge is sorely perplexing.

As tubular and cup- and vase-shaped specimens continue to grow, more oscula open into the depression, whose lip increases in height. Whether the depression observed in a particular specimen is vase-, cup- or tube-shaped will be due to the past relative longitudinal and circumferential growth rates at the lip. The relationship between these two growth rates will, in turn, be dependent upon the hydrodynamic requirements of the sponge. It is important to realize that such cups and tubes are not internal spaces of the sponge. They lie beyond the last point of the exhalant aquiferous system at which movements of exopinacocytes alone can regulate water velocity. Events within the depressions are not under the control of the sponge except in so far as the pump activities may vary the water movement in the depression or the whole wall of the depression may collapse or contract. This point is important for the definition of the term osculum, for which Hartman and Reiswig (1975) doubt that there can be a final consensus. To examine aquiferous systems in taxonomy without being able to distinguish an osculum unequivocally is to lose sight of data which can help to solve the perennial problem of the relationship of sponge form to genotype and to environment (see esp. Burton, 1932; Borojevič, 1967).

There is little difficulty in asserting that, in sponges with ultimate exhalant apertures of more than 1 cm diameter, the apertures are not oscula but vents emptying cloacae fed by oscula (cf. Burton, 1963; Borojevič et al., 1968). But ultimate exhalant apertures of less than 1 cm diameter are not necessarily oscula (see Figs. 5, 8). Some environmental conditions may induce particular genotypes to couple their oscula into cloacae at very small sizes, while other genotypes may delay oscular coupling into a cloacal condition either indefinitely, or until they attain a much greater mass (Fig. 10). Careful consideration of the aquiferous architecture of many specimens of different sizes—initially grouped on the basis of other characteristics—most easily offers data for delimiting interspecific boundaries. Other sources of such data are the expensive and technically difficult procedures of laboratory culture and biochemical comparison.

It might appear, from the foregoing arguments, that sponges' aquiferous units are fairly simply recognizable. That may well be so in many preserved specimens, but in the living sponge the matter is far more complicated. Terminal exhalant ducts of neighbouring modules apparently always interconnect, and in sponges with little oscular grouping one can observe readily the temporary closure of single oscula, whose neighbours then act as exits for the otherwise closed off module. In addition, in many sponges the incurrent ostia open directly into vestibules from whence water passes through volumes of sponge emptied through several oscula. Furthermore, the aquiferous modules are often ephemeral in life. Sponges frequently reorganize their aquiferous systems. Some even do this daily, closing their ostia and oscula and thoroughly reorganizing parts of their aquiferous ducting. New oscula may arise in the old sites, or at different positions. In other forms day-to-day reorganization may be only slight, with major reconstructions occurring seasonally (Reiswig, 1971, 1973, 1974). In no other organisms do there exist units like the aquiferous modules, recognizable as individual units in preserved specimens but showing such fugacity in life. On the one hand aquiferous modules have only a notional existence as individuals in the life of the sponge. On the other hand, in the taxonomists' specimens aquiferous modules are usually detectable, and their forms and disposition are often ultimately linked with structures rich in potential taxonomic information—the skeleton.

2. *The Skeleton*

The almost exclusive function of the sponge skeleton is the support of the aquiferous system. Consequently, it would be expected that the architecture of one will reflect clearly the architecture of the other. This is by no means always so, although oscular papillae, chimneys, vent walls and inhalant papillae may often have distinctive spicule complements or patterns of articulation e.g. *Siphonodictyon*, *Cliona* and *Polymastia* (Demospongia), Pharetronida (Calcarea) (see Fig. 13).

In very many silicious sponges the skeleton consists of a more or less regular three-dimensional network constructed on a very much smaller scale than the aquiferous modules. Thus, in *Ophlitaspongia seriata* the spongin and spicule network has sides of approximately 200–300 μm and the spicules of the framework are about 75 to 135 μm long, while the mean distance between oscula is approximately 6 mm and the skeletal presence of oscula is not easily detectable (Fry, 1970b). The high pressure choanocyte chambers are approximately 20 μm in diameter and occur in numbers in the spaces of the network, slung therefrom by fibrils of spongin. In other Poecilosclerida microscleres occur as additional minute (e.g. 15 μm) spicular components within the network.

In many of the Calcarea, although the oscular rim frequently is supported by strongly distinctive arrays of spicules, the main skeleton reflects the repetition of choanocyte chambers. These, being large (e.g. up to 130 μm by 2000 μm in

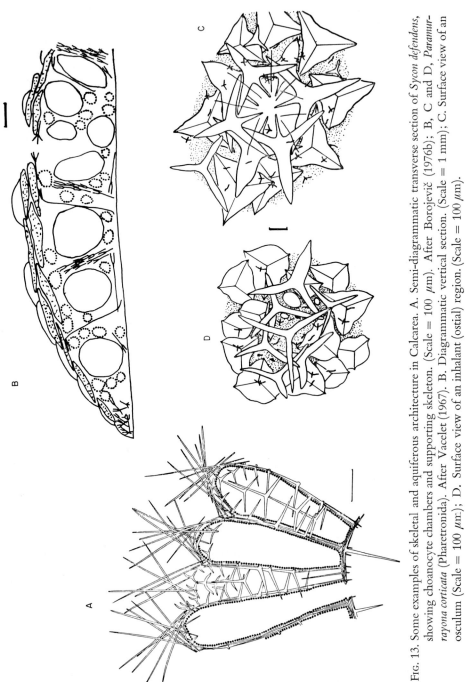

FIG. 13. Some examples of skeletal and aquiferous architecture in Calcarea. A. Semi-diagrammatic transverse section of *Sycon defendens*, showing choanocyte chambers and supporting skeleton. (Scale = 100 μm). After Borojević (1976b); B, C and D, *Paramurrayona corticata* (Pharetronida). After Vacelet (1967). B. Diagrammatic vertical section. (Scale = 1 mm); C. Surface view of an osculum (Scale = 100 μm); D. Surface view of an inhalant (ostial) region. (Scale = 100 μm).

Scypha okadai (Sasaki, 1941)) and working under low pressure, require direct support for their walls (see Fig. 13).

In some of the more massive silicious sponges the skeleton is fairly readily divisible into a general, large scale support system and a smaller scale interstitial system, this subdivision being superimposed on the skeletal differences between inhalant and exhalant regions. This situation is particularly well exemplified in the globular Tetractinellida (Fig. 14c), in many of which there is further skeletal subdivision into cortical and medullary components. Such a cortical/medullary skeletal subdivision is also apparent in some upright branching forms, as occur in the Axinellida, where spicules may be grouped into central coring bundles surrounded by a three dimensional network.

The sponge taxonomist is, therefore, faced with a difficult problem. The skeleton of a sponge to be sampled will reflect to a variable extent both the repetition of aquiferous units and the distribution of gravitational and other forces throughout the sponge. Also, particular portions of individual units of the aquiferous system will have variably distinctive skeletons. Consequently, samples taken with indifference to the disposition of aquiferous modules may well omit a number of skeletal features of taxonomic importance.

CONCLUSIONS

Oscular stratagems are diverse even within a single genus (e.g. *Haliclona*) and within a single ontogeny (e.g. *H. perleve*, *G. phlegraei*) and each specimen's aquiferous system must therefore be examined with care. The fruits of that examination are not only taxonomically valid data but also data which themselves describe the response of the sponge's aquiferous system to the environment in which it grew (Fig. 15).

Contrary to the belief of some, "spicules tell all" is not an accepted tenet of sponge taxonomy (*vide* esp. Burton, 1931, 1932; Lévi, 1960; Borojevič, 1967a; Bergquist and Sinclair, 1973). Skeletons and their component spicules provide a wealth of data, it is true, but the systematic and taxonomic value of the data are hard won by painstaking consideration of the disposition of aquiferous units. The disposition of aquiferous units largely determines form and it is important that we should bear in mind Burton's (1932) opinion of the overwhelming significance of external form in assigning specimens to species. There is strong evidence for the frequent occurrence of ontogenetic form changes in sponges but we have little clear idea which ontogenetic form changes are open to any particular genotype. We can, undoubtedly, achieve better knowledge in this respect by considering the relationship between environ-

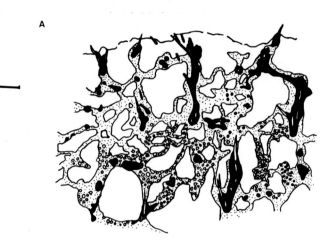

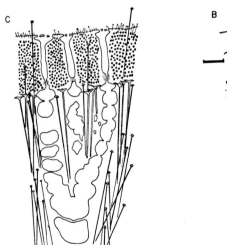
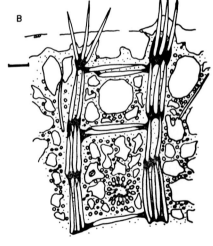

FIG. 14. Examples of skeletal and aquiferous architecture in some demosponges. A. Portion of a thin vertical section of an encrusting specimen of *Ophlitaspongia seriata*, showing the spongin (black) and spicule skeleton, irregular aquiferous spaces and subspherical choanocyte chambers (Scale = 100 μm); B. *Mycale richardsoni*. Portion of a vertical section (semi-diagrammatic) showing the irregular aquiferous spaces, isodictyal spicular skeletal architecture and subspherical choanocyte chambers (Scale = 100 μm), after Bakus (1966); C. Portion of a radial section of the globular sponge *Geodia muelleri* s. Arndt, 1935, showing the radially arranged medullary tracts of spicules, the special cortical skeleton of spherastral spicules, and ostial pits. After Vosmaer (1932).

5. Taxonomy, the Individual and the Sponge

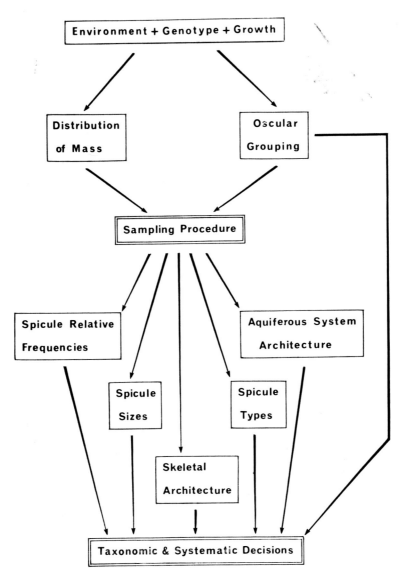

Fig. 15. The rôle of sponges' oscular grouping and aquiferous system architecture in taxonomic practice and systematic decision-making.

ments and the performances of aquiferous modules. Once we have that knowledge—and recent researches (see Harrison and Cowden, 1976; Reiswig, 1971, 1973, 1974; Rützler, 1970, 1971, 1974; Vogel, 1974) indicate that the necessary data are becoming available, then Burton's dictum can be tested. I suspect that it will be found to be an overgeneralization but, in just such a sense as the concepts of coloniality and individuality served in the past to unearth new data, so will a consideration of the relationship between hydrodynamic regimes, ontogeny of form, and the coupling of units of the aquiferous system lead us to a much clearer understanding of the phenotypic and genotypic determination of form within species.

For these reasons aquiferous units or modules, which are ephemeral and functionally overlapping in life, have a very real identity—an operational individuality—for the taxonomist.

Of their physiological, genetic, or more metaphysical individuality, sidestepping both Aristotle and Plato we may comment with Brien (1967) "Des spongologistes les plus avertis en doutent parfois".

ACKNOWLEDGEMENTS

I wish to express my gratitude to my wife, Patricia, for her help with the drawings; also to Shirley Stone for her assistance with literature and specimens and for untying the nomenclatorial knots which impeded this—as any—work on sponges.

SELECT BIBLIOGRAPHY

As has been indicated in the text, the literature on this topic referring to sponges is vastly extensive. Consequently, I have been highly selective in compiling the following list of references; it is to be hoped that the selection is not too eclectic.

BAGBY, R. M. (1966). The fine structure of myocytes in the sponges *Microciona prolifera* (Ellis and Solander) and *Tedania ignis* (Duch. and Mich.). *J. Morphol.* **118**, 167–182.

BAKUS, G. J. (1966). Marine poeciliscleridan sponges of the San Juan Archipelago, Washington. *J. Zool., Lond.* **149**, 415–531.

BEKLEMISHEV, V. N. (1964). "Osnovy Sravnitelnoy Anatomii Bespozvonochnykh" vol. I Promorphologiya. Nauka, Moscow.

BERGQUIST, P. R. and HARTMAN, W. D. (1969). Free amino acid patterns and the classification of the Demospongiae. *Mar. Biol.* **3**, 247–268; 1–14.

BERGQUIST, P. R. and HOGG, J. J. (1969). Free amino acid patterns in Demospongiae; a biochemical approach to sponge classification. *Cah. Biol. Mar.* **10**, 205–220.

BERGQUIST, P. R. and SINCLAIR, M. E. (1973). Seasonal variation in settlement and spiculation of sponge larvae. *Mar. Biol.* **20**, 35–44.

BIDDER, G. P. (1923). The relation of a form of a sponge to its currents. *Q. Jl microsc. Sci.* **67**, 293–323.

BIDDER, G. P. (1937). The perfection of Sponges. *Proc. Linn. soc. London.* **149**, 119–146.

BOROJEVIČ, R. (1966). Étude expérimentale de la différenciation des cellules de l'éponge au cours de son développement. *Devl Biol.* **14**, 130–153.

BOROJEVIČ, R. (1967a). Importance de l'étude de la répartition écologique pour la taxonomie des éponges calcaires. *Helgoländer wiss. Meeresunters.* **15**, 116–119.

Borojević, R. (1967b). Spongiaires d'Afrique du Sud, (2) Calcarea. *Trans. r. Soc. S. Afr.* **37**, 183–226; p. XVI, 1–25.

Borojević, R., Fry, W. G., Jones, W. C., Lévi, C., Rasmont, R. and Sarà, M. (1968). Mise au point actuelle de la terminologie des eponges (A reassessment of the terminology for sponges). *Bull. Mus. natn. Hist., nat. Paris* (2) **39**, 1224–1236.

Boury-Esnault, N. (1974). Structure et ultrastructure des papilles d'éponges du genre *Polymastia* Bowerbank. *Arch. Zool. exp. gen.* **115**, 141–165.

Bowerbank, J. S. (1857). On the vital powers of the Spongiadae. *Rept. Brit. Ass. Adv. Sci.* **1856**, 438–451.

Brien, P. (1967). Les éponges: leur nature métazoaire: leur gastrulation; leur état colonial. *Annls Soc. r. zool. Belg.* **97**, 197–235.

Brien, P. (1968). The sponges or Porifera. *In* "Chemical Zoology" (M. Florkin and B. T. Scheer, eds), Vol. 2. pp. 1–30. Academic Press, London and New York.

Bromley, G. R. and Tendal, O. S. (1973). Example of substrate competition and phobotropism between two clionid sponges. *J. Zool. Lond.* **169**, 151–155.

Burnet, F. M. (1969). "Self and Not Self. Book 1. Cellular Immunology" Cambridge University Press.

Burton, M. (1928). A comparative study of the characteristics of shallow-water and deep-sea sponges, with notes on their external form and reproduction. *J. Quekett microsc. Club* **15** (2), 49–70.

Burton, M. (1929). No. 1. Descriptions of South African sponges collected in the South African Marine Survey. Part II. The Lithistidae, with a critical survey of the desma-forming Sponges. *U.S. Afr. Fish. mar. Biol. Survey, Rep.* **7**, 1–12.

Burton, M. (1932). Sponges. *"Discovery" Rep.* **6**, 237–392.

Burton, M. (1949). Observation on littoral sponges, including the supposed swarming of larvae, movement and coalescence in mature individuals, longevity and death. *Proc. zool. Soc. London* **118**, 893–915.

Burton, M. (1965a). Sponges. Part I. The Third Kingdom? *Animals* **7** (16), 436–440.

Burton, M. (1965b). Sponges. Part II. Reproduction with and without sex. *Animals* **7** (16), 456–460.

Carter, H. J. (1848). Notes on the species, structure and animality of the freshwater sponges in the tanks of Bombay. *Ann. Mag. nat. Hist.* (2) **1**, 303–311.

Carter, H. J. (1857). On the ultimate structure of *Spongilla* and additional notes on freshwater Infusoria. *Ann. Mag. nat. Hist.* (2), **20**, 21–41.

Carter, H. J. (1869). On *Grayella cyathophora*, a near genus and species of sponges (sic). *Ann. Mag. nat. Hist.* (4) **4**, 189–197; VII.

Carter, H. J. (1871a). A description of two new Calcispongiae, to which is added confirmation of Prof. James-Clark's discovery of the true form of the sponge cell (animal) and an account of the polype-like pore-area of *Cliona corallinoides* contrasted with Prof. E. Haeckel's view on the relationship of the sponges to the corals. *Ann. Mag. nat. Hist.* (4) **8**, 1–27.

Carter, J. J. (1871b). Discovery of the animal of the Spongiadae confirmed. *Ann. Mag. nat. Hist.* (4) **7**, 445.

Carter, H. J. (1872). Proposed name for the sponge-animal, viz. "Spongozoon"; also on the origin of thread-cells in the Spongiadae. *Ann. Mag. nat. Hist.* (4) **10**, 45–51.

CARTER, H. J. (1874). On the Spongozoa of *Halisarca Dujardinii*. *Ann. Mag. nat. Hist.* (4) **13**, 315–316.

CURTIS, A. S. G. (1970). Problems and some solutions in the study of cellular aggregation. In "The Biology of the Porifera" (W. G. Fry, ed.), Symp. Zool. Soc. London. No. 25, 335–352. Academic Press, London.

DENDY, A. (1915). The biological conception of individuality. The President's Address. *J. Quekett microsc. Club* **12**, 465–478.

DUJARDIN, F. (1841). "Histoire naturelle des zoophytes. Infusoires, comprenant la physiologie et la classification des ces animaux et la manière de les étudier à l'aide de microscope" Librarie Encyclopédique de Rovet, Paris.

EFREMOVA, S. M. (1967). The cell behaviour of the freshwater sponge *Ephydatia fluviatilis*: a time-lapse microcinematography study. *Acta biol. hung.* **18**, 37–46.

ELLIS, J. (1766). On the nature and formation of sponges: in a letter from John Ellis, F.R.S. to Dr. Solander, F.R.S. *Phil. Trans. R. Soc. Lond.* **55**, 280–289.

FROST, T. M. (1976). Sponge feeding; a review with a discussion of some continuing research: In "Aspects of Sponge Biology" (F. W. Harrison and R. R. Cowden, eds.), 283–298. Academic Press, New York, San Francisco, London.

FRY, W. G. (1970a). Introduction, "The Biology of the Porifera" vol. 25, pp. xi–xix. *Zool. Soc. Lond. Symp.* Academic Press, London.

FRY, W. G. (1970b). The sponge as a population: a biometric approach. In "The Biology of the Porifera" (W. G. Fry, ed.), Symp. Zool. Soc. London No. 25, 135–162. Academic Press, London.

FRY, W. G. (1971). The biology of larvae of *Ophilitaspongia seriata* from two North Wales populations. In "IVth Europ. Mar. Biol. Symp." (D. J. Crisp, ed.) pp. 155–178.

FRY, W. G. (1973). The role of larval migration in the maintenance of an encrusting sponge population, in *7th Europ. Symp. mar. Biol., Neth. J. Sea Res.* **7**, 159–170.

GOOD, R. A. and FISHER, D. W. (eds) (1972). "Immunobiology" Sinauer, Stamford, Conn.

GRANT, R. E. (1826). Observations and experiments on the structure and functions of the sponge. *Edinb. phil. Jn.* **14**, 113–124.

HADŽI, J. (1963). "The Evolution of the Metazoa" Pergamon Press, Oxford, London, New York, Paris.

HADŽI, J. (1966). Uprašanje individualitete pri spužvah. *Slov. Akad. Znan. Umet.* **9** (4), 167–204.

HAECKEL, E. (1872). Die Kalkschwämme, vol. 1, (Genareller Thiel) "Biologica der Kalkschwämme" G. Reimer, Berlin.

HAECKEL, E. (1889). "Natürlicheschöpfungs-Geschichte" G. Reimer, Berlin.

HAECKEL, E. (1896). "Systematische Phylogenie, pt. 2. Systematische Phylogenie der wirbellosen Thiere (Invertebrata)" G. Reimer, Berlin.

HARRISON, F. W. and COWDEN, R. R. (eds) (1976). "Aspects of Sponge Biology" Academic Press, New York, San Francisco, London.

HARTMAN, W. D. (1958). Natural history of the marine sponges of southern New England. *Bull. Peabody Mus. nat. Hist.* **12**, 1–155.

HARTMAN, W. D. (1967). Revision of *Neofibularia* (Porifera, Demospongiae), a genus of toxic sponges from the West Indies and Australia. *Postilla*, **113**, 1–41.

HARTMAN, W. D. and GOREAU, T. F. (1970). Jamaican coraline sponges: their morphology, ecology and fossil relatives. *In* "The Biology of the Porifera" (W. G. Fry, ed.). Symp. Zool. Soc. London. No. 25, 205–243. Academic Press, London.

HARTMAN, W. D. and GOREAU, T. F. (1972). *Ceratoporella* (Porifera: Sclerospongiae) and the chaetetid corals. *Trans. Connecticut Acad. Arts. Sci.* **44**, 133–148.

HARTMAN, W. D. and GOREAU, T. F. (1975). A Pacific tabulate sponge, living representative of a new order of sclerosponges. *Postilla*, **167**, 1–14.

HARTMAN, W. D. and REISWIG, H. N. (1973). The individuality of sponges. "Animal Colonies" (R. S. Boardman, A. H. Cheetham and W. A. Oliver, Jr., eds), pp. 567–584. Dowden, Hutchinson and Ross Inc., Stroudsburg, Pa.

HIBBERD, D. J. (1975). Observations on the ultrastructure of the choanoflagellate *Codosiga botrytis* (Ehr.) Saville-Kent with special reference to the flagellar apparatus. *J. Cell Sci.* **17**, 191–210.

HUMPHREYS, T. (1970). Biochemical analysis of sponge cell aggregation. *In* "The Biology of the Porifera" (W. G. Fry, ed.) Symp. Zool. Soc. London. No. 25, 325–334. Academic Press, London.

JAMES CLARK, H. (1866). Conclusive proofs of the animality of the ciliate sponges and of their affinities with the Infusoria Flagellata. *Am. Jl. Sci.* (2) **42**, 320–324.

JAMES CLARK, H. (1868). On the Spongiae Ciliatae as Infusoria Flagellata, or, observations on the structure, animality, and relationship of *Leucosolenia botryoides*, Bowerbank *Ann. Mag. nat. Hist.* (4) **1** (2), 133–142.

JOHNSTON, G. (1842). "A History of British Sponges and Lithophytes" W. H. Lizars, Edinburgh.

KILIAN, E. F. (1952). Wasserströmung und Nahrungsaufnahme beim Süsswasserschwamm *Ephydatia fluviatilis. Z. Vergl. Physiol.* **34**, 407–447; 1–24.

KILIAN, E. F. (1964). Zur biologie der einheimischen Spongilliden, Ergebnisse und Probleme. *Zool. Beiträg.* N.F. **10**, 85–159.

KOLTUN, V. M. (1959). Kremnerogovii Gubki severnikh i dalnyevostochnikh morei SSR (otrad Cornacuspongida). *Akad. Nauk SSSR. Opred. fauna SSSR* **67**.

KOLTUN, V. M. (1966). Cheteirechlychevii Gubki severnikh i dalnyevostochnikh morei SSSR. (Otrad Tetraxonida). *Akad. Nauk SSSR, Opred. fauna SSSR*, **90**, 1–112.

LEADBEATER, B. S. C. and MORTON, C. (1974a). A light and electronmicroscopical study of the choanoflagellates *Acanthoeca spectabilis* Ellis and *A. brevipoda* Ellis. *Arch. Microbiol.* **95**, 279–92.

LEADBEATER, B. S. C. and MORTON, C. (1974b). A microscopical study of a marine species of *Codosiga* James-Clark (Choanoflagallata) with special reference to the ingestion of bacteria. *Biol. J. Linn. Soc. Lond.* **6** (4), 337–347.

LENTZ, T. L. (1966). Histochemical localization of neurohumors in a sponge. *Jl. exp. Zool.* **162**, 171–180.

LÉVI, C. (1960). Les démosponges des Côtes de France. I. Les Clathriidae. *Cah. Biol. mar.* **1**, 47–87.

LIEBERKÜHN, N. (1856). Zusätze zur Entwickelungsgeschichte der Spongilliden. *Arch. Anat. Physiol.* **1856**, 496–514.

MACKIE, G. O. (1963). Siphonophores, bud colonies, and superorganisms. *In* "The Lower Metazoa; Comparative Biology and Physiology" (E. C. Dougherty, ed.), pp. 329–337. Univ. Calif. Press. Berkeley.

MERGNER, H. (1964). Über die Induktion neuer Oscularrohre bei *Ephydatia fluviatilis*. *Wilhelm Roux Arch. Entwmech. Org.* **155**, 9–128.
MERGNER, H. (1970). Ergebnisse der Entwicklungsphysiologie bei Spongilliden. In "The Biology of the Porifera" (W. G. Fry, ed.), Symp. Zool. Soc. Lond. No. 25, 365–397. Academic Press, London.
MINCHIN, E. A. (1896). Suggestions for a natural classification of the Asconidae. *Ann. Mag. nat. Hist.* (6) **18**, 349–362.
MINCHIN, E. A. (1900). Sponges, In "A Treatise on Zoology Pt. 2, (E. Ray Lankester, ed.), pp. 1–178. Adam and Charles Black, London.
MOSCONA, A. A. (1968). Cell aggregation: properties of specific cell-ligands and their role in the formation of multicellular systems. *Devl Biol.* **18**, 250–277.
MACLENNAN, A. P. (1970). Polysaccharides from sponges and their possible significance in cellular aggregation, In "The Biology of the Porifera" (W. G. Fry, ed.), Symp. Zool. Soc. London. No. 25, 299–324. Academic Press, London.
PARKER, G. H. (1910). The reaction of sponges with a consideration of the origin of the nervous system. *J. exp. Zool.* **8**, 3–41.
PARKER, G. H. (1914). On the strengths of the volume of the water currents produced by sponges. *J. exp. Zool.* **16**, 443–446.
PAVANS DE CECCATTY, M. (1958). La mélanisation chez quelques éponges calcaires et siliceuses: ses rapports avec le système reticulohistiocytaire. *Arch. Zool. exp. Gen.* **96**, 1–51.
PAVANS DE CECCATTY, M. (1969). Les systèmes des activités motrices, spontanées et provoquées des éponges: *Euspongia officinalis* L. et *Hippospongia communis* Lmk. *C.r. hebd, Séanc. Acad. Sci. Paris.* Sér D. **269**, 596–599.
PAVANS DE CECCATTY, M. (1971). Effects of drugs and ions on a primitive system of spontaneous contractions in a sponge (*Euspongia officinalis*). *Experientia* **27**, 57–91.
PAVANS DE CECCATTY, M., THINEY, Y. and GARRONE, R. (1970). Les bases ultrastructurales des communications intercellulaires dans les oscules de quelques, éponges In "The Biology of the Porifera" (W. G. Fry, ed.), Symp. Zool. Soc. London. No. 25, 449–466. Academic Press, London.
PERTY, J. A. M. (1852). "Zur Kenntnis kleinster Lebensformen nach Bau, Funktionen, Systematik, mit Specialverzeichniss der in der Schweiz beobachteten" Verlag von Jent & Reinert, Brno.
PROSSER, C. L., NAGAI, T. and NYSTROM (1962). Oscular contractions in sponges. *Comp. Biochem. Physiol.* **6**, 69–74.
REISWIG, H. M. (1971). *In situ* pumping activities of tropical Demospongiae. *Marine Biology* **9** (1), 38–50.
REISWIG, H. M. (1973). Population dynamics of three Jamaican Demospongiae. *Bull. mar. Sci.* **23** (2), 191–226.
REISWIG, H. M. (1974). Water transport, respiration and energetics of three tropical marine sponges. *J. exp. mar. Biol. Ecol.* **14**, 231–249.
REISWIG, H. M. (1975). The aquiferous systems of three marine Demospongiae. *J. Morphol.* **145** (4), 493–502.
RIDLEY, S. O. and DENDY, A. (1887). Report on the Monaxonida collected by HMS Challenger during the years 1873–76. Rep. scient. Results Voy. Challenger **20**, i–lxviii, 1–275.

ROITT, I. (1975). "Essential Immunology" Blackwell Scientific Publications, Oxford, London and Edinburgh.

RÜTZLER, K. (1970). Spatial competition among Porifera: solution by epizoism. *Oecologia (Berl.)* **5**, 85–95.

RÜTZLER, K. (1971). Bredin-Archbold-Smithsonian Biological Survey of Dominica: burrowing sponges, genus *Siphonodictyon* Bergquist, from the Caribbean. *Smithson. Contr. Zool.* **77**, 1–37.

RÜTZLER, K. (1974). The Burrowing Sponges of Bermuda, *Smithson. Contr. Zool.* **165**, 1–32.

SARÀ, M. (1974). Sexuality in the Porifera, *In Atti del XLII Convegno dell' U.Z.I.* (E. Vannini, ed.), Some Aspects of Sex Differentiation in Pluricellular Animals at a lower order of organisation: Porifera, Fresh-water Hydras and Planarians. *Boll. Zool.* **41**, 327–348.

SASAKI, N. (1941). On the changes occurring in various parts of the body, especially in the spicules in accordance with the increase of body length in the case of the calcareous sponge, *Sycon okadai* Hozawa. *Sci. Rep. Tohoku Univ.* (4) Biol. **16**, 365–382; 12 figs.

SAVILLE KENT, W. (1878a). Observations upon Professor Ernst Haeckel's group of the "Physemaria" and on the affinity of the sponges. *Am. Mag. nat. Hist.* (5) **1**, 1–17.

SAVILLE KENT, W. (1878b). Notes on the embryology of sponges. *Ann. Mag. nat. Hist.* (5) **2**, 139–156.

SCHMIDT, O. (1864). "Supplement der Spongien des adriatischen Meeres, enthaltend die Histologie und systematische Ergänzungen": Wilhelm Engelmann, Leipzig.

SCHULZE, F. E. (1887). Report on the Hexactinellida collected by H.M.S. Challenger during the years 1873–1876. *Rep. scient. Results Voy. Challenger. Zoology*, **21** (53), 1–514.

SIMONS, J. R. (1963). Sponges—the first republicans. *Australian Nat. Hist.* **14**, 194–197.

SIMPSON, T. L. (1963). The biology of the marine sponge *Microciona prolifera* (Ellis and Solander). I. A study of cellular function and differentiation. *J. exp. Zool.* **154** (1), 135–151.

SIMPSON, T. L. (1973). Coloniality among the Porifera. *In* "Animal Colonies: Development and Function through Time" (R. S. Boardman, A. H. Cheetham and W. A. Oliver, Jr., eds.), pp. 549–565. Dowden, Hutchinson and Ross, Stroudsburg, Pa.

SOLLAS, W. J. (1888). Report on the Tetractinellida collected by HMS Challenger during the years 1873–76. Rep. Scient. Results Voy. Challenger **25**, i–clxvi, 1–458.

STEMPIEN, M. F. Jr. (1970). Sponges. *Animal Kingdom* **7** (4), 2–7.

STORR, J. F. (1976). Field observations of sponge reactions as related to their ecology, *In* "Aspects of Sponge Biology" (F. W. Harrison and R. R. Cowden, eds), Academic Press, New York, San Francisco, London.

TOPSENT, E. (1888). Contribution à l'étude des clionides. *Archs Zool. exp. gen.* (2) **5** bis, suppl. 1–165; I–VII.

TUZET, O. (1970). La signification des poriferes pour l'évolution des métazoaires. *Z. zool. Syst. Evolflorsch.* **8** (2), 119–126.

TUZET, O., PAVANS DE CECCATTY, M. and PARIS, J. (1963). Les éponges sont-elles des colonies? *Arch. Zool. Gen. Exp.* **102**, 14–19.

VACELET, J. (1966). Les cellules contractiles de l'éponge cornée *Verongia cavernicola* Vacelet. *C.r. hebd. Séanc. Acad. Sci. Paris, Sér. D.* **263**, 1330–1332.

VACELET, J. (1967). Descriptions d'Eponges pharetronides actuelles des tunnels obscurs sous-récifaux de Tuléar (Madagascar). *Rec. Trav. Sta. Mar. Endoume. Suppl.* **6**, 37–62.

VAN DE VYVER, G. (1970). La non-confluence intraspécifique chez les spongiaires et la notion d'individu. *Annls Embryol. Morphogen.* **3**, 251–262.

VAN DE VYVER, G. (1971a). Analyse de quelques phénomènes d'histoincompatibilité intraspécifique chez l'éponge d'eau douce *Ephydatia fluviatilis* (Linné). *Archs Zool. exp. gén.* **112**, 55–62.

VAN DE VYVER, G. (1971b). Mise en évidence d'un facteur d'agrégation chez l'éponge d'eau douce *Ephydatia fluviatilis. Annls Embryol. Morphogen.* **4** (4), 373–381.

VITETTA, E. S. and UHR, J. W. (1975). Immunoglobulin receptors revisited. *Science* **189** (4207), 964–969.

VOGEL, S. (1974). Current-induced flow through the sponge *Halichondria bowerbanki. Biol. Bull.* **147**, 443–456.

VOSMAER, G. C. J. (1908). *Poterion* a boring sponge. *Proc. Sect. Sci. k. med. Akad. Wet.* **1908**, 37–41.

VOSMAER, G. C. J. (1932). The sponges of the Bay of Naples: Porifera Incalcarea. With analyses of genera and studies in the variation of species. *Capita zool.* **3** (1), 1–320.

WILLMER, E. N. (1960). "Cytology and Evolution" Academic Press, London and New York.

WINTERMANN, G. (1951). Entwicklungsphysiologische Untersuchungen an Süsswasserschwammen. *Zool. Jb.* (Anatomie) **71**, 427–486.

6 | Coloniality in the Scyphozoa: Cnidaria*

BERNHARD WERNER

Biologische Anstalt Helgoland, Zentrale Hamburg, Federal Republic of Germany

Abstract: Coloniality as it occurs in Cnidaria is represented in permanent intraspecific animal communities which are systems connected by development, morphology and anatomy, metabolism and reproduction. The solitary life habit is the primary one from which the colonial status has originated by progressive steps of evolution. In most cases coloniality is connected with sessile life which must be interpreted as one primary basis of the origin of coloniality. Morphologically, the stolonal colony with unbranched uniform members connected by a network of stolons is considered the most simple and basic form from which the more advanced types have evolved by the development of a stolonal plate and/or branching. Di- and polymorphism with division of labour are other well known phenomena correlated with coloniality of higher levels. Considering the anatomical structures and the physiology, all cnidarian colonies have common epithelial layers and a common gastrovascular system which continue through all members and provide a common system of growth and metabolism. All members of a colony have the same ancestor by sexual reproduction. Ontogenetically, a colony of sessile Cnidaria originates from the single primary polyp which develops from the planktonic larval planula after its attachment to the substratum. By means of asexual budding processes and/or division, the primary polyp gives rise to a colony. Also the multiform colonies of higher levels go through the mentioned basic status. Free-swimming or floating polymorphic colonies exist in the order Siphonophora, class Hydrozoa. Colonial scyphozoan polyps exist only in the basic order Coronatae. They can serve as models for the several steps of evolution which leads from the primary solitary life habit to the secondary colonial status. Due to differences of morphological structures the colonial species can be arranged in a series of progressive complexity. Corresponding differences of regenerative qualities can be demonstrated experimentally. Different sequences of reproductive activity between solitary and colonial species are correlated with differences in their vertical distribution. Thus, as a final conclusion, an ecological basis for the evolution of coloniality in Scyphozoa seems to be evident and is discussed in detail.

Systematics Association Special Volume No. 11, "Biology and Systematics of Colonial Organisms", edited by G. Larwood and B. R. Rosen, 1979, pp. 81–103. Academic Press, London and New York.

*Dedicated to Professor Dr. Dr. h.t. W. E. Ankel, Giessen for his 80th birthday.

INTRODUCTION

The Coelenterata have been defined as diploblastic Metazoa in which the intestine is the sole body cavity. In a modern system this lowest group of true Metazoa is classified as the two separate phyla Cnidaria and Ctenophora. All Ctenophora are solitary animals.

Coloniality as it occurs in the Cnidaria is represented in permanent intraspecific animal communities which are connected systems by development, morphology and anatomy, metabolism and reproduction. In this contribution coloniality will be dealt with only in this restricted sense. The other form of coloniality, the gregariousness of specimens of the same or different solitary species, occurs also in the Cnidaria. In some well observed cases it depends on the planktonic larval stages to react to special qualities of the substratum or to the attractive presence of young or adult specimens of the same species. Furthermore, true coloniality and gregariousness appear to occur together in some groups. This happens if the primarily single larval stages of colonial species attach to the substratum in aggregations.

The solitary life habit is the primary one from which colonial status has originated by progressive steps of evolution. In most cases coloniality is connected with sessile life which must be interpreted as one primary basis of the origin of coloniality.

The multiform phenomena of colonial growth in the Cnidaria, especially of the classes Hydrozoa and Anthozoa, are well known. Morphologically, the stolonal colony with uniform unbranched members connected by a network of stolons is considered the most simple and basic form from which the more advanced types have evolved by the development of a stolonal plate and/or branching. Di- and polymorphism with division of labour are other well known phenomena correlated with coloniality at higher levels.

All members of a colony have the same ancestor by sexual reproduction. Ontogenetically, a colony originates in the sessile Cnidaria from the single primary polyp which develops from the planktonic larval planula after its attachment to the substratum. By asexual budding processes the primary polyp gives rise to the colony. The multiform colonies of higher levels also pass through the same basic stages. Free-swimming or floating polymorphic colonies exist in the order Siphonophora (in the class Hydrozoa) and these colonies also develop from a single planktonic planula.

With regard to anatomical structures and physiology, all cnidarian colonies have common epithelial layers and a common gastrovascular system which are continuous through all members and provide a common system of metabolism (nutrition, digestion and distribution of reserve substances) and growth.

6. Coloniality in the Scyphozoa

The Scyphozoa, consisting of the orders Coronatae, Semaeostomeae, Rhizostomeae and Stauromedusae, represent the most basic class of the phylum Cnidaria (Werner, 1973). Colonial species exist only in the Coronatae. In the class Cubozoa (Werner, 1973, 1975, 1976), which is positioned systematically and evolutionarily between Scyphozoa and Hydrozoa, coloniality is unknown. In Hydrozoa, coloniality must be considered as the basic status of Recent groups. The solitary life habit of several smaller groups has re-evolved by secondary steps of evolution which are adaptations to special environmental conditions. In the Anthozoa, solitary and colonial species also exist and, as in the Scyphozoa, the solitary status represents the original one.

In this paper emphasis is laid on the evolution of coloniality in the Scyphozoa. As will be shown, knowledge of the scyphopolyps offers the great advantage that they can serve as models for the stages of evolution which lead from a primary solitary life habit to a secondary colonial status. The essential argument is based on: (a) the differences of morphological structures which can be arranged in a progressively more complex series; (b) the differences of regenerative qualities; (c) the differences of distribution compared with different sequences of reproductive phenomena of solitary and colonial species. Thus, as a final conclusion, an ecological basis for the evolution of coloniality in Scyphozoa seems to be evident and is discussed in detail. The argument agrees with the generally accepted theory that a solitary sexual sessile polyp was the stem form of all Recent Scyphozoa and other Cnidaria (Chapman, 1966; Thiel, 1966; Uchida, 1969; Werner, 1971b).

SOLITARY AND COLONIAL GROWTH OF SCYPHOZOAN POLYPS

Members of the order Coronatae represent the basic group of the class Scyphozoa (Werner, 1971b, 1973). The coronate medusae are produced by polyps of the genus *Stephanoscyphus* Allman, 1874 which is distinguished by a firm peridermal tube completely enclosing the soft body. By virtue of this primitive feature and its structural qualities *Stephanoscyphus* can be derived from the extinct fossil group Conulata which lived from the Cambrian to the Triassic and were the ancestors of the Recent Scyphozoa (Kiderlen, 1937; Knight, 1937; Werner, 1966, 1967b, Glaessner, 1971). This relationship to the extinct Conulata which gives *Stephanoscyphus* the status of a living fossil and missing link has been denied so far only by Koztowski (1968). This author also rejects the widely held view that the Conulata were the ancestors of the Recent Scyphozoa. However, the arguments presented by him are not convincing, as will be shown in another paper which is in preparation. Most Conulata had a solitary life habit (Moore and Harrington, 1956; Müller, 1963) but some colonial species are also known. In several cases it has proved difficult to identify true coloniality where young specimens have been found attached to the tubes of solitary adults.

The polyp of the higher Recent orders Semaeostomeae and Rhizostomeae, for which the scyphistomae are well known has, in most species, only a thin small basal cup of periderm. Loss of the firm peridermal tube has been compensated by the development of a thick cellular mesoglea (the middle layer between ectoderm and endoderm) giving the body strength and support. Coloniality is unknown in the Semaeostomeae and Rhizostomeae. It follows, and is evident in the description of colonial species of *Stephanoscyphus*, that, generally, coloniality of Scyphozoa is correlated with the possession of firm peridermal structures. According to present state of knowledge there exists no colonial scyphozoan species in which the polyp is naked or enveloped only partly by a reduced periderm. This is true also for the Scyphozoa of the order Stauromedusae which is an aberrant group combining polypoid and medusoid features. The Stauromedusae are stalked sessile solitary animals which are naked.

Species of *Stephanoscyphus* occur everywhere on the continental shelves, but some species live at greater depths, down to 7000 m (Kramp, 1959). The systematics of the Coronatae suffer from the well known fact that the sessile polyp and the free-swimming medusa of the same species were often found and described by different authors at different times and in different localities. In the older literature (Komai, 1936; Leloup, 1937; Kramp, 1959) only two solitary species are described: *Stephanoscyphus simplex* Kirkpatrick 1890 (which is thought with good reason by Kramp (1959) to be synonymous with *S. striatus*, *S. sibogae* and *S. bianconis*) and *S. corniformis* Komai, 1936. However, more recent observations of Naumov (1959) and rearing experiments by the present author (Werner, 1966, 1967a, 1971a, 1974) have shown that a number of different species is included in each of these two names. The culture experiments led to the conclusion that reliable identification of the species of *Stephanoscyphus* and the systematics of this genus can be based only on knowledge of the medusae produced by the respective polyp species. In several cases species of the same genus *Stephanoscyphus* gave rise to medusae ascribed by other workers to different genera and even different families. In such cases the polyps were given the name of the medusa according to the valid nomenclatorial rules.

The systematic summaries (Komai, 1936; Leloup, 1937; Kramp, 1959) include a few colonial species of *Stephanoscyphus*:

(a) *Nausithoë punctata* Kölliker, 1853 (syn. *S. mirabilis* Allman, 1874), Mediterranean Sea,

(b) *Linuche unguiculata* (Schwartz, 1788) (syn. *S. komaii* Leloup, 1937), Indopacific, Caribbean Sea,
(c) *S. allmani* Kirkpatrick, 1890, Indopacific,
(d) *S. racemosus* Komai, 1936, southeastern coast of Japan.

The species *N. punctata*, *S. racemosus* and *S. allmani* belong to the same family: Nausithoidae. *S. racemosus* is very closely related to *Nausithoë*; these species differ in the structure of the medusa which in *N. punctata* is a normal free-swimming medusa, but in *S. racemosus* is an eumedusoid with reduced structures and a very short life-span. The exact generic affinity of *S. allmani* remains unclear because its medusa generation is not known. The polyp of *Linuche unguiculata* (Fam. Linuchidae) has been described by Leloup (1937) as the new species *S. komaii*. However, by rearing experiments, I was able to demonstrate that the polyp of *S. komaii* produces the medusa known as *L. unguiculata* (unpublished), and that the valid name for both was *L. unguiculata*.

As in the other groups of the Cnidaria, colonies of coronate polyps originate ontogenetically from a single primary polyp which develops from a planktonic planula and grows by budding. The following short outline of morphology and life-history of coronate polyps begins with a solitary species and draws attention to those characteristics which constitute evolutionary stages. All described species have been reared in long-term laboratory cultures and a reasonable description of their structures can therefore be given.

1. Nausithoë sp.

The solitary species of *Nausithoë* (Fig. 1) has a firm peridermal tube which completely encloses the soft body except for the head. The long slender tube consists of chitin and is attached by a minute basal disc to the substratum. It broadens only slightly from the base to the distal aperture and in most cases has the shape of a slender horn. The anterior region of the soft body consists of a narrow collar and a mouth area encircled by a crown of numerous tentacles and extends beyond the tube. At the lower end of the collar is a small transverse groove in which the soft body and rim of the tube are intimately connected. Within this groove secretory activity of gland cells maintains longitudinal growth of the chitinous tube. In most species the size of the tube (i.e. the length of the specimen) amounts to an average of 10–20 mm. The final size of most species is limited because the growth rate diminishes in old specimens; but the growth of the tube is virtually unrestricted. This has been demonstrated by rearing experiments in which specimens of several species grew to a size of at least 30–50 mm with a maximum of about 70–80 mm.

Anatomically, the sac-like, thin-walled soft body exhibits the typical tetraradial scyphozoan structures with four interradial, gastric septa. A particular characteristic is the possession of an endodermal ring canal or ring sinus located

FIG. 1. *Nausithoë* sp., a solitary species from the Indian Ocean. Note the small attaching disc at the base and the endodermal gastric septa which are seen in the upper part through the transparent periderm. From Werner 1966, Fig. 4.

in the upper part of the body below the mouth field. The ring canal opens into the coelenteron by four short perradial canals or pores.

The production of medusae takes place by the well known process of strobilation, typical of scyphozoan polyps (Fig. 2). Polydisc strobilation produces transverse constrictions in the upper part of the body and the ensuing fission produces numerous small medusa larvae (ephyrae) which become free and grow to maturity. At the end of the strobilation phase the small basal part of the polyp body, the residuum, which is not involved in strobilation, regenerates a complete soft body in a short time. Another period of nutrition and vegeta-

FIG. 2. *Nausithoë* sp., same species as in Fig. 1, the beginning of strobilation. From Werner 1966, Fig. 10.

tive growth follows in which the polyp replaces the loss of tissue material caused by the medusa production. Provided there are favourable conditions, there exists a regular alternation of the two phases of vegetative growth and strobilation throughout the year. More details of structures and life-history are given by Werner (1966, 1967a, 1970a, b) and Chapman and Werner (1972).

In passing on to the comparative morphology of colonial species below, it must be emphasized that the shape and growth of colonies depend on one important characteristic which they share with solitary species, that is, the possession of firm peridermal structures enclosing and covering all of the soft body except the head.

2. *Nausithoë punctata*

The polyp of *Nausithoë punctata* represents an elementary form of coloniality in which the individual polyp branches but in a simple and often irregular way (Fig. 3). Well developed colonies of this species with numerous branched polyps

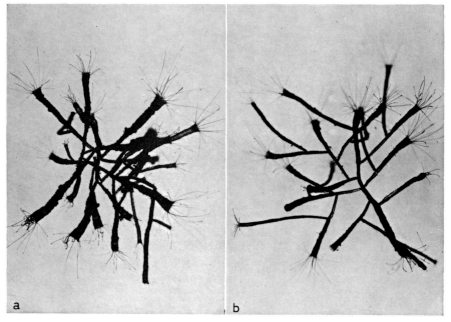

FIG. 3a, b. *Nausithoë punctata* Kölliker 1853, (a) an older and (b) a younger polyp colony, reared in the laboratory. From Werner 1970b, Fig. 2a, b.

are found living in sponges. The polyps' tubes and the stolonal connections within these colonies are confined to the interior of the host sponges, with only the anterior regions of the polyps extending beyond the surface of the sponges. Rearing experiments revealed that the polyp can also live freely in the normal way as it is not dependent on the host for normal growth and development. A distinct character linked to coloniality in this species is that the growth of the individual polyp is restricted to an average size of about 7·00 mm, the maximum being 12 mm. On the other hand, the colonial polyp of *N. punctata* shares a primitive feature with the solitary species by lacking an expanding stolonal plate (scyphorhiza). Therefore, branched polyps, with their limited growth and lack of a scyphorhiza are the essential features of coloniality in *N. punctata*.

3. *Linuche unguiculata*

The polyp of *Linuche unguiculata* (Fig. 4a, b) follows next in the evolutionary line deduced from morphological structures. In this species the colony com-

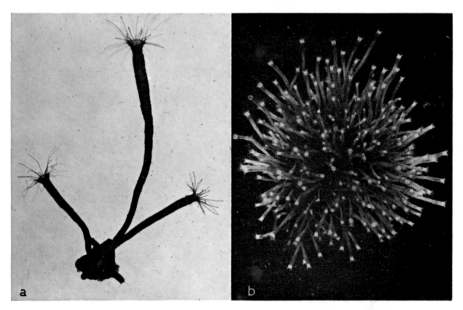

FIG. 4a, b. *Linuche unguiculata* (Schwartz 1788), (a) young colony demonstrating the basal scyphorhiza, (b) fully grown colony, 3 years old; reared in the laboratory. (a) From Werner 1973, Fig. 2a.

prises the scyphorhiza, a flat basal area of tissue with a fine peridermal cover and numerous unbranched polyps. The particular mode of growth, in which only individual unbranched polyps are produced by the enlarging scyphorhiza is evidently typical for colonies of *L. unguiculata*. This has been demonstrated by long-term rearing experiments in which colonies developed about 100 or more unbranched individuals. *L. unguiculata* shares the characters of unbranched polyps and unrestricted growth of individual polyps exceeding 20–30 mm with solitary species. These features, and the possession of a scyphorhiza, are thus characteristic of colonies of *L. unguiculata*.

4. Stephanoscyphus racemosus

The peak of colonial growth in coronate polyps is achieved by *Stephanoscyphus racemosus* (Fig. 5a, b). In this species the colonies are represented by large stems in which whorls of numerous polyps develop in the distal regions of the main vertical axis and of the lateral branches. As in *L. unguiculata*, the colonies of *S. racemosus* originate from a well developed scyphorhiza which consists of a coarse network of flat stolonal outgrowths. As shown in Fig. 5b, the size of the individual polyp is restricted to an average of 8·0 mm. Because of their extensive growth and budding, colonies of *S. racemosus* resemble small colonies of corals. Therefore colonial growth in *S. racemosus* combines characteristics which are diagnostic of colonies of both *N. punctata* and *L. unguiculata*: branched polyps, limited polyp growth, and possession of a scyphorhiza.

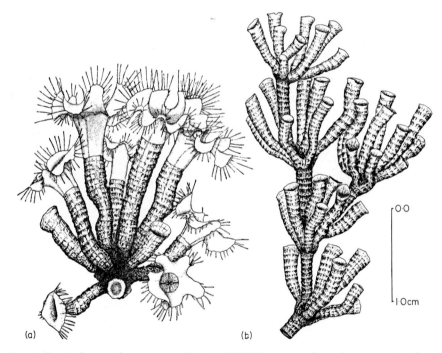

FIG. 5a, b. *Stephanoscyphus racemosus* Komai 1936, (a) young colony demonstrating the basal scyphorhiza, (b) part of a fully grown colony with empty tubes. From Werner 1970b, Fig. 4, 3.

In the foregoing survey of their structural characters colonial coronate polyps were described in a series of increasing complexity starting from a solitary species. It is of interest that in the series *Nausithoë* sp.–*N. punctata–Linuche unguiculata–Stephanoscyphus racemosus* the intermediate species share important characters with the solitary species. Although based only on the description of morphological structures the sequence appears to reflect an evolutionary sequence. From the viewpoint of systematics it seems to be important that solitary and colonial species exist within the same genus. This fact speaks for itself and suggests an evolutionary pathway. On the other hand coloniality also occurs in the related family of Linuchidae which differs in the structure of its medusae. One would expect that in the Coronatae, the sessile polyp generation in different members would have more structures in common than the medusa generation which evolved and radiated in different directions within the pelagic habitat.

REGENERATIVE QUALITIES IN SOLITARY AND COLONIAL SPECIES

The above conclusions appear to be borne out by observations on the regenerative qualities of the solitary and colonial species. The result of a simple experiment revealed that the species each possess different qualities which are correlated with the solitary or colonial habit.

Fig. 6 illustrates the results of an experimental comparison of growth habit in the species investigated (on the left) with their regenerative qualities (on the right). For the experiment a solitary polyp, or an individual polyp of a colonial species, was cut off near its base. In the solitary species (Fig. 6a) the detached upper part of the polyp closes the basal cut by contraction of the soft body walls with subsequent formation of a tissue sheet which secretes a thin external peridermal cover. No other activities of the upper part may be observed since it is unable to regenerate the lost basal part and to attach itself again to the substratum. Nevertheless, lying on the bottom, the upper part is able to grow and to strobilate in the normal manner. Additionally, it must be noted that the attached remaining stump (black), lacking the upper part, is still able to regenerate a new head (white) and to grow normally.

In *Nausithoë punctata* (Fig. 6b) the experiment reveals a situation which is partly similar to that of the solitary *Nausithoë* sp., in that the separated polyp is also unable to regenerate the lost basal part or to attach itself again to the substratum. On the other hand, regenerative qualities are present and active: at the basal hole of the separated polyp a new head is produced enlarging the nutritive abilities of the small new colony. Furthermore, the polyp is able to build up a large colony by lateral branching (white). Therefore, from the

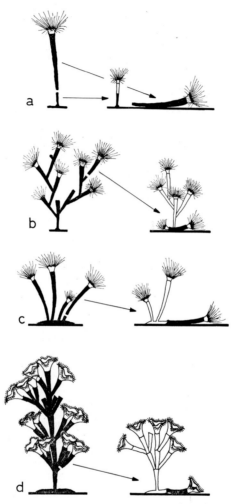

FIG. 6. Growth habit and regenerative qualities of solitary and colonial coronate polyps; (a) *Nausithoë* sp., (b) *Nausithoë punctata*, (c) *Linuche unguiculata*, (d) *Stephanoscyphus racemosus*, same species as in Fig. 1, 2 and 3–5. Left side: schematic patterns of growth habit; note the size of the solitary species and the individual polyps of the colonial ones. Right side: results of the experiment of cutting off a specimen near its base; black the separated polyp, white the regenerated structures which are in (a) the new head of the stump only, in (b–d) new colonies. The latter are formed in (b) from the separated polyp, in (c) and (d) from the new basal scyphorhiza. From Werner 1973, Fig. 1, slightly modified.

separated individual polyp a complete new colony may be produced which differs from the original one only by lack of the basal part and lack of the attachment disc, these having been formed by the primary polyp.

The same experiment gave a different result if carried out with an individual polyp of *Linuche unguiculata* (Fig. 6c). From the basal part of the separated polyp new undifferentiated cell material grows out which flattens and attaches to the substratum so forming a new normal scyphorhiza. New individual polyps (white) develop from the scyphorhiza. In this way a new normal colony originates from the separated polyp because it has a higher regenerative capacity.

In *Stephanoscyphus racemosus* (Fig. 6d) the result of the experiment is in principle the same as in *L. unguiculata*: from the basal hole of the separated polyp new tissue material grows out which attaches itself to the substratum and gives rise to a new scyphorhiza and to a new branched colony (white).

The results of these experimental observations lead to the conclusions that: (*a*) colonial species of *Stephanocyphus* have a greater regenerative capacity than solitary ones and, (b) coloniality in scyphozoan polyps correlates with higher regenerative qualities. Furthermore, certain differences between colonial species correlate with the position in the evolutionary sequence, derived above, from morphological structures, *Nausithoë punctata* being less advanced than *Linuche unguiculata* and *Stephanoscyphus racemosus*. Thus possession of a scyphorhiza allows replacement of accidentally lost basal parts, enlargement of the colony (in *Linuche*) and development of new branched colonies (in *S. racemosus*) by means of normal growth. The species *N. punctata*, does not have the capacity to regenerate a lost basal part, and is similar in this respect to the solitary species so illustrating physiological evolution in a single genus.

CONIALITY AND SEQUENCE OF REPRODUCTIVE ACTIVITIES

Colonial status signifies and implies an enlarged complex of developmental, morphogenetic and physiological potentials.

(a) The most basic prerequisite is the capacity to accumulate more reserve substances than are needed for the metabolism and reproduction of a specimen in its actual life condition and during its life span. In fact, a reserve has to be accumulated by a primary polyp in excess of its regular requirements before colonial budding can commence. The potential to accumulate such a surplus must have become genetically determined in certain solitary species at some earlier evolutionary stage.

(b) The second prerequisite is the ability to distribute accumulated reserves to the points where they are needed for branching and multiplying the morphological structures. The latter processes presume enzyme action to soften and resorb the chitinous material of the firm peridermal tubes.

(c) The same is true if the formation of basal stolons or a stolonal plate (scyphorhiza) is considered. As has been demonstrated experimentally (p. 91), the possession of a scyphorhiza is associated with greater regenerative qualities and there must have been a genetically fixed potential for the scyphorhiza to act as a distributor of reserve substances.

(d) The limited growth and genetically fixed size of an individual polyp is also a new feature which appears in association with coloniality. Once the maximum size of an individual is reached further reserves accumulated by nutritive and digestive processes are available for colony growth. It has been demonstrated that the limited growth of an individual polyp is correlated with its capacity to branch (in *Nausithoë punctata* and *Stephanoscyphus racemosus*).

Solitary species of *Nausithoë–Stephanoscyphus* can only be compared in a negative way. In particular they lack the ability to accumulate a reserve. As mentioned previously, the solitary species exhibit a regular alternation of the phases of vegetative growth and strobilation and all accumulated reserves are consumed during strobilation except those which are stored in the basal residuum and are therefore needed for regeneration of the soft body. Phase alternation is fixed genetically as it continues indefinitely in constant conditions, as has been verified in long-term culture experiments with numerous species over more than ten years. As an example, a solitary species of *Nausithoë* from the Caribbean Sea can strobilate about 7–8 or more times per year.

In contrast to solitary species, the colonial *Nausithoë punctata* and *Linuche unguiculata* strobilate only once per year. It was impossible to induce them to strobilate repeatedly in the same constant conditions and using the same culture methods which suggests that their strobilation activities are genetically fixed to one distinct short season. In *Stephanoscyphus racemosus*, laboratory cultures failed to induce strobilation at all. Therefore, a special investigation was carried out at Seto, Japan, using the natural stocks of this species which are abundant near the Seto Marine Laboratory (Werner, 1970b, 1973). The combined observations demonstrated that *S. racemosus* also has a genetically determined short period of strobilation which coincides with the time of the highest water temperature (see below). Thus coloniality of coronate polyps correlates with a single genetically determined season of strobilation activities. This would therefore explain how Recent colonial scyphopolyps are able to accumulate a surplus of reserve substances over the greater part of the year. It is this capacity which manifests itself by colonial growth.

VERTICAL DISTRIBUTION OF SOLITARY AND COLONIAL SPECIES

Consideration of the origin of coloniality consistently focuses on why solitary and colonial species differ in the sequence of their reproductive activities and how this difference could have evolved. As is generally known from modern genetics, the evolution of structures and organisms depends on the following basic factors: (a) mutations, (b) recombination of genes in bisexual organisms, (c) selection, (d) isolation, (e) random effects combined with fluctuations of populations. The factors c, d, e generally correlate with ecological factors. Since all organisms are affected by their environment to a greater or lesser degree, environmental conditions have always been important in evolution both now and in the past. It is therefore necessary to consider the ecological basis of evolution, by comparison of the environmental conditions of a habitat, with the ecology of its inhabitants and their special morphological structures and physiological adaptations. Attempting to answer our crucial question, we must consider not only morphology, physiology and life-history but also the ecology of both solitary and colonial scyphozoan polyps. If we can show that they live under different environmental conditions it should be possible to find out whether these differences correlate with the different evolutionary paths of descent.

This therefore leads to the problem of distribution and habitat conditions. According to present state of knowledge, most of the solitary species are found in deeper sublittoral and bathyal waters (Werner, 1970b, p. 12). Some solitary species also occur in the upper zones of the sublittoral (Moore, 1961; Brahm and Geiger, 1966) but a depth range of 50–200 m and deeper seems to be the favoured habitat.

The colonial species of coronate scyphopolyps, on the other hand, live in the upper sublittoral and must be considered to be true shallow-water forms. Up to now no colonial species has been detected in the deeper sublittoral nor in bathyal depths. In two species, *Linuche unguiculata* and *Stephanoscyphus racemosus*, occurrence in shallower waters is indicated by their symbiosis with unicellular brown algae (zooxanthellae) which need sufficient light for their development and growth. The zooxanthellae are located in the endodermal epithelia of *L. unguiculata* and *S. racemosus*, which therefore have a brown colour. Additionally, it must be noted that *Nausithoë punctata* is also a shallow-water species though it lacks symbiotic zooxanthellae. Although knowledge of the distribution of colonial polyps is fragmentary, the known data indicate that in coronate polyps coloniality correlates with occurrence in the higher levels of the upper sublittoral zone. This means that the colonial species live in an environment which undergoes significantly greater periodic and unpredictable changes than occur in the deeper more homogeneous habitats of the solitary species.

As mentioned above, the colonial species strobilate only once a year and have one genetically determined season for their reproductive activities. This in turn correlates with their vertical distribution pattern and it would seem fair to conclude that habitat conditions underly the evolution of coloniality, the physiological basis of which is the capacity to accumulate a surplus of resources.

In *Stephanoscyphus racemosus* vertical distribution is fairly well known and its correlation with seasonally restricted reproduction may be traced more closely. This species occurs on the eastern coast of Japan in the highest upper part of the sublittoral (Werner, 1970b, p. 12). The vertical depth range is from extreme low water level to about 5 m below this. Therefore, colonies are very much exposed to seasonal fluctuations in environmental conditions. This is particularly true for temperatures which vary from an average minimum of 14·1°C (9·8–18·6°C) in February to an average maximum of 27·7°C (24·8–29·8°C) in August (Werner, 1970b, p. 13). As described by Komai and Tokuoka (1939) and confirmed by the present author (see above) the strobilation phase is restricted to a short period in August. The species is apparently adapted to, and needs the highest water temperature of the year for strobilation.

A further remarkable feature is that in colonies of *S. racemosus*, only the oldest and most peripherally located polyps strobilate, whereas most other specimens fail to strobilate at all but show instead excessive vegetative growth and budding. In this species sexual reproduction is clearly suppressed partly in favour of colonial growth. This explains the interesting observation that, in contrast to all other species, *S. racemosus* could not be induced to strobilate in laboratory cultures and this also agrees with the general evaluation of this species as having reached the peak of coloniality in coronate polyps.

In comparison with the ecology of solitary species, it can be noted that they also have special temperature requirements, particularly for the process of strobilation. Requirements of different species sometimes differ because they correspond to the conditions of the original habitats. In the course of long-term culture experiments, a tropical sublittoral species of *Nausithoë* from the Indian Ocean (85 m depth, Fig. 1) was found to need a temperature of at least 22–24°C for its strobilation activities. An Atlantic species from greater depths (500 m) proved to require an optimum strobilation temperature of 15–17°C. Provided the temperature was constant, solitary species exhibited the above-mentioned alternation of the strobilation and vegetative growth phases.

DISCUSSION: THE EVOLUTION OF COLONIALITY IN SCYPHOPOLYPS

Knowledge of the phylogeny of the lower invertebrates is incomplete because the palaeontological record has many gaps. There are often no means of elucidating evolutionary relationships other than by comparison of structures and life-histories of Recent groups and inference of ancestral structural homologues. This applies to scyphozoan polyps for which one has to depend on comparison of morphological structures and their physiological implications.

Generally, it can be concluded from the broadly similar features of (a) growth and form of peridermal structures, (b) morphology and anatomy of the soft body and (c) life-history, in both the solitary and colonial species (Chapman and Werner, 1972), that coronate polyps had common ancestors. The idea that the colonial forms must be derived from solitary ancestors gains sufficient support and evidence from a study of ontogeny: in their development all colonial species pass through the stage of a single sexually produced primary polyp.

The Scyphozoa, being the basic class of Cnidaria, represent an obviously conservative group. It is in the most primitive order, the Coronatae that coloniality is known, and according to present state of knowledge there exist only a few colonial species. This offers the advantage that the evolutionary steps from solitary to colonial status can be traced in a fairly simple and clear approach.

1. Coloniality in coronate polyps correlates with firm supporting peridermal structures. This characteristic of Recent species can be thought with good reason to be inherited from their ancestors, the extinct fossil Conulata (Fig. 7). It must be mentioned by way of comparison that the polyps of the other metagenetic scyphozoan orders Semaeostomeae and Rhizostomeae, also have budding capacities. The semaeostome scyphistoma gives rise to short non-attaching stolons the tips of which develop into detaching secondary polyps. The rhizostome polyp, on the other hand, asexually produces small lateral buds, which then detach as planuloids and attach themselves to the substratum after a short free-swimming period; they, too, subsequently develop into small polyps, but in a different way. Why did their ancestors fail to incorporate their budding capacities genetically and thus develop into permanent colonies? It seems quite reasonable to assume that they were unable to undergo this evolutionary change because they evolved along another pathway in which they first lost the peridermal tubes which enclose and support the soft body. As mentioned previously, the ancestors of the Semaeostomeae and Rhizostomeae compensated for this by the development of a thick cellular mesoglea. With regard to the classes Hydrozoa and Anthozoa, we become aware of the fact

that coloniality is again correlated with the possession of supporting peridermal structures or skeletons.

2. Colonial species have been shown above to constitute an evolutionary

FIG. 7. *Conularia* sp., the extinct fossil Conulata were the ancestors of the Recent Scyphozoa. From Werner 1971b, Fig. 6.

series of progressive complexity. The argument gains special support from the observation that evolution from solitary to colonial life habit can be followed through intermediate stages within the same genus, or within closely related genera (*Stephanoscyphus, Nausithoë*) and families (Nausithoidae, Linuchidae). The most significant characteristics of the series are: (a) branching capacity; (b) limited size of individual polyps (these two characteristics appear to be obligatorily combined); (c) the possession of a basal scyphorhiza. A special form of coloniality is realized in *Linuche unguiculata* which has unbranched polyps of unlimited growth size. With respect to these features, *L. unguiculata* demonstrates a close relationship to its solitary ancestors.

3. In the further search for internal and external conditions which have effected and punctuated evolutionary changes, colonial status has been shown experimentally to correlate with greater regenerative qualities. It is of special interest that in this respect *Nausithoë punctata* partly resembles a solitary species in that individual polyps are unable to replace their lost basal parts and to attach themselves again to the substratum. The limited capacity of *N. punctata* obviously corresponds to the lack of a scyphorhiza, a structure which gives a higher potential for colony formation.

4. (a) An essential physiological process of colonial species is the capacity to accumulate a surplus of reserve substances which are not consumed by metabolism and reproduction. (b) Observations on strobilation activities revealed that coloniality appears to correlate with seasonally limited reproductive activities. There can be no doubt that there is a causal relationship between (a) and (b) as argued in an earlier section. (c) From an ecological viewpoint it has also been shown that Recent colonial species are very shallow-water forms which live in the upper sublittoral, where there are greater environmental fluctuations than in the deeper-water habitats favoured by solitary species.

Critical evaluation of these combined observations seems to offer a plausible explanation of the origin of coloniality in scyphozoan polyps. The final conclusion remains of course hypothetical but can be assumed to be in accordance with these facts. As mentioned before, the Japanese species *Stephanoscyphus racemosus* repesents the peak of coloniality in scyphozoan polyps and its special qualities can be interpreted as spotlights which illuminate the dark pathway of evolution.

In *S. racemosus* correlation of budding capacity with seasonally restricted reproductive activities is particularly obvious. The other facts, that the period of reproduction coincides with maximum water temperature and that the polyp lives in the highest part of the sublittoral close to the limit of the eulittoral, contain the ecological key to its evolution.

It is well known that reproduction and earliest stages of development represent the most vulnerable and sensitive phases in the life-history of invertebrates and even of the lower groups of vertebrates. With respect to these phases which depend on very particular environmental conditions, temperature has been proved to be one of the most important factors. This is especially true for the Cnidaria. In some well analysed cases (Werner, 1958, 1962), seasonal occurrence and local distribution of several hydroid species has been interpreted satisfactorily in terms of their temperature requirements and the temperature dependence of both the polyp and medusa generations. Of special significance, the polyp generation has proved to be eurythermal in its vegetative phase but stenothermal in its propagative phase of medusa budding. The medusa generation was also stenothermal.

Corresponding temperature requirements are also important for the strobilation of coronate scyphopolyps: solitary species strobilate indefinitely in constant ambient temperatures. Why is *Stephanoscyphus racemosus* able to reproduce only at the season of the highest water temperature (about 25–30°C)? It appears quite reasonable to assume that the ancestors of our Recent subtropical species lived in the tropics, i.e. in a climate in which this range of water temperature prevailed over the entire year. Solitary ancestors of *S. racemosus* would have strobilated throughout the year as do the Recent solitary species. In comparison, a Recent tropical species of a rhizostome polyp, *Cassiopea andromeda*, also exhibits a regular alternation of vegetative growth and strobilation phases (Ludwig, 1969) as described here for *Stephanoscyphus*.

It may be supposed that the solitary ancestors of *S. racemosus* migrated into the neighbouring northern regions with their seasonally changing temperatures, where they had to adjust to more exacting conditions. The local temperatures allowed them to grow vegetatively during the greater part of the year but restricted the temperature-sensitive sexual reproduction phase to the one short warmest period comparable in temperature range to the original habitat. The resources needed for continuous strobilation would have remained mostly unused and could accumulate instead as a surplus available for asexual budding. This eventually became a genetically determined adaptation to temperate conditions and thereby developed into true colonial growth.

This hypothesis agrees with the generally accepted idea that selection and processes of evolution are accelerated in areas of greater environmental fluctuations (Rensch, 1954). Therefore, the special conditions of the new habitat, especially the temperature regime, can account for the progressive evolution of the migrating ancestors of *S. racemosus*. Finally, the essential quality of coloniality (the capacity to accumulate a surplus of resources, correlated with a seasonally limited reproduction) can be interpreted as a physiological adaptation to changed environmental conditions.

SUMMARY

1. The morphological structures of one solitary and three colonial species of scyphopolyps of the basic order Coronatae are described. The coronate polyps

are distinguished by peridermal structures which completely envelop the soft body except the head. Generally, coloniality in Scyphozoa is correlated with the presence of firm peridermal structures.

2. These polyps constitute a series of progressive complexity which starts with a solitary species and demonstrates evolutionary stages. The significant characteristics are ramification, limited growth of an individual polyp and possession of a scyphorhiza.

3. As demonstrated experimentally, coloniality is correlated with greater regenerative qualities.

4. With respect to reproduction the solitary species exhibit a regular alternation of the vegetative growth and strobilation phases. Coloniality, however, is correlated with a single seasonally restricted and genetically determined short period of strobilation.

5. Deeper zones of the sublittoral and the bathyal are the favoured habitats of the solitary species. The colonial species, on the other hand, are true shallow-water forms which live in the upper zones of the sublittoral where there are significant periodic or irregular fluctuations of most environmental conditions.

6. Physiologically, the most significant phenomena of coloniality are the capacities: (a) to accumulate a surplus of resources which are not consumed by metabolism and sexual reproduction, (b) to distribute the resources to points of budding, branching and replication of the morphological structures, which also presumes (c) the ability to soften and dissolve the peridermal chitin by enzyme action.

7. Correlation of these capacities with the sequence of reproductive activities and with depth distribution offers an ecological explanation of the origin of coloniality. As a final conclusion, the coloniality of scyphopolyps can be interpreted as a physiological adaptation of the solitary ancestors to changed environmental conditions.

REFERENCES

BOARDMAN, R. S., CHEETHAM, A. H. and OLIVER, W. A., Jr (eds) (1973). "Animal Colonies: Development and Function through Time" Dowden, Hutchinson and Ross: Stroudsburg, Penn.

BRAHM, C. and GEIGER, S. R. (1966). Additional record of the scyphozoan *Stephanoscyphus simplex* Kirkpatrick. *Bull S. Calif. Acad. Sci.* **65**, 47–52.

CHAPMAN, D. M. (1966). Evolution of the scyphistoma. *In* "The Cnidaria and their Evolution" (W. J. Rees, ed.), Symp. Zool. Soc. London. No. 16, 51–75. Academic Press, London.

CHAPMAN, D. M. and WERNER, B. (1972). Structure of a solitary and a colonial species of

Stephanoscyphus (Scyphozoa, Coronatae) with observations on periderm repair. *Helgoländer wiss. Meeresunters.* **23**, 393–421.

GLAESSNER, M. F. (1971). The genus *Conomedusites* Glaessner and Wade and the diversification of the Cnidaria. *Paläont. Z.* **45**, 7–17.

KIDERLEN, H. (1937). Die Conularien, über Bau und Leben der ersten Scyphozoa. *N. Jb. Miner. Geol. Pal. BeilBd.* **77** B, 113–169.

KNIGHT, J. B. (1937). *Conchopeltis* Walcott, an Ordovician genus of Conularida. *J. Paleont.* **11**, 186–188.

KOMAI, T. (1935). On *Stephanoscyphus* and *Nausithoë*. *Mem. Coll. Sci. Kyoto Univ.* (Ser. B) **10**, 289–339.

KOMAI, T. (1936). On another form of *Stephanoscyphus* found in waters of Japan. *Mem. Coll. Sci. Kyoto Univ.* (Ser. B.) **11**, 175–183.

KOMAI, T. and TOKUOKA, Y. (1939). Further observations on the strobilation of *Stephanoscyphus*. *Mem. Coll. Sci. Kyoto Univ.* (Ser. B) **15**, 127–133.

KOZTOWSKI, R. (1968). Nouvelles observations sur les conulaires. *Acta palaeont. polon.* **13**, 497–535.

KRAMP, P. L. (1959). *Stephanoscyphus* (Scyphozoa). *Galathea Rep.* **1**, 173–185.

LELOUP, E. (1937). Hydropolypes et scyphopolypes recueillis par C. Dawydoff sur les côtes de l'Indochine française. *Mém. Mus. r. Hist. nat. Belg.* (Sér. 2) **12**, 1–73.

LUDWIG, F.-D. (1969). Die Zooxanthellen bei *Cassiopea andromeda* Eschscholtz 1829 (Polypstadium) und ihre Bedeutung für die Strobilation. *Zool. Jb. Anat.* **86**, 238–277.

MOORE, D. R. (1961). The occurrence of *Stephanoscyphus corniformis* Komai (Scyphozoa) in the western Atlantic. *Bull. mar. Sci. Gulf. Caribb.* **11**, 319–320.

MOORE, R. C. and HARRINGTON, H. J. (1956). Scyphozoa. In "Treatise on Invertebrate Paleontology. Pt. F: Coelenterata" Univ. of Kansas Press, Lawrence (Kansas) Repr. 1963.

MÜLLER, A. H. (1963). "Lehrbuch der Paläozoologie. Bd. 2: Invertebraten. T.1: Protozoa, Mollusca (1). 2. Aufl." G. Fischer, Jena.

NAUMOV, D. V. (1959). [Artliche Verschiedenheiten der Polypengeneration der Coronatae]. (Russ.). *Dokl. Acad. Nauk SSSR* **126**, 902–904.

RENSCH, B. (1954). "Neuere Probleme der Abstammungslehre: Die transspezifische Evolution" F. Enke, Stuttgart.

THIEL, H. (1966). The evolution of Scyphozoa. A review. In "The Cnidaria and Their Evolution" (W. J. Rees, ed.), Symp. zool. Soc. London, No. 16, 77–117. Academic Press, London.

UCHIDA, T. (1969). The interrelationships of scyphozoan class. *Bull. biol. Stn Asamushi* **13**, 247–250.

WERNER, B. (1958). Die Verbreitung und das jahreszeitliche Auftreten der Anthomeduse *Rathkea octopunctata* M. Sars, sowie die Temperaturabhängigkeit ihrer Entwicklung und Fortpflanzung. *Helgoländer wiss. Meeresunters.* **6**, 137–170.

WERNER, B. (1962). Verbreitung und jahreszeitliches Auftreten von *Rathkea octopunctata* (M. Sars) und *Bougainvillia superciliaris* (L. Agassiz) (Athecatae-Anthomedusae). *Kieler Meeresforsch.* **18**, 55–66.

WERNER, B. (1966). *Stephanoscyphus* (Scyphozoa, Coronatae) und seine direkte Abstammung von den fossilen Conulata. *Helgoländer wiss. Meeresunters.* **13**, 317–347.

WERNER, B. (1967a). Morphologie, Systematik und Lebensgeschichte von *Stephanoscy-*

phus (Scyphozoa, Coronatae) sowie seine Bedeutung für die Evolution der Scyphozoa. *Zool. Anz.* (Suppl.) **30**, 297–319.

WERNER, B. (1967b). *Stephanoscyphus* Allman (Scyphozoa, Coronatae), ein rezenter Vertreter der Conulata? *Paläont. Z.* **41**, 137–153.

WERNER, B. (1970a). Weitere Untersuchungen über die Entwicklungsgeschichteron *Stephanoscyphus* (Scyphozoa, Coronatae) und sein Bedeutung für die Evolution der Scyphozoa. *Zool. Anz.* (Suppl.) **33**, 159–165.

WERNER, B. (1970b). Contribution to the evolution in the genus *Stephanoscyphus* (Scyphozoa, Coronatae) and ecology and regeneration qualities of *Stephanoscyphus racemosus*. *Publs Seto mar. biol. Lab.* **18**, 1–20.

WERNER, B. (1971a). *Stephanoscyphus planulophorus* n.spec., ein neuer Scyphopolyp mit einem neuen Entwicklungsmodus. *Helgoländer wiss. Meeresunters.* **22**, 120–140.

WERNER, B. (1971b). Neue Beiträge zur Evolution der Scyphozoa und Cnidaria. *Acta Salmant.* (*Ciencias*) **36**, 223–244.

WERNER, B. (1973). New investigations on systematics and evolution of the class Scyphozoa and the phylum Cnidaria. *Publs Seto mar. biol. Lab.* **20**, 35–61.

WERNER, B. (1974). *Stephanoscyphus eumedusoides* n.spec. (Scyphozoa, Coronatae) ein Höhlenpolyp mit einem neuen Entwicklungsmodus. *Helgoländer wiss. Meeresunters.* **26**, 434–463.

WERNER, B. (1975). Bau und Lebensgeschichte des Polypen von *Tripedalia cystophora* (Cubozoa, class. nov., Carybdeidae) und seine Bedeutung für die Evolution der Cnidaria. *Helgoländer wiss. Meeresunters.* **27**, 461–504.

WERNER, B. (1976). Die neue Cnidarierklasse Cubozoa. *Verh. Dtszool. Ges.* 1976, 230.

7 | Development of Coloniality in Hydrozoa

K. W. PETERSEN

Zoological Museum, University of Copenhagen, Copenhagen Ø, Denmark

Abstract: It must be assumed that Hydrozoa developed from Cubozoa through simplification of the polyp and modifications to the process of medusa formation. The most primitive condition is found in the hydrozoan subclass Narcomedusae in which the whole larva metamorphoses into a medusa, thus resembling conditions in Cubozoa. In many present-day Narcomedusae the polyps are ectoparasitic on other medusae. In most species the primary larva buds off a few other larvae, forming a small cluster of polyps which all metamorphose into medusae. In some species the primary larva develops into a "stolo prolifer" which buds other polyps laterally. These lateral polyps metamorphose while the primary larva remains at the polyp stage. A condition similar to this may be the origin of the lateral medusa bud formation found in the other hydrozoan subclasses.

Trachymedusae probably evolved from Narcomedusae but their life-history has been changed to direct development as an adaptation to holoplanktonic life. The Limnomedusae (Olindiidae) which are related to Trachymedusae have simple solitary or primitively colonial polyps which develop lateral, planula-like polyp buds and medusa buds. The subclasses Thecata and Athecata, which likewise have lateral medusa bud formation, must have developed from, or be related to primitive Limnomedusae. In the Thecata, evolution from very primitive forms cannot be followed but this can be done in some detail in Athecata, as follows.

1. Capitata

In the Moerisiidae, the most primitive Athecata, the polyp has a whorl of moniliform tentacles around the middle part of the body. These polyps are either solitary and attached by a basal disc, or primitively colonial due to the primary polyp being able to produce another polyp with reversed polarity from its aboral end. This second polyp can remain attached to the primary polyp by a thin tube which acts as a primitive stolon.

From forms evolved from the Moerisiidae through the acquisition of oral tentacles, two evolutionary lines can be traced which mainly differ in the development of the two tentacle whorls. In the Tubulariida, the original aboral whorl remains as one whorl which

---

Systematics Association Special Volume No. 11, "Biology and Systematics of Colonial Organisms", edited by G. Larwood and B. R. Rosen, 1979, pp. 105–139. Academic Press, London and New York.

may eventually become reduced while the oral tentacles tend to spread downwards over the hydranth. In the Zancleida, the oral tentacles remain as one whorl while the aboral tentacles spread upwards over the hydranth.

In the Tubulariida, the families Euphysidae, Corymorphidae, Acaulidae and Myriothelidae retain the solitary habit, and the reverse budding from the aboral end can still be seen in *Euphysa* and *Vannuccia*. The line leading to the families Tubulariidae, Margelopsidae and Paracorynidae is interesting because they retain the "undecided" solitary/primitively colonial condition found in Moerisiidae. In the Tubulariidae, *Ectopleura* forms stolonal colonies but *Tubularia* is solitary and has retained the basal disc. Of the two other families in this line, the Margelopsidae have evolved a solitary, pelagic polyp, but the Paracorynidae form a peculiar, polymorph colony evidently developed from the hydranth head of tubulariids. Of two side branches from the tubularioid line, the Halocordylidae have perfected the monopodial colony with terminal hydranth, but the Corynidae and related families form either stolonal or racemose colonies or, as in the Solanderiidae, form colonies with a chitinized skeleton externally covered by coenosarc.

In the Zancleida the primitive forms, Cladocorynidae and Zancleidae, have stolonal colonies, but the Teissieridae (*Teissiera* and *Rosalinda*) develop a skeletal colony with a skeleton consisting of trabeculae contained within an encrusting hydrorhiza. The colony in *Teissiera* is polymorphic, with long dactylozooids bearing numerous capitate tentacles, and shorter gastrozooids which also carry the medusa buds. The pelagic Velelloidea, forming compact colonies with a large, central gastrozooid surrounded by gonozooids and dactylozooids, represent a separate evolutionary line within the Zancleida.

2. Filifera

It must be assumed that Filifera evolved from forms close to primitive Moerisiidae. In the primitive family Clavidae, which retains the scattered tentacle whorl found in moerisiids, the colonies are stolonal or sympodial. Two evolutionary lines can be traced from stolonal, primitive Filifera, both characterized by the restriction of the moerisiid tentacle whorl to two or, usually, one whorl under a fairly long hypostome. One line leads to the Pandeida, within which the Hydractinioidea develop polymorphic colonies with an encrusting chitinose or calcareous skeleton, and the Bougainvillioidea and Pandeoidea form racemose colonies. The other line leads to the Eudendriida, which likewise form racemose colonies, but differ from all other Athecata in their trumpet-shaped hypostome and their characteristic gonophores.

INTRODUCTION

The palaeontological record contains so little information on most cnidarian groups other than Anthozoa that their phylogeny has to be reconstructed from a careful analysis of Recent forms. For the higher taxa this has been done recently by Werner (1973), but attempts to clarify taxonomy within the class Hydrozoa seem to have been hampered by the traditional separation of hydrozoan specialists into workers dealing exclusively with medusae and those studying only the hydroids. The family-level classification of hydromedusae proposed by Russell (1953) and Kramp (1961) was a significant advance, as was Rees' (1957) discussion of the evolutionary trends in capitate hydroids. Many aspects will, however, continue to be overlooked so long as hydroids and medusae are, for all practical purposes, treated as two separate animal groups. At genus level we have suffered from the effects of our inability to create a single classification for the two "generations". This last difficulty arises from the disagreement as to the value of free medusae against fixed gonophores as a generic character. Even though a number of authors, such as Broch (1916), Kramp (1935) and Russell (1953), have clearly demonstrated that the degree of reduction of the medusa cannot be used to separate genera of otherwise similar hydroids, other authors, such as Rees (1957), Brinkmann-Voss (1970) and Millard (1975), have continued to use the possession of medusae or fixed gonophores as generic characters. What has probably contributed to this is the inability of the former group of workers to propose a classification in which hydroids with fixed gonophores could be classified in their mainly medusa-based genera. This led both Russell and Kramp to advocate the retention of a dual classification in which hydroids with free medusae and with fixed gonophores were placed in the same genus, while the medusae produced by these hydroids could be classified in one or more separate medusa genera. The attitude adopted by both groups of workers seems illogical and does not lead to any real solution of the problem.

In the present paper, the phylogeny of the hydrozoan subclass Athecata will be traced and, based on studies for a monographic revision of the group, it will be shown that following the principles laid down by Broch, Kramp and Russell, a single classification for hydroids and medusae can be achieved through a reevaluation of the characters used in both hydroid and medusa classification.

INTERRELATIONSHIPS OF THE CNIDARIAN CLASSES

In his very thorough analysis of the evolution of the cnidarian classes Werner (1973) concluded that the stem form of all Recent Cnidaria is a solitary, sessile, tetramerous polyp, that the Anthozoa are apparently an early offspring from this common ancestor, and that another evolutionary line leads to the Scyphozoa, Cubozoa and Hydrozoa. Werner's concept certainly seems to be the most logical interpretation of the facts known to us. The taxonomic consequence of the acceptance of this concept must be the classification of the phylum Cnidaria into two subphyla: **Anthozoa**, where the polyp is the normal, sexual adult and a medusa has never been developed, and **Medusozoa** subphylum nov., in which a medusa, whether free or reduced, is the normal sexual adult, and in which the polyp must be regarded as a larval stage. Medusozoa thus comprises the extinct class Conulata and the Recent classes Scyphozoa, Cubozoa and Hydrozoa.

PHYLOGENY OF THE HYDROZOA

In Scyphozoa the medusae are developed through strobilation, a transformation of the oral end of the polyp. In the Cubozoa the whole polyp metamorphoses into the medusa. Although both scyphozoan and cubozoan polyps reproduce asexually to form secondary polyps, only certain species of *Stephanoscyphus* Allman, 1874, belonging to the scyphozoan order Coronata, have become colonial. The Hydrozoa evolved from the Cubozoa at a relatively late geological period through simplification of the polyp and modifications of the process of medusa formation.

In many of the more important characters the Narcomedusae are the most primitive of the Hydrozoa. Certain species have a direct development, probably as a result of an adaptation to oceanic life. In other species the ontogeny resembles that of the Cubozoa in that the larvae may reproduce asexually to produce secondary larvae, but in turn all of them metamorphose into medusae. In some *Cunina* larvae described by Kramp (1957), the primary larva acts as a "stolo prolifer" from which other larvae arise by budding and, while these secondary larvae metamorphose into medusae, the primary larva in a cluster remains in the polypoid stage. The Narcomedusae are similar to Scyphomedusae and cubomedusae, and different from other hydromedusae, in having the umbrella margin interrupted by clefts so that it forms marginal lobes or lappets; their broad stomach may give rise to peripheral pouches, but radial canals and a continuous ring canal are absent; the gonads develop on the stomach periphery, but they are situated in the ectoderm, whereas scyphozoan gonads are always endodermal. Based on comparative embryological studies Goette (1907) concluded that the narcomedusan bell could not be homologous with that of other Hydromedusae. This is no doubt too strict an interpretation, but the fact remains that both the ontogeny and the morphology of the Narcomedusae places the group as intermediate between the Cubozoa and the rest of the Hydrozoa.

In all other subclasses of the Hydrozoa, medusae are produced through lateral budding from a polyp, never by metamorphosis of the whole larva. Medusae produced through this process of lateral budding have entire, circular umbrella margins, their stomachs are normally narrow and they possess narrow radial canals and a continuous ring canal. It is impossible to judge whether this lateral budding of medusae represents an entire innovation or whether it is a modification of the lateral polyp-bud formation already found in the Cubozoa, during which a secondary polyp metamorphoses into a medusa during the development of the bud. The details of the formation of this medusa bud differ from those in the development of the young Narcomedusae, mainly in the formation of the subumbrellar cavity through the development of a "glochenkern" (Goette, 1907).

The subclass which in certain structural characters, such as statocysts, seems closest to the Narcomedusae, is the Trachymedusae. Unfortunately, this group offers no evidence in the problem of modification to the polyp and formation of the medusa, as the Trachymedusae have a direct development, from the fertilized egg to an actinula which gradually grows into a medusa. This embryology may well be an adaptation to oceanic, holoplanktonic life, parallel to what we know from Scyphomedusae like *Pelagia noctiluca* (Forsskål, 1775) and a number of Narcomedusae, because the Trachymedusae agree with the higher Hydrozoa in the structure of the umbrella.

Of the remaining subclasses, the Limnomedusae seems to occupy a basal position. Its medusae have the same type of statocyst as the Narcomedusae and Trachymedusae, and its polyps have in some species retained the formation of planula-like polyp buds which must be regarded as primitive. We know the hydroid in seven of the 13 genera. In *Craspedacusta* Lankester, 1880; *Gonionemus* A. Agassiz, 1862; *Limnocnida* Günther, 1893; *Olindias* F. Müller, 1861; and *Vallentinia* Browne, 1902, the polyp is solitary, without tentacles or with three to six tentacles, and with asexual reproduction through planula-like polyp buds which develop on the lower part of the body wall. In *Craspedacusta* and *Limnocnida* the polyp buds can fail to liberate, and as a result a small, colony-like cluster is formed. In *Monobrachium* Mereschkowsky, 1877, and *Aglauropsis* Müller, 1865, the hydroid is colonial and forms linear or reticulate stolons from which new polyps are budded. The hydroid has only one tentacle. Medusa buds are borne at the base of the polyp in *Aglauropsis* and on the stolons in *Monobrachium*.

The subclasses Siphonophora, Thecata and Athecata probably originated independently from forms close to the primitive Limnomedusae. In the first two of these subclasses we do not have more than a few indications of this relationship, so the transition from primitive solitary hydrozoans to these specialized colonial forms cannot be followed. In the Athecata we have a number of transitional forms, which makes it possible to follow the evolution in reasonable detail.

SUBCLASS ATHECATA (ANTHOMEDUSAE)

The Subclass Athecata was divided by Kühn (1913) into the Capitata and the Filifera, mainly on the characters of the tentacles in the hydroids. When Rees (1958) returned the former limnomedusan family Moerisiidae to the Athecata he regarded it as the most primitive family in the Capitata. However, it must be realized that the tentacles of moerisiid hydroids correspond to the aboral tentacles of other capitate hydroids, whose oral tentacles are an acquisition unique to this group (cf. the development of tentacles in actinulae of the Tubulariidae and Acaulidae, and the position of medusa buds over the tentacle whorl in the Moerisiidae and in *Sphaerocoryne* Pictet, 1893). There are reasons to assume that the tentacles in the Filifera correspond to the moerisiid tentacles and to the aboral tentacles in the Capitata (see p. 131).

Turning to the athecata medusae, the quadrate stomach with perradial pouches found in all moerisiid meduase, is not found in the Capitata, except perhaps in the form of swellings in the radial canals as seen in *Dipurena* McCrady, 1857, and *Zanclea* Gegenbaur, 1856 (Rees, 1966). A quadrate stomach with perradial extensions is, however, found among the Filifera in the families Pandeidae, Bougainvilliidae and Hydractiniidae, and the tentacle bulbs of moerisiid and pandeid medusae are morphologically very similar.

Because these moerisiid characters point to both the Capitata and Filifera, it seems reasonable to assume that forms which basically possess the characters of the Moerisiidae are ancestral to both groups. The family is here referred to the Capitata, principally because the combination of nematocyst types agrees with that found in the Tubulariida.

1. Order Capitata

Our ideas about relationships within the Capitata have changed considerably over the past 30 years. While Kramp (1949) in his discussion of the origin of the Corymorphidae regarded the colonial Corynidae as the most primitive family, Garstang (1946) viewed the solitary Corymorphidae as the ancestral group. This concept was strongly supported by Rees (1957), who based his conclusions on an assessment of the morphology and embryology of both hydroids and medusae. In the following year Rees (1958) became aware of the similarity in structure between newly liberated moerisiid medusae and medusae of the Tubulariidae and Corynidae and of the similarities between the hydroids of *Moerisia lyonsi* Boulenger, 1908 and *Euphysa aurata* Forbes, 1848. As mentioned above, he referred the Moerisiidae to the Capitata, regarding it as the most primitive family of this group.

The Capitata can conveniently be subdivided into three suborders (Fig. 1). The most primitive is the Moerisiida, with one more or less scattered whorl of tentacles which, judging from the position of the medusa-budding zone among or over the tentacles, must correspond to the aboral tentacles of other capitate hydroids. In the remaining two suborders the hydroids have acquired a whorl of oral tentacles but the arrangement of tentacles in these two whorls differs in the two suborders. In the Tubulariida, the two whorls are either nearly equally developed, even though the oral whorl shows a tendency to duplication, or the oral whorl is elaborated and spreads down over the body of the hydranth while the original, aboral whorl tends to be reduced. By contrast in the Zancleida although the oral whorl is present, it is the aboral whorl which is elaborated and spreads upwards over the body of the hydranth.

Moerisiida subordo nov.
The suborder Moerisiida contains only one superfamily of two families. Of these, the Moerisiidae must be regarded as the most primitive. The hydroid has either a single whorl of moniliform tentacles or moniliform tentacles scattered over the upper, middle part of the hydranth. It has retained the planula-like polyp buds found in the Cubozoa and Limnomedusae and it is solitary or very primitively colonial.

The suborder Moerisiidae is usually divided into five genera, based on characters of both hydroids and medusae. Of these, *Halmomises* von Kennel, 1891, is known only from the medusa and is retained only because the sole species of the genus is so inadequately described that it cannot be distinguished from other species. *Tiaricodon* Browne, 1902, has one species and is well known in the medusa stage, but the hydroid has not been found. In the remaining three genera, both hydroids and medusae are known. *Moerisia* Boulenger, 1908 and *Ostroumovia* Hadzi, 1928, have very similar hydroids and have been kept distinct on a single medusa character, namely globular tentacle bulbs in *Moerisia* and clasping, adnate tentacle bulbs in *Ostroumovia*. Examination of material of all the species in the two genera, including medusae from Boulenger's original material of *M. lyonsi*, shows that in both genera the tentacle bulbs originate as round "sarsiid" bulbs, but that during their growth they do not increase in length and the adaxial part of the bulb develops more than the abaxial, pushing the bulb and tentacle towards the exumbrellar side of the margin. Fully developed tentacles in both genera spring from the abaxial part of a low bulb clasping the umbrella margin. There are no evident reasons for keeping *Moerisia* and *Ostroumovia* separate and *Ostroumovia* must thus be considered a junior synonym of *Moerisia*. The fifth genus, *Odessia* Paspaleff, 1937, has only one

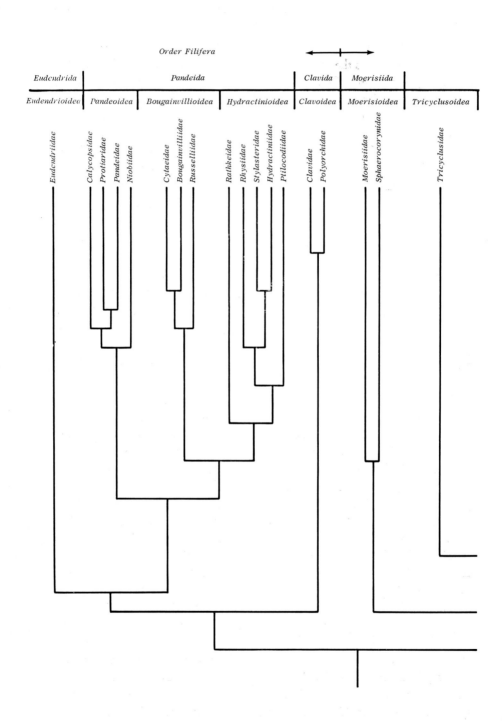

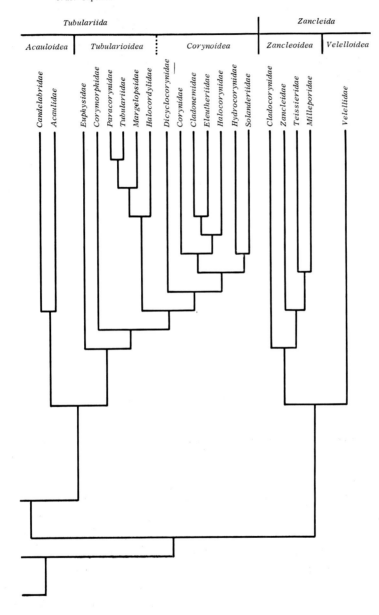

Fig. 1. Phylogeny in the Athecata. A diagram of the relationships of the families with an indication of the proposed classification. The figure should not be interpreted as a cladogram.

species. It resembles *Moerisia* in all characters except in the arrangement of the nematocyst armature on the tentacles, which are not moniliform but have the nematocysts arranged in clasps in the medusa and in adoral warts in the hydroid. The polyp of *Odessia maeotica* (Ostroumoff, 1896) is solitary and attached to the substrate by a perisarc-covered pedal disc. *Moerisia hori* (Uchida and Uchida, 1929) and *M. inkermanica* Paltschikowa–Ostroumova, 1925, are also solitary, attached by a pedal disc and 4–10 podocysts, i.e., elliptical, convex, perisarc-

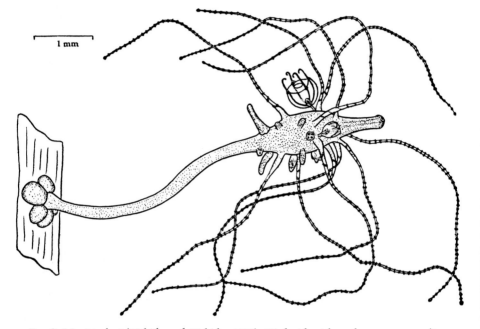

FIG. 2. *Moerisia hori* (Uchida and Uchida, 1929). Hydroid with podocysts surrounding the pedal disc (from Uchida and Nagao, 1959).

covered discs connected to the pedal disc (Fig. 2). After the death of the original polyp, each of these podocysts gives rise to a new polyp. In both species the podocysts may arise from stolon-like outgrowths from the base of the polyp so that the total impression is that of a branched hydrorhiza with adhesive discs. This latter condition is the normal one in the hydroid of *M. lyonsi* Boulenger, 1908 (Fig. 3). From the perisarc-invested lower part of the polyp other long, filiform tubes arise, each bearing a hydranth at its distal end. Asexual reproduction takes place not only at the ends of these stolons, but also through lateral, planula-like buds on the hydranths and through transverse fission of the basal part of the hydranth, whereby a series of spherical bodies is formed each of which later develops into a new polyp.

In an undescribed species of *Moerisia* from Mauritius the filiform outgrowths from the basal part of the polyp also resemble stolons, and both polyp and medusa buds are formed on these stolons.

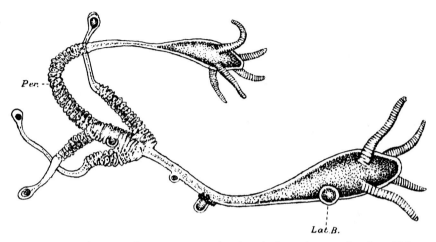

FIG. 3. *Moerisia lyonsi* Boulenger, 1908. Hydroid with the podocysts placed on filiform outgrowths from the base of the polyp and with a secondary polyp formed at the distal end of another outgrowth. Lat. B., lateral polyp bud. Per., perisarc (from Boulenger, 1908).

The second family in the Moerisioidea is the Sphaerocorynidae, fam. nov. (type genus *Sphaerocoryne* Pictet, 1893). *Sphaerocoryne bedoti* Pictet, 1893, *S. multitentaculata* (Warren, 1908) (Fig. 4) and the closely related *Linvillea agassizi* (McCrady, 1857) have up to now been regarded as belonging to the Corynidae, presumably because the hydroids possess capitate tentacles and the medusae possess ocelli. In all three species the hydranths are vasiform with a conical proboscis; the tentacles are capitate, placed in a more or less scattered whorl around the proximal third of the hydranth, and the medusa buds are borne on short pedicels among, or distal to, the tentacles. The hydranths are placed on relatively long, branched or unbranched stems covered by a firm perisarc which terminates below the hydranth. The medusae have a bell-shaped umbrella, with or without nematocyst tracts. Their manubrium is more or less cruciform in cross-section, and the mouth is without lips. The gonads cover the perradial surfaces of the manubrium, with deep interradial furrows. The four tentacle bulbs, which each bear an abaxial ocellus, clasp the umbrella margin. The tentacles have a terminal nematocyst cluster and oblique nematocyst patches along their length.

The arrangement of the tentacles in a zone around the basal third of the hydranth, the position of the medusa-budding zone among or distal to the tentacles, the perradial position of the gonads in the medusa and the occurrence

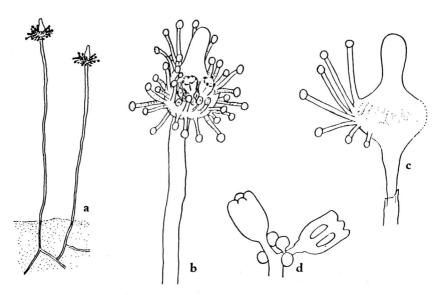

Fig. 4. *Sphaerocoryne multitentaculata* (Warren, 1908). a. Two hydranths arising on long, perisarc-covered stems from creeping stolons embedded in a sponge. b, c. Hydranths showing the position of tentacles around the lower part of the body and of medusa buds in a zone distal to the tentacles. d. Blastostyle with a cluster medusa buds (from Yamada and Konno, 1973).

of stenotele and desmoneme nematocysts in both hydroid and medusa are definite indications that these three species cannot be included in the Corynidae. They show instead a definite relationship to the Moerisiidae, in which the moniliform tentacles and the medusa-budding zone are positioned in the same way as in the hydroids. In the medusa generation, the tentacle bulbs of *Sphaerocoryne* and *Linvillea* bear a strong resemblance to the tentacle bulbs of young moerisiid medusae, and the perradial development of the gonads is also reminiscent of the gonad development in Moerisiidae.

As with the Moerisiidae, the Sphaerocorynidae lack an oral tentacle whorl, but the tentacles of the existing whorl have become capitate. The ability to produce lateral planula-like polyp buds is lost, the perisarc covers not only the very base of the polyp but reaches to the base of the hydranth, and primitive, creeping stolons have been developed.

Tubulariida, subordo nov.

As a consequence of my opinion that the evolutionary lines leading to the Acauloidea, Tubularioidea and Corynoidea have originated from a common moerisiid stock, these superfamilies have here been united in the new suborder Tubulariida. This suborder comprises those groups in which the original aboral tentacles are retained in the more primitive forms but tend to be reduced in the derived forms, while the oral tentacles are retained, either as a single whorl, or as several closely set whorls, or they tend to spread down over the hydranth body. Desmoneme and stenotele nematocysts are always present.

Tricyclusoidea

This superfamily was erected by Rees (1957) for the curious little hydroid *Tricyclusa singularis* (Schulze, 1876). This hydroid is solitary and attached by a pedal disc covered by a gelatinous perisarc which reaches to the base of the vasiform hydranth proper. The original, aboral tentacles are arranged in two whorls around the middle of the hydranth, but in addition to these tentacles an oral whorl of capitate tentacles is present. The fixed gonophores are budded from the area between the two whorls of aboral tentacles and polyp buds are produced from an area just below the basal tentacle whorl. *Tricyclusa* seems to be a survivor of a primitive group of the Capitata which has acquired the whorl of oral tentacles but has otherwise retained many of the characteristics of the Moerisioidea (Fig. 5).

Acauloidea

This superfamily comprises two families, the Acaulidae and the Candelabridae, both of which contain solitary species. In the Acaulidae a pointed aboral end is encased in a gelatinous perisarc which attaches the hydroid to muddy bottoms by means of filaments. In the Candelabridae, species either have a basal, chitinized perisarcal tube with stout anchoring filaments (Fig. 6), or the perisarc is reduced to adhesive discs at the tip of the anchoring filaments (see Manton, 1941). Both these types of attachment point directly back to the conditions in the Moerisiida.

Tubularioidea

Early Tubularioidea, derived from forms close to the Moerisioidea, which, in addition to the aboral tentacles, acquired a whorl of oral tentacles, cannot have been solitary as presumed by Rees (1957) but has probably been at an intermediate stage halfway between solitary and primitively colonial, just as in the Recent moerisiids.

At some point in the tubulariid development—when the hydranth had a morphology close to the present *Hypolytus* Murbach, 1899, i.e. with moniliform tentacles in both whorls and incipient parenchyma formation in the hydrocaulus

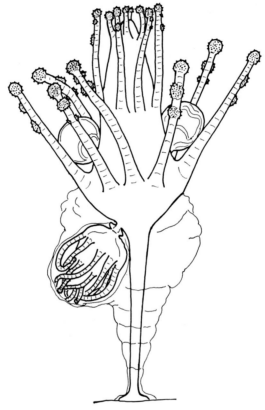

Fig. 5. *Tricyclusa singularis* (Schulze, 1876). A specimen with medusa buds between the aboral tentacles and a polyp bud on the lower part of the hydranth (redrawn from Schulze, 1876).

although still a moerisiid type mode of attachment—the Euphysidae split off to develop their special type of solitary polyp in which the hydrocaulus is attached to the loose perisarcal sheet by papillae on the hydranth just below the aboral tentacle whorl.

The main corymorphid-tubulariid line further developed the stem parenchyma as a support for larger hydranths; the long blastostyles may also have begun to develop at this stage. From this point the group evolved in two directions. One line, leading to the Corymorphidae, developed the solitary polyp as a soft-bottom form. This occurred through a further development of the stem parenchyma and the development of the special parenchymatous diaphragm structure in the hydranth. Corymorphids are further characterized

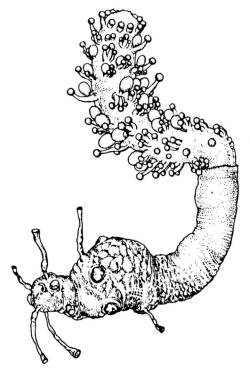

FIG. 6. *Monocoryne gigantea* (Bonnevie, 1898), a candelabrid hydroid with basal perisarc and stout anchoring filaments (from Rees, 1957).

by the numerous rooting filaments on the aboral end of the hydrocaulus, filaments which are probably homologous with stolons.

The second line, leading to the Tubulariidae and Halocordylidae, and giving rise at an early point to the corynid group of families (Corynoidea), developed the basal disc or stolon type of attachment to firm substrates. In this line the mesolamella has become continuous over the tentacle bases, separating the tentacular endoderm from that of the gastric cavity. Also in the tubulariid and halocordylid line the relatively large hydranths acquired a supporting structure in the form of a parenchymatous cushion under the lower tentacle whorl, but this cushion is structurally very different from the diaphragm of the corymorphids. At a later stage of evolution, this line split, so leading to the present-day Tubulariidae and Halocordylidae. In the tubulariid line the tentacles in both whorls eventually became filiform, the stem parenchyma with central and peripheral canals developed, though the colony form remained primitive. From the tubulariid line the Margelopsidae evolved, developing a solitary pelagic hydranth which in *Margelopsis* Hartlaub, 1897, is reminiscent of the pelagic form of *Moerisia pallasi* (Derzhavin, 1912), while *Pelagohydra* Dendy, 1902, is much more specialized. In the halocordylid line the capitate oral tentacles spread downwards to form two or three whorls of imperfectly capitate tentacles, and the colonial habit was developed to such an extent that pennariid colonies are among the most highly organized in the Athecata.

The Tubulariidae includes, among others, the present hydroid genus *Tubularia*, which contains species with sessile gonophores and also species liberating medusae of the two medusan genera *Ectopleura* and *Hybocodon*. Often, as in Fraser (1944) and Brinkmann-Voss (1970), the species with free medusae are regarded as belonging to two separate genera and the species with sessile gonophores to a third, a good example of the unsatisfactory state of the art in hydrozoan taxonomy where the degree of reduction of the medusa generation, contrary to all evidence, is used to delimit genera.

From the morphology of the gonophores it is quite clear that in certain species such as *Tubularia indivisa* Linnaeus, 1758, *T. ceratogyne* Perez, 1920, and *T. regalis* Boeck, 1859, the gonophore represents a reduced asymmetrical medusa, while in species such as *T. crocea* (L. Agassiz, 1862) and *T. larynx* Ellis and Solander, 1786, the gonophore is reduced from a symmetrical medusa. This alone should be sufficient indication that the species with fixed gonophores do not form a natural genus of their own but should be transferred to one or the other of the medusa-producing genera. Theoretically, this could be done on the characters afforded by the gonophores, but the hydroid generation also offers good systematic characters on which a distinction between two genera can be based.

Tubularian hydroids are either solitary or colonial. In the solitary species the attachment varies from a small, perfectly circular disc in *T. cornucopia* Bonnevie, 1897, through lobed discs in species like *T. indivisa* and *T. regalis*, to tuber-like attachments in the sponge-dwelling species *T. prolifera* (L. Agassiz, 1862) and *T. unica* (Browne, 1902). When the hydrocaulus of most of the disc-attached species reaches a certain height, one or more stolon-like tubes covered by thick, rigid perisarc grow out from the hydrocaulus well above the basal disc. They continue to grow downwards, adhering to the stem, and may be straight or entwining. On reaching the substrate these tubes may continue to grow over the surface of it for a shorter or longer distance, thus anchoring the growing hydroid firmly and stiffening the thin lower part of the stem. Distal to the region from which these supporting tubes originate, the stems are straight, with smooth perisarc. At times it may seem as if *T. indivisa*, *T. ceratogyne* and *T. regalis* form small colonies of entwined stems but a closer examination will show that these "colonies" consist of a number of separate individuals which have settled close to each other; their stems and supporting tubes have become entwined during growth and there is never any connection between these individuals. The solitary species either produce free medusae of the medusa genus *Hybocodon* L. Agassiz, 1862, or have fixed gonophores of the asymmetrical type. These species form a natural genus, *Tubularia* Linnaeus, 1758, with *Tubularia indivisa* L. as the type species.

The colonial species form a series within which the stolons vary from a pattern very close to that seen in *Moerisia lyonsi*, present in *Ectopleura wrighti* (this is a nom. nov. for *Acharadia larynx* Wright, 1863, as the species name *larynx* becomes a secondary homonym of the combination *Ectopleura larynx* (Ellis and Solander, 1786) when these two species are treated as congeneric) and *E. dumortieri*, to the more or less regular tripod arrangement seen in *E. ochracea* L. Agassiz, 1862, *E. larynx*, *E. crocea*, etc. In the colonial species the young polyp sends out stolons over the substrate, and from these stolons new hydranths arise at fairly regular intervals. In the species with a tripod arrangement of the stolons these usually bifurcate at the point where a hydranth arises, so that each stem seems to arise from a tripod of stolon (Mackie, 1966). Since this happens at the budding point of each stem, the stolons will soon grow over each other as a dense tangle complicated by the formation of new colonies on stolons and stems of the parent colony. The hydrocauline perisarc of the colonial species is annulated or wavy at the base and at intervals along the length of the stems. These species either produce free medusae of the medusa genus *Ectopleura* or have fixed gonophores of the symmetrical type. They form a natural genus for which the name *Ectopleura* L. Agassiz, 1862, has priority with *E. dumortieri* (van Beneden, 1844) as the type species.

In addtion to these two genera the Tubulariidae consists of: the solitary *Zyzzyzus solitarius* (Warren, 1906), a sponge-dwelling species with a thick hydrocaulus, tuber-like attachment and very thin perisarc; an undescribed genus and species with very thick, irregular perisarc and an attachment reminiscent of certain species of *Candelabrum* de Blainville, 1830.

The Margelopsidae is clearly derived from the tubulariid line through the loss of the hydrocaulus. *Margelopsis* Hartlaub, 1897, has a solitary pelagic hydranth structurally similar to the head of tubulariid hydroids, and its larval development includes an actinula stage. In *Pelagohydra* Dendy, 1902, the aboral region of the hydranth has developed into a float through a specialization of the parenchyma and the aboral tentacles have spread out over this float.

The Halocordylidae is a small family of hydroids closely related to the Tubulariidae. The hydroids form upright, branching colonies of a very definite bush-like or feather-like form, springing from a creeping stolon.

The curious family Paracorynidae contains only one species, *Paracoryne huvei* Picard, 1957. *Paracoryne* forms a small, plate-like colony of three types of individuals: gastrozooids, gonozooids and dactylozooids. The gastrozooids are very small, with up to four whorls of tentacles of which the most oral are short and capitate and the ones further down the hydranth are longer and more filiform. The gonozooids are even smaller than the gastrozooids and carry a cluster of sessile gonophores at their distal ends. The colonies are dioecious. Around the periphery of the colony are placed a number of dactylozooids which are about ten times as long as the other types of individuals in the colony. The basal plate of the colony is made up of two layers: an upper layer of normal coenosarc from which the gastrozooids and gonozooids arise, vascularized by anastomosing endodermal canals, and a basal layer of large, parenchymatous cells which are continuous with the endodermal core of the dactylozooids. The presence of an actinula larva, and the kinds of nematocyst which are present, place *Paracoryne* close to the Tubulariidae. Bouillon (1974a) advances the interesting theory that the *Paracoryne* colony can be derived from the head of a normal tubulariid hydranth (Fig. 7).

Corynoidea

The Moerisiida and the Tubulariida discussed previously, have a nematocyst armature which always bear stenoteles and desmonemes, both in the hydroid and in the medusa. The corynid group of families, which I believe originates from primitive tubulariid–halocordylid stock, differs from this pattern in having both kinds of nematocyst in the medusa, but only stenoteles in the hydroid. The hydroids in this group of families have capitate oral tentacles which can be present in one whorl or spread down over the hydranth, either as scattered tentacles or in whorls. The aboral tentacles are filiform and there is a pronounced tendency to reduce this whorl. Within the family group there is considerable evolution both in the medusa generation and in the hydroid, especially with regard to colony development. From various characters, such as the supporting mesogloeal lamella, which is continuous over the tentacle bases and separates the endoderm of the gastric cavity from that of the tentacles, it is clear that the corynid group of families is derived from the Tubularioidea of the tubulariid–halocordylid line. Maintaining the Corynoidea as a separate superfamily is thus based on convenience of classification rather than on phylogenetic considerations.

The family Corynidae affords a good example of the difficulties encountered

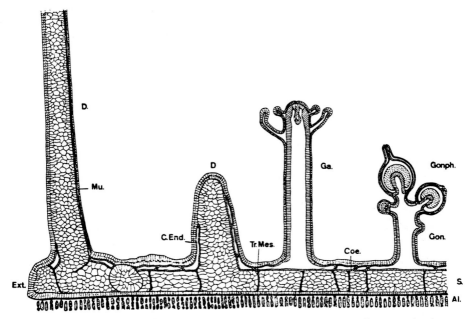

FIG. 7. Diagram of the structure of a colony of *Paracoryne huvei* Picard, 1957. Al., alga. C. End., endodermal canal. Coe., coenosarc. D., dactylozooid. Gon., gonozooid, Gonph., gonophore. Mu., spiral muscle. S., parenchyma. Tr. Mes., mesogloea (from Bouillon, 1974a).

in creating a single classification to include both hydroids and medusae. Russell (1953) summarized the situation as far as the northern species are concerned, concluding that although most of the necessary facts were available, it was not possible to create a single classificatory system for members of this family. Using the present dual classification the Corynidae consists of the genera listed in Table I.

The genus *Dicyclocoryne* (and *Bicorona*) is characterized by a whorl of 4–7 capitate oral tentacles and an aboral whorl of 6–21 capitate tentacles. The gonophores are borne between or slightly below the aboral tentacles. In *Dicyclocoryne filamenta* (Annandale, 1907) the colony consists of a creeping stolon from which hydranths spring at intervals. Sometimes the hydranths bear a lateral polyp. In *Bicorona elegans* Millard, 1966, the colony is an erect, branched monopodial hydrocaulus with terminal hydranths. Since the only character of any generic

TABLE I. Genera of the Corynidae

Hydroids	Medusae
Coryne Gaertner, 1774	Fixed gonophore
Syncoryne Ehrenberg, 1834	*Sarsia* Lesson, 1843
Staurocoryne Rotch, 1872	Fixed gonophore
Stauridiosarsia Mayer, 1910	*Stauridiosarsia* Mayer, 1910
Dipurena McCrady, 1857	*Dipurena* McCrady, 1857
Bicorona Millard, 1966	Fixed gonophore
Dicyclocoryne Annandale, 1915	*Dicyclocoryne* Annandale, 1915

value separating these two genera is the production of free medusae in *Dicyclocoryne*, even though *Bicorona* has fixed gonophores, we can safely unite these two species under the first genus. The number and arrangement of the tentacles places *Dicyclocoryne* as a primitive corynid which points directly back to an origin from tubulariid forms. The medusa is sarsiid in most characters but lacks ocelli on the tentacle bulbs. Both hydroid and medusa characters indicate that the creation of a new family, Dicyclocorynidae, fam. nov., is indicated for these species.

The remainder of the corynids is presently grouped in the medusa genera *Sarsia*, *Stauridiosarsia* and *Dipurena*, and in the hydroid genera *Coryne*, *Syncoryne*, *Staurocoryne*, *Stauridiosarsia* and *Dipurena*. As clearly indicated by Russell (1953), it is impossible to merge the two systems into one as they stand today. This would indicate that either the hydroid or the medusa system or both are erroneous and based on characters without actual significance.

Let us first consider the medusae. *Sarsia* and *Dipurena* are today separated through the number of rings, which are formed by the gonads surrounding the manubrium, while *Stauridiosarsia* is distinguished from *Sarsia* by the presence of filiform tentacles in its hydroid. Hartlaub (1907) subdivided the genus *Sarsia* into an "*eximia* group" and a "*tubulosa* group". In the "*eximia* group" the manubrium is short and completely encircled by gonad and its endoderm is digestive throughout, i.e., the whole manubrium equals the stomach proper. In the "*tubulosa* group", the manubrium is divided into a distal stomach and a long, narrow, nondigestive part, only the latter carrying the gonad. If we continue this train of thought we find that the medusae under discussion can be divided into three groups on characters pertaining to manubrium and gonad: one group comprising among others *Sarsia eximia* (Allman, 1859), *Stauridiosarsia producta* (Wright, 1858) and *Sarsia prolifera* Forbes, 1848, in which the whole manubrium corresponds to the stomach proper and in which the gonad encircles the whole stomach, leaving only the mouth rim and a short proximal portion free of gonad. A second group consists of *Sarsia tubulosa* (M. Sars, 1835) and *S. princeps* (Haeckel, 1879), in which the stomach proper is a distal swelling on the manubrium while the longer, narrow part of the manubrium is nondigestive. The gonad encircles only this nondigestive part, leaving the stomach proper and a short proximal part free of gonad. Finally, a third group comprises the *Dipurena* species and *Sarsia gemmifera* Forbes, 1848; here the manubrium is similar to that of *S. tubulosa* and *S. princeps* but the gonad is developed on the stomach proper with, in many cases, additional rings of gonad around parts of the nondigestive part of the manubrium.

If we now try to compare this grouping of the medusae with the corresponding hydroids, we find that hydroids of the medusae of the first-mentioned group have their tentacles arranged in alternating whorls, and the medusa buds are developed either in the axils of the tentacles or in a whorl replacing the lowest whorl of capitate tentacles. They may or may not have a basal whorl of filiform tentacles.

Hydroids of medusae in the second group have their capitate tentacles scattered over the distal two thirds of the hydranth and the medusa buds are developed under the tentacle-bearing zone.

Hyroids of the third group are either of the primitive corynid type, which has only the oral whorl of capitate tentacles in addition to an aboral whorl of filiform tentacles, and whose medusa buds are situated just distal to the filiform whorl; or they are of a more advanced corynid type, which, in addition to the oral whorl of capitate tentacles, has scattered capitate tentacles which have spread downwards into the medusa-bud producing zone, where the buds are situated in between but independent of the tentacles. These species may or may not have an aboral whorl of filiform tentacles.

Using the foregoing characters, it is now possible to place in their proper genus the hydroids whose medusae are unknown or reduced to a fixed gonophore, and similarly, to refer medusae whose hydroids are not known to a genus. The proposed classification is shown in Fig. 8.

With regard to colony structure, *Dipurena* forms creeping, stolonal colonies and many of the species inhabit sponges. *Coryne* contains species with creeping stolonal colonies in addition to species with upright racemose colonies, and *Sarsia* usually forms upright, racemose colonies.

Of the remaining families in the Corynoidea, the Cladonemidae and the Eleutheriidae have developed very specialized medusae, while the hydroids have remained rather primitive, with an oral whorl of capitate tentacles and a whorl of filiform aboral tentacles which are, however, often reduced. Hydroids of both families are characterized by the concentration of the gland cells around the mouth in a cup-shaped depression (Bouillon, 1963, 1966, 1971). The hydroid colony in both families is of the creeping, stolonal type.

The Halocorynidae occurs on Bryozoa and forms creeping colonies with

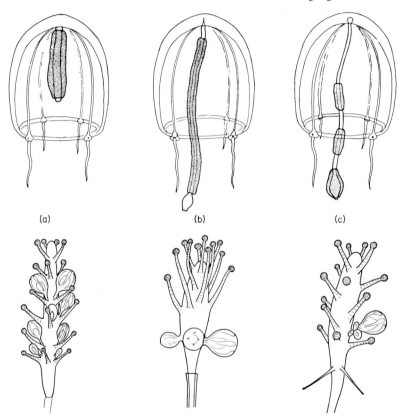

FIG. 8. Proposed classification of corynid hydroids and medusae. a, genus *Coryne* Gaertner, 1774. b, genus *Sarsia* Lesson, 1843. c, genus *Dipurena* McCrady, 1857. For further details see text.

anastomosing stolons. The hydroids are dimorphic, with gastrozooids devoid of tentacles and dactylozooids with uniserial nematocyst clusters. The medusa is figured by Bouillon (1974b).

The two last families in this group have very specialized colony forms. The Hydrocorynidae has a strongly developed, chitinized skeleton similar to that in *Hydractinia* van Beneden, 1841, while the Solanderiidae has developed upright colonies with a curious ectodermal skeleton consisting of chitinized skeletal parts covered by ectoderm and provided with tubes of endoderm (Vervoort 1966).

Zancleida, subordo nov.

The third suborder of the Capitata, the Zancleida, must, like the Tubulariida, be derived from hydroids close to the Moerisiida, which in addition to the aboral tentacles have acquired a whorl of oral tentacles. In the Zancleida the aboral

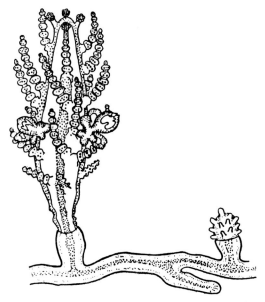

FIG. 9. *Asyncoryne ryniensis* Warren, 1908, a primitive member of the Zancleoidea with moniliform aboral and capitate oral tentacles (after Warren, from Rees, 1957).

tentacles tend to spread up over the body of the hydranth while the oral tentacles remain as a single whorl around the mouth. The medusae are very characteristic, with the exumbrellar nematocysts confined to a specialized tissue situated as oval, club-shaped or elongated patches over the tentacle bulbs, and with stalked cnidophores on the marginal tentacles. While the hydroids and medusae belonging to the Tubulariida possess desmonemes and stenoteles in addition to other nematocyst types, the Zancleida are characterized by the presence of macrobasic euryteles and stenoteles. The Zancleida contains two superfamilies, the Zancleoidea and the Velelloidea.

Zancleoidea

In the most primitive of the Zancleoidea the oral and aboral tentacles differ in structure. In *Asyncoryne ryniensis* Warren, 1908, the scattered aboral tentacles are moniliform and the oral whorl capitate (Fig. 9). The medusa buds, which may develop into free medusae, are situated in clusters on short blastostyles between the moniliform tentacles. Mature buds show the exumbrellar nematocyst patches and the tentacular cnidophores characteristic of the superfamily (Bouillon, 1974b). The colony consists of a creeping stolon on which hydranths are borne at intervals. At times a branch of the hydrorhiza may turn upwards and the end of it develops a hydranth, which is reminiscent of the colony form in *Moerisia lyonsi* and *Ectopleura wrighti*.

Lobocoryne travancorensis Mammen, 1963, has fleshy aboral tentacles scattered over the body and four very short capitate oral tentacles. The aboral tentacles have the nematocysts placed in a number of capitulae along the lateral sides. *Cladocoryne littoralis* (Mammen, 1963) and *C. floccosa* Rotch, 1871, have respectively one or several whorls of long aboral tentacles with stalked nematocyst knobs in addition to the oral whorl of short, capitate tentacles. The colonies of these three species are of the stolonal type with hydranths arising at intervals along a creeping stolon. It is hardly justifiable to place these four hydroids in two families, the Asyncorynidae and the Cladocorynidae, as they appear to form a continuous series, so I prefer to transfer *Asyncoryne* Warren, 1908 to the Cladocorynidae.

In the Zancleidae, the oral and aboral tentacles are of the same general structure, capitate in *Zanclea* Gegenbaur, 1856, or nearly filiform in *Pteroclava* Weill, 1931. The Zancleidae seem to have a common origin with the Cladocorynidae, the capitate or filiform aboral tentacles probably being derived from the moniliform type. Both *Zanclea* and *Pteroclava* form creeping, stolonal colonies. The medusae, of which in *Pteroclava* only newly liberated ones are known, have the exumbrellar nematocyst patches characteristic of the superfamily, together with simple circular mouths, and two or four marginal tentacles with abaxial cnidophores; they lack ocelli.

The family Teissieridae Bouillon, 1978 (type genus *Teissiera* Bouillon, 1974), at present contains three species, *T. milleporoides* Bouillon, 1974b (Fig. 10), *T. australe* Bouillon 1978 and *T. medusifera* Bouillon 1978. The colonies known only in *T. milleporoides* are dimorphic, composed of short, stout gastrozooids with about 17 scattered, short capitate tentacles, and long, slender dactylozooids without a mouth, each bearing about 65 long, capitate tentacles. The two types of polyps arise from a stolonal plate made up of anastomosing endodermal canals covered by ectoderm. The ectoderm secretes a thin outer periderm and a thicker basal plate with irregularly toothed spines which penetrate the coenosarc. Medusa buds arise at the base of the gastrozooids. These buds develop into free medusae with a simple, circular mouth, four radial canals and two tentacular bulbs having long tentacles provided with abaxial cnidophores. Each of the four exumbrellar nematocyst patches carries an ocellus at its upper end. The

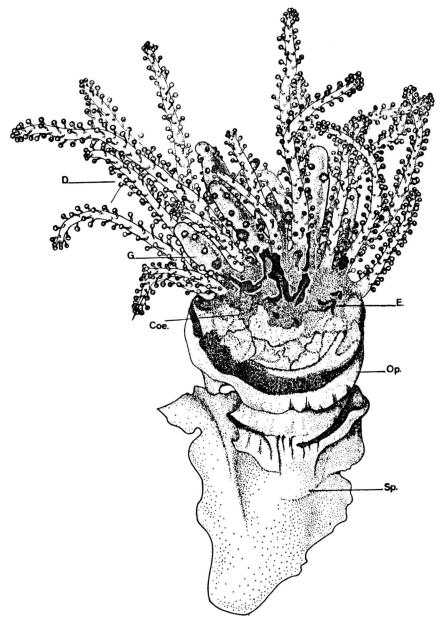

Fig. 10. Colony of *Teissiera milleporoides* Bouillon, 1974, growing on the operculum of polychaete *Spirobranchus tetraceros*. B.M., medusa bud. Coe., coenosarc. D., dactylozooid. E., spine. G., gastrozooid. Op., operculum of Sp., *Spirobranchus* (from Bouillon, 1974b).

dimorphism, the encrusting spiny stolonal plate in the hydroid, and the presence of ocelli in connection with the exumbrellar nematocyst patches in the medusa, separate *Teissiera* from the Zancleidae and made it necessary to create a new family for this genus. The presence of scattered capitate tentacles in both types of polyps in the Teissieridae indicates a derivation from the Zancleidae.

As demonstrated by Bouillon (1974b) the genus *Rosalinda* Totton, 1949, has many characters in common with *Teissiera*. The colonies have a stolonal plate consisting of a basal perisarc with elevated ribs covered by ectoderm with worm-like tubes made up of an external layer of ectoderm and an internal layer of endoderm. In *R. williami* Totton, 1949, the elevated ribs reach the surface of the stolonal plate layer of endoderm giving it a fine honeycomb-like structure, while in *R. incrustans* (Kramp, 1947) the ribs are less elevated, giving the plate a reticulate structure (Vervoort, 1966). The hydranths, which arise singly from the stolonal plate, have 30–60 scattered capitate tentacles. The gonophores are unknown. The stolonal plate of *Rosalinda* has a structure close to that of *Teissiera* except that the structure is not covered by an outer perisarc. This last fact, together with the absence of polymorphism in the hydranths, makes it questionable whether *Rosalinda* should be included in Teissieridae or should form a separate family.

The family Milleporidae forms calcareous colonies, covered on the outside with ectoderm and containing a meshwork of coenosarc tubes which secrete the calcareous skeleton. The hydranths, which are situated in pores in the stolonal basal plate, are of two types: short, plump gastrozooids with four to six very short capitate tentacles around the mouth, and long, slender, mouthless dactylozooids with a number of scattered capitate tentacles. Small, abortive medusae are produced in special chambers in the stolonal plate. They lack velum and tentacles but have four or five nematocyst patches on the exumbrellar side of the margin. Gonads develop on the manubrium. In the character of the polymorphism and in the structure of the two types of polyp, the Milleporidae resemble the Teissieridae, and the skeletal stolonal plate also occurs in *Teissiera* and *Rosalinda*. These two genera and *Millepora* Linneaus, 1758, possess a nematocyst type unique to them, the macrobasic mastigophore, making the hypothesis that they are related even more plausible.

Velelloidea, superfam. nov.

In the Velellidae the nematocyst types present in their medusae are like those of the Zancleoidea, but the family should probably be regarded as a separate evolutionary line derived from primitive Zancleoidea, and thus be placed in a separate superfamily.

2. Order Filifera

The phylogeny of the Filifera is more difficult to trace than that of the Capitata, mainly because the morphology of the polyp has undergone relatively few changes within the order.

The Clavida is the most primitive suborder. Its filiform tentacles in the polyp are scattered over the upper middle part of the body, as is also seen in certain Moerisiidae. The Clavida are clearly derived from primitive forms close to Moerisiida, but the Recent Clavida cannot be used as a starting point in filiferan development to the same extent as can the Moerisiida in capitate phylogeny. The nematocyst types in both hydroid and medusa are microbasic euryteles and desmonemes. From the Clavida it is, however, possible to trace two lines leading to the Pandeida and to the Eudendriida. In both these lines there is a development towards concentrating the tentacles in one whorl somewhat below the mouth, but oral tentacles as we know them in the Capitata are not developed. In the Pandeida the general shape of the hydroid body remains unchanged, and the general pandeid polyp is spindle-shaped with a conical hypostome and a single whorl of filiform tentacles. In the medusa, the mouth rim is elaborated, either as large, folded lips in the Pandeoidea, drawn out as mouth arms in the Hydractinioidea, or provided with proper oral tentacles in the Bougainvillioidea. The nematocyst types in Pandeida are microbasic euryteles and desmonemes, as in the Clavida.

In the Eudendriida the hydranth is large, with a wide, trumpet-shaped hypostome. The gonophores, which are always fixed sporosacs, are borne by hydranths which are reduced to varying degrees. The male gonophores consist of a number of round chambers, while each of the female gonophores consists of a spadix curving around a single large egg. The nematocysts are microbasic euryteles and sometimes also atrichous isorhizas or macrobasic euryteles, but never desmonemes.

Clavida, subordo nov.

The Clavida contains only one superfamily. Clavoidea, superfam. nov., with hydroids in which the filiform tentacles are scattered over the body of the hydranth and in which the gonophores are borne just under the tentacles on the hydranth, or on the hydrocaulus, or on the stolons. The medusae have a more or less quadrate stomach, a mouth with four lips armed with a continuous row of nematocysts, and numerous marginal tentacles with ocelli.

In the family Clavidae, *Clava* Gmelin, 1791 and *Rhizogeton* L. Agassiz, 1862, both have stolonal colonies in which the creeping, reticulate hydrorhiza gives rise to sessile, naked hydranths. Both genera have fixed gonophores, borne in *Clava* on the hydranth and in *Rhizogeton* on the stolons. *Merona* Norman, 1865, also has stolonal colonies, but these differ from those of the two preceding genera in being polymorphic, with normal clavid gastrozooids, short, mouthless gonozooids and very small, thin nematophores. The genus *Corydendrium* van Beneden, 1844, forms unique stolonal colonies in which the stolons grow up in fascicles covered by a common perisarc. Two species have free medusae and the others have fixed gonophores. *Cordylophora* Allman, 1844, forms erect, racemose colonies. Only two species of the Clavidae (*Corydendrium*) produce free medusae. They have four simple radial canals, a quadrate stomach with four lips with a continuous row of nematocyst clusters along the rim, interradial gonads, and numerous solid marginal tentacles with adaxial ocelli.

It is possible that the medusa family Polyorchidae, in which the hydroid is unknown, should be placed in the Clavoidea. The medusae have sausage-shaped gonads which hang down from the junction between the manubrium and radial canals, abaxial ocelli on the numerous marginal tentacles, and radial canals with a tendency to develop side branches.

Pandeida, subordo nov.

The Pandeida includes the majority of the families of the Filifera. In this suborder the hydroid retains a club-shaped body with a conical hypostome, and the tentacles tend to move up into a single whorl around the hypostome. The medusae of the Pandeida have some basic features, such as the quadrate stomach with perradial extensions in common with the medusae of the Moerisiida.

The Pandeoidea, superfam. nov., is mainly based on medusa characters. It brings together the families in which the medusae have four simple or crenulated unarmed lips and hollow marginal tentacles with the bases either adnate to the exumbrella or widened to a long, hollow bulb. The hydroids are known only in the family Pandeidae. They possess the general morphology common to the suborder. Their colonies are usually of creeping stolonal form, with single hydranths growing on thin stems which can occasionally bear one or two lateral hydranths. Some species form a rhizocaulus in which bundles of stolons bearing nearly sessile hydranths grow up from the substrate. Medusa buds are developed on the hydranth stems or on the stolons. The medusae have large, crenulated lips and quadrate stomachs with adradial or interradial gonads which may continue along the radial canals. Their tentacle bulbs can easily be derived from the moerisiid tentacle bulb.

The medusa family Pandeidae as understood at present contains a number of genera which should be removed to families of their own. This is true for

Protiara Haeckel, 1879; *Halitiara* Fewkes, 1882; *Paratiara* Kramp and Damas, 1925; and *Cnidotiara* Uchida, 1927. Russell (1953) placed *Protiara* in a subfamily of its own, and this should be raised to the rank of family, containing the above-mentioned genera. The genus *Niobia* Mayer, 1900, with watch-glass shaped umbrella, a stomach without peduncle, with interradial gonads and four simple lips, with bifurcating radial canals and having marginal tentacle bulbs which develop into new medusae, clearly also requires its own family, the Niobiidae, fam. nov.

The medusa family Calycopsidae is placed in the Pandeoidea, firstly because of its hollow tentacles of which the basal parts, which are not widened to a bulb, are adnate to the umbrella margin, and secondly because of the quadrate manubrium with four simple or crenulated lips. In these characters the medusae seem close to the primitive pandeoid stock.

Of the two remaining superfamilies in the Pandeida, the Hydractinioidea Bouillon, 1978, contains the polymorph filiferan families Hydractiniidae, Ptilocodiidae, Rhysiidae and Stylasteridae. They have been discussed recently by Bouillon (1978), so it is sufficient to mention that the Rhysiidae has a colony with a creeping stolon covered by perisarc. The Ptilocodiidae has either creeping perisarc-covered stolons or a naked basal coenosarc. The Hydractiniidae has either perisarc-covered stolons which may fuse to form a basal plate from which the upper perisarc tends to disappear, leaving the ectoderm naked, but pierced by horny spines arising from the basal perisarc, or it has an elaborate calcareous skeleton covered by naked coenosarc. Finally, the Stylasteridae forms porous calcareous colonies criss-crossed by stolonal tubes and covered by naked coenosarc. The medusae in the Hydractinioidea have the mouth lips drawn out and sometimes branched, and provided with terminal clusters of nematocysts. The family Rathkeidae may belong in this superfamily.

The Bougainvillioidea, superfam. nov., contains the medusa-based families Cytaeidae, Bougainvilliidae and Russelliidae, in which the medusa has a set of oral tentacles, sometimes branched, inserted above the mouth. The hydroid in the Cytaeidae arises singly from a creeping stolon, surrounded at the base by a short peridermal cup. The medusa buds are developed singly from the stolons. In the Bougainvilliidae the genera *Balella* Stechow, 1919, *Clavopsella* Stechow, 1919 and *Silhouetta* Millard and Bouillon, 1973, have the tentacles arranged in two to four whorls concentrated around the hypostome, but the remaining genera have only one tentacle whorl. The simplest colony form in the Bougainvilliidae has erect, unbranched or slightly branched stems with a terminal hydranth. Most species however have erect, much branched stems of the racemose type which may be fascicled as stolons grow up along the stem. Both stolons and stems are covered by perisarc which terminates below the hydranths in *Clavopsella* and *Silhouetta*, or continues around the hydranth as a pseudohydrotheca. In *Koellikerina* Kramp, 1939 and *Bimeria* Wright, 1859, the pseudohydrotheca forms a sheath around the base of each tentacle. The hydroid of the Russelliidae is not known.

Eudendriida, subordo nov.

The Eudendriida contains only the superfamily Eudendrioidea, superfam. nov., with the single family Eudendriidae. This family differs from all other filiferan hydroids in the shape of the hydranth body, which is urn-shaped with a trumpet-shaped hypostome, and in the cnidome, which always contains microbasic euryteles, occasionally macrobasic euryteles and atrichous isorhizas, but never desmonemes. None of the species has free medusae. Reproduction is through fixed sporosacs borne on the hydranths below the tentacles. Male sporosacs usually consist of a row of bulbous chambers and female sporosacs a spadix curving around a single egg. The hydranths bearing the gonophores may be normally developed or reduced in size and in the number of tentacles. In the genus *Myrionema* Pictet, 1893, the hydranth is long and slender and bears the tentacles in several closely set whorls around the trumpet-shaped hypostome. By contrast, in *Eudendrium* Ehrenberg, 1834, the hydranth is broadly urn-shaped and the tentacles are arranged in a single whorl. In both genera however, the colony is upright, often profusely branched, and with terminal hydranths. The stem may be fascicled and is covered by a firm perisarc which terminates at the base of the hydranth.

DEVELOPMENT OF COLONIALITY IN THE ATHECATA

As will appear from the preceding taxonomic sections, I consider the Athecata to be derived from solitary hydroids with a perisarc-covered pedal disc. In the Recent Moerisiidae this condition has largely been retained. *Odessia maeotica* has what must be the most primitive type of pedal disc. Species such as *Moerisia hori* and *M. inkermannica* have, in addition to the pedal disc, a number of elliptical perisarc-covered podocysts connected with the pedal disc. These podocysts can be placed immediately around the pedal disc or arise from filiform outgrowths at the base of the polyp. In *Moerisia lyonsi* the podocysts are always placed on the distal end of such filiform tubes while other tubes may bear secondary hydranths at their distal ends. The podocysts are known to develop into new polyps once the polyp forming them has died off. It is therefore tempting to assume that development is inhibited by the original polyp and that the podocysts furthest removed from the polyp can develop into secondary polyps if the concentration of the inhibiting agent is sufficiently weak. In this way a primitive form of coloniality might have originated.

In most moerisiids the polyp may detach itself from the pedal disc and form a bluntly pointed aboral end. In this phase the polyp can move around and later attach itself and form a new disc.

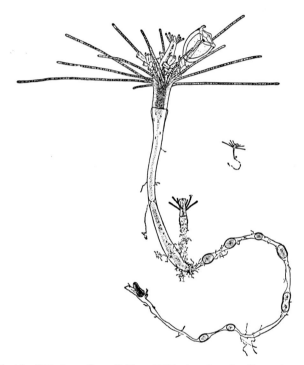

FIG. 11. Hydroid of *Euphysa farcta* (Miles, 1937) showing the formation of secondary polyps through transverse fission of the aboral end (from Miles, 1937).

The variation in the structure of the basal part of the polyp found in Moerisiidae can also be traced within the Tubulariida. Polyps of the Acaulidae, Euphysidae and Corymorphidae have a pointed aboral end similar to that found in detached moerisiid polyps; and in the Euphysidae the ability to produce new polyps by transverse fission of the aboral end, as found in *Moerisia lyonsi*, has been retained (Fig. 11). The simple solitary polyp with a pedal disc occurs in the Tricyclusidae and in the tubulariid genus *Tubularia*. A somewhat modified form of the attachment with podocysts on tube-like outgrowths is found in the Candelabridae. The colony form found in *Ectopleura wrighti* (Fig. 12) is easily derived from the primitive colonies formed by *Moerisia lyonsi*, and the tripod arrangement of the stolons in other *Ectopleura* species has developed from this colony form. The most primitive colonial condition in the Corynoidea consists of creeping

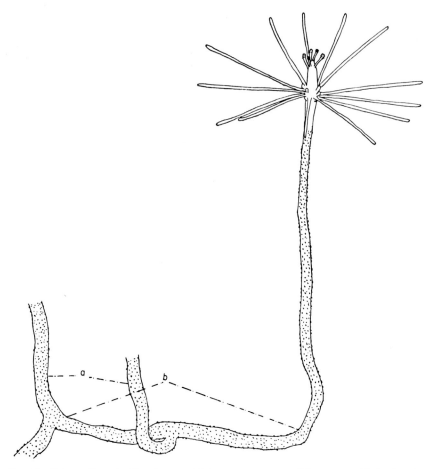

FIG. 12. Colony form in *Ectopleura wrighti* nom. nov. a, upright stems with hydranths. b, part of stem acting as stolon (from Brinckmann-Voss, 1970).

stolonal colonies very similar to those of *Ectoplerua wrighti*. This colony form is also the most primitive condition within the suborders Zancleida, Clavida and Pandeida. This kind of creeping stolonal colonial form is probably the origin of the various specialized rhizocaulome formations found in the Corynoidea (Hydrocorynidae and Solanderiidae), the Zancleoidea (Teissieridae and Milleporidae) and the Hydractinioidea (Hydractiniidae and Stylasteridae), and of the simpler upright rhizocaulome colonies of certain Pandeidae.

Within certain groups in the Tubulariida, in the Pandeida and in all the Eudendriida, the stolonal colonies, in which unbranched hydranths arise on shorter or longer stems along the length of the stolons, have developed further through the addition of lateral branching from the stem of the primary polyp, so that a racemose colony, i.e. a monopodial hydrocaulus with terminal hydranth, is formed.

The curious colony formation in the Paracorynidae has already been mentioned (p. 122), and the Velellidae may well be another example of a colony formation of this type.

Within the Athecata, colony formation has led to polymorphism only in the filiferan superfamily Hydractinioidea, in one species of the Clavoidea and in a single species of the Corynoidea. Dimorphism, where nutritive polyps and polyps bearing gonophores differ in size and sometimes in structure, is more widespread, occurring in the Corynoidea, Zancleoidea and Eudendrioidea.

SUMMARY

Following Werner's (1973) analysis of the interrelationships of the cnidarian classes, the phylum Cnidaria is subdivided into two subphyla, the Anthozoa and the Medusozoa (subphylum nov.). In a discussion of the class Hydrozoa, the subclass Narcomedusae is regarded as the most primitive group with the Trachymedusae and Limnomedusae derived from it. The subclasses Siphonophora, Thecata and Athecata probably originated independently from forms close to primitive Limnomedusae. In a review of the taxonomy of the Athecata, this subclass is divided into two orders, the Capitata and the Filifera, both of which are derived from the Moerisioidea.

The Capitata are grouped in three new suborders: the Moerisiida, with a scattered whorl of aboral tentacles; the Tubulariida, in which the aboral tentacles are concentrated in one whorl which is sometimes reduced while the oral tentacles tend to duplicate and spread down over the hydranth; and the Zancleida, in which the aboral tentacles spread up over the hydranth while the oral tentacles remain as one whorl. As a consequence of this classification, the genera *Sphaerocoryne* and *Linvillea* are referred to the Moerisiida, forming the Sphaerocorynidae (fam. nov.), and the genus *Dicyclocoryne* is designated the type genus of a family of primitive Corynoidea, the Dicyclocorynidae (fam. nov.). Within the Tubulariidae and Corynidae the genera are revised to give a single classification for hydroids and medusae. The Zancleida is divided into two superfamilies, the Velelloidea (superfam. nov.), and the Zancleoidea, containing the families Cladocorynidae, Zancleidae, Teissieridae, and Milleporidae.

The Filifera have not acquired the oral tentacles found in most capitate hydroids but have only the moerisiid aboral tentacles. The order is grouped in three new suborders: the Clavida, in which the tentacles remain scattered; the Pandeida, comprising families with fusiform hydranths in which the tentacles move up to form a whorl under the conical proboscis; and the Eudendriida, with vasiform hydranths in which the tentacles move up to form one or more closely-set whorls under the trumpet-shaped proboscis. While the Clavida and Eudendriida contain only one superfamily each, the Pandeida are grouped in three superfamilies: the Hydractinioidea, comprising the families in which the medusae have the mouth rim drawn out into arms, and in which the hydroid colonies are polymorphic; the Bougainvillioidea (superfam. nov.), comprising the families in which the medusae have solid oral tentacles inserted above the mouth; the Pandeoidea (superfam. nov.), grouping together the families with simple or folded lips.

The family Pandeidae is divided into three: the Protiaridae, the Pandeidae and the Niobiidae (fam. nov.).

Based on the taxonomic review it is shown that the solitary polyp with an attachment through a pedal disc and podocysts forms the starting point for colonial development in the Athecata. Attachment types clearly derived from this moerisiid type are found scattered among the Tubulariida and it is shown that the stolonal colony form is probably derived from the moerisiid type podocyst attachment with the podocysts placed on filiform outgrowths, while the other colony forms in the Athecata, rhizocauline and racemose hydrocauli, can be derived from the stolonal type.

REFERENCES

BOUILLON, J. (1963). Les cellules glandulaires des hydroïdes. *C. r. hebd. Séanc. Acad. Sci. Paris* **256**, 1617–1620.

BOUILLON, J. (1966). Les cellules glandularies des hydroïdes et hydroméduses. Leur structure et la nature de leurs sécrétions. *Cah. Biol. mar.* **7**, 157–205.

BOUILLON, J. (1971). Sur quelques hydroïdes de Roscoff. *Cah. Biol. mar.* **12**, 323–364.

BOUILLON, J. (1974a). Sur la structure de *Paracoryne huvei*, Picard 1957 (Coelenterata, Hydrozoa, Athecata). *Mém. Acad. r. Belg. Cl. Sci.* **18** (3), 1–45.

BOUILLON, J. (1974b). Description de *Teissiera milleporoides*, nouveau genre et nouvelle espèce de Zancleidae des Seychelles (hydrozoaires; athécates-anthoméduses), avec une révision des hydroïdes "Pteronematoidea". *Cah. Biol. mar.* **15**, 113–154.

BOUILLON, J. (1978). Sur un nouveau genre et espèce de Ptilocodiidae et la nouvelle super-famille Hydractinoidea (Hydroida, Athecata). *Steenstrupia*, in press.

Brinkmann-Voss, Anita (1970). Anthomedusae/Athecatae (Hydrozoa, Cnidaria) of the Mediterranean. Part I Capitata. *Fauna Flora Golfo Napoli* **39**, 1–96.
Broch, H. (1916). Hydroida (Part I). *Dan. Ingolf Exped.* **5** (6), 1–66.
Fraser, C. McLean (1944). "Hydroids of the Atlantic coast of North America." Toronto University Press.
Garstang, W. (1946). The morphology and relations of the Siphonophora. *Q. Jl microsc. Sci.* **87** (2), 103–193.
Goette, A. (1907). Vergleichende Entwicklungsgeschichte der Geschlechtsindividuender Hydropolypen. *Z. wiss. Zool.* **87**, 1–335.
Hartlaub, C. (1907). Craspedote Medusen. 1. Teil. 1. Lief.: Codoniden und Cladonemiden. *Nord. Plankt.* **6** (12), 1–135.
Kramp, P. L. (1935). *Corydendrium dispar*, a new athecate hydroid from Scandinavian seas, with remarks on classification. *Göteborgs K. Vetensk.-o. vitterhSamh. Handl.* Ser. V, B, **4** (11), 1–15.
Kramp, P. L. (1949). Origin of the hydroid family Corymorphidae. *Vidensk. Meddr dansk naturh. Foren.* **111**, 183–215.
Kramp, P. L. (1957). Hydromedusae from the Discovery collections. *"Discovery" Rep.* **29**, 1–128.
Kramp, P. L. (1961). Synopsis of the medusae of the World. *J. mar. biol. Ass. U.K.* **40**, 5–469.
Kühn, A. (1913). Entwicklungsgeschichte und Verwandtschaftsbeziehungen der Hydrozoen. 1. Teil: Die Hydroiden. *Ergebn. Fortschr. Zool.* **4**, 1–284.
Mackie, G. O. (1966). Growth of the hydroid *Tubularia* in culture. *In* "The Cnidaria and their evolution" (W. J. Rees, ed.), Symp. zool. Soc. London. No. 16, 397–412. Academic Press, London.
Manton, S. M. (1941). On the hydrorhiza and claspers of the hydroid *Myriothela cocksi* (Vigurs). *J. mar. biol. Ass. U.K.* **25**, 143–150.
Millard, N. A. H. (1975). Monograph on the Hydroida of southern Africa. *Ann. S. Afr. Mus.* **68**, 1–513.
Rees, W. J. (1957). Evolutionary trends in the classification of capitate hydroids and medusae. *Bull. Br. Mus. nat. Hist., Zool.* **4** (9), 455–534.
Rees, W. J. (1958). The relationships of *Moerisia lyonsi* Boulenger and the family Moerisiidae with capitate hydroids. *Proc. zool. Soc. Lond.* **130** (4), 537–545.
Rees, W. J. (1966). The evolution of Hydrozoa. *In* "The Cnidaria and their evolution" (W. J. Rees, ed.), Symp. zool. Soc. London. No. 16, 199–222. Academic Press, London.
Russell, F. S. (1953). "The medusae of the British Isles." Cambridge University Press, Cambridge.
Vervoort, W. (1966). Skeletal structure in the Solanderiidae and its bearing on hydroid classification. *In* "The Cnidaria and their evolution" (W. J. Rees, ed.), Symp. zool. Soc. London. No. 16, 373–396. Academic Press, London.
Werner, B. (1973). New investigations on systematics and evolution of the class Scyphozoa and the phylum Cnidaria. *Publs Seto mar. biol. Lab.* **20**, 35–61.

ADDENDUM

(*Submitted July, 1976, after the Symposium*)

Colonies of Colonies in *Physalia*

P. F. S. CORNELIUS

Department of Zoology, British Museum (Natural History), London, England

It is arguable that the highest level of coloniality in the Hydrozoa, and possibly in the Coelenterata, is seen in *Physalia physalis*. As is well known *Physalia* is a "colony" comprising several kinds of "individuals". The "colonies" themselves also aggregate—albeit passively—in suitable conditions, forming vast swarms extending sometimes for hundreds of kilometres, and sometimes so dense that they partially obscure the surface of the sea (for example the great density reported in *The Marine Observer* (1976) Vol. 46, p. 61). The local impact of the long tentacles (up to about 20 m) on the sub-surface environment must be considerable. Passive though the aggregations may be, it would seem likely that they have some influence on the *Physalia* populations. The influence might be positive (for example in promoting spawning) or negative (for example in depleting food sources); but information seems lacking.

Aggression between the separate "colonies" has apparently not been recorded from *Physalia*, and in a planktonic environment it is hard to see how spacing out could be achieved by a passively floating species. Nevertheless, if aggression is in fact absent then it is arguable that in forming non-aggressive aggregations of "colonies", *Physalia* exhibits a high level of coloniality. The swarms, or "colonies of colonies", when they occur achieve a dominance of the local water mass to a degree unusual among the Coelenterata.

8 | Co-ordination of Behaviour in Cnidarian Colonies

G. A. B. SHELTON

Department of Zoology, University of Oxford, England

Abstract: Solitary and colonial species from two cnidarian classes, the Anthozoa and the Hydrozoa, are compared using ultrastructural, behavioural and electrophysiological data.

In the Anthozoa, there are similarities in the conduction systems found in both subclasses (Octocorallia and Hexacorallia). In addition to the nerve net is a second system (similar to the slow conduction systems described in sea anemones). Both systems are used in the colonial control of behaviour.

Hydrozoan colonies also have multiple conduction systems involving both nervous and non-nervous elements. Several pacemaker systems may be present. Complexity of behaviour is mirrored by the underlying electrical activity.

Though certain other phyla of colonial animals, notably the Bryozoa, may show similar patterns of behaviour to those found in the Cnidaria, electrophysiological evidence indicates marked differences in the control mechanisms.

The advantages of coloniality for the different groups are discussed.

INTRODUCTION

Lepers, slime moulds, ants and corals have all been described as living in "colonies" and it seems that this mode of life can provide significant selective advantages, but the exact definition of a colony is not easy. Wells (1973) suggests that "Animal colonies are groups of individuals either structurally bound together or structurally separated but bound by behaviour." A flock of birds could be classed in this category. To define a colony it is necessary to consider the meaning of the term "individual" but to define "individual" proves an equally difficult conundrum. The reader is urged to read Mackie (1963) on the

---

Systematics Association Special Volume No. 11. "Biology and Systematics of Colonial Organisms", edited by G. Larwood and B. R. Rosen, 1979, pp. 141–154. Academic Press, London and New York.

concept of the individual. He quotes Bertalanffy as saying "only in the consciousness of ourselves as beings different from others are we immediately aware of individuality that we cannot define rigidly in the living organisms around us." Mackie concludes that "Individuality is not simply a morphological concept nor an ecological one but is equally genetical and physiological." The situation is further confused by some of the more advanced colonies, e.g. siphonophores which may behave in such a co-ordinated way as to take on a "colonial individuality" (Mackie, 1963). The fact is that the quest for a totally satisfactory definition of either "individual" or "colony" is fraught with problems. Mackie rightly urges us away from semantic arguments toward a consideration of scientific evidence. In this paper, clues to the nature of a colony are taken from morphological studies and from electrophysiological and behavioural analysis of members of the phylum Cnidaria. Of particular interest are the two important colonial classes—Anthozoa and Hydrozoa. Comparisons are made between solitary and colonial species. This approach yields important data on the extent to which zooids are independent and the extent to which they are integrated as a colonial organism. From this follows a discussion of the advantages of coloniality for the different groups.

ANTHOZOA

1. Structure

The Anthozoa are divided into two subclasses: the Hexacorallia, e.g. sea anemones and corals (solitary and colonial) and the Octocorallia, e.g. gorgonians and soft corals (colonial). Fig. 1 (a) shows a diagrammatic vertical section of a typical solitary sea anemone (Actiniaria). The tentacles are simple, usually arranged in several cycles around the mouth. The outside is covered with a single layer of ectoderm which may contain gland cells, nematocysts, spirocysts, nerve cells, etc. A layer of mesogloea separates the ectoderm from the endodermal epithelium. Leading from the mouth into the gastrovascular cavity is the pharynx which is held in place by complete mesenteries (Fig. 1(d)). The inner margins of the mesenteries are elongated at the base to form mesenterial filaments, used in the digestion of food. The pedal disc usually secretes an adhesive which secures the animal to the substratum. Figs. 1(b) and (e) show corresponding sections through a typical madreporarian coral polyp. There are clear similarities in the general organization of mesenteries, tentacles, pharynx etc. between the Actiniaria and the Madreporaria. A most important difference, however, is the presence in the Madreporaria of a rigid calcium carbonate theca and a number of calcium carbonate sclerosepta. This imposes severe constraints on the kinds of movement possible. Solitary individuals may attain oral disc diameters of centimetres or tens of centimetres. The individuals in hermatypic colonies are often much smaller but may collectively form massive colonies

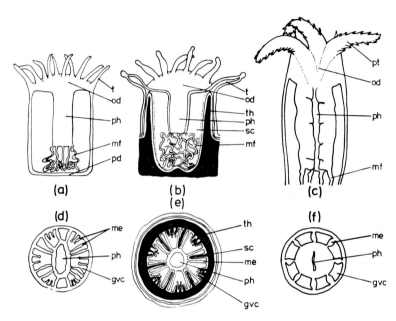

FIG. 1. Structure of anthozoans. Diagrammatic vertical sections through (a) a sea anemone, (b) a madreporarian polyp, (c) an autozooid of an octocoral. (d), (e) and (f) are the corresponding transverse sections. Abbreviations: gvc—gastrovascular cavity, me—mesenteries, mf—mesenterial filament, od—oral disc, pd—pedal disc, ph—pharynx, pt—pinnate tentacle, sc—scleroseptum, t—tentacle, th—theca. See text for further details.

which comprise thousands of individuals. The third pair of diagrams (Figs 1(c) and (f)) shows vertical and transverse sections respectively through a polyp (an autozooid) of an octocoral. In the Octocorallia, only eight mesenteries are present. Each of the eight tentacles is pinnate. There is no massive calcium carbonate theca for support and the hydrostatic skeleton may take on a more important role in this respect. Spicules, which may be horny or calcareous, are also involved in mechanical support and can form massive fused structures in, for example, the gorgonians. Though there is no evidence for division of labour and polymorphism within species belonging to the subclass Hexacorallia, many examples may be found in the subclass Octocorallia. In *Pennatula phosphorea*, for instance, the colony develops by branching from a highly modified individual and contains two other sorts of zooids: autozooids (concerned with feeding) and siphonozooids (concerned with maintaining hydrostatic pressure within the colony).

2. Behaviour

A sea anemone such as *Calliactis parasitica*, one of the most extensively-studied species, shows two main sorts of behaviour. First, there are symmetrical responses, e.g. protective polyp withdrawal or general oral disc expansion, and secondly, there are many of the more interesting asymmetrical responses e.g. column bending, local tentacle bending, mouth opening, peristalsis, locomotion and aggression. The Actiniaria are usually regarded as solitary rather than colonial species. Dicquemare (1775), however, showed that *Metridium* when reproducing asexually by pedal laceration often remained attached to one or more of its offspring and that there appeared to be functional (presumed nervous) connection between them. This was shown by the simple experiment of lightly touching one individual and observing the almost simultaneous withdrawal of others connected to it. This sort of "colonial" protective withdrawal is found in many soft and hard corals. "Monsters" possessing two or more oral crowns are also sometimes produced. There is usually close coordination of behaviour between them.

It is possible for anemones to show colonial behaviour although not physically connected in any way. *Anthopleura elegantissima* is an intertidal species which lives attached to rock faces. Genetically identical individuals cluster together in clones but there is always a clear area of rock between one clone and a genetically different one. This is maintained by inter-clonal "aggression" (Francis, 1973). The example is an interesting case of colonial "co-operation" *despite lack of physical connections between polyps.*

Among the Madreporaria, interspecific aggression between neighbouring colonies has been described (Lang, 1971). In this case, mesenterial filaments are used against adjacent colonies in the competition for space. Coral colonies show a number of other behaviour patterns. Colonial co-operation during sediment shedding, in for example, *Porites* has been described by Hubbard (1973). The use of mucous feeding nets in, for example, agariciid colonies has recently been reported by Lewis and Price (1975, 1976).

Perhaps one of the most widely-known behaviour patterns found in corals is the protective withdrawal response. Horridge (1957) found that in *Favia*, for example, electrical or mechanical stimulation of one polyp could lead to the retraction of all the polyps in the colony. *Porites* and other perforate species

8. Cnidarian Colonies: Behaviour Co-ordination

could not be made to give whole-colony responses but many polyps surrounding the point of stimulation could be made to retract. Shelton (1975a, d) has investigated the electrophysiology of these and related responses (see below).

The Octocorallia show many of the responses which have been described for the solitary and colonial Hexacorallia, such as polyp retraction and local feeding behaviour. In the examples which have been studied so far, for instance *Alcyonium digitatum* (Horridge, 1956), *Pennatula phosphorea* (Shelton, 1975c) and *Renilla köllikeri* (Anderson and Case, 1975), a retraction response can always be made to affect all the polyps in a colony. The exception to this is the Gorgonacea which seem to show little colonial co-ordination of behaviour and even the most vigorous stimulation (e.g. cutting the colony in half) does not cause polyp retraction to travel far from the point of stimulation. Many octocorals show a number of other colonially-integrated behaviour patterns. In the Alcyonacea, e.g. *Veretillum cynomorium* (Buisson, 1971), *Renilla köllikeri* (Parker, 1920) or *Pennatula phosphorea*, the colony undergoes regular expansions and contractions which involve changes in the volume of fluid inside the gastrovascular cavity of the colony. Waves of peristalsis pass over the colony and these facilitate burrowing. Burrowing requires that the whole colony acts as a single individual and poses interesting electrophysiological problems which require further investigation. In the same category are escape responses of sea pens and the co-ordinated movements of the arms bearing autozooids in genera such as *Pennatula*.

One of the other well-known responses found in the Octocorallia is that of phosphorescence (Parker, 1920; Nicol, 1958). Again, this is a response which spreads in "waves" all over the colony and can follow electrical or mechanical stimulation of a single autozooid. It may possibly have a protective function following harmful stimuli.

The study of the behaviour of cnidarian colonies provides valuable data on the "individuality" of each polyp and the degree of "coloniality" of the colony. Investigation of responses such as the spread of polyp retraction or phosphorescence has often been undertaken as an indirect method of studying nervous co-ordination. Recent advances in electrophysiological techniques, particularly the use of suction electrodes, have enabled a more direct approach. This has shown two important points: the overt spread of behaviour is not always a reliable guide to the underlying electrical activity, and similar patterns of behaviour can be co-ordinated in different ways (see below).

3. Electrophysiology

For comparative purposes, it is necessary to know about the control of behaviour in solitary species. Fig. 2(a) shows an electrical recording made with a suction

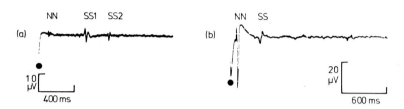

FIG. 2. Electrophysiological recordings from Hexacorallia. (a) The sea anemone *Adamsia palliata*. Following a single electrical stimulus ●, three pulses may be recorded from a tentacle: NN—nerve net pulse, SS1–SS1 pulse, SS2–SS2 pulse. (b) The coral *Eusmilia fastigiata*. A similar recording from this species revealed two conduction systems: NN—nerve net, SS—slow system. In this species, both systems conducted without decrement between polyps.

electrode from a single tentacle of the sea anemone *Adamsia palliata*. One 10 V electrical stimulus was applied to the column of the animal via a second suction electrode. Three responses are visible and these are labelled NN, SS1 and SS2. They represent activity in three different conduction systems similar to those found in *Calliactis parasitica* (McFarlane, 1969a, 1969b, 1970). Conduction velocity, threshold to electrical stimulation and the regions of the animal in which the systems are found are different for each one and allow ready distinction to be made between them. Activity in all three systems spreads without decrement, that is, it through-conducts. The nerve net (NN) is primarily an excitatory system and at different frequencies of stimulation, different muscle groups are activated to contract. The SS1 (slow system one) is an ectodermal system, though it is not yet clear whether the ectodermal cells themselves conduct the pulses (Shelton, 1975b). SS1 activity seems to inhibit ectodermal muscles (McFarlane and Lawn, 1972). The SS2 is an endodermal system and has a role in the inhibition of endodermal muscles and in the inhibition of nerve net pacemaker activity (McFarlane, 1974). Both slow systems are so called because of the slow conduction velocity (10 cm s^{-1} or less) and the long, slow duration of the pulses.

When the electrical activity in a solitary anemone is compared with that in a madreporarian coral (Fig. 1(b)), there are some points of similarity. Fig. 1(b) shows a corresponding recording made from *Eusmilia fastigiata*. This species may form colonies containing many individuals though usually live tissue connections are such that the colony is divided up into groups of 5–10 polyps. Nerve net pulses through-conduct to all connected polyps. A second conduction system (SS) is present which appears to be similar to the SS1 of sea anemones. Unlike the nerve net, which has a uniformly short refractory period, the slow system has a short refractory period in tentacles and oral disc (60 ms) but a much longer one in inter-polyp regions (several seconds) (Shelton and McFarlane, 1976a, 1976b). This means that low frequency slow system activity is conducted between polyps, but high frequency pulses are only conducted within polyps. A polyp thus has some independence of action whilst the colony still receives a low level of information concerning the activities of individuals. Slow system activity evokes polyp expansion and may cause mucus release. Shelton (1975d) has shown that there may be colonial control of mouth opening and release of mesenterial filaments. Mouth opening could be under slow system control. Control of nematocyst discharge threshold is another possible function of slow systems (McFarlane and Shelton, 1975).

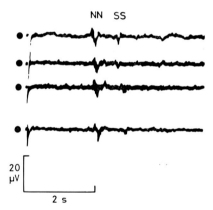

FIG. 3. Electrophysiological recordings from Octocorallia. Four electrical stimuli ● displayed one below the other. Stimuli to the outside of a colony of *Pennatula phosphorea*. A recording from the oral disc of an autozooid indicates the presence of two colonial conduction systems: NN—nerve net, SS—slow system.

In the Octocorallia, a colonial nerve net and at least one slow system are present (Fig. 3). As with the Hexacorallia, the nerve net (which is through-conducting) controls polyp retraction (Shelton, 1975c; Anderson and Case, 1975). Phosphorescent responses may also be under nerve net control. It is not known what functions the slow system serves. They may well be similar to those of the slow system of *Eusmilia*. It has been suggested (Shelton, 1975c) that it may influence the pumping activities of siphonozooids.

The electrophysiology of the Anthozoa which have been studied so far has shown considerable similarity throughout both subclasses. At least two or three conduction systems are present and in colonial species there is evidence for varying degrees of integration of behaviour at a colonial level.

HYDROZOA

Behaviour and Electrophysiology

This aspect of the Hydrozoa has been recently reviewed by Mackie (1973). A great deal of electrophysiology has been carried out on this class and most species which have been examined by means of suction electrodes have given large, easily recorded electrical pulses. Some typical examples are shown in Fig. 4. Several conduction systems are usually present (Josephson, 1974) and may

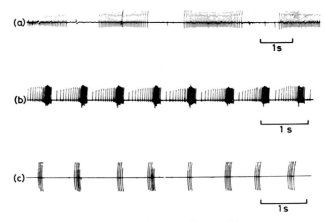

FIG. 4. Electrophysiological recordings from Hydrozoa. (a) Concert activity in *Velella*. (b) Concert sequence in *Tubularia*. (c) Concert-like sequence in a fed *Hydra*. From Fields and Mackie (1971). Reproduced by permission of the Journal of the Fisheries Research Board of Canada.

involve both nervous and neuroid (epithelial) components. One or more pacemakers may be present, for example, those controlling concert activity of the tentacles of *Tubularia* (Fig. 4).

Electrophysiology has been a useful tool in providing data in the controversy concerning the colonial or solitary nature of the chondrophore, *Velella* (Fields and Mackie, 1971). At first sight, this appears to be a hydrozoan colony. Behavioural evidence, however, suggests that it could really be an individual. Concerts of tentacle flexion and related muscle contractions closely resemble those found in, for example, *Tubularia*, a solitary hydrozoan. Electrophysiological data (Fields and Mackie, 1971) and morphological evidence (Fig. 5) tend to support the view that *Velella* is an individual and not a colony, despite its

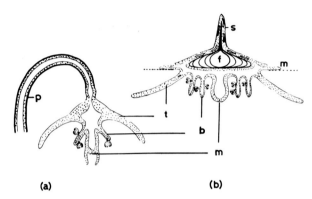

FIG. 5. Suggested homologous anatomical features between a sessile tubularian hydroid (left) and the chondrophore body plan (right) showing that the chondrophore plan may be equivalent to a single polyp. Abbreviations: b—blastostyle, f—float chamber, m—mantle, mouth, p—perisarc, s—sail, t—tentacle. From Fields and Mackie (1971). Reproduced by permission of the Journal of the Fisheries Research Board of Canada.

appearance. The Hydrozoa, of course, is a very plastic group and one could never be certain, just on the basis of physiological data, of the relationships between species but if Fields and Mackie are correct, then here we have an individual which looks like a colony. This contrasts with groups such as the Siphonophora which, though colonies, are so highly integrated that they behave as individuals.

There is good evidence for colonial integration of behaviour in *Obelia* (Morin and Cooke, 1971), *Hydractinia* (Stokes, 1974), *Proboscidactyla* (Spencer, 1974) and in *Millepora* (de Kruijf, 1975). Fulton (1961) has shown, however, that in *Cordylophora* movements which appear to originate locally are really co-ordinated over the whole colony. This highlights the need for combined behavioural and electrophysiological experiments.

Though my research experience has been mainly on Anthozoa, Dr Ian McFarlane and I have made some recordings of electrical activity from hydrozoans (Fig. 6). Recordings made from the pneumatosaccus of *Physalia* show a single system which gives large pulses and which is probably neuroid. There is no evidence for nerve fibres in this region (Mackie, 1965). As one might expect, recordings from fishing tentacles (Fig. 6(b)) show more complex activity involving at least two systems. There seems to be some independence of action of different fishing tentacles.

Nanomia cara is known to be very well co-ordinated as a colony (Mackie,

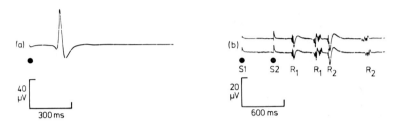

FIG. 6. Electrophysiology of *Physalia*. (a) Recording from the pneumatosaccus following a single electrical stimulus ●. A single, large pulse (probably neuroid) was recorded. (b) Two-channel recording from a fishing tentacle. Two electrical stimuli were given to the tentacle—S1 and S2. Four responses were recorded—R1, R1 to S1 and R2, R2 to S2. At least two conduction systems present.

1963). Small stimuli only cause a few nectophores to contract but a large stimulus to the anterior nectophores and float leads to co-ordinated reverse swimming by the colony, whilst a similar stimulus to the posterior of the colony leads to co-ordinated forward swimming.

Figure 7 shows electrophysiological recordings made from four points on a

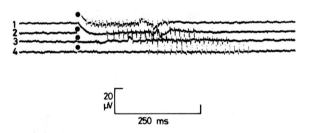

FIG. 7. Electrophysiology of the bryozoan *Membranipora membranacea*. Four-channel recording. Electrodes 1, 2, 3 and 4 were placed at progressively increased distances from the point of stimulation. Following a single electrical stimulus ●, a burst of pulses was produced. Note the reduction in frequency with the distance conducted.

colony of the bryozoan *Membranipora membranacea*. This is of interest here because *Membranipora* shows a similar pattern of spread of polypide retraction following electrical or mechanical stimulation (Thorpe et al., 1975a, 1975b) to that shown by the madreporarian *Porites*. It is clear, however, that a quite different control mechanism is involved in the two species. Whereas in *Porites* a single electrical stimulus only leads to a single response in the nerve net, in *Membranipora* a single electrical stimulus leads to a burst of pulses in the nerve plexus. It emphasizes once again the importance of studying behaviour in conjunction with electrophysiology and the value which both have in the study of animal colonies.

ADVANTAGES OF COLONIALITY—CONCLUSIONS

Cnidarian colonies show both local responses from individual polyps and varying degrees of colonial co-ordination. In, for example, the Octocorallia and the Hydrozoa, the colonial mode of life has been exploited such that the colony has become more than just a collection of individuals. At the highest level (e.g. Siphonophora), colonies have become so highly integrated that they have been rightly termed "superorganisms" (Mackie, 1963). Many advantages have been suggested for coloniality. Obvious benefits accrue from rapid asexual reproduction and there can be advantages in aggregation if, for example, this aids the individuals in the competition for a suitable substratum. A large biomass can be achieved without any increase in the size of individuals thus avoiding possible structural difficulties. Indeed, great skeletal strength can be achieved in a united colonial structure. In some Octocorallia and Hydrozoa, polymorphism with division of labour between different polyp-types, has been exploited. This raises the colony to a higher level of integration and necessitates communication and co-operation between polyps. Many new sorts of behaviour become possible with the development of new polyp-types. Knight-Jones and Moyse (1961) have suggested that an organized colony enables optimum spacing of individuals without overcrowding and if one part grows into an unfavourable region it can be resorbed and used elsewhere in the colony. Finally, colonial conduction systems are useful in the event of an attack, e.g. by a fish (Mackie, 1973) since other polyps can be warned and withdrawn before the danger reaches them.

ACKNOWLEDGEMENTS

I am most grateful to the Royal Society for the award of a John Murray Travelling Studentship and to Dr R. C. Brace for his criticism of the manuscript.

REFERENCES

ANDERSON, P. A. V. and CASE, J. F. (1975). Electrical activity associated with luminescence and other colonial behaviour in the pennatulid *Renilla kollikeri*. *Biol. Bull. mar. biol. Lab./Woods Hole* **149**, 80–95.

Buisson, B. (1971). Les activités rythmiques comportementales da la colonie de *Veretillum cynomorium* (Cnidaire Pennatulidae). *Cah. Biol. mar.* **12**, 11–48.
Dicquemare, Abbé. (1775). A second essay on the natural history of sea anemonies. *Phil. Trans R. Soc. Lond.* **65**, 207–248.
Fields, W. G. and Mackie, G. O. (1971). Evolution of the Chondrophora: evidence from behavioural studies on *Velella. J. Fish. Res. Bd Can.* **28**, 1595–1602.
Francis, L. (1973). Clone specific segregation in the sea anemone *Anthopleura elegantissima. Biol. Bull. mar. biol. Lab. Woods Hole* **144**, 64–72.
Fulton, C. (1961). The development of *Cordylophora. In* "The Biology of Hydra and of Some Other Coelenterates" (H. Lenhoff and W. F. Loomis, eds), pp. 287–295, University Press Miami, Coral Gables, Florida.
Horridge, G. A. (1956). A through-conducting system co-ordinating the protective retraction of *Alcyonium* (Coelenterata). *Nature, Lond.* **178**, 1476–1477.
Horridge, G. A. (1957). The co-ordination of the protective retraction of coral polyps. *Phil. Trans R. Soc. Lond.* B **240**, 495–528.
Hubbard, J. A. E. B. (1973). Sediment-shifting experiments: a guide to functional behavior in colonial corals. *In* "Animal Colonies" (R. S. Boardman, A. H. Cheetham and W. A. Oliver, eds), pp. 31–42, Dowden, Hutchinson and Ross Inc., Pennsylvania.
Josephson, R. K. (1974). The strategies of behavioral control in a coelenterate. *Am. Zool.* **14**, 905–915.
Knight-Jones, E. W. and Moyse, J. (1961). Intraspecific competition in marine animals. *Symp. Soc. exp. Biol. Cambridge* **15**, 72–95.
Kruijf, H. A. M. de (1975). General morphology and behaviour of gastrozooids and dactylozooids in two species of *Millepora* (Milleporina, Coelenterata). *Mar. Behav. Physiol.* **3**, 181–192.
Lang, J. C. (1971). Interspecific aggression by scleractinian corals. 1. The re-discovery of *Scolymia cubensis* (Milne Edwards & Haime). *Bull. mar. Sci.* **21**, 952–959.
Lewis, J. B. and Price, W. S. (1975). Feeding mechanisms and feeding strategies of Atlantic reef corals. *J. Zool. Lond.* **176**, 527–544.
Lewis, J. B. and Price, W. S. (1976). Patterns of ciliary currents in Atlantic reef corals and their functional significance. *J. Zool. Lond.* **178**, 77–89.
McFarlane, I. D. (1969a). Two slow conduction systems in the sea anemone *Calliactis parasitica. J. exp. Biol.* **51**, 377–385.
McFarlane, I. D. (1969b). Co-ordination of pedal disc detachment in the sea anemone *Calliactis parasitica. J. exp. Biol.* **51**, 387–396.
McFarlane, I. D. (1970). Control of preparatory feeding behaviour in the sea anemone *Tealia felina. J. exp. Biol.* **53**, 211–220.
McFarlane, I. D. (1974). Excitatory and inhibitory control of inherent contractions in the sea anemone *Calliactis parasitica. J. exp. Biol.* **60**, 397–422.
McFarlane, I. D. and Lawn, I. D. (1972). Expansion and contraction of the oral disc in the sea anemone *Tealia felina. J. exp. Biol.* **57**, 633–649.
McFarlane, I. D. and Shelton, G. A. B. (1975). The nature of the adhesion of tentacles to shells during shell-climbing in the sea anemone *Calliactis parasitica. J. exp. mar. Biol. Ecol.* **19**, 177–186.
Mackie, G. O. (1963). Siphonophores, bud colonies and superorganisms. *In* "The Lower Metazoa, Comparative Biology and Physiology" (E. C. Dougherty, Z. N. Brown,

E. D. Hanson and W. D. Hartman, eds), pp. 329–337, University of California Press, California.

MACKIE, G. O. (1965). Conduction in the nerve-free epithelia of siphonophores. *Am. Zool.* **5**, 439–453.

MACKIE, G. O. (1973). Coordinated behaviour in hydrozoan colonies. *In* "Animal Colonies" (R. S. Boardman, A. H. Cheetham and W. A. Oliver, eds), pp. 95–118. Dowden, Hutchinson and Ross Inc., Pennsylvania.

MORIN, J. G. and COOKE, I. M. (1971). Behavioural physiology of the colonial hydroid *Obelia. J. exp. Biol.* **54**, 707–722.

NICOL, J. A. C. (1958). Observations on the luminescence of *Pennatula phosphorea* with a note on the luminescence of *Virgularia mirabilis. J. mar. biol. Ass. U.K.* **37**, 551–563.

PARKER, G. H. (1920). Activities of colonial animals. II. Neuromuscular movements and phosphorescence in *Renilla. J. exp. Zool.* **31**, 475–515.

SHELTON, G. A. B. (1975a). Electrical activity and colonial behaviour in anthozoan hard corals. *Nature, Lond.* **253**, 558–560.

SHELTON, G. A. B. (1975b). The transmission of impulses in the ectodermal slow conduction system of the sea anemone *Calliactis parasitica* (Couch). *J. exp. Biol.* **62**, 421–432.

SHELTON, G. A. B. (1975c). Colonial conduction systems in the Anthozoa: Octocorallia. *J. exp. Biol.* **62**, 571–578.

SHELTON, G. A. B. (1975d). Colonial behaviour and electrical activity in the Hexacorallia. *Proc. R. Soc. Lond.* B **190**, 239–256.

SHELTON, G. A. B. and MCFARLANE, I. D. (1976a). Electrophysiology of two parallel conducting systems in the colonial Hexacorallia. *Proc. R. Soc. Lond.* B **193**, 77–87.

SHELTON, G. A. B. and MCFARLANE, I. D. (1976b). Slow conduction in solitary and colonial Anthozoa. *In* "Coelenterate Ecology and Behavior" (G. O. Mackie, ed.), pp. 599–607, Plenum Publishing Corp, New York.

SPENCER, A. N. (1974). Behavior and electrical activity in the hydrozoan *Proboscidactyla flavicirrata* (Brandt). 1. The hydroid colony. *Biol. Bull. mar. biol. Lab. Woods Hole* **145**, 100–115.

STOKES, D. R. (1974). Physiological studies of conducting systems in the colonial hydroid *Hydractinia echinata*. 1. Polyp specialisation. *J. exp. Zool.* **190**, 1–18.

THORPE, J. P., SHELTON, G. A. B. and LAVERACK, M. S. (1975a). Electrophysiology and co-ordinated behavioural responses in the colonial bryozoan *Membranipora membranacea* (L). *J. exp. Biol.* **62**, 389–404.

THORPE, J. P., SHELTON, G. A. B. and LAVERACK, M. S. (1975b). Colonial nervous control of lophophore retraction in cheilostome Bryozoa. *Science* **189**, 60–61.

WELLS, J. W. (1973). What is a colony in anthozoan corals? *In* "Animal Colonies" (R. S. Boardman, A. H. Cheetham and W. A. Oliver, eds), pp. 29–30, Dowden, Hutchinson and Ross Inc., Pennsylvania.

ADDENDUM

This paper was written in April 1976. Since then several interesting and relevant papers have been published.

ANDERSON, P. A. V. (1976). An electrophysiological study of mechanisms controlling polyp retraction in colonies of the scleractinian coral *Goniopora lobata. J. exp. Biol.* **65**, 381–393.

KRUIJF, H. A. M. de. (1976). The effect of electrical stimulation and some drugs on conducting systems of the hydrocoral *Millepora complanata. In* "Coelenterate Ecology and Behavior" (G. O. Mackie, ed.), pp. 661–670. Plenum Publishing Corp, New York.

McFARLANE, I. D. (1978). Multiple conducting systems and the control of behaviour in the brain coral *Meandrina meandrites* (L.). *Proc. R. Soc. Lond.* B **200**, 193–216.

SHELTON, G. A. B. (1978). *Lophelia pertusa* (L.): electrical conduction and behaviour in a deep-water coral. *J. mar. biol. Ass. U.K.* (in press).

9 | On Some Aspects of Coloniality in Permian Corals

J. FEDOROWSKI

*Laboratory of Palaeontology, Department of Geology,
A. Mickiewiczs University, Poznań, Poland*

Abstract: A collection of lower Permian rugose corals from Sierra Diable, southwestern Texas, U.S.A., contains numerous well preserved colonies with protocorallites, together with solitary corals. I believe that all these specimens belong to one species.

In contrast to the only two species previously studied, the ontogeny of the protocorallites shows zaphrentoid increase in septa. Offsetting (budding) is lateral and shows a number of interesting characters, including channels and basal structures, which are briefly discussed.

Emphasis is placed on two particular aspects of astogeny: a) Offsetting and the problem of maturity; b) growth and pattern in fasciculate colonies. Relatively immature specimens may show offsetting and more mature specimens may lack it. Some remain solitary even though they are ontogenetically advanced, and morphologically mature single corallites within colonies may similarly be found without offsets. It is concluded that capacity for coralite increase should not be identified with morphological maturity. Colony growth patterns are varied and include an approximate colony wide synchronization of offsetting, interpreted here as a consequence of both genetic and environmental factors.

INTRODUCTION

Different aspects of coloniality in the extinct zoantharian rugose corals have been studied by many workers (e.g. Koch, 1883, 1896; Smith, 1913; Smith and Ryder, 1926, 1927; Oliver, 1960, 1968; Rozkowska, 1960; Jull, 1965, 1967, 1973; Fedorowski, 1965, 1970; Fedorowski and Jull, 1976; Ulitina, 1973, 1974, and many others). The present paper is based in part on some of my recently published investigations of new species of *Heritschioides* from the Permian of Texas (Fedorowski, 1978). Some new ideas concerning maturity in rugosans

---

Systematics Association Special Volume No. 11, "Biology and Systematics of Colonial Organisms", edited by G. Larwood and B. R. Rosen, 1979, pp. 155–171. Academic Press, London and New York.

and their ability to produce colonies are introduced here to supplement that paper.

The collection studied contains more than twelve hundred silicified specimens all of which were etched from a few blocks of limestone. The blocks came from locality USNM 728e in Sierra Diable, south-western Texas, and from one bed of limestone 60 cm thick. Although all specimens belong to the same thanatocoenosis they probably had a variety of living conditions. This is indicated by the shapes of specimens, by the presence or absence of attachment processes (talons) and by other external characters.

Rapid sedimentation of the limestone (Dr G. A. Cooper, personal communication) and the close morphological similarities between all specimens indicate that they belong not only to the same species but most probably also to a single population. This is believed to be so even though growth form may be colonial, quasi-colonial or solitary. These different growth forms and different stages to be observed in blastogeny (the appearance of offsets) and in astogeny (the development of colonies) allow a palaeobiological interpretation to be made.

TYPES OF COLONIALITY IN RUGOSANS

Growth form is usually a constant feature in rugose coral genera. In many cases it is used as a generic character (Smith and Ryder, 1926; Minato and Kato, 1965a,b; Fedorowski and Gorianov, 1973, etc.). Spassky (1965) even used it as a basis for division of all rugosans into two independent, parallel groups: Solitaria and Associata. This division has been discussed recently (Fedorowski, 1978) as highly improbable and artificial. It may be stated, however, that although the majority of rugose corals are either solitary or colonial the taxonomic value of this feature is usually not suprageneric and in almost all the families distinguished so far there are both solitary and colonial forms. There are also a few species in which the corallum may be either solitary or colonial. These composite species may occur both in genera which are normally solitary or in those which are usually colonial. Colonial genera may be grouped according to their mode of asexual reproduction (budding, referred to here, as by other rugosan specialists, as offsetting), which may be lateral or peripheral, and according to their degree of coloniality, which may result in quasi-colonies (Fagerstrom and Eisele, 1966) or incipient colonies (Fedorowski, 1970) or normal colonies. In normal colonies only lateral increase successfully provides large and well developed colonies.

Quasi-colonies (Pl. 1, 12; Pl. 2, 5) are protocorallites that have produced lateral offsets even though this capacity is weak and was probably induced by environmental factors. Offsets in such colonies are not able to reach complete morphological maturity. They may be produced only a few times during the development of the protocorallite. *"Craterophyllum" verticillatum* Barbour is the best known species in which quasi-colonies are common. *Timania rainbowensis* Rowett is in my opinion a second example of such a species and *Heritschioides* n. sp. is another. It may also be possible that very weak colonies of *Neokoninckophyllum kansasense* (Miller and Gurley) illustrated by Cocke (1970) are also examples of quasi-colonies. These few records may indicate:

1. Quasi-coloniality is not a chance development but a controlled process. It has been found in *Heritschioides* n. sp. that offsets in quasi-colonies were produced in the same way as those in regular colonies of this species. They have been called "lost structures" (Fedorowski, 1978) because non-environmental factors prevented them from reaching morphological maturity. Environmental conditions were the same for both a parent polyp and its progeny but only the parent was able to develop continuously. Records are known also (Pl. II, 5) where the "lost structure" died shortly after being separated from the parent polyp, although at the time of a direct connection between these individuals it was itself healthy enough to start asexual reproduction.

2. Quasi-coloniality may appear both in solitary species of *Timania* and *"Craterophyllum"* and in colonial species of *Heritschioides* and is considered (Fedorowski, 1978) to be either a progressive feature in solitary species or a rudimentary feature in colonial ones. Whatever it is it seems to remain halfway between the solitary and fully colonial conditions in rugosans, as none of the asexually produced protopolyp's progeny are able to form healthy offsets. The investigations by Fagerstrom and Eisele (1966) of quite a large collection of *"Craterophyllum" verticillatum* and by Fedorowski (1978) of *Heritschioides* n.sp. show that there are groups of specimens within populations of some colonial or solitary species which develop lost structures and groups of specimens which never do so. The second of these groups predominates in generally solitary species. In mainly colonial species there is also a third, predominating group of specimens which form normal colonies. In species in which all three groups are present (e.g. in *Heritschioides* n.sp.) there is no significant difference in the morphology of mature specimens belonging to particular groups. The probable genetic control of these groups will be discussed later in this paper.

A phenomenon similar to quasi-coloniality has also been discussed briefly by Gardiner (1902) in some specimens of *Flabellum rubrum* (Scleractinia). He referred to specimens which produce "buds" on the peripheral parts of calices. Gardiner, however, did not discuss the nature of these "buds" or their genetic significance. He supposed that they might have been developed from eggs. Semper (1872) who had also observed this phenomenon separated such "budding" specimens into the different genus and species *Blastrotrochus nutrix*.

Incipient colonies are protocorallites that have produced peripheral offsets which are not able to reach morphological maturity and do not themselves reproduce asexually. Incipient colonies, typical early colonies, are so far known only within solitary genera and in *Heliophyllum halli* which contains solitary and weakly colonial corals as well. They have been described or illustrated in *Tabulophyllum schlueteri* (Peetz), Bulvanker, 1958; *T. rotundum* Spassky, 1960; *Clisaxophyllum ava atetsuense* Minato and Nakazawa, 1957; *Bothrophyllum dobroljubovae* (Fomichev), Yakovlev, 1965; *Ceratophyllum eifeliense* Fedorowski, 1967; *Spirophyllum geminum* Fedorowski, 1970, and in some other species. Peripheral increase which leads to the appearance of such incipient colonies was considered by Fedorowski and Jull (1976) as a kind of multiplied rejuvenescence.

PLATE I

All corallites belong to *Heritschioides* n. sp., Bone Spring Fm. (Lower Permian) of Sierra Diable, Texas, U.S.A., locality USNM 728e.

1. Specimen USNM 196583 × 2·4. Calicular view of solitary, juvenile individual.
2. Specimen USNM 196584 × 2·4. Start of offsetting in a very early ontogeny (after approximately 2 mm of growth of a protocorallite); a—calicular view, b—side view.
3. Specimen USNM 196593 × 2·4. Start of offsetting in a neanic protocorallite (optimum stage for offsets) a—calicular view, b—side view.
4. Specimen USNM 196602 × 1·6. Start of a few offsets in an ephebic protocorallite; a—side view, b—calicular view.
5. Specimen USNM 196659 × 1·6. a—calicular view, b—side view of a large solitary specimen attached to another specimen of the same species.
6. Specimen USNM 196591 × 2·4. A protocorallite with strong attachment to bryozoan colony and with two offsets produced very early in ontogeny.
7. Specimen USNM 196596 × 1·6. A protocorallite producing four offsets in the neanic (optimum) stage.
8. Specimen USNM 196603 × 1·6. Solitary corallite showing a few rejuvenescences.
9. Specimen USNM 196633 × 1·6. A protocorallite with strong talon and with first offsets produced at the ephebic stage.
10. Specimen USNM 196569 × 1·6. Very late offsetting from a protocorallite; a—side view, b—calicular view, c—two young offsets.
11. Specimen USNM 196630 × 1·6. A solitary corallite attached to an inner surface of a brachiopod shell.
12. Specimen USNM 196636 × 1·6. A quasi colony.
13. Specimen USNM 196619 × 1·6. A protocorallite with only a single offset developed at the ephebic stage.

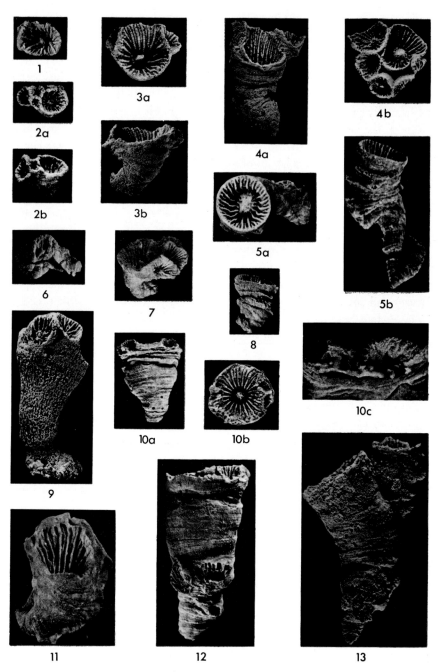

PLATE I (Caption on facing page)

Normal colonies are very seldom created in this way and even then they are never large and contain only a very limited number of asexually produced generations (e.g. *Entelophyllum articulatum* or *Kodonophyllum truncatum*). Ulitina (1974) has also observed that peripheral increase may appear as an incidental mode of reproduction at the periphery of laterally offsetting colonies. These peripherally produced offsets are never able to reach morphological maturity although they may be regularly increased as explained by Jull (1965) who found peripherally increased individuals in colonies of Australian *Lithostrotion*, a genus which normally produces lateral offsets. This, as well as some observations on very weak incipient colonies (Fedorowski, 1967, 1970), may indicate that this type of early colony formation and this type of increase are in many cases only a vestige of the solitary condition. It seems to be largely a response to unfavourable or unusual environmental conditions.

Yakovlev (1965) hypothesized that in the case of *Bothrophyllum dobroljubovae* coloniality was a kind of specific adaptation. He wrote: "At the end of their life these corals were probably not able to reproduce normally, sexually (by larvae) and so produced offsets. The parent polyp died during this process. Transfer to a budding mode of reproduction allowed the species to continue and, as such, was a character of specific adaptation." In other words Yakovlev considered coloniality in *B. dobroljubovae* as a gerontic character. If this were correct, one would expect corallites producing offsets to be more advanced morphologically than those which do not. This is seen in some, but not all, examples of this species (compare Yakovlev, 1965; Fomichev, 1953).

Heliophyllum halli is an example of a species in which not only solitary corals

---

PLATE II

All corallites belong to *Heritschioides* n. sp., Bone Spring Fm. (Lower Permian) of Sierra Diable, Texas, U.S.A., locality USNM 728e.

1. Specimen USNM 196584 × 1·6. A succession of three generations of offsets; p—protocorallite, 1–3—first to third asexually produced generations of corallites.
2. Specimen USNM 196610 × 1·6. p—protocorallite, 1,2—two successively produced generations of offsets.
3. Specimen USNM 196580 × 1·6. p—protocorallite, 1,2—first and second generation of asexually produced corallites, 1a—second interval of reproduction of first asexually produced generation of corallites.
4a, b. Specimen USNM 196598 × 1·6. p—protocorallite, 1,1a—its two intervals of reproduction.
5. Specimen USNM 196620, nat. size. A quasi colony.
6. Specimen USNM 196626 × 1·6. An example of the effects of environmental influence on a colony: simultaneous rejuvenescence and offsetting of different corallites.

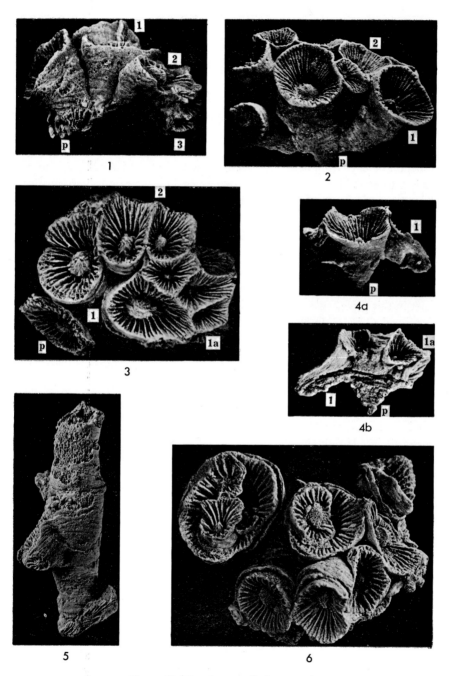

PLATE II (Caption on facing page)

and incipient colonies are found but also small normal colonies of both branching and massive (astreoid) forms. The classic study by Wells (1937) on variation in form and growth within this species and the recent study by Sorauf and Oliver (1976) on septal carinae and microstructure also provide data concerning the nature of coloniality in this species. In contrast to Wells, who considered almost all of the variants observed as being purely ecologically controlled, Sorauf and Oliver (p. 331) concluded that "... the differences observed in all populations may be genetically controlled rather than being a simple function of an environment (facies)". Further study is envisaged by Sorauf and Oliver to establish the species question and the linkage from solitary forms to branching and massive colonies in this species.

Unfortunately, records concerning incipient coloniality are rare and insufficient to establish any more general and certain rules. On recent knowledge of this phenomenon it is possible only to suppose that a number of solitary rugose corals possessed a predisposition to reproduce asexually in only certain extreme living conditions. A possible genetic basis for this predisposition is discussed later in this paper.

Normal Colonies. In these at least two asexually produced generations of offsets occur. This is of course only on arbitrary distinction, but it exists in most of the examples investigated so far and gives an impression of a kind of barrier between the ability to produce just a single generation of underdeveloped offsets and the capacity to begin a colony. On the other hand, this apparently small difference between early coloniality and normal coloniality makes it more obvious that the process of achieving colonial status might have been repeated many times and quite independently in different groups of rugosans. Normal coloniality is too complex to be discussed further here.

CONTROL OF COLONY FORMATION

As has been already considered, each type of coloniality or early coloniality shows obvious regularities in blastogeny and in astogeny. It has also been suggested that extinction of progeny in early colonies is dependent on non-environmental factors.

The genetic control of the formation of normal colonies is unquestionable and only phenomena such as the existence of budding and non-budding individuals within colonies, and of types and patterns of colonies may be discussed in either environmental or genetic aspects. The problem is rather different in the case of quasi-colonies however and even more so in incipient colonies which are always associated with either generally solitary or with generally colonial species as extraordinary occurrences. They also exhibit certain features attributable to environmental factors (e.g. rejuvenescences, deep narrowings, etc.). In my opinion genetic control of both the appearance and extinction of lost structures in quasi-colonies and of offsets in incipient colonies is the only way to explain this phenomenon. The process of budding involves deep changes of part of a parent polyp body and its skeleton. These changes are introduced step by step in a regular manner common to all specimens within the species and lead to the development of more or less well formed young individuals. These individuals may stop developing but comparable ontogenetic stages are found both among offsets of incipient or quasi-colonies and in offsets of normal colonies of the same species (e.g. *Heritschioides* n. sp.). It is highly unlikely that such a regularity in so important a process as reproduction would be controlled solely by random external factors.

Although direct genetic study is not possible in palaeontology, information from biological genetic studies has interesting implications, for example, the random frequency of particular genes within a population. There are genes observed which appear only once in one thousand samples or even once in 250 000. Although these observations were made mainly on man (Montague, 1963; Harris, 1974) they probably hold for primitive animals also. Studies of Gooch and Schopf (1972) or Ayala *et al.* (1973) on polymorphism of proteins in certain deep sea animals or the similar frequency of polymorphic loci genes in man and other species (e.g. *Limulus*) stated by Lewontin (1974) may indicate that the similarity in various frequencies of particular genes in a human and animal populations is real.

In the case of those rugose coral species in which early coloniality occurs together with a solitary condition and/or normal coloniality, it may be supposed that a gene (or gene complex) determining coloniality is one which appears irregularly. Frequency of these genes (or gene complexes) is differentiated in particular species and, consequently, there are species in which early colonies or normal colonies appear quite frequently, and species in which these appearances are extremely rare. It is also possible that in most cases coloniality in such a "complex" species has a lower rank character and may be easily suppressed by other characters. It may develop in some extraordinary environmental circumstances, which induce its appearance, or it may develop as a gerontic character. The more strongly the gene complex determining coloniality is developed, and of necessity the higher the rank it possesses, the higher the degree of coloniality it can achieve. In the case of very differentiated complexes of such genes there may appear such variable species as *Heliophyllum halli* or *Heritschioides* n.sp.

The sample of *Iovaphyllum*, discussed from Frasnian deposits of Belgium

(Lecompte, 1970) and Poland (Różkowska, in press), shows also that an opposite direction of development is possible. This generally colonial (astreoid and aphroid) genus may be solitary in some circumstances. Although Lecompte (1970) considered this solitary condition only as a function of environment, Różkowska (in press) found that both solitary corals and small colonies may coexist. The genes relating to the solitary condition may have existed in European species and populations of this genus, though seemingly not in American ones, as no solitary forms of this genus have been found there. This gene might have permitted solitary growth in particular environmental circumstances.

OFFSETTING AND INDIVIDUAL MATURITY

1. Appearance of the first offset in the protopolyp

Since the ability to reproduce is normally a function of maturity in individuals the timing of the first offset in protopolyps may be formally correlated with this factor. The collection of Lower Permian *Heritschioides* n. sp. corallites from Sierra Diable (south-western Texas) provided much information concerning the stage of protocorallite development at which the protopolyps first began to offset. In different corallites initiation of offsetting occurred at very different ontogenetic stages. In some corallites the first offset appears after as little as 2 mm of protocorallite growth (Pl. I, 2a, b). Protocorallites are very juvenile in character when their polyp offsets are so early. They have no dissepimentarium, the arrangement of septa is zaphrentoidal and their axial structure is underdeveloped. The pseudocolumella is usually only slightly upstanding. All these features are characteristic of an early neanic stage. The protopolyp is the only individual within the colony with the capacity to produce offsets at such a young ontogenetic stage. It reaches full morphological development only after considerable further growth.

The next point in the ontogenetic series is marked by those protocorallites whose polyps started asexual reproduction when they possessed one or several rings of dissepiments and reasonably well developed axial structures, with lamellae and tabellae present. The diameter of such corallites is much smaller than that of cylindrical corallites, and all those characters are at the introductory stage of development (Pl. I, 3a–b, 7; Pl. II, 2, 4a–b). Consequently, such specimens are supposed to be at the neanic stage. Protocorallites whose polyps did not begin to offset until full morphological development are also present in the collection. They do not differ in morphology and measurements from other large individuals of the species (Pl. I, 4a–b, 9, 10a–b, 13).

In addition to corallites showing offsetting there are also solitary corallites in the collection (Pl. I, 1, 5a–b, 8, 11). Different ontogenetic stages of these corallites are present, there being juvenile as well as mature individuals in this group. Since the species is generally colonial these solitary corallites may be

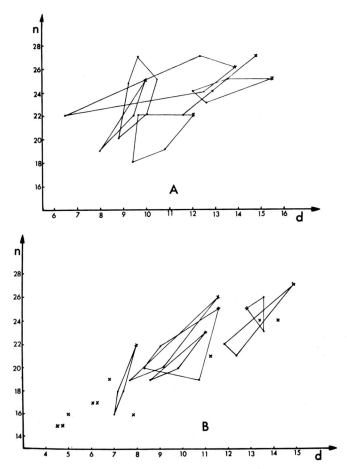

FIG. 1. Diagrams of "n" (number of septa) to "d" (diameter in mm) ratios. Points corresponding to particular corallites of the same colony are united by lines. Protocorallites of particular colonies, and those which have only started to produce their first offsets (diagram B) are marked by crosses. Diagram A—n/d ratio measured beneath calices. Diagram B—n/d ratio measured close to calicular margins.

thought of either as corallites in which the capacity to produce offsets is not yet realized, or as individuals lacking those genes which determine colonial growth. Whichever is true, they do not differ morphologically when similar stages of growth are compared and the supposed genetic difference is purely speculative. In Fig. 1 the ratio of number of septa to diameter for some solitary corallites has been plotted together with variation diagrams for corallites of

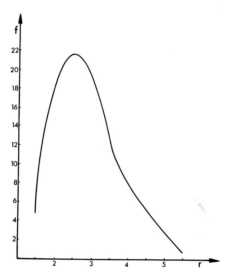

FIG. 2. Frequency (f) in particular classes of n/d ratio (r) for protocorallites which have started to produce their first offsets only.

small colonies emphasizing their similarity within the range of *Heritschioides* n. sp. as well as their actual measured differences.

Forty-five samples of protocorallites of different ontogenetic stages whose polyps had started to produce their first offsets were measured for statistical analysis. This number included all the very young and all the mature protocorallites present in the collection. All sampled specimens evidently died at their respective development stages because of extrinsic factors (storms?) as they do not show any pathological or similar changes. Figure 2 shows that, in spite of artificial over-exaggeration of the extreme groups, neither very young nor mature protocorallites predominate but rather those which started their asexual reproduction at the neanic stage. This pattern is also true for all larger colonies with preserved protocorallites. They were not included here or in the plots as continued growth of protocorallites and further offsets make only their young-stage diameters available for measurement. The juvenile character of the dominant group in which protocorallites have only just started asexual reproduction is also clearly marked morphologically.

All the protocorallites discussed reach one of their most important characters

—the capacity to produce offsets—at quite different stages of morphological development. Is it then reasonable to treat this character as diagnostic for maturity? Judging from the foregoing observations, there is no simple relationship between the timing of offsets and the morphological maturity of protocorallites. Although the majority of polyps in the observed protocorallites started to produce offsets quite early in ontogeny, they are differentiated by their measurements and their morphology, which is overemphasized by protocorallites starting to produce offsets in more advanced stages of ontogeny and by solitary corals. Such wide variation in such an important character as asexual reproduction must have been genetically determined. It is also possible that not only the capacity to produce offsets was coded, but also its potential. These genetic tendencies with the influence of extrinsic factors, which suppressed or emphasized them, resulted in the observed pattern within the present species. The inconstant frequency of the gene determining coloniality discussed previously may also be important in the case of solitary corallites.

There is no published definition of, or observations on sexual maturity in protopolyps. Therefore it seems necessary to distinguish two kinds of maturity of protocorallites:

(a) Morphological maturity, that is the stage of development at which they reach the highest degree of individual morphological development, but do not start to degenerate gerontically.

(b) Sexual maturity, that is the stage of development at which they develop the capacity to produce offsets; this character is individual in some species and may never develop at all in some specimens of "composite" species.

Another regular feature of early stages of coloniality observed in the present species is less well documented but preliminary investigations suggest that the extreme (very young and mature) protocorallites do not generate healthy and well developed colonies. This may mean that very early asexual reproduction is beyond the capacity of an organism at its earliest growth stage. Offsetting at an early stage evidently leads to an early extinction of the protopolyp, which is physically not able to provide itself and its offsets with adequate nutrition. It leads also to the extinction of the progeny because these are not adequately supported by the parent and, at the same time, cannot yet live independently. In the case of mature polyps the supply of nutrients should be adequate for all components of the initial stage of a colony, but nevertheless such colonies died very early in most of the observed specimens.

Since the process of asexual reproduction is generally regarded here as being genetically determined it is supposed that such extreme cases result from a simultaneous action of both genetic and environmental factors. Either the environment induced the onset of asexual reproduction too early (in the case of very young polyps) or coloniality was a subordinate character and, as such, was induced by special circumstances (in the case of mature protopolyps). In this second case offsetting may also be treated as a gerontic feature. "Lost structures" or the combination of "lost structures" and healthy offsets in simple colonies are commonly produced.

There is an optimism period in the ontogeny of a protopolyp, which is most suitable for the initiation of a healthy colony. This varies according to individual polyps but in *Heritschioides* n. sp. one particular interval is dominant for the species. It may be called the first reproductory interval. Unfortunately I have no data concerning other species for comparison. In most small colonies of *Heritschioides* n.sp. there is also a second reproductory interval observed (Pl. II, 2, 3) and in rare cases a third one (Pl. II, 1). The lack of large colonies did not allow further studies on this problem.

2. Appearance of second and next asexual generations

As was stated in the previous section, protopolyps of the great majority of observed protocorallites produced their first offset at the neanic stage. Supposing this to be a valid generalization, one can observe that the first asexually produced generation reached its capacity to reproduce at a later stage of its development than the protopolyp. The corallites of this generation do not show any clearly juvenile characters, but again, their polyps almost always begin their asexual reproduction before reaching complete morphological maturity. The second asexual generation also starts to reproduce before acquiring all the characteristics of fully mature cylindrical corallites. Unfortunately, here too the lack of large colonies made further observations impossible.

It seems to be most probable, however, that the morphological maturity of asexually produced polyps, as in protopolyps, does not relate in any direct way to their ability in turn to produce offsets, at least in early astogeny. In general it seems to relate to the actual generation of offsets or at least to the stage of astogeny (early or late). As with protocorallites, the morphological maturity of asexually produced corallites within colonies should be distinguished from the capacity to produce offsets by their polyps. Furthermore the moment when offsets are produced varies individually here, even within colonies though it usually falls within the range of a particular reproductory interval.

The relationship between maturity and the capacity to produce offsets is very important in the Rugosa. Only corallites at similar growth stages can be compared and only the comparison of mature corallites gives us an opportunity to establish intraspecific variability, boundaries of species, subspecies, etc. The problem of maturity is especially important in colonial species, where variability within colonies is inferred from intraspecific variability.

CONCLUSIONS

1. In spite of species in which growth form is constant, solitary, early colonial and sometimes also fully colonial forms can all occur together within certain single species.

2. Early coloniality, like the solitary and fully colonial conditions in the Rugosa, is determined by genes, the frequency of which is differentiated in some species. As a result of this, colonies or early colonies may vary in the frequency of their occurrence within particular populations.

3. Complexes of genes determining coloniality are more or less fixed and possess higher or lower rank in the genetic system of an individual. Consequently there are more or less well developed colonies or early colonies within the same species or population.

4. Although in most of the cases discussed there seems to be a general tendency to pass from the solitary form into early coloniality or coloniality, the opposite trend is also possible (e.g. *Iovaphyllum* in Europe).

5. Environment plays an important role in inducing offsets, especially in the case of early colonies. It does not control the process of blastogeny and does not cause the start of asexual reproduction unless this is also genetically encoded.

6. Ability to produce offsets is not a function of morphological maturity either in protopolyps or in the first few asexually produced generations. On the contrary, they offset for the first time at neanic or late neanic stages of ontogeny.

7. Acceleration of offsetting, observed in early stages of astogeny, seems to be important as a form of lateral colony expansion enabling it to occupy the largest possible living space.

ACKNOWLEDGEMENTS

I wish to express my warmest thanks to Drs G. Arthur Cooper and Richard E. Grant for providing the specimens for study and to Dr Brian Rosen who read and corrected the manuscript. Finally I thank Mr Antoni Pietura for taking the photographs.
This study was supported by Smithsonian Institution Grant RF-60100.

REFERENCES

AYALA, F. J., HEDGECOCK, D., ZUMWALT, G. S., VALENTINE, J. W. (1973). Genetic variation in *Tridacna maxima*, an ecological analog of some unsuccessful evolutionary lineages. *Evolution* **27**, 177–191.

BULVANKER, E. Z. (1958). Devonskie chetyrekhluchevye korally okrain Kuzneckogo basseina. WSEGEI. 1–212.

COCKE, J. M. (1970). Dissepimental rugose corals of Upper Pennsylvanian (Missourian) rocks of Kansas. *Univ. Kansas Paleont. Contrib. Art.* **54** (Coelenterata 4). 1–67.

FAGERSTROM, J. A. and EISELE, C. R. (1966). Morphology and classification of the rugose coral *Pseudozaphrentoides verticillatus* (Barbour) from the Pennsylvanian of Nebraska. *J. Paleont.* **40**, 595–602.

FEDOROWSKI, J. (1965). Lower Permian Tetracoralla of Hornsund, Vestspitsbergen. *Studia geol. pol.* **17**, 1–173.

FEDOROWSKI, J. (1967). A revision of the genus *Ceratophyllum* Gürich, 1896 (Tetracoralla). *Acta palaeont. pol.* **12**, 213–222.

FEDOROWSKI, J. (1970). Some Upper Viséan columnate tetracorals from the Holy Cross Mountains (Poland). *Acta palaeont, pol.* **15**, 549–613.

FEDOROWSKI, J. (1978). Some aspects of coloniality in rugose corals. *Palaeontology* **21**, 177–224.

FEDOROWSKI, J. and GORIANOV, V. B. (1973). Redescription of tetracorals described by E. Eichwald in "Palaeontology of Russia". *Acta palaeont. pol.* **18**, 3–70.

FEDOROWSKI, J. and JULL, R. K. (1976). Review of blastogeny in Palaeozoic corals and description of lateral increase in some Upper Ordovician rugose corals. *Acta palaeont. pol.* **21**, 37–78.

GOOCH, J. L. and SCHOPF, T. J. M. (1972). Genetic variability in the deep sea: relation to environmental variability. *Evolution* **26**, 545–552.

HARRIS, H. (1974). Common and rare alleles. *Sci. Progr. Oxford* **61**, 495–514.

JULL, R. K. (1965). Corallum increase in *Lithostrotion*. *Palaeontology* **8**, 204–225.

JULL, R. K. (1967). The hystero-ontogeny of *Lonsdaleia* McCoy and *Thysanophyllum orientale* Thomson. *Palaeontology* **10**, 617–628.

JULL, R. K. (1973). Ontogeny and hystero-ontogeny in the Middle Devonian rugose coral *Hexagonaria anna* (Whitfield). *In* "Animal colonies" (R. S. Boardman, A. H. Cheetham and W. A. Oliver, Jr., eds), pp. 59–68. Stroudsburg.

JULL, R. K. (1974). The rugose corals *Lithostrotion* and *Orionastaea* from Lower Carboniferous (Viséan) beds in Queensland. *Proc. roy. Soc. Queensland* **85**, 57–76.

KOCH, G. v. (1883). Die ungeschlechtliche Vermehrung der Palaeozoischen Korallen. *Palaeontographica* **29**, 325–348.

KOCH, G. V. (1896). Das Skelett der Steinkorallen. Eine morphologische Studie. Festschrift f. Gegenbauer. **2**, 251–276.

LECOMPTE, M. (1970). Die Riffe im Devon der Ardennen und ihre Bildungsbedingungen. *Geologica et Palaeontologica* **4**, 25–71.

LEWONTIN, R. C. (1974). Population genetics. *Ann. Rev. Genet.* **7**, 1–17.

MINATO, M. and KATO, M. (1965a). Waagenophyllidae. *J. Fac. Sci. Hokkaido Univ.* (4) **12**, 1–241.

MINATO, M. and KATO, M. (1965b). Durhaminidae (tetracorals). *J. Fac. Sci. Hokkaido Univ.* (4) **13**, 11–86.

MINATO, M. and NAKAZAWA, K. (1957). Two Carboniferous corals from Okayama Prefecture. *Trans. Proc. palaeont. Soc. Japan.* n.s., **25**, 17–20.
OLIVER, W. A., Jr. (1960). Inter- and intracolony variation in *Acinophyllum*. *Bull. geol. Soc. America* **71**, 1937–1938.
OLIVER, W. A., Jr. (1968). Some aspects of colony development in corals. *J. Paleont.* **42**, 16–34.
RÓŻKOWSKA, M. (1960). Blastogeny and individual variations in tetracoral colonies from the Devonian of Poland. *Acta palaeont. pol.* **5**, 3–64.
RÓŻKOWSKA, M. (1979). New Frasnian tetracorals in Poland (in press).
SMITH, S. (1915). The genus *Lonsdaleia* and *Dibunophyllum rugosum* (McCoy). *Q. J. Geol. Soc.* **71**, 218-272.
SMITH, S. and RYDER, T. A. (1926). The genus *Corwenia* gen. nov. *Ann. Mag. nat. Hist.* (9) **17**, 149–159.
SORAUF, J. E. and OLIVER, W. A., Jr. (1976). Septal carinae and microstructure in Middle Devonian *Heliophyllum* (Rugosa) from New York State. *J. Paleont.* **50**, 331–343.
SPASSKY, N. YA. (1960). Devonskie chetyrekhluchevye korally Rudnogo Altaya. *Paleontologicheskie obosnovanije stratigrafii paleozoya Rudnogo Altaya.* **3**, 1–143.
SPASSKY, N. YA. (1965). Devonskie chetyrekhluchevye korally SSR sistematika, paleoekologia, stratigraficheskoe i geograficheskoe znachenie. Avtoreferat dissertacii na soiskanie uchenoi stepeni doktora geologo-mineralogicheskihk nauk. Leningradskiy Gorniy Institut. pp. 1–41.
ULITINA, L. M. (1973). Razvitie kolonii *Phillipsastraea hennahi. Paleont. Zhurnal* **1**, 97–102.
ULITINA, L. M. (1974). Rost massivnykh kolonii rugoz (na primiere *Hexagonaria* Gürich, 1896). *In* "Ancient Cnidaria" (B. S. Sokolov, ed.) Vol. 1, pp. 172–179. Publishing House "Nauka", Siberian branch, Novosibirsk.
WELLS, J. W. (1937). Individual variation in the rugose coral species *Heliophyllum halli* E. & H. *Paleontogr. Amer.* **2**, 5–22.
YAKOVLEV, N. N. (1965). Interesnye sluchai pochkovaniya odinochnykh kamennougolnykh korallov Rugosa iz Donbassa. *Paleont. Zhurnal.* **1**, 147–148.

10 | Coloniality in the Lithostrotionidae (Rugosa)

JOHN R. NUDDS

*Department of Geological Sciences, University of Durham, England**

Abstract: During the evolution of the Lithostrotionidae lineage there was a gradual change towards higher levels of colonial integration. The corallum evolved from fasciculate (dendroid and phaceloid) to cerioid, then to astraeoid, thamnastraeoid and aphroid, and finally to a situation where it is composed of dissepiments and tabulae only, a condition which is described in the paper as "indivisoid". The changes that took place as each new growth habit evolved are described and the significance of these changes in terms of soft part morphology and level of colonial integration is discussed.

It is suggested that the fasciculate colonies show only a low level of integration, that fusion of coral tissue first takes place at the cerioid stage and that in later types of growth habit there is also communication between the gastrovascular cavities of adjacent corallites. The degree of communication increases from astraeoid to thamnastraeoid and through aphroid, until in the indivisoid coralla the colony exhibits one large intercommunicating enteron.

INTRODUCTION

The Lithostrotionidae is an extinct family of rugose corals which are common fossils in Carboniferous rocks. They make their first appearance in the Arundian Stage and become extinct in the Namurian. As the lineage evolved the most significant change was in the growth form of the colony such that the level of colonial integration gradually increased.

In early stages of phylogeny the colony was fasciculate, and evolved from dendroid to phaceloid. Later colonies were massive and evolved from cerioid, through astraeoid and thamnastraeoid to aphroid and finally to indivisoid (Figs 1, 2). The advantage of studying an extinct group of corals is that the

---

Systematics Association Special Volume No. 11, "Biology and Systematics of Colonial Organisms", edited by G. Larwood and B. R. Rosen, 1979, pp. 173–192. Academic Press, London and New York.

* Present address: Department of Geology, Trinity College, Dublin, Republic of Ireland.

complete evolution of the colony can be evaluated and trends recognized. The disadvantages are that no direct comparisons can be made with living members of the same family or even the same order as far as soft part morphology is

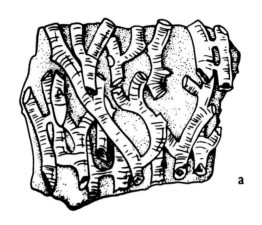

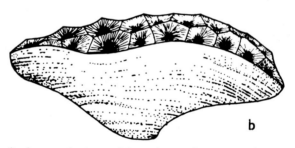

Fig. 1. Longitudinal external views of fasciculate and massive colonies; (a) fasciculate colony (dendroid) with separated corallites, (b) massive colony (cerioid) with corallites in contact. Fig. 1a from the "Treatise on Invertebrate Palaeontology", courtesy of The Geological Society of America and The University of Kansas Press. Fig. 1b from "Animal Colonies" (R. S. Boardman, A. H. Cheetham and W. A. Oliver, eds), courtesy of Dowden, Hutchinson and Ross, Inc.

concerned. We can only assume that the rugosan polyp was similar to the polyp of the present day order Scleractinia. A description of the salient features of the soft part morphology of this group is given below (after Wells, 1956).

For the purpose of this paper a colony is defined as an association of individuals living together for their mutual benefit and connected to each other cellularly or by a common skeleton.

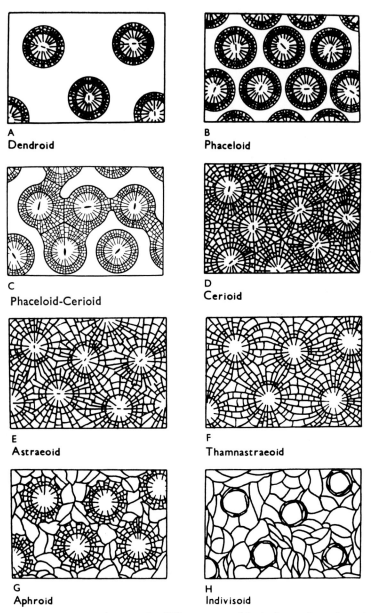

FIG. 2. Transverse sections showing the different types of fasciculate and massive colonies; (A–B) fasciculate, (A) dendroid, (B) phaceloid, (C) intermediate between fasciculate and massive, (D–H) massive, (D) cerioid, (E) astraeoid, (F) thamnastraeoid, (G) aphroid, (H) indivisoid.

THE ANTHOZOAN POLYP

The typical anthozoan polyp is illustrated in Fig. 3. It is basically cylindrical in shape, terminated at the top by the oral disc, and at the bottom by the basal disc; the sides of the cylinder being the column wall. This body wall is often considered two-layered (ectoderm and endoderm), but these are separated by a jelly-like mesogloea. In the centre of the oral disc is a mouth surrounded by tentacles and leading through a gullet or stomodaeum into the gastrovascular cavity, which occupies the majority of the interior of the cylinder including the interior of the tentacles. The gastrovascular cavity is divided by two series of partitions. The first consists of a radial system of vertical folds in the basal disc which rise over and secrete the radial calcareous septa of the skeleton. The second series of partitions is also radial consisting of mesenteries, the chief organs of digestion and excretion. These are folds of endodermal tissue projecting from, and attached to, the interior of the oral disc and the column wall. They extend into the gastrovascular cavity and commonly their lower margins are free. The ectoderm of the polyp can be divided into two regions; first, the exposed surface of the oral disc, tentacles and column wall, and secondly the calicoblast layer which includes all ectoderm that is directly in contact with skeleton (i.e. basal disc and lower parts of the column wall). Such a polyp sits in a depression at the top of the skeleton, known as the calice, which may or may not be bounded by a wall. In many species the column wall of the polyp is folded over the lip of the calice and extends down the outside of the corallite for a variable distance. This tissue which lies outside the calice is known as the edge zone and the gastrovascular cavity is continuous over the corallite wall into this edge zone (Fig. 3a). In colonial corals the edge zone may be extensive and eventually fuse with the edge zone of the adjacent polyp. In this case the connecting tissue between the two polyps lying outside the calices is termed coenosarc. The calicoblast layer occurs on the underside of the edge zone and the coenosarc, and secretes skeletal plates collectively termed coenosteum; the term coenenchyme is used collectively for both connecting tissue and skeleton.

At the point where the edge zone folds over the lip of the calice calcium carbonate may be secreted in which case a corallite wall is built. This wall is entirely secondary in nature and its method of formation varies in different groups; a common type is a septotheca formed by thickening of the peripheral ends of the septa (Fig. 3a). In many colonies the lower margins of the edge zone over this corallite wall are free and as the polyp grows upwards the edge zone rises and the cells of its lower margin secrete a thin sheath of calcium carbonate over the whole surface of the skeleton. This is the epitheca; a primary wall deposit (Figs 3b, 4b).

10. Coloniality in the Lithostrotionidae

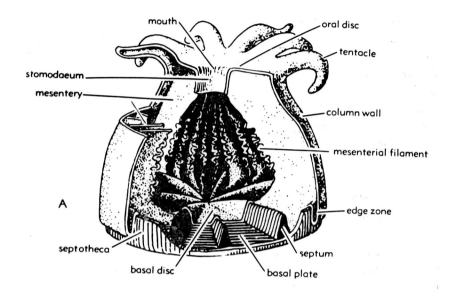

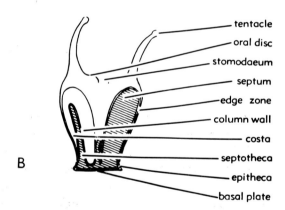

Fig. 3. Illustrations of the anthozoan polyp showing the relationship of the soft tissue to the skeleton. From the "Treatise on Invertebrate Paleontology", courtesy of The Geological Society of America and The University of Kansas Press.

In polyps with an extensive edge zone extending a considerable distance down the corallite wall the secretion of the epitheca lags behind the secretion of the other elements of the calice (Fig. 3b). In fasciculate colonies whose polyps have an edge zone continuing right down the branch and fusing with the edge zone of the adjacent polyp (Fig. 4a) the lower margins of this edge zone are clearly no longer free. As the polyp grows, the lower margins of the edge zone do not rise up the corallite wall and so no epitheca is deposited. In these colonies the coenosarc may contain extensions of the mesenteries and folds of the body wall and if so it will secrete costae, which are continuations of the septa connecting with the septa of the adjacent polyp to which it is fused. However, coenosarc may not contain such mesenterial extensions and in this case the coenosteum may be solid or consist wholly of dissepiments (small supporting plates of the skeleton).

FASCICULATE (DENDROID AND PHACELOID) COLONIES

The fasciculate colony is a branching system of corallites with the polyps situated at the open end of each branch, so that it appears like a number of solitary corals joined at their basal ends (Fig. 1a). In the earlier dendroid colony the branches are well separated whereas in the later phaceloid coralla the branches are much closer together and subparallel (Figs 2a, b).

These types of colony are exhibited by many species of the Lithostrotionidae. The first species to appear in this family is *Lithostrotion martini* and during the initial stages of the phylogeny of this species, in the lower part of the Arundian Stage, the corallum is dendroid. Later coralla of *L. martini*, in the upper part of the Arundian, the Holkerian, Asbian and Brigantian stages, are phaceloid. Similarly *L. pauciradiale*, which appears in the Asbian and is an indirect descendant of *L. martini*, exhibits a phaceloid colony.

Although no members of the Lithostrotionidae are solitary it is important to evaluate the changes that occurred when the fasciculate colony evolved from the solitary state as the Lithostrotionidae lineage must have been derived from a solitary ancestor even if indirectly. The most important change occurring in the evolution of the fasciculate growth habit is in the nature of reproduction. The ability to reproduce asexually is vital to the development of a colonial organism. Boardman *et al.* (1973) state that solitary corals can reproduce asexually, but the fasciculate coral has developed the ability to prevent its asexual buds from separating from the parent skeleton. Thus the important development from simple to fasciculate colonies is not only the ability to reproduce asexually, but also the continued adherence of the buds to the parent after reproduction has taken place so that a colony can be initiated (see Fedorowski, this volume).

To determine the level of colonial integration in the fasciculate species of the Lithostrotionidae we may draw an analogy with present day fasciculate scleractinian corals; such colonies fall into two main categories. The first of these contains the vast majority of the present day fasciculate scleractinians and is characterized by the edge zone of each polyp extending over the open top of the corallite right to the base of the corallite branch, as coenosarc, and fusing with

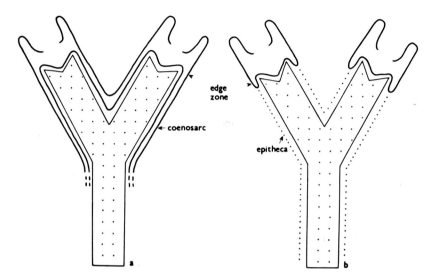

FIG. 4. Diagrammatic illustrations showing the relationship of the polyps to the skeleton in fasciculate colonies. Stippling represents skeleton. (a) Polyps are in cellular connection; the edge zone extends down the sides of each branch as coenosarc, (b) polyps separated cellularly; the edge zone is narrow and secretes an epitheca.

the coenosarc on the adjacent corallite branch, so that all polyps are connected by coenosarc (Fig. 4a). The second group contains only a few extant scleractinians and is characterized by a narrow edge zone or no edge zone at all so that the polyps are not cellularly connected (Fig. 4b).

It has already been shown that there is also a difference in the skeleton of such colonies in that corallites of the second group are coated by a thin epitheca (Fig. 4b), whereas those with coenosarc extending to the base of the corallite (Fig. 4a) do not develop an epitheca (Wells, 1956). The fasciculate species of the Lithostrotionidae (and all other fasciculate rugose corals) belong to the second category, possessing a well developed epitheca. Thus once the asexually produced daughter polyps appeared, although they remained connected to the parent by a fused skeleton, they were separated cellularly and were entirely independent biologically.

The rugose fasciculate colony therefore indicates a low level of coloniality as every individual was entirely independent. The polyps were not connected by any soft tissue, but only at their proximal ends by their previously deposited skeleton (Fig. 4b). This type of colony is little different from a number of solitary corals attached to the same block of substrate in that the individuals do not rely or depend on each other at all. Coates and Oliver (1973) state that if there is any co-operation between such polyps it is only that which results from extrinsic factors acting upon a cluster of polyps and not from any biological integration.

There are two main advantages of this type of colony. First, each new individual polyp that was produced did not have to find hard ground substrate to initiate the process of skeleton building. Secondly, by continuing the skeletal growth of earlier individuals, the polyps are able to commence growth away from the substrate and so reduce the risk of sediment deposition fouling the colony. As Coates and Oliver state (1973) all coralla tend to lift polyps away from the substrate and up into the water column and this is reflected in the mound shape of many fossil coral reefs. A third significant point is that as every individual polyp in the colony was produced asexually from the same original protopolyp (the first formed polyp after a period of sexual reproduction), then every individual polyp will be genetically identical. Thus at this stage the polyps develop the potential to become cellularly fused with their neighbours. This feature can only occur when the fusing cells contain an identical genetic make up. It is therefore the essential genetic feature of the cerioid and more complex colonial corals.

There is little difference in the level of colonial integration between dendroid and phaceloid coralla. The advantages of polyps growing close together might be first, the prevention of food being lost from the colony in the wide spaces between the corallites, and secondly the colony growing on the sea bed takes up less space for which there was probably strong competition. Moreover, the closer packing of the corallites in the phaceloid colony is an important beginning of a move towards a higher level of integration.

CERIOID COLONIES

Many of the fasciculate Lithostrotionidae species have evolutionary descendants which are massive, and the first massive colony to appear is of the cerioid type, which evolves directly from the phaceloid state. For example *Lithostrotion martini* gives rise directly to the cerioid *L. araneum*, which appears in the Holkerian Stage, while *L. pauciradiale* gives rise to *L. decipiens*, which appears soon after *L. pauciradiale* in the Asbian Stage.

The skeletal change occurring when the ceroid colony evolves is the fusion of the epithecae of adjacent corallites along their whole length instead of just at the base of the corallites (Figs 1b, 2d). The corallites cease to be cylindrical and free from their neighbours, but become polygonal and in contact with all the surrounding corallites although separated by epitheca.

The development of the cerioid colony has often been thought to be the result of compaction (obviously not mechanical squashing, but competition for space within the colony), i.e. as new buds appear there is no room for their circular outline and they are forced into a polygonal shape. This view is taken by Coates and Oliver (1973) who suggest that the cerioid corallum is merely a continuation of the trend of gradual closer packing of corallites and is thus little more than a compact phaceloid corallum.

However, close examination of some partly phaceloid/partly cerioid coralla which are the transitional phylogenetic stages between *L. pauciradiale* (phaceloid) and *L. decipiens* (cerioid) shows that the corallite centres do not become closer and closer together until their margins are in physical contact. Instead it can be seen that the corallite centres in cerioid coralla remain at a similar distance as in the phaceloid coralla. Serial sections of intermediate coralla show that while the corallite centres stay at a uniform distance apart the corallites actually extend their peripheral tissues laterally until they meet the adjacent corallite (Figs 2c, 5). This method of formation of the cerioid colony is reflected in the fact that cerioid corallites have a larger diameter than the corallites of their direct fasciculate ancestors. (Corallites of *L. decipiens* can be up to 6 mm in diameter; corallites of *L. pauciradiale* are rarely more than 3·5 mm in diameter.) If lack of space was responsible for the cerioid colony then the reverse would be true.

Now, because Coates and Oliver (1973) believe the cerioid corallum to be only a compact phaceloid corallum they suggest that the corallites of a cerioid corallum were not occupied by polyps connected to each other by cellular tissue. They assume that the polyps were probably still separate individuals, although in physical contact with other polyps on all sides. However, bearing in mind the actual method of formation of the cerioid colony, it now becomes difficult to explain why two polyps should actually grow towards each other in the partly cerioid coralla, and why their skeletal tissue should be deposited in complete apposition (Figs 2c, 5) if they were not achieving cellular connection. Lack of space can no longer be given as a reason for the close proximity of corallites in a cerioid corallum.

Coates and Oliver's argument (1973) is based on a supposition that any corallites separated from each other by epithecae could not be occupied by polyps connected to each other by edge zone. Further evidence can be obtained by studying the transitional phylogenetic stages between a cerioid species (*L. decipiens*) and its astraeoid descendant (*Orionastraea phillipsii*). *O. ensifer* forms an evolutionary link between these two species and is partly cerioid and

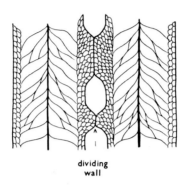

Fig. 5. Longitudinal section of a colony intermediate between fasciculate (phaceloid) and massive (cerioid). The corallites grow towards each other by extensions of the dissepimentarium. Temporarily they exhibit the features of a cerioid colony, growing side by side with their epithecae fused together forming a dividing wall.

partly astraeoid in that it possesses inconsistent traces of a wall structure between some of its corallites. Moreover in this species the wall structure definitely consists of some primary epitheca. Now it is generally accepted that the polyps of a completely astraeoid corallum do have cellular connection (see later text). Thus, if Coates and Oliver's supposition is correct the cellular connection in partly astraeoid coralla must suddenly fail between those polyps whose corallites possess an inconsistent epitheca. Again this seems unlikely and it is more probable that cellular connection can occur over the top of the epitheca and that the fusion of the distal hard parts in a cerioid corallum points to a similar fusion of the soft tissue which was secreting those hard parts; at this stage the polyps may have realized the potential for fusion that was present in the fasciculate colony.

Assuming then that cellular connection does occur in the cerioid corallum, it is still difficult to evaluate whether this was a simple fusion of the outer cells of the edge zone or whether there was a higher level of integration and that within the fused zone there was a continuation of the gastrovascular cavity over the dividing wall. The latter case is probably true in many of the cerioid scleractinian colonies where the septa continue over the wall into the adjacent corallite, as this is direct evidence that the folds of the basal disc (the floor of the gastrovascular cavity) were continuous from one polyp to another. However, in the cerioid species of *Lithostrotion* the septa of adjacent corallites are separated by an aseptate space along the top of the corallite wall so that the gastrovascular cavities may not have been continuous (Fig. 6). Secondly such continuation of

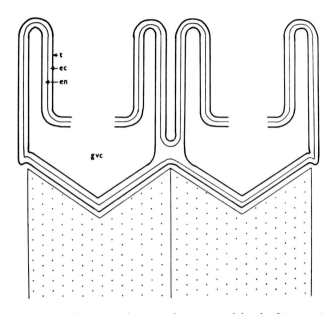

FIG. 6. Diagrammatic illustration showing the suggested level of integration between the polyps of a cerioid colony and their relationship to the skeleton. Stippling represents skeleton. Adjacent corallites have the cells of their outer wall fused, but the gastrovascular cavities are not continuous. t = tentacles, ec = ectoderm, en = endoderm, gvc = gastrovascular cavity.

gastrovascular cavities would have to take place near the base of the interior of the polyp, beneath the lower free edges of the mesenteries. In the cerioid species of *Lithostrotion* the calice is deep and the epitheca is a high ridge; there would have been little depth to the fusion of the cells. The proximal soft parts may not have become fused until the calice shallowed and the epitheca became reduced in the later types of corallum.

Whichever of these possibilities are true, fusion of the cells does allow for nourishment to be transferred from one polyp to another across the colony and this is an important adaptive advantage. The cerioid coralla therefore exhibit a higher level of integration regarding exchange of nutrients. Such a connection between the corallites utilizes all the available space, and ensures that no food particles fall between the corallites. Also the development of a cerioid colony reduces the surface area of exposed skeleton and therefore reduces the possibility of death to the colony by a fouling organism settling on such exposed skeleton (Jackson, this volume).

ASTRAEOID AND THAMNASTRAEOID COLONIES

In these colonies the walls separating corallites disappear so that the septa of one corallite continue to the adjacent corallite. In the thamnastraeoid coralla the septa of two adjacent corallites are actually confluent (Fig. 2f); in astraeoid coralla they merely abut against each other (Fig. 2e).

In the Lithostrotionidae the cerioid *Lithostrotion decipiens* gives rise to *Orionastraea ensifer*, which appears in the Brigantian Stage and which is intermediate between cerioid and astraeoid. Slightly higher in the Brigantian *O. ensifer* evolves through *O. phillipsii* to *O. tuberosa*, which may be either astraeoid or thamnastraeoid.

It is also apparent that the calices of these four species become gradually shallower as evolution progresses. Now, most of the vertical skeletal plates of a corallite are secreted by the soft tissue underneath a point in which there is a sharp upfold in that tissue (Kato, 1963). Fusion of two adjacent polyps takes place on top of the dividing wall and so the fused tissue is sharply upfolded over this wall (Fig. 6). As the calices shallow this upfold would become gradually flatter so that less calcium carbonate is secreted and the dividing wall disappears. When the corallite wall disappears the septa continue into the adjacent corallite. This is direct evidence that the radial upfolds of the basal disc (the floor of the gastrovascular cavity) were continuous so that adjacent polyps had communicating enterons (Fig. 7). This communication between gastrovascular cavities can only occur when the calices become sufficiently shallow for the proximal soft parts of the polyp to fuse, because communication of the gastrovascular cavities must occur below the level at which the mesenteries are attached to the column wall.

This is a further move towards a higher level of integration regarding exchange of nutrients. In thamnastraeoid coralla, the septa of adjacent corallites are not only continuous, but also confluent and therefore so are the radial upfolds in tissue at the base of the gastrovascular cavity. Coates and Oliver (1973) suggest that as the septa in astraeoid colonies are not confluent the radial upfolds (and therefore the mesenteries) also were not confluent and that the gastrovascular cavities did not communicate. However, while it is true that the radial upfolds and mesenteries of astraeoid coralla are not confluent, continuation of the gastrovascular cavity could still occur at a level immediately above the folds in the basal body wall. If this is correct then the soft part morphology would suggest that there is little significant difference between astraeoid and thamnastraeoid coralla. However, the thamnastraeoid coralla would have had a greater continuity of the gastrovascular cavity.

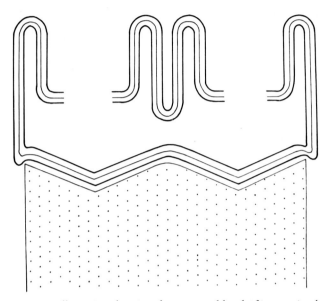

Fig. 7. Diagrammatic illustration showing the suggested level of integration between the polyps of astraeoid and thamnastraeoid colonies and their relationship to the skeleton. Stippling represents skeleton. The calices are shallower than the cerioid colony and the basal parts of the polyps are fused so that the gastrovascular cavities are continuous. For expanation of the soft tissue see Fig. 6.

The increased communication in such astraeoid and thamnastraeoid coralla is an obvious advantage over the cerioid colony. Moreover the dividing wall between two corallites of a cerioid colony is a vestigial feature derived from the fasciculate colony; its loss will mean less energy involved in colony building and more calcium carbonate available for secretion elsewhere.

The loss of wall is related to a complete change of emphasis from the individual corallite to the colony as a whole. The epitheca defining the boundary of the corallite is replaced by a holotheca which covers only the exposed sides and base of the colony and therefore defines the boundaries of the colony.

APHROID AND INDIVISOID COLONIES

The aphroid coralla are characterized by a peripheral retreat of the thamnastraeoid septa, so that the septa of adjacent corallites are no longer continuous. The resulting areas between corallites are filled by irregular dissepiments. In an aphroid colony the septa have retreated, but are still present (Fig. 2g) and this is seen in *Orionastraea edmondsi* which evolves directly from the thamnastraeoid members of *O. tuberosa* higher in the Brigantian. (*O. edmondsi* is restricted to one cyclothem limestone, the Jew Limestone, of the Brigantian of the Alston Block in Northern England.) In most rugose lineages this is the most advanced state of coloniality achieved, but in the Lithostrotionidae a further development of colony form occurs in *O. indivisa* when the septa retreat completely and are absent altogether. The colony is built entirely of tabulae and dissepiments and the term "indivisoid" is here proposed to refer to this type of colony (Fig. 2h). This species is a direct descendant of *O. edmondsi* and occurs in the Brigantian Tyne Bottom Limestone, immediately above the Jew Limestone in Northern England.

In the development of an aphroid colony Coates and Oliver (1973) point out that as the septa are lacking in the peripheral parts, direct evidence of polyp integration is lacking. However, in the Lithostrotionidae lineage the aphroid species clearly evolved from thamnastraeoid ancestors and we can assume that they also had gastrovascular cavities intercommunicated.

Coates and Oliver (1973) believe that the presence of dissepiments between the corallites indicates that coenosarc began to develop between the polyps of the aphroid colony. I do not consider that such dissepiments necessarily imply the development of coenosarc, but before discussing this it is necessary to define the term "coenosarc". Vaughan and Wells (1943) state, "the coenosarc is that part of a polyp that lies outside the skeletal wall". The term "coenosarc" is therefore only strictly applicable to connecting tissue in colonies whose corallites possess a definite wall structure. In the astraeoid, thamnastraeoid, aphroid and indivisoid coralla of rugose corals a corallite wall is absent, and we must therefore interpret "coenosarc" as being that tissue lying outside a hypothetical "boundary" of the polyp; this "boundary" being homologous with the corallite wall in a cerioid corallum. Obviously direct evidence of the position of this "boundary" in other types of corallum is lacking, but in some lineages its position can be inferred. If the corallites of the ancestral cerioid species are of similar dimensions (e.g. number of septa, diameter of tabularium) to the corallites of the astraeoid, thamnastraeoid, aphroid and indivisoid species we can assume that the position of the "boundary" in the astraeoid and other species lies the same distance from the corallite centre as does the corallite wall in the cerioid species. Thus, if the astraeoid, thamnastraeoid, aphroid and indivisoid corallites are spaced at greater intervals than the cerioid corallites we can infer that coenosarc has developed.

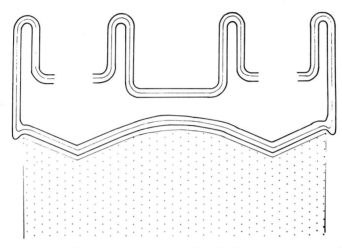

FIG. 8. Diagrammatic illustration of the suggested level of integration between the polyps of aphroid, indivisoid and some late thamnastraeoid colonies and their relationship to the skeleton. Stippling represents skeleton. The polyp heads become more separated by lateral extensions of the connecting tissue so that coenosarc develops. For explanation of the soft tissue see Fig. 6.

To determine the colony form in which coenosarc first developed the number of corallites in a given area was counted in *Lithostrotion decipiens* (cerioid), *Orionastraea tuberosa* (astraeoid and thamnastraeoid), *O. edmondsi* (aphroid) and *O. indivisa* (indivisoid) in which corallite dimensions are similar. This analysis showed that in astraeoid and most thamnastraeoid coralla the spacing of corallites is similar to cerioid coralla and so no additional tissue between polyps was developed. However, in the aphroid and indivisoid coralla the corallites are spaced at much wider intervals. Thus additional tissue was developed between the polyps of these coralla (Fig. 8). It should, however, be pointed out that some of the later thamnastraeoid coralla also had their corallites more widely spaced and hence coenosarc did develop in some thamnastraeoid colonies also. This means that the areas of dissepiments between the corallites of an aphroid and indivisoid colony do not necessarily correspond exactly to the position of the coenosarc as such dissepiments are absent in thamnastraeoid coralla.

When the coenosarc first developed in the late thamnastraeoid coralla its base was folded into radial partitions and secreted septa. However, in aphroid and indivisoid coralla the basal layer of the coenosarc ceased to be folded into the radial partitions and therefore septa were not secreted in the areas between polyps. The coenosarc secreted coenosteum of dissepiments only. This peripheral flattening of the basal body tissue means that the endoderm was reduced in total surface area so that the capacity of digestion was also reduced. However, more connecting tissue was present in the aphroid and indivisoid coralla which presumably had the reverse effect of increasing surface area so that overall digestive capacity was not reduced. The reduction in the partitions of the basal tissue also has the advantage that the interior of the polyp is more open and in indivisoid coralla it is completely open and the colony has one large gastrovascular cavity. Integration regarding exchange of nutrients is therefore now at a high level.

Beklemishev (1970) states that the original role of coenosarc was for providing organic contact and exchange of matter between zooids, but in the Lithostrotionidae and in most other rugose lineages organic connection was achieved in the cerioid corallum before coenosarc developed; its development in the aphroid corallum merely maintains organic connection.

According to Beklemishev (1970) the coenosarc in colonial organisms may have a further role when differentiation of organs common to the whole colony takes place from it; the organs are built from individuals which are necessarily polymorphic. Polymorphism in colonial organisms is not necessarily as complex as this; there may just be a differentiation into feeding and reproductive individuals. However, no evidence of polymorphism is seen in the Lithostrotionidae nor in any other rugosan lineage; integration in this direction clearly does not proceed in rugose corals.

However, there is some evidence that the coenosarc may have developed beyond its role of exchange of matter between polyps. Hubbard (1973) suggests that colonies with coenosarc extending across the flat upper surface may have been more efficient in removing sediment (and therefore also waste products) from the calicular surface. The coenosarc is at a lower level on the surface than the polyp heads and waste can be channelled through these "valleys" without falling into the polyp mouths (Fig. 8). Some aphroid colonies of *Orionastraea* show the calices as distinct hummocks on the calicular surface, with the coenosteum as lower valleys. Assuming that this has the same effect on waste removal, such aphroid coralla show integration in a direction other than exchange of nutrients as the entire colony is invovled in this waste removal system.

COMPARISONS WITH OTHER GROUPS

Other groups of rugose corals exhibit similar types of corallum to the Lithostrotionidae and probably therefore progressed along a similar cycle of increasing integration. However, exact lineages have yet to be determined in many of these groups.

The main difference between rugose corals and the scleractinian corals seems to be in the direction along which colony form evolves, and this is related to the difference between the rugose and scleractinian fasciculate colony. In the rugose fasciculate colony polyps are not connected to each other by living tissue (Fig. 4b), but in scleractinians the edge zone in most fasciculate colonies extends to the base of the corallite branch where it fuses with edge zone of the adjacent polyp (Fig. 4a). Coenosarc and cellular connection are thus achieved in most fasciculate scleractinian colonies. The coenosarc of such a colony secretes coenosteum on the outside of the corallite wall and this coenosteum consists of costae, which are external extensions of the septa. Some scleractinian colonies advance from this stage and the coenosarc at the base of the branch starts to secrete a coenosteum of costae or dissepiments actually in the space between the two corallite walls so that this space is infilled and the colony develops a flat upper surface. Such a colony is termed plocoid (Fig. 9a). The coenosarc thus differs in lying not directly on the outside of the corallite wall, but across the top of the colony. If this latter type of colony now evolves to lose its wall it will automatically produce an astraeoid or aphroid colony (depending on whether the coenosarc was secreting costae or dissepiments) without passing through a cerioid stage.

It is apparent that there are fewer cerioid genera in the scleractinians than in the rugose corals and this may be due to the fact that cellular connection is achieved in the scleractinians without passing through a cerioid stage. In the scleractinians the cerioid colony is replaced by the plocoid colony which does not occur in the rugose lineages. Thus the scleractinian fasciculate colony shows a higher level of integration regarding exchange of nutrients than the rugose fasciculate colony in which there is never cellular connection.

A second type of colony which occurs in the scleractinians and tabulate corals is the ramose colony which is dendroid in appearance, but instead of each branch consisting of just one individual, the sides of the branch also open to polyps (Fig. 9b). It is possible that many ramose coralla form when the flat calicular surface of any masssve colony is projected into "finger-like" extensions. It is also possible, however, that some ramose colonies form directly from the scleractinian fasciculate colony by polyps budding asexually from the coenosarc extended down the corallite wall. Obviously such a colony cannot evolve from

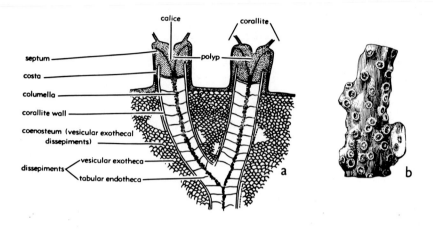

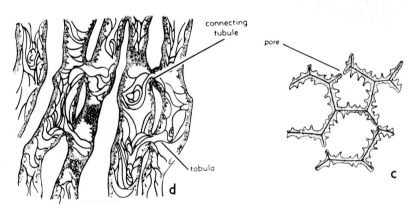

Fig. 9. Variations in colony form in the scleractinian and tabulate corals. (a–b) Scleractinian colonies, (a) plocoid, (b) ramose; (c–d) tabulate colonies, (c) cerioid colony with mural pores, (d) fasciculate colony with hollow connecting processes (compare with Fig. 5b of connecting processes in rugose corals which are not hollow, but divided by a wall). From the "Treatise on Invertebrate Paleontology", courtesy of The Geological Society of America and The University of Kansas Press.

the rugose fasciculate colony as it does not possess coenosarc outside its corallite wall.

Many tabulate corals are similar to rugose corals in the level of coloniality, but one feature seen persistently in the tabulates is the presence of mural pores in cerioid coralla (Fig. 9c), and hollow connecting processes in fasciculate coralla (Fig. 9d). As their name suggests mural pores are perforations in the dividing wall between two corallites; the connecting processes achieve a similar end as each wall of two adjacent corallites is perforated and connection is by a hollow tube. These structures occur in the wall of the calice and Coates and Oliver (1973) suggest that they probably provided communication of the gastrovascular cavities. Thus some tabulate fasciculate and cerioid coralla show a higher level of integration regarding exchange of nutrients than the rugose equivalents.

Many rugose fasciculate colonies do possess structures which appear superficially to be identical to the tabulate connecting processes, but close examination shows that the process in rugose fasciculate corals is always terminated by an epitheca and so there is never direct communication between polyps through such processes (Fig. 5). They merely represent an intermediate stage between a fasciculate colony and a cerioid colony, the process being a lateral extension of the peripheral parts of the polyp on one side only. These occur during the trend towards a cerioid colony when the polyps extended laterally all around their circumference and fused with the adjacent polyp to produce the polygonal cerioid corallite.

Compared to other coelenterates the corals as a whole show only a low level of integration. Polymorphism of individuals is not known in the rugose corals and only one instance has been recorded in the scleractinians. Fowler (1887) described feeding and reproductive individuals in *Madrepora durvillei*. However the (hydrozoan) siphonophores not only show polymorphism, but developed even higher levels of integration by building colonial organs from polymorphic individuals.

CONCLUSION

Beklemishev (1970) discussed the development of a colony in terms of a cycle which starts with an individual, progresses to the initiation of a colony and ends when the colony is so integrated as to be regarded again as an individual, albeit of a higher order.

Table I illustrates the progression, with evolution, of the Lithostrotionidae lineage along such a cycle. This table applies to most rugose lineages, but not to the scleractinians or the tabulates. The level of integration of the lineage can be considered as being low in comparison to other groups of organisms, as all the changes taking place during phylogeny are more beneficial to the individual than to the colony as a whole.

TABLE I. Progression of the Lithostrotionidae lineage

1. Solitary individual	Simple species (distant ancestor of *Lithostrotion*)
2. Colonial organism	
a. Polyps separated in space; connected only by skeleton	Fasciculate species
b. Polyps connected by fusion of the outer cells	Cerioid species
c. Confluent body cavities between polyps	Astraeoid and Thamnastraeoid species
d. Development of coenosarc between polyps	Aphroid and Indivisoid species
e. Polymorphism	—
f. Organs constructed from polymorphic individuals	—
3. Higher order individual	—

REFERENCES

BEKLEMISHEV, W. N. (1970). "Principles of Comparative Anatomy of Invertebrates" Oliver and Boyd, Edinburgh and University of Chicago Press.

BOARDMAN, R. S., CHEETHAM, A. H. and OLIVER, W. A. (eds) (1973). "Animal Colonies" Dowden, Hutchinson and Ross, Inc., Pennsylvania.

COATES, A. G. and OLIVER, W. A. (1973). Coloniality in Zoantharian corals. In "Animal Colonies", (R. S. Boardman, A. H. Cheetham and W. A. Oliver, eds), pp. 3–27. Dowden, Hutchinson and Ross, Inc., Pennsylvania.

FOWLER, G. H. (1887). The anatomy of Madreporaria, II, *Madrepora*. *Q. Jl microsc. Sci.* **17**, 1–16.

HUBBARD, J. A. E. B. (1973). Sediment-shifting experiments: a guide to functional behavior in colonial corals. In "Animal Colonies" (R. S. Boardman, A. H. Cheetham and W. A. Oliver, eds), pp. 31–42. Dowden, Hutchinson and Ross, Inc., Pennsylvania.

KATO, M. (1963). Fine skeletal structure in Rugosa. *J. Fac. Sci. Hokkaido Univ.*, ser. 4, *Geology and Mineralogy* **9**, 571–630.

MAKIE, G. O. (1963). Siphonophores, bud colonies, and superorganisms. In "The Lower Metazoa, Comparative Biology and Physiology" (E. C. Dougherty, ed), pp. 329–337. Univ. California Press, Berkeley, California.

VAUGHAN, T. W. and WELLS, J. W. (1943). Revision of the suborders, families and genera of the Scleractinia. *Spec. Pap. geol. Soc. Am.* **44**, 1–363.

WELLS, J. W. (1956). Scleractinia. In "Treatise on Invertebrate Paleontology, pt. F, Coelenterata" (R. C. Moore, ed), pp. 328–444. Geol. Soc. America and Univ. Kansas Press.

11 | Some Problems in Interpretation of Heteromorphy and Colony Integration in Bryozoa

PATRICIA L. COOK

*Department of Zoology, British Museum (Natural History)
London, England*

Abstract: The enormous diversity in colony form in Bryozoa is accompanied by an equal diversity in level of colony integration. In some colonies, zooids function almost as an aggregation of solitary animals. In others, zooids contribute to distinct colony functions. Some functions can be inferred from the morphology of zooids, others can only be demonstrated in living specimens. In all cases, series of increasing integration can be observed or inferred.

Differences in morphology of tentacle crowns, or differences in position or behaviour only, are now known to produce colony-wide patterns of water currents involved in both feeding and cleaning. Similar differences in male zooids form a distinct series of increasing integration. Zooid modifications which produce brood chambers are very complex and do not involve homologous structures. One possible series showing increasing integration may be inferred from the number of member zooids contributing to the formation of a brood chamber.

Other heteromorphic zooids, generally described as avicularia, also show gradients of morphological modification and astogenetic position. Knowledge of their behaviour and individual function is slight, but some observations have been made which indicate that they also show series of integration among colonies.

INTRODUCTION

Bryozoan colonies exhibit an enormous range of diversity in overall structure and in the form of member zooids. In some colonies, the zooids function in many ways autonomously, like an aggregation of solitary animals; in other colonies the zooidal functions are partially, at least, subordinated to colony control and may contribute to colony-wide functions.

Systematics Association Special Volume No. 11, "Biology and Systematics of Colonial Organisms", edited by G. Larwood and B. R. Rosen, 1979, pp. 193–210. Academic Press, London and New York.

The bryozoans discussed here belong to the Cheilostomata and Ctenostomata. Cheilostome colonies have member zooids with variously calcified body walls, a distally placed orifice through which the crown of ciliated tentacles is everted, and a cuticular flap, the operculum, which covers the orifice. Colonies of ctenostomes are very similar in many ways, but the zooids have uncalcified, cuticular body walls and a simpler orifice. In both groups zooids intercommunicate through pores in the body walls which are plugged with special, interdigitating cells. Nervous impulses and nutrients are transmitted through the pores from one zooid to another (see Ryland, this volume).

Recently, Boardman and Cheetham (1973) analysed some morphological characters of bryozoan colonies and arranged them in series of states which indicated increasing colony control, or integration. Included amongst the characters they examined were the nature and relationships of the body walls, the types and degree of zooidal intercommunication, and the types and arrangement of heteromorphic zooids.

It is thus possible to look at colonies in several ways and to assess their qualities by arranging the observations in different series of integrational gradients. This often reveals not only gaps in information, but problems of recognition and interpretation. In study of living colonies particularly, problems arise when attempts are made to correlate morphological integration with colony-wide function. Some of these problems are discussed here.

OVERALL INTEGRATION AND HETEROMORPHY

Examples of colonies at either end of a gradient of overall integration, from virtual zooid autonomy to almost complete colony control, are easy to demonstrate. Colonies like those of the cheilostome genus *Aetea* (see Marcus, 1937), have a low degree of integration, which is reflected in all their characters. These colonies have a discrete, uniserial budding pattern and most of each zooid body wall is in direct contact with the environment. Zooids are monomorphic and communicate with each other through a single interzooidal pore. Each zooid feeds and generally functions very like a solitary animal. At the other end of the gradient, one of the most integrated forms of colony is found in the cheilostome family Cupuladriidae, where colony control is reflected in morphology, function and behaviour. The free-living discoid or cup-shaped colonies have a rigidly regular budding pattern, and only about one sixth of each zooid body wall is in direct contact with the environment. Zooids intercommunicate through several pores in three ways, with each other distally and laterally, and with a basal colony-wide coelome (see Håkansson, 1973).

This high degree of communication is necessary as about one half of the zooids are non-feeding heteromorphs which apparently require transfer of nutrients from the feeding zooids. Some of the reactions of these heteromorphs are apparently colony-wide (see p. 200), and it is easy to recognize and regard each colony as a single entity.

Heteromorphy itself presumes some degree of colony integration, but first there must be a base-line from which heteromorphs may be recognized. Colonies result from asexual budding and all member zooids within a colony are assumed to be genetically uniform (see Ryland, this volume). Analogous structures, such as tentacle crowns, are also assumed to be homologous among colonies. All colonies have some zooids, usually the majority, which have tentacle crowns and an actual or potential feeding function. At any one period, however some zooids will be passing through the regular stages, universal among bryozoans, of degeneration or regeneration of feeding and other organs.

The feeding zooid (autozooid) is therefore utilized as a base-line, and zooids with differences in morphology not attributable to cyclic ontogenetic changes or astogenetic factors are regarded as heteromorphic, with functions presumed to be additional or alternative to feeding. Four types of heteromorph are briefly considered here. They pose problems both of recognition, and of the degrees of integration they express in morphology, function and behaviour.

HETEROMORPHIC TENTACLE CROWNS—EXCURRENT CHIMNEYS

The occurrence, in actively feeding colonies, of water circulation patterns which ensure the unimpeded flow of feeding and cleaning currents has been described by Banta et al. (1974) and Cook (1977). In some colonies small differences in behaviour or morphology of the tentacle crowns of feeding zooids may reflect apparently sophisticated, colony-wide function. Groups of zooids in which some of the tentacles are lengthened, and in which the tentacle crowns are held at a constant angle to the zooid frontal wall, form "active chimneys". These are excurrent water outlets allowing free flow of water away from distinct, regularly spaced areas of the colony surface. The chimney zooids often surround a non-feeding zooid, and the elongated tentacles are produced distally, laterally or proximally in those zooids which are respectively proximal, lateral and distal in position to the central, non-feeding zooid. The production (by budding through at least three zooid generations), and presumed maintenance (though cyclic degeneration and regeneration of tentacle crowns) of such regularly spaced groups of zooids indicates considerable colony control of zooid morphology and function. The zooids involved are however, feeding zooids, and are indistinguishable from other autozooids when the tentacles are retracted.

In other colonies, zooids are produced which for a time have no feeding function. These zooids may accommodate brooding larvae, and in the ctenostome, *Alcyonidium nodosum* (O'Donoghue and de Watteville, 1944), which encrusts the shells of the gastropod, *Comminella papyracea* (B.M. 1963.3.30.12) are budded in raised, regularly spaced groups. Such zooids form "passive

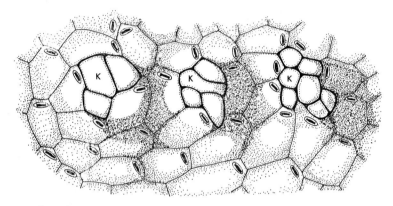

FIG. 1. *Alcyonidium* sp. encrusting *Tegella verrucosa*, Panama, B.M. 1975.9.22.1. Autozooids and groups of kenozooids (K), developed on protuberances of gastropod shell as a result of crowding. × 0·25.

chimneys" in that they provide a focus for outward water currents because they themselves are producing no inward current. Permanently non-feeding zooids, also budded in raised groups, form passive chimneys in the cheilostome genus *Hippoporidra*. The non-feeding zooids are strongly heteromorphic (see p. 197), and are budded in regular, cyclically repeated patterns throughout colony growth. This suggests a high degree of colony integration (see Cook, 1977).

Groups of heteromorphic, non-feeding zooids may have a fortuitous occurrence but still perform the same function. A species of the ctenostome genus *Alcyonidium*, encrusting shells of the gastropod, *Tegella verrucosa* (B.M. 1975.9.22.1), has a somewhat random and discrete budding pattern, and produces raised groups of non-feeding kenozooids (zooids with body walls but no orifices). The gastropod shell itself has regularly spaced protuberances, and the kenozooids tend to occur at these points, as a result of crowding of zooid buds where the substratum is restricted (see Fig. 1). Although these groups of kenozooids almost certainly act as passive chimneys, they are the result of microenvironmental influences and a high degree of integration is not inferred for this species. Thus occurrence of chimneys as part of the budding pattern may be under varying degrees of colony control, but all the zooids involved contribute to an integrated, colony-wide function. The degrees of integration may belong to several series and some may be recognizable only in living colonies. If the zooids forming the chimneys differ from autozooids in skeletal morphology, information on colony growth patterns is also needed for interpretation of their degree of integration.

HETEROMORPHIC TENTACLE CROWNS—MALE ZOOIDS

A somewhat less complex series of heteromorphic tentacle crowns is known, although incompletely, in male zoooids.

Spermatozoa are emitted from the tips of all or some of the tentacles, and the tentacle crowns exhibit distinct behaviour patterns during emission. In many colonies, male zooids are in all other respects exactly like autozooids (see Silén, 1966, 1972). In *Hippopodinella*, male zooids have modified, non-feeding tentacle crowns with a reduced number of tentacles and their behaviour patterns are distinct (see Gordon, 1968). After cyclic degeneration and regeneration, these zooids have "normal" tentacle crowns and function as autozooids, from which they do not differ skeletally in any way. Zooids interpreted as male zooids, which do differ skeletally from autozooids, are known in several cheilostome genera. Generally they have smaller orifices than autozooids, and may occur near female, brooding zooids (as inferred for *Trypostega claviculata* (Hincks), see Plate II, 1), or in patterned groups or series, as in *Adeonella* (see Harmer, 1957). In *Hippoporidra*, the male "cortical" zooids (see Cook, 1968a), have modified, non-feeding tentacle crowns, distinct behaviour patterns, heteromorphic skeletons and rigidly patterned positions in raised groups which may also act as passive chimneys (see p. 196).

Emission of spermatozoa may be epidemic in some species (see Silén, 1966), but even if it is an autonomous function in others, the association of male zooids with brooding zooids or with water current patterns indicates a high degree of colony control and requires investigation. Recognition of male zooids in colonies with lower degrees of integration may be possible only in living specimens. Interpretation of heteromorphic zooids which may be males is complicated by the possibility of more than one function, and by the difficulty in establishing their sexual function during ontogenetic periods when spermatozoa are not produced.

AVICULARIA

In the Cheilostomata, heteromorphic zooids in which the orifice and operculum (termed the mandible) are enlarged in comparison with the rest of the zooid, are interpreted as "avicularia". The term was established for highly modified zooids, but an enormous diversity of zooids which appear to fall within this definition is known. Some are budded apparently at random, are little differentiated from autozooids and are capable of feeding. These heteromorphs have a high degree of autonomy, as in *Crassimarginatella similis* (see Cook, 1968b and see Fig. 2a, Plate III, 1), which has occasional "avicularian" zooids that differ very little from autozooids. At the other end of the range avicularia may be budded by one zooid, in a rigidly patterned position, and be highly modified and incapable of feeding, which itself suggests colony control in their budding and perhaps their function, as for example, in the Cupuladriidae (Fig. 2h).

The morphological diversity of avicularia suggests a corresponding diversity in function and behaviour. In fact, although behaviour has been observed in many colonies, the function of a large number of avicularia is unknown or difficult to interpret.

One series which can be constructed shows an integrational gradient depending upon the degree of avicularian autonomy. This series poses some interesting problems because where the function is known, it is not apparently correlated with behaviour, irritability, morphology or integration (see Fig. 2).

For example, the large, vicarious avicularia of *Stylopoma* (Fig. 2d) tend to occur irregularly in depressed areas of the colony surface, where sediments accumulate, and produce water currents which clean the zooidal surfaces as the spatulate mandibles are closed. Similarly shaped avicularia in *Labiporella* (Fig. 2b), which occur with greater frequency and regularity, have not been seen to function at all. In addition, in some species of this genus, these avicularia are capable of feeding. The long, setiform mandibles of the avicularia of *Escharina porosa* (Smitt), (Fig. 2e) make violent sweeping movements on stimulation, and presumably function defensively; whereas those of *Scrupocellaria* (Fig. 2g), which have a similar morphology, make continuous, cleaning move-

---

FIG. 2. Overall integrational gradient of avicularian heterozooids, based on autonomy (low degree of integration) to colony control (high degree of integration). a–c, autonomous avicularia (at right), all capable of feeding. a, *Crassimarginatella similis*, b, *Labiporella dipla*, c, *Steganoporella magnilabris*. d, vicarious avicularium (at right), not capable of feeding. *Stylopoma*. e–g, dependent avicularia, budded from one autozooid, not capable of feeding. e, *Escharina porosa*, f, *Bugula calathus*, g, *Scrupocellaria*. h, avicularia and zooids developed within a colony-wide coelome, *Cupuladria*.

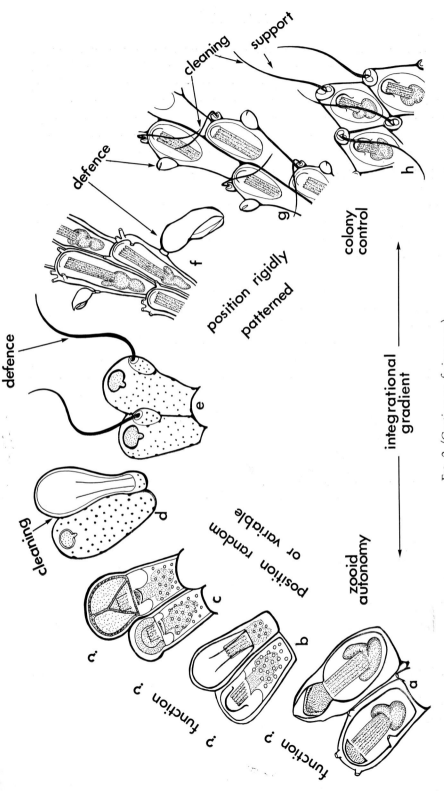

Fig. 2. (Caption on facing page)

ments, which are automatic. In the Cupuladriidae (Fig. 2h) similar setiform mandibles clean on stimulation and support the colony at the periphery (see Cook, 1963). The "bird's head" avicularia of *Bugula* (Fig. 2f) move automatically, and the mandibles shut both on stimulation and automatically. They are defensive in function (see Kaufmann, 1971), as are the lateral avicularia of *Scrupocellaria* (Fig. 2g) which however, only react on stimulation. The massive, toothed mandibles of *Steganoporella* (Fig. 2c), which could be inferred to be defensive from their morphology, do not apparently have a defensive function and, in most species, are capable of feeding (see Banta, 1973).

There are several complex series of integration inherent in the budding patterns and specialization of avicularia. In some cases, avicularian function is known to be colony-wide, as in the supporting peripheral setiform mandibles of the Cupuladriidae. All the supporting setae may move simultaneously to shake sediment off the colony surface (see Cook, 1963) suggesting that behaviour patterns too, may be colony controlled in this highly integrated family. In such cases, the possibility of relatively sophisticated nervous communication among zooids is a strong one.

Although many avicularia are easily recognized, interpretation of their function requires more observation of living colonies, and correlations of morphology with presumed function need testing. In addition, the relationships of patterned positions of avicularia and their known autonomous reactions with possible colony-wide function require investigation.

BROODING ZOOIDS AND BROOD CHAMBERS

Brooding in cheilostomes is extra-coelomic. Ova are laid through a pore at the base of the distal tentacles into a brood chamber (see Silén, 1945). Brood chambers may be membranous diverticula housed within zooid body walls. Such brooding zooids are frequently polymorphic, with larger orifices than those of autozooids (e.g. *Tropidozoum cellariiforme* Harmer, see Plate I, 1). Most brood chambers are external and have calcified walls.

Recent work has shown that a basically globular structure with an opening close to a zooidal orifice has evolved in many ways. Even brood chambers

---

PLATE I

1. *Tropidozoum cellariiforme* Harmer. Siboga Stn 144, East Celebes, B.M. 1964.3.10.1. Autozooids and brooding zooid with enlarged orifice (⊙), ×70.
2. *Usica mexicana* Banta. Gulf of California, B.M. 1966.3.2.6. Autozooids and heteromorphic zooids with enlarged orifices (⊙), ×40.

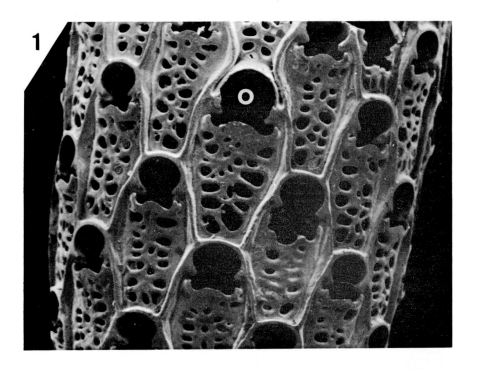
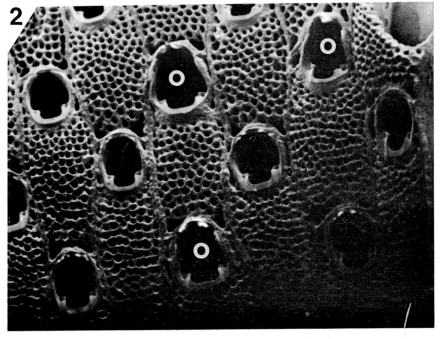

PLATE I. (Caption on facing page)

which are very similar in appearance may thus not prove to be composed of homologous structures or to have the same ontogeny.

One series showing increasing integration can be constructed by considering only how many zooids are involved in formation of a brood chamber, not the homologies of the structures themselves (see Fig. 3).

For example, in *Aetea*, *Reptadeonella* and *Thalamoporella* (Figs. 3a, b, c) the zooid which produces the ovum (the "maternal zooid"), also broods it, either in an internal diverticulum or in external structures produced during ontogeny. In *Tendra* (Fig. 3d) the embryos are brooded in structures produced wholly by the next sequentially budded zooid distal to the maternal zooid. In *Crassimarginatella* (Fig. 3f) the brood chamber has a double wall and is produced partly by the maternal zooid (inner lamina) and partly by the distal zooid (outer lamina) (see Harmelin, 1973). A similar type of development occurs in some species of *Scrupocellaria* (Fig. 3e) and in *Doryporella alcicornis* (O'Donoghue and O'Donoghue, 1923, and see Plate II, 2). In *Trypostega claviculata* (Hincks), the zooid distal to the maternal zooid is a kenozooid (see Levinsen, 1909) which in its entirety forms the outer layer of the brood chamber (see Plate II, 1). In *Catenicula* (Fig. 3g) two modified, distally budded kenozooids form the brood chamber (the "synoecium" of O'Donoghue, 1924 and O'Donoghue and de Watteville, 1944). In *Bugula neritina* (Linnaeus) the brood chamber is a modified kenozooid itself budded from the sequentially budded distal zooid (see Woollacott and Zimmer, 1972, Fig. 3h and Ryland, this volume). This type of development is similar to that seen in *Trypostega venusta* (Norman), where the brood chamber apparently arises as a frontal bud from a heteromorphic (?male) zooid which is regularly budded distal to each autozooid and female zooid. In *Tremogasterina* (Fig. 3i), only two zooids, the maternal and distal, are usually involved, but as many as three zooids may act together to produce the brood chamber. The inner lamina is budded by the maternal zooid, the outer lamina arises from extensions of the frontal calcified walls of the distal zooids, both laminae growing into an extrazooidal (i.e. colony-wide) frontal coelomic cavity. This type of brood chamber is apparently a complex type of frontally budded interzooidal kenozooid, and is similar to that found in *Aspidostoma* (see Plate III, 2-5). Here there is a heavily calcified internal cryptocyst lamina which is covered in life by a cuticular frontal wall and hypostegal coelome. Brood chambers develop from a thin, unilaminar expansion of the frontal (upper) part of the continuous lateral/distal zooidal walls. Septulae in the cryptocyst, which presumably connect the hypostegal and visceral coelomic cavities, are present both proximally (in the distal part of the maternal zooid) and distally (in the proximal part of the distal zooid). The lamina thus grows into a cuticular

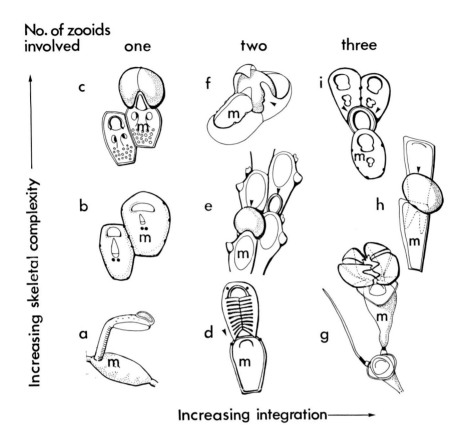

FIG. 3. Overall integrational gradient of brood chambers, based on the number of contributory zooids. a–c, brood chamber produced by maternal zooid (m). a, *Aetea*, small external, cuticular sac, b, *Reptadeonella*, dimorphic brooding zooid (right) with "internal" brood chamber, c, *Thalamoporella*, brood chamber (right) with calcified walls. d–f, brooding contributed to by two zooids. d, *Tendra*, chamber formed by zooid distal to maternal zooid, e, *Scrupocellaria*, outer layer of brood chamber contributed to by zooid distal to maternal zooid, f, *Crassimarginatella*, inner layer of brood chamber produced by maternal zooid, outer layer by distal zooid (after Harmelin, 1973). g–i, brood chamber produced by maternal zooid and two sequential zooids. g, *Catenicula*, "synoecium" produced by two modified kenozooids, h, *Bugula neritina*, brood chamber a kenozooid budded by zooid distal to maternal zooid, i, *Tremogasterina*, inner layer of brood chamber produced by maternal zooid, outer layer by two distal zooids, all growing into extrazooidal coelomic space.

sac which is an expansion of the hypostegal coelomes of two sequentially budded zooids, and the brood chamber which results is also apparently an interzooidal kenozooid. Later ontogenetic calcification in *Aspidostoma* may result in continuous thickening of the outer surface of the cryptocyst of the distal zooid and the outer surface of the brood chamber, which may become totally immersed.

Brood chambers are produced sporadically in some species, or in distinct zones or groups in others. Their development may take place either early in the ontogeny of the zooids concerned (e.g. *Doryporella alcicornis*, see Plate II, 2), or may not occur until a much later stage (e.g. *Aspidostoma giganteum*, see Plate III, 2-5). There is strong evidence for substantial colony control where two or three zooids act together to produce a brood chamber, or where zones of brood chambers are produced simultaneously. Whether there are any correlations between the numbers of contributory zooids, the onset of brood chamber ontogeny, its relationships to the ontogeny and astogeny of the zooids, the nature of structures involved, and the function of the colony as a unit remains to be investigated.

CONCLUSIONS

Bryozoa are complex and diverse and often difficult to observe alive for long periods. There has therefore been a natural tendency to argue from the particular to the general, and to assume that morphological similarity indicates both structural homology and similarity of function. In addition, because zooids are small, observations of autonomous function, which can often only be made at high magnifications, have not usually been related to possible colony-wide functions which would only be observable at low magnifications.

Integrated, colony-wide functions may be much more common in bryozoan colonies than hitherto suspected. Analysis of colony morphology is one of the first steps in discovering the degrees of integration present in, for example, zooidal intercommunication and in heteromorph patterning. An efficient inter-

---

PLATE II

1. *Trypostega claviculata* (Hincks). Queen Charlotte Is., B.M. 1886.3.6.35. Autozooids, ?male zooids (arrowed), maternal zooid (M) and brood chamber (B). Broken brood chamber showing formation from kenozooid distal to maternal zooid at right, ×60.
2. *Doryporella alcicornis* (O'Donoghue & O'Donoghue). Vancouver Is. region, B.M. 1963.12.30.39. Autozooids and two maternal zooids with brood chambers developing very early in zooidal ontogeny, ×125. (1) inner lamina of brood chamber developing from maternal zooid (2) later stage showing development of outer lamina derived from frontal wall of distal zooid.

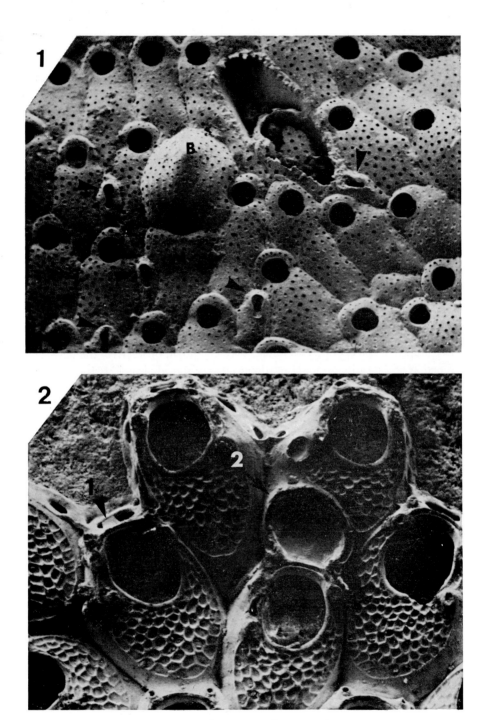

PLATE II. (Caption on facing page)

communication system, through pores or colony-wide coelomes, or both, seems to be an essential feature of highly integrated colonies. A high degree of interzooidal communication may mean a similar degree of nervous organization (see Ryland, this volume).

The correlation between highly integrated morphology and colony-wide functions is not, however, a simple one. For example, the presence of active chimneys seems to show a high degree of integration, but has relatively little morphological expression. On the other hand, high morphological integration is certainly positively correlated with colony-wide behaviour patterns and functions in the Cupuladriidae.

Colony control may be expressed in the budding patterns of autozooids and heterozooids, as for example, in the regeneration of regular, discoidal colonies from small fragments consisting of only a few complete zooids in the Cupuladriidae (see also Ryland, this volume). The functions of other, often regularly repeated groups of autozooids and heterozooids ("cormidia"), have not been observed in many species. Although autonomous behaviour patterns and functions of the members of a cormidium may be known or inferred, the possible function of the entire group in relation to that of the whole colony is generally not known (see also Ryland, this volume).

Assumptions that the type of heteromorph and its function can be recognized from morphology need testing. It is interesting to note that although the function of many "avicularia" is not known, their morphology has generally been well investigated, whereas the overwhelmingly demonstrable function of brood chambers seems to have obscured a similar degree of analysis of their morphology and development.

Recognition of male zooids is complicated by the cyclic ontogeny of the

---

PLATE III

1. *Crassimarginatella similis* Cook. Ghana, B.M. 1969.9.5.86. Autozooids and one avicularian zooid with modified oral region (arrowed), × 30.

2–5. *Aspidostoma giganteum* (Busk.). Challenger Stn 135, Nightingale Is., B.M. 1887.12.9.646. Showing late development of brood chambers in zooidal ontogeny, × 36.

2. Stage 1, autozooids and maternal zooids (m) with minute expansions of distal/lateral walls (arrowed).

3. Stage 2, maternal zooid with expanded lamina forming base of brood chamber, developing between maternal zooid septulae and distal zooid septulae (arrowed). Stage 3, partially developed brood chamber at left.

4. Stage 4, nearly complete brood chamber at left. Stage 1, minute expansion at right (arrowed).

5. Stage 5, complete brood chamber.

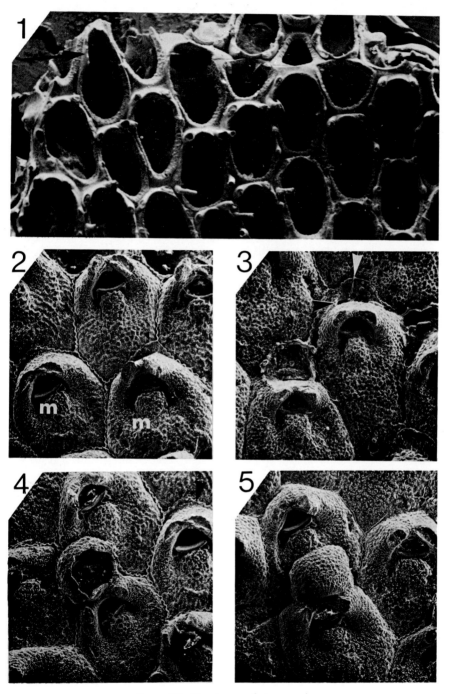

PLATE III. (Caption on facing page)

reproductive products, even where there are differences between male zooids and autozooids in tentacle crown and/or skeletal morphology. It is very probable, for example, that the small zooids of *Trypostega* are males, even though Marcus (1938) found no spermatozoa in *T. venusta*.

Sometimes the differences among heteromorphs of known differing function are very small. For example, male zooids in *Adeonella* (see Harmer, 1957) differ from autozooids only in the slightly smaller size of the orifices. Brooding zooids in *Tropidozoum* (see Harmer, 1957) differ principally in the slightly enlarged orifices (see also Plate I, 1). Avicularia in *Crassimarginatella similis* differ only in the slightly enlarged orifice and modified operculum (see Plate III, 1). These last heteromorphs are inferred to be "avicularia" only because brood chambers are already known in the species. Banta (1973) has discussed a somewhat similar problem in the "B-zooids" (avicularia) of *Steganoporella*. In the absence of evidence provided by specialized morphology, behaviour, or by reproductive products it may be almost impossible to recognize or interpret some heteromorphs. Those known in *Uscia mexicana* Banta are of unknown function (see Banta, 1969), and have orifices and opercula differing only slightly from those of the autozooids (see Plate I, 2). These heteromorphs are very similar in skeletal appearance to those of *Tropidozoum* and *Crassimarginatella* mentioned above, and could be variously interpreted as male, female, brooding or avicularian zooids. In addition, the heteromorphic zooids of *Adeonella*, *Tropidozoum*, *Crassimarginatella* and *Uscia* are all capable of feeding. The possibility that their position in the colony structure might be connected with some additional or alternative function such as water current production or storage of nutrients tends to be obscured by assumptions of a single function inferred from available morphology.

More observations of living colonies are urgently needed in order that some of these problems may be resolved. This does not mean, however, that information gained from the study of dead Recent colonies or from fossil colonies cannot contribute a great deal (see Taylor, 1975). Once they have been recognized from detailed morphological comparisons, autozooids and heterozooids occurring in regular patterns may suggest possible functions within colonies. These may then be tested in homologous or analogous cases from living specimens.

The adaptive advantages of integration have been examined in detail in very few forms. Highly integrated colonies such as those of *Hippoporidra* and the Cupuladriidae are associated with specialized environments usually unsuitable for colonization by other bryozoans (see Cook, 1968c), and both their structural morphology and colony-wide functions can be correlated with these environmental conditions. Although their extreme specialization may itself be a barrier to expansion into other ecological niches, the free-living, cup-shaped or discoid colonies particularly have had a long record of abundance from the late Cretaceous to Recent times (see Håkansson, 1973, 1975). There are also distinct indications that increasing integration of colony structure (and perhaps of colony-wide functions) was correlated with rapid expansion of the cheilostomes during the late Cretaceous (see Boardman and Cheetham, 1973). This does not mean that colonies with a low degree of integration are necessarily either primitive or unsuccessful, although, like the specialized colonies, their ecological niches may be restricted. In fact, many colonies show a high degree of integration in some characters and a low degree in others.

Although study of integration in relation to systematics is in its early stages, consideration of integrative states of characters raises questions, and provides new ideas and observations on morphology which may prove to have systematic value. In addition it can only help to discover more about Bryozoa and more about the phenomenon of coloniality in general.

ACKNOWLEDGEMENTS

I should like to thank Dr J. Rosewater (United States National Museum) for presenting specimens of *Alcyonidium* sp., Dr J. S. Ryland (University College, Swansea) for technical criticism and Mr P. J. Chimonides (British Museum, Natural History) for the S.E.M. photography.

REFERENCES

Banta, W. C. (1969). *Uscia mexicana*, new genus, new species: a watersiporid bryozoan with dimorphic autozooids. *Bull. Sth Calif. Acad. Sci.* **68**, 30–35.
Banta, W. C. (1973). Evolution of avicularia in cheilostome Bryozoa. In "Animal Colonies" (R. S. Boardman, A. H. Cheetham and W. A. Oliver, eds), Dowden, Hutchinson and Ross, Stroudsburg.
Banta, W. C., McKinney, F. K. and Zimmer, R. L. (1974). Bryozoan monticules: excurrent water outlets? *Science, N.Y.* **185**, 783–784.
Boardman, R. S. & Cheetman, A. H. (1973). Colony dominance in stenolaemate and gymnolaemate Bryozoa. In "Animal Colonies" (R. S. Boardman, A. H. Cheetham and W. A. Oliver, eds), Dowden, Hutchinson and Ross, Stroudsburg.
Cook, P. L. (1963). Observations on live lunulitiform zoaria of Polyzoa. *Cah. Biol. mar.* **4**, 407–413.
Cook, P. L. (1968a). Observations on living Bryozoa. *Atti. Soc. ital. Sci. nat.* **108**, 155–160.
Cook, P. L. (1968b). Polyzoa from west Africa. The Malacostega Pt. 1. *Bull. Br. Mus. nat. Hist. (Zool.)* **16** (3) 113–160.
Cook, P. L. (1968c). Bryozoa (Polyzoa) from the coasts of tropical west Africa. *Atlantide Rep.* **10**, 115–262.

Cook, P. L. (1977). Colony-wide water currents in living Bryozoa. *Cah. Biol. mar.* **18**, 31–47.

Gordon, D. P. (1968). Zooidal dimorphism in the polyzoan *Hippopodinella adpressa*. *Nature, Lond.* **219**, 633–634.

Håkansson, E. (1973). Mode of growth of Cupuladriidae (Bryozoa Cheilostomata). *In* "Living and Fossil Bryozoa" (G. P. Larwood, ed), Academic Press, London and New York.

Håkansson, E. (1975). Population structure of colonial organisms. A palaeoecological study of some free-living Cretaceous bryozoans. *Docums Lab. Géol. Fac. Sci.* Lyon H. S. **3**, 385–399.

Harmelin, J.-G. (1973). Les bryozoaires des peuplements sciaphiles de Mediterranée: le genre *Crassimarginatella* Canu (chilostomes Anasca). *Cah. Biol. mar.* **14**, 471–492.

Harmer, S. F. (1957). The Polyzoa of the Siboga Expedition. Pt. IV, Cheilostomata Ascophora II. *Siboga Exped.* **28**d, i–xv, 641–1147.

Kaufmann, K. W. (1971). The form and functions of the avicularia of *Bugula*. *Postilla* **151**, 1–26.

Levinsen, G. M. R. (1909). "Morphological and systematic studies on the cheilostomatous Bryozoa" Nationale Forfatterers Forlag Copenhagen.

Marcus, E. (1937). Bryozoarios marinhos brasileiros I. *Bolm Fac. Filos. ciênc. Univ. S. Paulo Zool.* **1**, 3–224.

Marcus, E. (1938). Bryozoarios marinhos brasileiros II. *Bolm Fac. Filos. ciênc. Univ. S. Paulo Zool.* **2**, 1–196.

O'Donoghue, C. H. (1924). The Bryozoa (Polyzoa) collected by the SS "Pickle". *Rep. Fish. mar. biol. Surv. Un. S. Afr. Spec. Rep.* **X**, 1–63.

O'Donoghue, C. H. and O'Donoghue, E. (1923). A preliminary list of Polyzoa (Bryozoa) from the Vancouver Island region. *Contr. Can. Biol. Fish.* n.s. **1**, 143–201.

O'Donoghue, C. H. and de Watteville, D. (1944). Additional notes on Bryozoa from South Africa. *Ann. Natal Mus.* **10**, 407–432.

Silén, L. (1945). The main features of the development of the ovum, embryo and ooecium in the ooeciferous Bryozoa. *Ark. Zool.* **35A**, **17**, 1–34.

Silén, L. (1966). On the fertilization problem in the gymnolaematous Bryozoa. *Ophelia* **3**, 113–140.

Silén, L. (1972). Fertilization in the Bryozoa. *Ophelia* **10**, 27–34.

Taylor, P. D. (1975). Monticules in a Jurassic bryozoan. *Geol. Mag.* **112**, 601–606.

Woollacott, R. and Zimmer, R. L. (1972). Origin and structure of the brood chamber in *Bugula neritina* (Bryozoa). *Mar. Biol. Berlin* **16**, 165–170.

12 | Structural and Physiological Aspects of Coloniality in Bryozoa

J. S. RYLAND

Department of Zoology, University College of Swansea, Wales

Abstract: Bryozoans are colonial and the zooids in any colony are genetically uniform and organically linked. In some taxa, intercommunication by confluence of body cavities had led to a substantial degree of colonial integration, apparently at the expense of restricting the evolutionary potential of the constituent zooids. Greatest adaptive radiation has therefore occurred in the Cheilostomata, in which the retention of zooidal individuality has been facilitated by development of the funiculi into a colonial network for metabolite distribution. The departure from cylindrical shape and the ability to mineralize walls may also have contributed significantly to the diversity of zooidal evolution in this order. Thus, while polymorphism is a feature of bryozoans generally, its most spectacular manifestation is in the Cheilostomata. The morphs are here generally grouped in replicated cormidia, of size and complexity varying from taxon to taxon, which effectively replace the autozooids as the fundamental unit of the colony. Especially elaborate are the cormidia of the Cribrimorpha and Scrupocellariidae. Behavioural integration of zooids has been accomplished by means of a colonial nervous system in some species, but other instances of apparently co-ordinated behaviour are not yet properly understood. Some can be related to, though only partially explained by, ageing processes and the existence of physiological gradients. Bryozoan colonies may thus be complex, displaying co-ordinated responses of zooids and a high degree of metabolic integration, but their sessile habit has inhibited the evolution of finite colony form, which has been approached only in a few, very specialized species.

INTRODUCTION

Bryozoans (ectoprocts of some authors) are abundant and diverse, and constitute one of the more successful colonial phyla. However, tending to be both small and complex, their particular evolutionary trends and adaptations to the

---

Systematics Association Special Volume No. 11, "Biology and Systematics of Colonial Organisms", edited by G. Larwood and B. R. Rosen, 1979, pp. 211–242. Academic Press, London and New York.

colonial habit are less well known than those of cnidarians, for example. Some of these facets of coloniality will be explored from a biological perspective in this paper. It has been written intentionally to introduce bryozoan coloniality to non-bryozoologists, as much as for specialists. Only a fairly basic knowledge of anatomy and classification has been assumed, such as may be obtained from my introductory text *Bryozoans* (Ryland, 1970).

Bryozoans are, probably without exception, colonial animals as generally understood (Beklemishev, 1969). As such: (1) the colony consists of zooids possessing some degree of individuality; (2) the colony arises by asexual budding from a founder zooid (ancestrula); (3) the zooids are organically linked; (4) there is a flow of metabolites between zooids during their lifetime; and (5) the formation, metabolism and behaviour of the member zooids harmonize with each other to some extent (Beklemishev, 1969). With these points in mind we can consider Beklemishev's statement (1969), that: "The basic problem of colonial structure is the relationship between the individuality of separate zooids and the individuality of the colony."

When Remane (1936, 1938) described the remarkable meiofaunal species *Monobryozoon ambulans* he regarded it as a solitary bryozoon, and this assertion is still being repeated. However, as recognized by Hyman (1959), each "individual" is in reality a colony consisting of a functional autozooid and several stolons. Though interzooidal septa were not illustrated by Remane (1936), Franzén (1960) clearly showed in his description of *Monobryozoon limicola* that each stolon is a zooid separated from the parent autozooid by a septum. Thus the assertion that bryozoans are colonial needs no qualification, although it should be noted the Ott (1972) has suggested recently that *M. ambulans* and *M. bulbosum* Ott may differ quite fundamentally in this respect from *M. limicola*.

POLYMORPHISM

One of the commonest manifestations of colonial development is polymorphism: the presence of several different forms of zooid within the colony. Polymorphism is variably developed within the three bryozoan classes, being absent in the Phylactolaemata, of sparse occurrence in the Stenolaemata, but of major importance in the Gymnolaemata, especially in the order Cheilostomata. Polymorphs here occur both in great abundance and in their maximum diversity. A contemporary review of bryozoan polymorphism is that of Silén (1977). Briefly, typical feeding zooids are termed *autozooids*. In non-cheilostomes the principal morphs are linear, polypide-less structures (*kenozooids*) making up attachment rhizoids or connecting stolons, and embryonic brood-chambers (*gonozooids*). Special dwarf zooids (*nanozooids*) are present in certain stenolaemates, notably *Diplosolen obelia* (Johnston). The single tentacle of the nanozooid polypide apparently helps to keep the colony surface clean (Silén and Harmelin, 1974).

Of cheilostome polymorphs, the obviously specialized avicularia and vibracula are morphologically well known (e.g., Silén, 1938; and brief reviews by Hyman, 1959; Ryland, 1970), even though both their evolution and their function are matters for debate (Banta, 1973; Schopf, 1973; Cook, this volume). Less well established as polymorphs are mural spines and their derivatives, perhaps including ooecia.

"Spines" in bryozoans have varied origins and homologies. Thus the slender *processus spiniformes* of the Crisiidae comprise a series of kenozooids (Borg, 1926); the tubercles at the autozooidal corners in *Membranipora membranacea* (L.), when studied comparatively with the "reticulum zooids" of *Conopeum*, seem clearly to be minute kenozooids (Jebram, 1968); while the marginal spines of *Electra verticillata* (Ellis and Solander) are described simply as hollow projections of the gymnocyst (Bobin, 1968). Jebram (1973) has questioned Bobin's view of *Electra* spines, but the latter's interpretation receives support from Lutaud's (1973) study of the nervous system in *Electra pilosa* (L.), in which the nerve supplying the mural rim simply branches into each spine. Different again are the internal, or cryptocyst, denticles seen, for example, in various membraniporines (Cook, 1968a). Most interest is centred, however, on true marginal spines present in so many cheilostome genera. Our information here still depends almost exclusively on the researches of Silén (1942). He showed that such spines: (1) terminate (at least in many cases) not with a point, but with a flat, uncalcified area (comparable to the anascan frontal membrane); (2) are hollow, with the lumen characteristically constricted to a communication pore at the point of origin from the parent zooid; and (3) may replace a zooid of a different polymorph, e.g., an autozooid in *Scruparia* sp. and an avicularium in *Tricellaria ziczac* Silén. There seems little doubt that his view is correct, and *Callopora*-type spines are polymorphs. This applies, therefore, to such modified spines as the scutum in Scrupocellariidae and costae in Cribrimorpha. The formation of such spines was described by Harmer (1902) and, in more detail, by Levinsen (1909: 9, Pl. IV, fig. 2) and is mentioned again by Harmelin (1973). The spines in the ctenostomatous genus *Flustrellidra* are different, each being the chitinous extension of a superficially placed kenozooid (Silén, 1947).

The brood chambers in cheilostomes are of the general type known as ovicells. I have elsewhere (Ryland, 1976) suggested that it is useful to distinguish between the "ovicell", which is the complete structure including its internal or closing membranes, and the "ooecium", which is the hooded part. Such ooecia develop in more than one way (see Cook, this volume). The independent ooecium of *Bugula neritina* (L.) arises as a single rudiment from the proximal gymnocyst of the distal zooid (Calvet, 1900; Woollacott and Zimmer, 1972), not from the maternal zooid. It forms a bilaminar hood, of variable extent, opening towards the maternal zooid. The two layers of the hood, respectively entooecium and ectooecium, are separated by a flat lumen which is continuous, through a communication pore, with the coelom of the distal zooid. There can be little doubt that the *Bugula* ooecium is a zooid morph, though its further homologies have not been determined.

In many other Anasca, including (from personal observation)*Scrupocellaria scabra* (van Beneden), the ooecium arises as a double rudiment or pair of flat knobs from the region of the distal wall of the maternal zooid. The knobs fuse, and the combined plate extends to form once again a hood opening towards the maternal zooid (Silén, 1944a; Harmelin, 1973). As in *Bugula*, the ooecium is bilaminar, with the entooecium and the ectooecium separated by a flat lumen. Cook (this volume) states that the ooecium of *Scrupocellaria* arises from the distal zooid, although Silén (1944c) shows the ooecial lumen in *S. scabra* communicating with the coelom of the maternal zooid. This is said also to be the case in *Callopora* (Silén, 1944a). In Ascophora the ooecial rudiment is usually single (Harmer, 1957) and the finished structure more complex, although even in anascans the ooecium may form partly from the maternal zooid and partly from the distal zooid (Harmelin, 1973). In the sertellid *Reteporellina evelinae* Marcus the ooecium has been shown by Banta (1977) to develop from the distal zooid.

In his study on *Euthyroides episcopalis* (Busk), Harmer (1902) suggested that the ooecium was formed "by the fusion of a pair of greatly expanded oral spines, the bases of which should communicate with the fertile zooecium on each side of the operculum". In this species, the complete ooecium bears a median keel representing the line of fusion of the two halves. Levinsen (1902) expressed a similar view with regard to the bivalvular ooecia of *Catenaria parasitica* (*Alysidium parasiticum* Busk). Silén rather tentatively extrapolated this view to the typical cheilostome ooecium: while such an interpretation remains at the level of an hypothesis, it still seems most reasonable to consider that the ooecium frequently comprises one or a fused pair of zooid polymorphs. In the Cretaceous cribrimorph *Leptocheilopora tenuilabrosa* Lang the ooecium was either formed by or covered with a series of spines (Larwood, 1962).

The remaining major zooid morph in cheilostomes is the septulum, dietella or pore-chamber: i.e., the structure associated with an interzooidal communication pore. Our understanding of the role of communication pores during the budding and astogeny of Gymnolaemata stems mainly from the study by Silén (1944b), although the structures involved had earlier been well described by Levinsen (1909). When such pores occur in an outer wall they lead simply to a pore-chamber or dietella. A distinction is usually made between dietellae, in which the inner wall extends to the base of the parent zooid, and septula, in which it does not; but, as Levinsen (1909) himself recognized, such a distinction is quite arbitrary. Moreover, the term septulum logically can be applied only to the perforated inner wall of the chamber, not to the entire structure (see Gordon and Hastings, 1979). Where there is a chamber, irrespective of whether it meets the basal wall or not, it is here termed a dietella. A dietella is thus initially a zooid morph: a kenozooid of unfulfilled potential which may, under appropriate circumstances, continue development to form a zooid of a different kind, usually an autozooid. The dietella in this way then becomes a small part of the new zooid (p. 219).

CORMIDIA

Cormidia are groupings of zooids, in which at least two morphs are represented, that recur regularly within the colony. In the strict sense cormidia may be described as "colonies within colonies" (Beklemishev, 1969) capable of performing most of the vital functions of the colony. Beklemishev recognized the occurrence of cormidia in bryozoans, notably in the Scrupocellariidae, but without perhaps appreciating their full complexity. Since their consideration by Boardman and Cheetham (1973) was also rather cursory, some discussion of cheilostome cormidia, in the light of the polymorphism already described, is apposite.

It appears that the "unit" in most cheilostomes is not in fact an autozooid at all, but a cormidium of varying complexity. This is true, as implied above, even in such an apparently simple form as *Membranipora*. Notably excepted from this generalization, however, are the Electridae. In this family there are no ovicells and, according to Bobin's (1968) analysis of *Electra verticillata*, the spines are simple outgrowths from the gymnocyst. The first cheilostomes to appear in the fossil record, of Upper Jurassic (Pohowsky, 1973) and Lower Cretaceous age, are *Pyripora*-like and classified in, or related to, the Electridae. In *Pyripora* and *Pyriporopsis* there are no spines, but in *Rhammatopora* spines, as evidenced by their bases, were present (Thomas and Larwood, 1960). *Wilbertopora*, also early Cretaceous, placed by Cheetham (1954) in the Calloporidae, has both occasional marginal spines outside the mural rim (Cheetham, 1976) and ooecia.

The typical unit in Recent calloporids seems to be a cormidium. Marginal spines are present, typically—as in *Callopora lineata* (L.)—as a complete ring around the frontal area, and both ooecia and adventitious avicularia are usual. In view of the apparently "primitive" position of the Calloporidae, and their central position in the adaptive radiation of the Cheilostomata (Silén, 1942), it follows that increasingly complex cormidia are the norm in this order. If the presence of a "tata" or *Callopora*-like ancestrula be accepted as valid evidence, even the cellularines must have evolved from calloporid stock, for *Bicellariella* (Smitt, 1868) and several species of *Bugula* (e.g., *B. turrita* Marcus, 1938 [not Desor]) have a horn-shaped ancestrula in which the horizontal frontal area is surrounded by spines. Also in *Callopora* there is a morphological link with the Cribrimorpha through species such as *C. rylandi* Bobin and Prenant in which the marginal spines have been transformed into a series of costae.

Some of the most obviously complex cormidia are found in the cellularine family Scrupocellariidae. A typical unit frequently comprises the autozooid, frontal adventitious avicularium (sometimes internally placed according to Harmer, 1923), lateral adventitious avicularium, basal vibraculum, attachment rhizoid arising from the vibracular chamber, a series of marginal spines—including the scutum and a group on each distal angle of the autozooid—and the ooecium. Although all these zooids clearly form a morphologically integrated unit, there is little information concerning the degree of physiological co-ordination. This is equally true in the Bugulidae, in which it is not known whether the pedunculate avicularium is physiologically integrated with its bearing autozooid.

Accepting the premises: (1) that spines are polymorphs, and (2) that ontogeny and astogeny here recapitulate phylogeny, some of the structurally most highly developed cormidia are found in the Cribrimorpha. A young colony of *Callopora rylandi* shows the transition from spines to costae in the zone of astogenetic change (Fig. 36B in Ryland and Hayward, 1977). The costae of *Membraniporella nitida* (Johnston) also develop from spines, as noted by Hincks (1880) and Harmer (1902). The same method of formation is seen in undisputed cribrimorphs such as species of *Cribrilina* (Harmer, 1902; Norman, 1903) and *Cribrilaria* (according to Norman's account, though not Harmer's). The success of and diversity attained by the cribrimorphs during the Cretaceous is apparent from Larwood's (1962) monograph.

The frontal shield of cribrimorphs may then, as Harmer (e.g., 1930) proposed, in certain instances have led to an imperforate frontal wall by complete fusion of costae: such ascophorans were referred to as Spinocystidea (or Polyspinocystidea) by Silén (1942) and represent the peak of structural integration for this line of evolution. *Exechonella, Triporula* (Cook, 1967), *Tremogasterina* (Powell and Cook, 1967) and *Chorizopora* (Powell, 1967) have been suggested as probable examples.

The two instances here discussed represent the bryozoan ultimate in two evolutionary lines. In *Scrupocellaria* the associated zooids in the cormidium belong essentially to different morphological types, specialized to perform particular functions. In the cribrimorph or spinocystidean, a group of similar morphs has developed in the same way to produce what is effectively a component part of the main autozooid.

ASTOGENY AND COLONY INDIVIDUALITY

The bryozoan colony arises by asexual budding from the ancestrula. This process, in relation to colony differentiation, complexity and morphological integrity, has been fully explored by Boardman and Cheetham (1969, 1973), Boardman *et al.* (1969) and Cheetham (1973). Little can usefully be added in a short review, although attention might be drawn to those curious instances in which a zooid of one morph regenerates inside one of a different morph: for example, an avicularium inside an autozooid and vice versa, several examples of which were cited by Levinsen (1907). In consequence, two points only will be discussed: the genetic homogeneity of the colony (heretofore assumed but never demonstrated) and the physiological individuality of a colony ("self" and "non-self") as indicated by the ability or inability of homospecific colonies to fuse.

Considerable evidence demonstrating genetic homogeneity within a single bryozoan colony has recently been obtained by Mr J. P. Thorpe, during a study of population genetics using enzyme electrophoresis. He has established that enzyme polymorphism occurs at rather a low level in bryozoans. Nevertheless, in 183 experiments involving polymorphic loci, replicate pieces from any colony invariably contained the same allele. The main species used for the experiments were *Alcyonidium gelatinosum* (L.) *A. hirsutum* (Fleming) and *Bugula turbinata* Alder, with a smaller number of results from *Bugula flabellata* (Thompson in Gray), *Cryptosula pallasiana* (Moll), *Membranipora membranacea* and *Pentapora foliacea* (Ellis and Solander). Thus, although it would be difficult— if not impossible—to prove genetic homogeneity within a colony, no exceptions were discovered in an extensive experimental sample. Mr Thorpe's findings will be fully published elsewhere.

Data on fusion compatibility in bryozoans are, regrettably, at present non-quantitative. There is nothing comparable with the observations of Karakashian and Milkman (1967), who recorded 78 fusions (6·2%) in 1262 contiguous borders of the compound ascidian *Botryllus schlosseri* (Pallas), or with the experiments of Theodor (1976) who obtained 11 fusions (0·7%), 40 semi-rejections (2·7%) and 1428 total rejections (96·6%) out of 1479 surviving allografts of the gorgonian *Eunicella stricta* (Bertoloni). However, autosyndrome (fusion between parts of the same colony: Knight-Jones and Moyse, 1961) appears normal in bryozoans, occurring for example in *Alcyonidium* spp., *Flustrellidra hispida* (Fabricius) and *Membranipora membranacea* (from personal observations), *Beania discordermae* (Ortmann) (Silén, 1944b), and *Camptoplites* spp., in which rhizoids may link up with a neighbouring branch (Hastings, 1943). Conversely, homosyndrome is rare, and only the following records have been located: *Camptoplites atlanticus* Hastings (Fig. 40 and p. 460 in Hastings, 1943: young colonies are attached by their rhizoids to the pore-plates of a well-grown colony; it seems likely that all the young colonies so attached would have been siblings derived from the well-grown colony by self-fertilization); *Membranipora hyadesi* Jullien (Moyano, 1967), *M. membranacea* and *Parasmittina trispinosa* (Johnston) (Stebbing, 1973); and in colonies of *P. nitida* (Verrill) derived from the same maternal parent (Humphries, 1979) and cyclostomes *Diplosolen obelia* (Johnston) and *Microecia suborbicularis* (Hincks) (in Harmelin, 1974) as the fused colonies in each pair were of about the same size, they may have been genetically identical twins, for development is polyembryonic (Harmer, 1893)). I have not noticed homosyndrome in mosaics of *Alcyonidium hirsutum*. *A. polyoum* (Hassall) or *Flustrellidra hispida*.

INTRA-COLONY CONTINUITY

In the most primitive and non-integrated bryozoan colonies, individual zooids possibly did not communicate through their soft tissues. This was apparently the situation in Palaeozoic cyclostomes (Brood, 1973; Boardman and Cheetham, 1973), although the zooids in post-Palaeozoic stomatoporids are linked internally through pores. In many other Palaeozoic stenolaemates the walls separating adjacent zooids were non-perforate, but the epithelia and coeloms were continuous around the end of the vertical zooid walls. This condition was found in both trepostomes and cryptostomes and is characteristic of Recent cyclostomes such as *Heteropora* and *Hornera*. More typically in cyclostomes adjacent zooids are in communication through mural pores, which are simply small round holes through the interzooidal walls. The zooidal epithelia are continuous through these pores, but there is no association between pores and funicular tissue (Borg, 1926), presumably because the polypide is enclosed by the membranous sac.

In branching phylactolaemates the zooids are separated only by incomplete septa, and there is no interzooidal association of funiculi. As the colony has evolved into a more compact form, the septa have tended to disappear altogether. The interzooidal pores of gymnolaemates differ from those of the other two classes, first, in being closed by cellular plugs and, second, by their association with the funiculus. They are also skeletally more complex in that the pore area (pore-plate or septulum) may be uniporous or multiporous. Moreover, in lateral walls, which are essentially double formations, the pore-plate is associated with a small chamber or dietella between the two elements of the wall.

The significance of the interzooidal communications in the astogeny of various gymnolaemates was elucidated by Silén (1944b), and it was mainly in this study that the relationships between dietellae, the parent zooid and the budded (or daughter) zooids was established. Thus in *Electra pilosa* or *Callopora dumerilii* (Audouin), for example, the terminal dietella of any particular zooid is in reality the proximal part of the next zooid in the same series; similarly, if a lateral zooid bud develops, the originating chamber—be it purely in the lateral wall or in the latero-basal angle—constitutes the most proximal part of the daughter zooid. A dietella, as a discrete kenozooid, will be present in such cases only until the daughter zooid develops. In certain positions, as around the ooecia of *Hippothoa*, the dietellae normally remain as kenozooids. In parallel and contemporaneously developing series of zooids, separated by duplex walls, the pore structures arising in the distal half of one zooid open into the proximal half of the adjacent zooid. Irrespective of how the fusion is achieved (the accounts of Silén, 1944b, and Banta, 1969, differ somewhat), the dietellae have only a transient existence, though it is perhaps a matter for debate whether or not such communicating lateral dietellae should be regarded as kenozooids.

Since kenozooidal dietellae do not represent a polymorph fundamentally different from the autozooids in a colony, it is hardly surprising to find that they were present in the earliest cheilostomes, such as the Jurassic *Pyriporopsis* (Pohowsky, 1973) and the various Lower Cretaceous genera mentioned above (Boardman and Cheetham, 1973; Banta, 1976).

THE NERVOUS SYSTEM AND CO-ORDINATED BEHAVIOUR

The presence or absence of a colonial nervous system in bryozoans has for long been a matter of controversy, which even now has been satisfactorily resolved in only a few well-studied species of Anasca. Certainly, the nervous system within a gymnolaemate zooid is extremely complex. This is not the place for a full description as I have given a short review elsewhere (Ryland, 1976), and a detailed and authoritative account has been prepared by Lutaud (1977). Briefly, the cerebral ganglion is dorsally (anally) situated between the base of the lophophore and the pharynx, and is laterally extended as a peripharyngeal ring (not necessarily a complete ring according to Gordon, 1974). Sensory and motor nerves from the peripharyngeal ring can be traced into the tentacles.

Three distinct areas can be recognized in the ganglion of cheilostomes: (1) a distal region associated with the peripharyngeal ring and sensory nerves from the lophophore; (2) a central portion, which may serve for general cerebral co-ordination, to which lead the pair of cystidial nerves which constitute part of the colonial nervous system (Lutaud, 1969); and (3) twin proximal cell aggregations associated with the non-lophophoral nerves of the polypide. The latter have a paired origin on each side of the proximal cell clusters, although the two roots join (as the "compound tentacle sheath nerve") during their pathway along the tentacle sheath. They are functionally "mixed" and incorporate the sensory nerves from the frontal membrane, including the operculum (Lutaud, 1973).

In anascans such as *Electra pilosa* the colonial nervous system comprises a pericystidial ring from which branches lead into the dietellae and transverse-wall pores, plus the two main nerves (mentioned above) which connect the ring with the ganglion (Hiller, 1939; Lutaud, 1969). Although the nature of the nerve junctions in the dietellae has not been established, it is evidently through the latter that the nervous systems of contiguous zooids are associated. The arrangement makes interzooidal nervous co-ordination possible, and Hiller (1939) reported that zooids frequently react in groups to the stimulation of a single frontal membrane. However, Marcus (1926a) had earlier established in *E. pilosa* and other species that mechanical stimulation of an expanded tentacle crown caused withdrawal of that crown but no other. Unfortunately, Marcus' generalization that each zooid's behaviour is not influenced by that of its neighbours has been more widely disseminated than Hiller's, establishing a quite erroneous impression of the level of behavioural co-ordination within a bryozoan colony.

Hiller's (1939) observations have been substantiated and elaborated in a recent series of experiments on *Membranipora membranacea* (Thorpe et al., 1975). It was confirmed that mechanical or electrical stimulation of an expanded lophophore resulted solely in retraction of that lophophore, but that similar stimulation of the frontal membrane, especially in the region of the orifice, led to the immediate withdrawal of all the expanded lophophores near to the point of stimulation.

The area affected was usually roughly diamond shaped, elongated in correspondence with the longitudinal axis of the zooids, to a maximum extent of about 100 × 50 mm. Co-ordinated withdrawal responses have also been reported in the phylactolaemate *Cristatella mucedo* Cuvier (Mackie, 1963; but with no details).

Thorpe et al. found that *M. membranacea* zooids responded to a mechanical stimulus of about 10 mg on the frontal membrane. A cut in the colony inhibited propagation of the response, although there was limited spread around the ends of the cut. Electrical recording revealed two types of pulse: one of these, recorded whenever a retractor muscle contracted, was of long duration and high amplitude; the other pulses were of short duration (~ 3 ms) and low amplitude ($\sim 10\,\mu V$), and produced in bursts with a peak frequency sometimes in excess of 200 Hz. The conduction velocity of these pulses was about 100 cm s^{-1} in directions parallel to the long axis of the zooids and about half of this at right angles to it. These rates of conduction are the same as those for the spread of lophophore retractions from a point of stimulation. The authors argue that the physiological properties of the system are consistent with those that would be predicted from the linkage of the cystidial nerve plexi demonstrated histologically by Hiller (1939) and Lutaud (1969).

Activity in *Membranipora* and many other marine bryozoans consists mainly of movements of the polypide in relation to protrusion and retraction of the lophophore. The co-ordination of such movements involves intrazooidal sensory nerves leading to the proximal part of the ganglion, and motor nerves arising from the same part. Neighbouring zooids influence the ganglion (and vice versa) through the pericystidial/colonial system, the component nerves of which join cells in the centre of each ganglion. The position of the lophophore in any zooid thus depends on the integration within the ganglion of input from the zooid's own sensory nerves and from the colonial system.

A superficial network, reputedly nervous, has been reported in some stoloniferans (Marcus, 1926a; Bronstein, 1937; Lutaud, 1974) and phylactolaemates (Gerwerzhagen, 1913; Marcus, 1926b). It is obviously morphologically quite different from the colonial nervous system of anascans, and nothing is known of its physiology.

Various other patterns of behaviour, apparently involving colony-wide co-ordination, have been reported in bryozoans. One example is the movement of whole *Cristatella mucedo* colonies from a well lighted to a shaded part of an aquarium (Marcus, 1926b). The mechanics of movement in this species are not understood, but it seems fairly clear (Ryland, 1977) that the behavioural response is one of high photokinesis: activity is greater in the light than in the dark, so that if the wanderings of a colony take it into a dark place it tends to stay there. Thus, although the behavioural response to light is clearly colony-wide, there is at present no evidence that this can be expressed through co-ordinated and directed movement.

The best known example of apparently co-ordinated movements by individual zooids within a colony concerns the vibracula of *Caberea*. It was first recorded by Darwin (1839, Journal for 19 May, 1834) and elaborated upon by Hincks (1880): "In this genus the entire dorsal surface of the branch is covered by the vibracula, and the movements of the setae are synchronous; they act together with perfect regularity—the whole company on a branch swinging to and fro at the same moment, and as if under a common impulse. We can hardly doubt that there must be some intercommunication between the nerve centres of the individual vibracula, on which these combined movements depend." However, the movements were re-examined by Silén (1950), who reached the conclusion that the co-ordination was entirely mechanical: one seta could not move without disturbing at least the one immediately distal to it, and the apparent synchrony was in reality a very rapid chain of independent responses. Simultaneous, and possibly co-ordinated, movements of vibracular setae also occur in *Cupuladria* (see Cook, this volume).

Another example of possibly integrated movement investigated by Silén (1950) was the bending and straightening of branches in *Kinetoskias* spp. The colony in this deep-water genus is of specialized form, with a tall straight stem anchored at its proximal end by rhizoids. The branches of autozooids arise as a bush at the distal end of the stem, the zooids facing outwards. Silén showed that the relaxed position for the zooid is that which results in the branches being downcurled (see Smitt, 1868). Depression of the frontal membrane in each zooid raises the hydrostatic pressure of the coelomic fluid, which in turn causes the flexible distal half of the basal wall to straighten. Silén assumed that such straightening followed eversion of the polypide. The movements of the zooids are said not to be integrated, but it is obvious that the branches would not stiffen and lift unless all or most of the zooids were involved: perhaps there may, after all, be some degree of co-ordination.

The observation of Mackie (1963) that in the phylactolaemate *Plumatella fungosa* (Pallas) "the lophophores over wide regions become orientated by muscular action so that they point in the same direction" seems possibly to imply co-ordination (unless it is simply a response to streaming water). A further example of static co-ordination concerns the exhalant "chimneys" reported in *Membranipora membranacea*. According to Banta et al. (1974), the everted polypides of little groups of zooids, of constant position in the colony, lean away from one another, creating a central space. Water which has been filtered through the expanded and interdigitated lophophores over a considerable area of colony needs an exhalant channel, which the chimneys provide. How the position of these chimneys is determined, and why the everted polypides of the associated zooids lean in the direction they do remain unexplained (see Cook, 1977 and this volume, for further discussion). Evidently, while certain aspects of co-ordination within colonies are becoming better understood, others have yet to be satisfactorily explained.

A more elaborate system of chimneys may be seen in species of *Hippoporidra*, which form colonies commensally with hermit crabs. The colony surface is mammillate, with each hummock capped by a group of small-orificed zooids, presumably males (see Cook, 1968b; and from analogy with *Hippopodinella adpressa* (Busk), described by Gordon, 1968b). In a probably undescribed species from Beaufort, North Carolina, each of these zooids has four, rigidly held, upwards directed, unciliated tentacles. Ovicellate zooids are restricted to areas between the mammillae, wherein lophophores comprise twelve ciliated tentacles spread as a radially symmetrical funnel. Where the feeding zooids border the mammillae the expanded lophophores display bilateral symmetry, with the tentacles towards the mammilla being greatly extended and little outcurved: this posture appears to be usual around exhalant outlets (Cook, 1977).

A striking feature of the Beaufort specimens is the mottled colour pattern, white blotches on a brown background or vice versa, the disposition of pigment exactly corresponding with the topography of the surface. The mammillae thus stand out clearly from the remaining area. In *H. senegambiensis* (Carter) a similar pattern may be observed (Cook, 1964, 1968), and sections display radiating bands of pigment (Carter, 1882), deposited as the colony thickens by frontal budding. *Hippoporidra* colonies of this or a similar species from the Gulf of Mexico are also patterned (F. J. S. Maturo, personal communication). *Hippoporidra* thus provides a remarkable example of a bryozoon in which the grouping of zooid morphs into filtration areas and outlets is paralleled by the distribution of pigment. A full account of the Beaufort *Hippoporidra* is in preparation.

THE FUNICULAR SYSTEM

The funiculus is a filamentous structure composed of fibrillar mesenchyme cells. In phylactolaemates it forms an unbranched cord extending from the stomach caecum proximally to the ventral body wall (see, for example, Ryland,

1970), while in cyclostomes it may be even less developed, failing to reach the zooid wall (Borg, 1926; Nielsen, 1970). The funiculus of ctenostomes is better developed, the main polypidial strand being linked to the stolonal cord via the interzooidal septum (e.g. Ryland, 1970). Its greatest development, however, is found in cheilostomes, where the main polypidial trunk (e.g., in *Bugula*: Calvet, 1900) or trunks (e.g., in *Membranipora*: Lutaud, 1961) have lateral extensions to the interzooidal pores

The funiculus in the stolon of a ctenostome such as *Bowerbankia* extends right to the apex and is thus already present as the septa form during growth (Brien and Huysmans, 1938), although it apparently breaks as the septum develops, its two free ends then associating with the cells plugging the pore (Bobin, 1962). In *Membranipora membranacea* and *Watersipora nigra* (Canu and Bassler) the longitudinal funiculi also precede the transverse walls in formation (Lutaud, 1961; Banta, 1969). In both, however, it seems that the pore (or multiporous septulum) forms at the point of transection of the funiculus by the septum. The association of funicular cells with the pore-cells in *Bowerbankia imbricata* (Adams) has been described at the level of optical microscopy (Bobin, 1958a, 1962, 1964; review 1977) but awaits investigation by electron microscopy.

The arrangement of cells at the interzooidal pore in gymnolaemates is consistent but moderately complex. The structure in its simplest form is found in a ctenostome such as *Alcyonidium hirsutum* (Gordon, 1975). A simple pore in the inner wall is marginally lined by an annular "cincture cell" (or, possibly, cells) of U-section, the two flanges being applied to the wall. The space delimited by the cincture cell is plugged by dumb-bell shaped "special cells", narrow waisted in the pore itself; this plug is then capped on each side by squamous "limiting cells". In cheilostome transverse walls the pore (or each pore of a multiporous septulum) is lined by a cincture (Banta, 1969), now seen as part of a cincture cell (Gordon, 1975) and plugged by one or more dumb-bell shaped special cells, each having its nucleus in the proximal dilatation. The limiting cells are cuboidal or columnar, arranged in one or two layers around the special cells.

In the lateral, duplex walls of *Membranipora membranacea* and *Watersipora nigra*, for example, there is a multiporous septulum on the parental ("abannular") side of the dietella and a large, rimmed opening on the non-parental ("annular") side. The cell types are the same as those found in a transverse wall pore, with the nuclei of the special cells located in the parental or abannular dilatation (Banta, 1969). In a stoloniferous ctenostome such as *Bowerbankia imbricata*, the structure corresponds to that of the single-pore type in simplex walls. Where autozooids adjoin stolons, the nuclei of the special cells are positioned in the autozooidal dilatation (Bobin, 1958a, b); in a growing stolon the special cell nuclei are placed in the proximal dilatation (Bobin, 1962), although their polarization may later change. The terminal membranes of the special cells are extensively folded and extended in slender prolongations which interdigitate with similar processes on the inner surface of the limiting cells. The abseptal face of the limiting cells has a similar configuration: in every instance the prolongations tend to be longer and more irregular than the microvilli of a normal brush-border (Gordon, 1975). It has thus become absolutely clear that the special cells of the pore must be instrumental in effecting the passage of metabolites from one zooid to another. They are manifestly polarized, with their nuclei positioned on the side from which nutrients would be expected to flow. Their apices are microvillous, and intimate contact is made with the funicular cells either directly or through the limiting cells.

When a new autozooid is produced in *Bowerbankia imbricata*, the main events involving the funiculus are as follows: The bud arises as a swelling which becomes temporarily separated from the bearing stolon by a curtain of mesenchyme cells and mucous strands. As the polypide rudiment is produced by cellular invagination in the bud apex, the septum starts to form as an annular involution of ectoderm on the bud side of the mucous curtain. The stolonal funiculus becomes linked to the mesenchyme of the curtain. As the septum constricts, like the closing of an iris diaphragm, with the cuticle being secreted between the ectodermal layers, the resulting pore is initially closed by a mucous plug; at this time the mesenchyme in the bud starts to form the autozooidal funiculus (Bobin, 1958a). During septum formation the polypide bud does not grow.

Mesenchyme cells in the bud elongate and move into the septal pore, displacing the mucous plug and differentiating into the special cells. During the immediately succeeding period the special cells undergo a phase of fluctuating polarity during which their nuclei may be attenuated, split or migrating from one side of the septum to the other. In many of the cells the nuclei are temporarily situated in the stolonal dilatation. This phase of reversed or fluctuating polarity coincides with the period of rapid growth and organogenesis of the polypide rudiment, and it seems that development of the polypide is not possible until: (1) the stolonal funiculus has become linked to the pore cells, and (2) the reverse polarity of the special cells facilitates metabolic inflow to the bud. By the time the polypide is complete and able to feed, the nuclei of the special cells have resumed their normal, and hence-forward fixed, position (Bobin, 1958a, 1962).

During the active life of the polypide there is evidently only a sparse movement of lipid through the autozooidal pore cells, but small oil globules are visible in the stolonal funiculus. At the stolonal septa in the proximity of an apex, the highest concentration of lipid is in the funiculus on the proximal side, but there are large globules in the proximal limiting cells. The proximal fringe of the special cells appears to be capturing globules, and fine droplets only are seen in the body of these cells and in their distal fringe. Large globules are present in the distal limiting cells (Bobin, 1971). The stolonal funiculus appears to store lipid as well as transport it. Although the impact from individual feeding zooids may be slight, the sum transport of lipid throughout the stolonal septa and into the growing tips seems to be considerable.

During polypide degeneration the autozooidal funiculus remains intact, and ends up attached to the brown body (Bobin, 1964). During this degenerative phase there is a massive efflux of lipid from the autozooid into the stolonal funiculus (Bobin, 1971). Globules of all sizes are abundant in both the autozooidal funiculus and the stolonal funiculus below the septum. Large globules can be seen in both groups of limiting cells, while the autozooidal fringe of the special cells is packed with small droplets. When degeneration is complete the new polypide starts to form: this is accomplished without any reversal of polarity in the special cells, so adequate reserves are presumably retained within the autozooid (Bobin, 1962). Bobin (1964) recognized that the stomach region of the gut of the living polypide was functioning as a nutrient store, and that the periodic discharge of lipid might well be associated with reproductive activity in general and vitellogenesis in particular. In an earlier paper on another stoloniferan, *Terebripora comma* Soule, it had been noted that the presence in a zooid of a developing embryo was invariably associated with a degenerating polypide or brown body (Bobin and Prenant, 1954). This was also very clearly shown in *Bowerbankia caudata* (Hincks) (*B. gracilis* Leidy) by Braem (1951), although, since the embryo is brooded in the atrium, it would be in any case impossible for the lophophore to be everted. In the encrusting ctenostome *Alcyonidium polyoum* several embryos develop simultaneously in a brood pouch which forms only after degeneration of the maternal polypide (Matricon, 1963); here also breakdown is concomitant with oocyte enlargement. The relationship between vitellogenesis and polypide degeneration will be further considered later in this paper.

AGEING, REGENERATION AND PHYSIOLOGICAL GRADIENTS

The ontogenetic or developmental changes seen in bryozoan zooids have recently received considerable attention (e.g. by Boardman *et al.*, 1969); but the ageing process does not, of course, terminate with the attainment of morphological maturity. Continuing changes may be associated with the thickening of the frontal wall, as in *Pentapora foliacea* (Balavoine, 1956; Hastings and Ryland, 1968), with polypide degeneration and replacement (review by Gordon, 1977) or with the reproductive cycle (see below). One manifestation of ageing, which has received too little attention, is regenerative ability.

When considering regeneration in bryozoans it is useful to make a distinction between within-zooid events, such as the cycle of brown body formation and polypide renewal, and colonial regeneration in which one or more whole zooids are involved. Admittedly, in stoloniferous ctenostomes such as *Walkeria uva* (L.) and species of *Bowerbankia* these situations are linked, for polypide degeneration may be associated with loss and replacement of the entire autozooid (Levinsen, 1907)—although papers already cited show that in *Bowerbankia* brown body formation is not necessarily followed by autotomy of the zooid (Bobin, 1958a, 1962, 1971).

The fronds of many erect, branching species routinely disintegrate during autumn, leaving only stolons to survive the winter. Nutritional reserves enable regenerative colonies to sprout from the stolons the following spring. This happens, for example, in *Crisia* (Harmer, 1891), *Bugula* (Numakunai, 1960; Eggleston, 1972) and *Zoobotryon* (see below). In many species special hibernacula are formed which, when the colony breaks up, propagate and disperse the parent colony. This is particularly characteristic of *Zoobotryon verticillatum* (Della Chiaje) (Zirpolo, 1924a), *Victorella pavida* Kent (Braem, 1951) and *Aetea sica* (Couch) (Simma-Krieg, 1969) among marine species and of *Paludicella articulata* Ehrenberg in fresh water (Harmer, 1913). Freshwater phylactolaemates, of course, produce statoblasts for the same purposes. It is interesting to note that when regeneration commences from stolons or rhizoids, neanic (or ancestruloid) zooids are formed first (Borg, 1926, for *Crisia*; Marcus, 1938, for *Bugula*).

Regeneration is not confined to overwintering propagules, however, but occurs readily in damaged colonies. Thus, the production of new zooids from pieces of colony has been described in *Zoobotryon verticillatum* (Zirpolo, 1924b), *Electra pilosa* and *Membranipora membranacea* (Bronstein, 1938, 1939), *Bugula neritina* (Abeloos, 1951) and *Eurystomella foraminigera* (Hincks) (Gordon, 1968). The experiments, though relatively few, seem to establish certain principles in relation to both age gradient and polarity.

The greatest body of work was that of Zirpolo (1924b) on the ctenostome *Zoobotryon verticillatum*. This species forms large, branching clumps of up to 0·5 m or more in length. The stolons, about 2 mm in width, form obvious internodes, each bearing the seried autozooids along their length and giving rise to two or more stolons of the subsequent order at their distal end. The first order stolon is attached to the substratum by rhizoids. Zirpolo cut segments of various length out of the first to fifth order stolons. Although the length of the piece had important consequences on the regenerative ability, it was the order from which the segment was cut that most influenced the result. After cutting, the segment at first went flaccid, as might be expected, but by the following day the cuts had sealed and turgour had been regained. In first order internodes regeneration occurred in pieces of middle length (18 mm), with a stolon emerging from one end after about two weeks; the shorter pieces died quickly, and the 27 mm segment died after a period of initial recovery. Second order branches regenerated more quickly, but again produced a single stolon from only one end. Third and fourth order branches regenerated much more readily, producing stolons from each end of the piece. One of the new stolons became an attachment rhizoid, so that a new, independent colony became established. Regenerative ability was most pronounced in fifth order branches, pieces of which often produced two or three stolons from each end.

The experiments of Bronstein (1939) and Gordon (1968) also show that the speed of wound repair and of new zooid formation is proportional to the youth of the damaged area: pieces of *Electra pilosa*, *Membranipora membranacea* and *Eurystomella foraminigera* from near the growing edge regenerated rapidly, older pieces regenerated slowly or not at all. Thus a square of *Eurystomella* cut from near the colony margin on 24 November (spring in the southern hemisphere, which is possibly important) had grown 58 new zooids by 1 February, whereas three similarly shaped but slightly larger pieces cut to include the ancestrula had produced only 29 zooids between them. Bronstein related the variations in regenerative ability to a physiological gradient within the colony. The detectable manifestation of this gradient was a change in refractive index of coelomic fluid from 1·3470 in young zooids to 1·3403 in older zooids (refractive index of sea water = 1·3400). Since the refractive index could be reduced to 1·3400 by treatment with trichloracetic acid, Bronstein assumed that the gradient reflected a concentration gradient of colloidal protein.

Zirpolo (1924b), unfortunately, paid rather little attention to the polarity of his experimental pieces, but the existence of polarity in another species was later demonstrated by Bronstein (1938). Small squares of *Membranipora membranacea* were cut out in such a way that no intact zooids (i.e., with feeding polypides) remained. The initial phase of regeneration was restricted to the distal side of the square. The area of regenerated growth increased in sigmoid manner to

12. Bryozoan Coloniality—Structure and Physiology

reach an asymptote after 50–100 h, by which time the food reserves in the square had presumably been exhausted. Somewhat similar results were obtained by Gordon (1968) with *Eurystomella foraminigera*, although he used a larger square which certainly contained several feeding polypides. After about six weeks 18 zooids were forming distally, five proximally, and five and seven on the two sides. These results are clearly explicable in terms of the polarity of the special cells in the pores of the transverse walls (*vide supra*): all available reserves and subsequently obtained food products would be transported to the distal margin of the square.

The orientation of a zooid budded from a cut or broken zooid, however, is orthogonal to the damaged edge, whether this accords with the polarity of the piece or not.

AGE-CORRELATED ZONES: SOMATIC AND REPRODUCTIVE CYCLES

A second manifestation of the ageing phenomenon is the existence across the colony of bands or zones of zooids in the same ontogenetic state, for example, as the zooids reach maturity, reproduce and senesce. Stach (1938) described one clear example in *Carbasea indivisa* Busk. In a well-grown colony he recognized: (1) the actively budding marginal zone; (2) a zone in which the zooids contained differentiating first polypides; (3) a zone of reproducing zooids with functional polypides, plus one of the following: spermatozoa, oocytes, ova or embryos in exetrnal ovisacs (the polypide not degenerating during embryogenesis); (4) a zone of post-reproductive zooids with brown bodies and regenerating polypides; (5) the oldest, polypide-less zooids, from which originate the attachment rhizoids.

Corrêa (1948) described a similar, but more complex, zonation in the fronds of *Bugula flabellata*, thus: (1) the budding zone; (2) the zone of polypide development; (3) the zone of oocyte enlargement and of ovicell formation; (4) the zone of small (second generation) oocytes and of preplacental embryos in ovicells; (5A) the zone of placental embryos and (on my interpretation of his account) degenerating polypides; and (6A) the zone of late embryos and regenerating polypides. Once the larva has been released and the new polypide is complete, a second ovum can be transferred to the ovicell (Silén, 1944a) and spermatozoa can presumably be released through the tentacles (Silén, 1972). The last three zones are then repeated (4B, 5B, 6B) for the second cycle of embryogenesis, and then repeated yet again (4C, 5C, 6C) in the third cycle. Corrêa found in the Bay of Santos that zooids produced, on average, three larvae in succession, and these events are diagrammatically summarized in Fig. 1B. Eggleston (1972) reported a similar zonation in *Bicellariella ciliata* (L.).

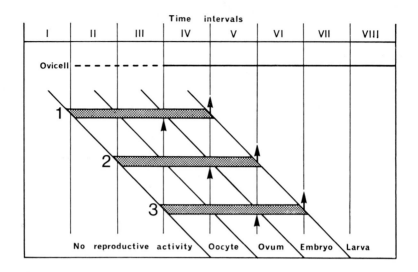

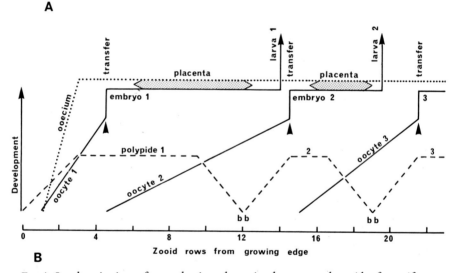

FIG. 1. Synchronization of reproductive phases in the maternal zooid of ooeciferous bryozoans. **A.** Three oogenetic and embryogenetic cycles in a non-placental cheilostome such as *Callopora dumerilii*: during successive time intervals the oocyte enlarges in the ovary; the ovum matures in the coelom; and, after transfer (first arrowhead), the embryo develops in the ovicell until the larva is released (second arrowhead). Based on the observations of Silén (1944a). **B.** Three oogenetic and two embryogenetic cycles in a placental cheilostome such as *Bugula flabellata*: during oocyte development no distinction has been made between enlargement in the ovary and maturation in the coelom; after transfer (first arrowhead), the embryo enlarges and develops in the ovicell; the stippled bar indicates the approximate duration of placentation. The polypide regresses during embryogenesis and it has been assumed that the replacement will be fully developed in time for the next transfer. (Note that the abscissa is not a time scale.) Based on the observations of Corrêa (1948). Abbreviation, bb, brown body.

Shortly before Corrêa's study, Silén in Sweden had found that individual zooids of *Callopora dumerilii* may also produce three or four successive larvae. As in *Bugula flabellata*, oocyte enlargement coincides with the formation of the ovicell, which led Silén (1944a) to postulate that the ovary secretes a stimulatory hormone. The reproductive sequence he described in *C. dumerilii* is as follows (Fig. 1A): (1) the first oocyte enlarges in the ovary while the ovicell is simultaneously developing: (2) the first oocyte is enlarging and maturing in the coelom (*Callopora* is non-placental and the ovum is large) while the ovicell is completing its development; the second oocyte is enlarging in the ovary; (3) the first ovum (presumably fertilized) has been transferred to the ovicell; the second oocyte has been ovulated into the coelom; the third oocyte is enlarging in the ovary; (4) the first larva has been released; the second ovum has been fertilized and transferred to the ovicell; the third oocyte has been ovulated. There is no indication that the reproductive and polypide renewal cycles are co-ordinated; indeed, Silén noted that if the polypides degenerated through adverse conditions, ova in the coelom disintegrated.

From a comparison of the succession of events in *Bugula* and *Callopora*, it is tempting to speculate that there is some physiological relationship between the incidence of vitellogenesis and of coincident polypide degeneration in the bearing zooid. The demonstration by Bobin (1971) that polypide degeneration in *Bowerbankia* results in a burst of released lipid is possibly highly significant in this context.

While there remains much to learn about the biology of the reproductive cycle, the points emphasized here are: (1) the high degree of co-ordination in *Bugula* between the three zooids involved in brooding, viz. the maternal zooid, the distal zooid and the ooecium. The inner vesicle of the ovicell, which forms the placenta (or embryophore, Silén, 1944c; Woollacott and Zimmer, 1975), develops from the maternal zooid inside the ooecium which has no direct connection with that zooid; (2) the co-ordinated timing of events during oogenesis and embryogenesis; (3) the co-ordination of the reproductive and polypide-renewal cycles in *Bugula*, with the inference that polypide degeneration may play a significant role in facilitating embryonic vitellogenesis through the placenta; and (4) that the timing of given events in the life of zooids of a particular age seems to be the same across the whole colony (or at least within any one frond of it). The wholly unresolved question is whether this similarity in timing is caused by regulation within the colony (i.e. the influence of zooids upon their neighbours) or by strictly identical ageing rates in zooids of any given age-group. Future work may help to answer this question.

DISCUSSION AND CONCLUSIONS

I have previously expressed the view that the three bryozoan classes were well separated by the Ordovician at least and that they have survived until the present day as independently evolving lines (Ryland, 1970). (A rather different scheme has recently been proposed by Jebram, 1973.) It may be necessary to assume that the most primitive bryozoans comprised "colonies" of non-communicating zooids, although such, on Beklemishev's criteria (see (3) on p. 212), would not be a colony as here understood but an aggregation of clonally related individuals.

The first class to achieve success, as indicated by fossil remains, was the Stenolaemata, probably represented by three orders: Cystoporata, Trepostomata and Cryptostomata. As can be seen from the initial budding pattern of tubular zooids (e.g., Fig. 17 in Ryland, 1970), a true colony is formed rather simply if the interzooidal septa remain a little distant from the terminal membrane. The zooids then remain organically joined through the resulting "hypostegal coelom", although it is difficult to envisage this confluence promoting a very efficient distribution of metabolites. Hypostegal coeloms were present in all three stenolaemate orders, although cystoporates apparently possessed interzooidal pores as well. Early representatives of the surviving order Cyclostomata apparently had no hypostegal coelom (although some existing divisions do) and must soon have developed interzooidal pores (cf. above). Obviously the characteristic gonozooid of cyclostomes could have evolved only in a true colony. The large number of embryos that develops in a gonozooid is indicative of an abundance of food, but Borg's (1926) conclusion that the nutritive tissue therein is derived from disintegrating follicle cells seems merely to beg the question of the original source. How are metabolites moved in quantity through a system dependent solely on open interzooidal pores?

Phylactolaemates, as evidenced by extant species, evolved a different pattern of budding, whereby the daughter polypide forms essentially within the parent zooid. Septa may or may not be laid down later, but the whole of phylactolaemate evolution reflects the lack of rigid compartmentalization.

Although Tavener-Smith (1969) has shown how the separated, chamber-like zooids of fenestellids may have been budded in a manner dependent upon a hypostegal coelom, the method of chamber replication characterizing gymnolaemates essentially demands the presence of interzooidal pores.

The problem of budding daughter zooids has thus been solved by bryozoans in basically three different ways: (1) the stenolaemate method, by which longitudinal septa are laid down behind, and spaced apart from, the terminal membrane (with interzooidal pores being a secondary development); (2) the phylactolaemate method, by which the daughter polypide forms inside the coelom of the mother zooid; (3) the gymnolaemate method, by which a chamber is separated from the mother zooid by a perforated, transverse septum, prior to the formation of the daughter polypide.

The form of the colony reflects to a considerable extent the differences in budding method: the common feature is simply the organic connection of zooids. It seems clear, however, from the geological history of the classes, that the three colony types have provided decidedly disparate evolutionary potential. The Stenolaemata have been enormously successful over a vast period of geological time, yet one by one the constituent orders have dwindled to extinction or insignificance. In all this time the form of the colony evolved, but that of the zooid did not. The Phylactolaemata, we may surmise, have not been notably successful; their colonies have become better integrated but scarcely more diversified, and the zooids (or polypides) have become larger but otherwise little changed. The Gymnolaemata, it is believed, had been present for many millions of years before they achieved real success, which followed quickly upon the origin of the Cheilostomata. Among bryozoans, only in cheilostomes can we see the logical fulfilment of the colonial habit: namely the evolution of both zooids and colonies on a significant scale. Why?

It seems likely that bryozoan adaptive radiation—one index of success—has depended largely upon a few basic attributes. Possibly the most important are: (1) the degree of zooidal compartmentalization and (2) the system of interzooidal communication; but also of likely importance are (3) zooid shape and (4) the ability to calcify. The last two will be discussed first.

1. Zooid shape

The primitive bryozoan shape was undoubtedly cylindrical, as evidenced by the simplest members of all three classes. Possibly this shape offers little evolutionary potential. Zooid form in stenolaemates, from the early Palaeozoic to the present, has varied rather little: the walls may be thick or thin and the chamber long or short, but a tubular zooid remains a tubular zooid, and polymorphs are few and not radically different from autozooids. In gymnolaemates, whether significantly or not, the almost explosive evolution of zooid form which characterizes the Cheilostomata was preceded by a change from cylindrical to squat shape.

2. Calcification

Neither of the major taxa (Phylactolaemata and Ctenostomata) with non-calcareous walls has been particularly successful, though both have certainly been long-lasting. The ability to mineralize the walls thus seems to be essential to evolutionary success.

3. Zooidal compartmentalization

Colonial evolution in the Phylactolaemata has been in the direction of morphological integration achieved by a reduction or even elimination of interzooidal walls. The degree of compartmentalization has been progressively diminished in parallel with the evolution of a communal coelom. Thus while the colony has become more individualistic (culminating with *Cristatella*), the only direction of zooidal evolution has been towards larger and more elaborate lophophores. With the elimination of the individual zooid, polymorphism becomes impossible.

Stenolamaetes have preserved the individuality of their zooids, so that their non-diversification must be related either to a restrictive shape (see above) or to the inefficient distribution of metabolites (see below).

Spectacular zooidal evolution and diversification into polymorphs have occurred only in cheilostomes, and the maintenance of zooidal integrity has been of the utmost importance in these processes. The advantages of compartmentalization would, however, be nullified by the absence of an efficient system for the transport of metabolites.

4. Interzooidal communication

Free interchange of coelomic fluid has been achieved by the breakdown of septa in phylactolaemates, and is similarly dependent on spatial continuity (at the surface or through open pores) in stenolaemates. Distribution of metabolites is seemingly by diffusion and by pressure changes associated with polypidial protraction/retraction and presumably rather inefficient, although this can hardly be true for the nutrient supply to the cyclostome gonozooid.

A funiculus is present in the zooid in all three classes. Primitively it seems to have linked the stomach caecum with the wall in the proximal part of the zooid, though for what reason is unclear. Certainly, in all classes, it supports the testis (Figs. 3, 16 and 21 in Ryland, 1970), perhaps channelling nutrient to it. In phylactolaemates the funiculus serves also to support the developing statoblasts, and the role of nutrient transfer from the caecum would again appear reasonable. Only in the Gymnolaemata has the funiculus become a colonial network rather than a zooidal strand. The evidence from Bobin's work, discussed on p. 225, suggests very strongly that the colonial funiculus provides the pathway for rapid colony-wide transport of metabolites in this class. The presence of such a system has provided the necessary *combination* of individualistic zooids and efficient interzooidal transport of metabolites, which (together, perhaps, with the ability to calcify) underlies the successful evolution and diversification of both zooid structure and colony form which eventuated only in the Cheilostomata.

Some of the trends towards greater individuality of the colony, in all three classes, have been noted by Beklemishev (1969). Clear peaks in this respect are the colonial brood-chambers of stenolaemates *Disporella* and *Lichenopora*, as described by Borg (1926); the hypertropheid epidermal cells and musculature that differentiate the colonial "foot" of *Cristatella mucedo*; and the colony-wide calcification found on the basal surface of the free-living cheilostome *Cupuladria biporosa* Canu and Bassler, as described by Håkansson (1973) and in the proximal, supporting part of *Tessaradoma boreale* (Busk) and in many other erect cheilostomes (Cheetham and Cook, in press). Moreover, propagation of colonies by lobulation is found in advanced phylactolaemates, such as *Lophopus cristallinus* (Pallas) and *Cristatella mucedo*, perhaps to improve feeding efficiency (Bishop and Bahr, 1973); however, propagation by fragmentation occurs in *Fredericella* (Bushnell, 1973). Lobulation also occurs in a relative of *Cupuladria*, *Discoporella umbellata* (Defrance), according to Marcus and Marcus (1962), and colony multiplication is found in ctenostomes such as *Monobryozoon* (Remane, 1936) and *Zoobotryon* (p. 227).

Recognition of "self" and "non-self" is hardly an attribute of individualistic colonies, since it is equally the property of individuals in a clone (as in anemones such as *Anthopleura elegantissima*: Francis, 1973a, b).

The clearance rate of suspension feeders is related to the size of the organism, tending to be high in small animals and low in larger ones. Thus Bullivant (1968) quoted the following rates in ml h^{-1} (mg dry wt)$^{-1}$: sponge *Halichondria panicea*, 6·7; tubeworm *Pomatoceros triqueter*, 7·6; polychaete *Chaetopterus variopedatus*, 0·4; oyster *Ostrea edulis* (adult), 6·6; mussel *Mytilus edulis*, 6·7; brine shrimp *Artemia salina*, 16; ascidian *Phallusia mamillata*, 0·2–1·7; rotifers *Brachionus* spp., 320–760; *Ostrea edulis* veliger, 56; cladoceran *Daphnia pulex*, 181; and copepod *Acartia clausi*, 0·02–37. For *Zoobotryon verticillatum*, in which colonies form tufts 0·5 m or more in length but the zooids measure but 0·5–1·0 × 0·25 mm, Bullivant himself found a clearance rate of 33·7. This is a high value, suggesting that it is the zooid, rather than the colony, that constitutes the metabolic unit. The contrary conclusion, however, seems at first sight to be implied by the experiments of Bishop and Bahr (1973) on *Lophopodella carteri* (Hyatt). These authors found that clearance rate per zooid, y, expressed in

ml min^{-1} was inversely proportional to colony size, x, expressed as number of zooids thus:
$$y = 25 + 181\ x^{-1}.$$

The explanation given by Bishop and Bahr, though, is not inconsistent with Bullivant's (1968) conclusion, since they suggest that the inverse proportionality is a result of the close proximity of lophophores, causing muutal interference with feeding currents and hence a reduction in the volume filtered by each one.

Thus, despite their undoubted sophistication in respect to polymorphism and metabolite transport, which reaches its apogee in the Cheilostomata, individuality in bryozoans remains largely with the replicated unit (zooid or cormidium). Zooids may certainly display co-ordinated activities, though the extent to which nervous control is general in Bryozoa remains unclear. It is equally uncertain whether physiological ageing phenomena are regulated at colony level. While colony form has undergone great evolutionary change in some taxa, the level of integration at which the colony essentially becomes an individual has never been reached and rarely approached.

This indubitably is a consequence of a sessile mode of life, with the colony permanently attached to a firm substratum. The most individualistic colonies are those of soft substrata (e.g. *Kinetoskias, Monobryozoon*), including the little deep-sea (500 m and more) ctenostomes such as *Metalcyonidium gautieri* and *Pseudalcyonidium bobinae* recently described by d'Hondt (1976), and those of partially motile habit (e.g., *Cristatella, Cupuladria* and *Discoporella*). Continuance of the sessile habit explains why, despite having zooidal and cormidial organizations of a complexity equalling that of advanced cnidarians, no bryozoan has achieved a finite colony form matching the supreme individuality of *Porpita, Velella* and some siphonophores.

REFERENCES

ABELOOS, M. (1951). Morphogénèse des colonies du bryozoaire *Bugula neritina* L. *C.r. hebd. Séanc. Acad. Sci., Paris* **232**, 654–656.

BALAVOINE, P. (1956). Sur deux bryozoaires de la région nord de Saint-Malo. *Bull. Lab. mar. Dinard*, **42**, 35–40.

BANTA, W. C. (1969). The body wall of cheilostome Bryozoa. II. Interzooidal communication organs. *J. Morph.* **129**, 149–170.

BANTA, W. C. (1973). Evolution of avicularia in cheilostome Bryozoa. *In* "Animal colonies: development and function through time" (R. S. Boardman, A. H. Cheetham and W. A. Oliver, eds). Dowden, Hutchinson and Ross, Stroudsburg.

BANTA, W. C. (1976). Origin and early evolution of cheilostome Bryozoa *In* "Bryozoa 1974—Proceedings of the third congress" (S. Pouyet, ed.). *Docums Lab. Géol. Fac. Sci. Lyon*, H.S. **3**, 565–582.

Banta, W. C. (1977). Body wall morphology of the sertellid cheilostome bryozoan, *Reteporellina evelinae. Am. Zool.* **17**, 75–91.
Banta, W. C., McKinney, F. K. and Zimmer, R. L. (1974). Bryozoan monticules: excurrent water outlets? *Science, N.Y.* **185**, 783–784.
Beklemishev, W. N. (1969) "Principles of comparative anatomy of invertebrates: 1. Promorphology" (translated from the 3rd Russian edition by J. M. MacLennan; ed. Z. Kabata). Oliver and Boyd, Edinburgh.
Bishop, J. W. and Bahr, L. M. (1973). Effects of colony size on feeding by *Lophopodella carteri* (Hyatt). In "Animal colonies: development and function through time" (R. S. Boardman, A. H. Cheetham and W. A. Oliver, eds). Dowden, Hutchinson and Ross, Stroudsburg
Boardman, R. and Cheetham, A. H. (1969). Skeletal growth, intracolony variation, and evolution in Bryozoa: a review. *J. Paleont.* **43**, 205–233.
Boardman, R. S. and Cheetham, A. H. (1973). Degrees of colony dominance in stenolaemate and gymnolaemate Bryozoa. In "Animal colonies: development and function through time" (R. S. Boardman, A. H. Cheetham and W. A. Oliver, eds). Dowden, Hutchinson and Ross, Stroudsburg.
Boardman, R. S., Cheetham, A. H. and Cook, P. L. (1970). Intracolony variation and the genus concept in Bryozoa. Proceedings of the North American Paleontological Convention, Part C, pp. 294–320.
Bobin, G. (1958a). Structure et genèse des diaphragmes autozoéciaux chez *Bowerbankia imbricata* (Adams). *Archs Zool. exp. gén.* **96**, 53–100.
Bobin, G. (1958b). Histologie des bourgeons autozoéciaux et genèse de leurs diaphragmes chez *Vesicularia spinosa* (Linné) (bryozoaire cténostome). *Bull. Soc. zool. Fr.* **83**, 132–144.
Bobin, G. (1962). Histogenèse des diaphragmes septaux stoloniaux et valeur des rosettes chez les vesicularines (bryozoaires, cténostomes). *Archs Zool. exp. gén.* **101**, Notes et revue 1, 14–42.
Bobin, G. (1964). Cytologie des rosettes de *Bowerbankia imbricata* (Adams). *Archs Zool. exp. gén.* **104**, 1–44.
Bobin, G. (1968). Morphogenèse du termen et des épines dans les zoécies d'*Electra verticillata* (Ellis et Solander) (bryozoaire chilostome, Anasca). *Cah. Biol. mar.* **9**, 53–68.
Bobin, G. (1971). Histophysiologie du système rosettes-funicule de *Bowerbankia imbricata* (Adams) (bryozoaire cténostome). Les lipides. *Archs Zool. exp. gén.* **112**, 771–792.
Bobin, G. (1977). Interzooecial communications and the funicular systems. In "The Biology of Bryozoans" (R. M. Woollacott and R. L. Zimmer, eds). Academic Press, New York and London.
Bobin, G. and Prenant, M. (1954). Sur un bryozoaire perforant (*Terebripora comma* Soule), trouvé en Méditerranée. *Archs Zool. exp. gén.* **91**, Notes et revue 3: 130–144.
Borg, F. (1926). Studies on Recent cyclostomatous Bryozoa. *Zool. Bidr. Upps.* **10**, 181–507.
Braem, F. (1951). Über *Victorella* und einige ihrer nächsten Verwandten, sowie über die Bryozoenfauna des Ryck bei Greifswald. *Zoologica, Stuttg.* **102**, 1–59.
Brien, P. and Huysmans, G. (1938). La croissance et le bourgeonnement du stolon chez les Stolonifera. *Annls Soc. r. zool. Belge* **68**, 13–40.
Bronstein, G. (1937). Étude du systéme nerveux de quelques bryozoaires gymnolémides. *Trav. Stn biol. Roscoff* **15**, 155–174.

BRONSTEIN, G. (1938). Note sur la croissance résiduelle des fragments de zoarium chez un bryozoaire *Membranipora membranacea* (L.). *C.r. hebd. Séanc. Soc. biol.* **126**, 65–68.

BRONSTEIN, G. (1939). Sur les gradients physiologiques dans une colonie de bryozoaires. *C.r. hebd. Séanc. Acad. Sci., Paris* **209**, 602–603.

BROOD, K. (1973). Palaeozoic Cyclostomata (a preliminary report). In "Living and Fossil Bryozoa" (G. P. Larwood, ed.), pp. 247–256. Academic Press, London.

BULLIVANT, J. S. (1968). The rate of feeding of the bryozoan, *Zoobotryon verticillatum*. *N.Z. Jl mar. freshw. Res.* **2**, 111–134.

BUSHNELL, J. H. (1973). The freshwater Ectoprocta: a zoogeographical discussion. In "Living and Fossil Bryozoa" (G. P. Larwood, ed.), pp. 503–521. Academic Press, London.

CALVET, L. (1900). Contributions à l'histoire naturelle des Bryozoaires ectoproctes marins. *Trav. Inst. Zool. Univ. Montpellier* **8**, 1–488.

CARTER, H. J. (1882). Remarkable forms of *Cellepora* and *Palythoa* from the Senegambian coast. *Ann. Mag. nat. Hist.*, Ser. 5, **9**, 416–419.

CHEETHAM, A. H. (1954). A new early Cretaceous cheilostome bryozoan from Texas. *J. Paleont.* **28**, 177–184.

CHEETHAM, A. H. (1973). Study of cheilostome polymorphism using principal components analysis. In "Living and Fossil Bryozoa" (G. P. Parwood, ed), pp. 385–409. Academic Press, London.

CHEETHAM, A. H. (1976). Taxonomic significance of autozooid size and shape in some early multiserial cheilostomes from the Gulf Coast of the U.S.A. In "Bryozoa 1974—Proceedings of the third congress" (S. Pouyet, ed.). *Docums Lab. Géol. Fac. Sci. Lyon.*, H.S. **3**, 547–564.

CHEETHAM, A. H. and COOK, P. L. (1979). General features of the class Gymnolaemata. *Treatise on invertebrate paleontology*. Bryozoa (revised). In press.

COOK, P. L. (1964). Polyzoa from west Africa. Notes on the genera *Hippoporina* Neviani, *Hippoporella* Canu, *Cleidochasma* Harmer and *Hippoporidra* Canu & Bassler (Cheilostomata, Ascophora). *Bull. Br. Mus. nat. Hist. (Zool.)* **12**, 1–35.

COOK, P. L. (1967). Polyzoa (Bryozoa) from west Africa: the Pseudostega, the Cribrimorpha and some Ascophora Imperfecta. *Bull. Br. Mus. nat. Hist. (Zool.)* **15**, 321–351.

COOK, P. L. (1968a). Polyzoa from west Africa: the Malacostega, part 1. *Bull. Br. Mus. nat. Hist. (Zool.)* **16**, 113–160.

COOK, P. L. (1968b). Observations on living Bryozoa. *Atti Soc. ital. Sci. nat.* **108**, 155–160.

COOK, P. L. (1968c). Bryozoa (Polyzoa) from the coasts of tropical West Africa. *Atlantide Rep.* **10**, 115–262.

COOK, P. L. (1977). Colony-wide water currents in living Bryozoa. *Cah. Biol. mar.* **18**, 31–47.

CORRÊA, D. D. (1948). A embryologia de *Bugula flabellata* (J. V. Thompson) (Bryozoa Ectoprocta). *Bolm Fac. Filos. Ciênc. S. Paulo.*, Zool. **13**, 7–71.

DARWIN, C. R. (1839). "Journal of researches into the natural history and geology of the countries visited during the voyage of H.M.S. 'Beagle' round the world, under the command of Capt. Fitzroy, R.N." Murray, London.

D'HONDT, J.-L. (1976). Bryozoaires cténostomes bathyaux et abyssaux de l'Atlantique Nord. In "Bryozoa 1974—Proceedings of the third congress" (S. Pouyet, ed.). *Docums Lab. Géol. Fac. Sci. Lyon*, H.S. **3**, 311–333.

Eggleston, D. (1972). Patterns of reproduction in marine Ectoprocta of the Isle of Man. *J. nat. Hist.* **6**, 31–38.
Francis, L. (1973a). Clone specific segregation in the sea anemone *Anthopleura elegantissima*. *Biol. Bull. mar. Biol. Lab., Woods Hole* **144**, 64–72.
Francis, L. (1973b). Intraspecific aggression and its effect on the distribution of *Anthopleura elegantissima* and some related sea anemones. *Biol. Bull. mar. Biol. Lab., Woods Hole* **144**, 73–92.
Franzén, Å. (1960). *Monobryozoon limicola* n. sp., a ctenostomatous bryozoan from the detritus layer on soft sediment. *Zool. Bidr. Upps.* **33**, 135–147.
Gerwerzhagen, A. (1913). Beiträge zur Kenntniss der Bryozoen. *Z. wiss. Zool.* **107**, 3–345.
Gordon, D. P. (1968a). Growth, regeneration and population biology of cheilostomatous polyzoans. Unpublished thesis, University of Auckland.
Gordon, D. P. (1968b). Zooidal dimorphism in the polyzoan *Hippopodinella adpressa* (Busk). *Nature, Lond.* **219**, 633–634.
Gordon, D. P. (1974). Microarchitecture and function of the lophophore of a marine bryozoan. *Mar. Biol.* **27**, 147–163.
Gordon, D. P. (1975). Ultrastructure of communication pore areas in two bryozoans. *In* "Bryozoa 1974—Proceedings of the third congress" (S. Pouyet, ed.). *Docums Lab. Géol. Fac. Sci. Lyon*, H.S. **3**, 187–192.
Gordon, D. P. (1977). The ageing process in bryozoans. *In* "The Biology of Bryozoans" (R. F. Woollacott and R. L. Zimmer, eds), pp. 335–374. Academic Press, New York.
Gordon, D. P. and Hastings, A. B. (1979). Interzooidal communications in *Hippothoa* (Bryozoa, Cheilostomata). *J. nat. Hist.* (In press.)
Håkansson, E. (1973). Mode of growth of the Cupuladriidae (Bryozoa, Cheilostomata) *In* "Living and Fossil Bryozoa" (G. P. Larwood, ed.) pp. 287–298. Academic Press, London.
Harmelin, J. G. (1973). Les bryozoaires des peuplements sciaphiles de Méditerranée: le genre *Crassimarginatella* Canu (chilostomes Anasca). *Cah. Biol. mar.* **14**, 471–492.
Harmelin, J. G. (1974). Les bryozoaires cyclostomes de Méditerranée: écologie et systémique. Thèse de l'Université d'Aix-Marseille.
Harmer, S. F. (1891). On the regeneration of lost parts in Polyzoa. *Rep. Br. Ass. Advmt Sci.* **60**, 862–833.
Harmer, S. F. (1893). On the occurrence of embryonic fission in cyclostomatous Polyzoa. *Q. Jl microsc. Sci.* **34**, 199–241.
Harmer, S. F. (1902). On the morphology of the Cheilostomata. *Q. Jl microsc. Sci.* **46**, 263–350.
Harmer, S. F. (1913). The Polyzoa of waterworks. *Proc. zool. Soc. Lond.* **1913**, 426–457.
Harmer, S. F. (1923). On cellularine and other Polyzoa. *J. Linn. Soc. (Zool.)* **35**, 293–361.
Harmer, S. F. (1930). Polyzoa. *Proc. Linn. Soc. Lond.* **141**, 68–118.
Harmer, S. F. (1957). The Polyzoa of the Siboga expedition, Part 4, Cheilostomata Ascophora II. *Siboga Exped.* **28**d, 641–1147.
Hastings, A. B. (1943). Polyzoa (Bryozoa), I. *'Discovery' Rep.* **22**, 301–510.
Hastings, A. B. and Ryland, J. S. (1968). The characters of the polyzoan genera *Pentapora* and *Hippodiplosia*, with redescriptions of *P. foliacea* (Ellis & Solander) and *H. verrucosa* Canu. *J. Linn. Soc. (Zool.)* **47**, 505–514.

HILLER, S. (1939). The so-called "colonial nervous system" in Bryozoa. *Nature, Lond.* **143**, 1069–1970.

HINCKS, T. (1880). "A History of the British Marine Polyzoa" vol. 1, 2. Van Voorst, London.

HUMPHRIES, E. (1979). Selected features of astogenetic growth in *Parasmittina nitida*. Proc. 4th Congress on Bryozoa. (In press.)

HYMAN, L. H. (1959). "The Invertebrates, 5: Smaller Coelomate Groups" McGraw-Hill, New York.

JEBRAM, D. (1968). A cultivation method for saltwater Bryozoa and an example for experimental biology. *Atti Soc. ital. Sci. nat.* **108**, 119–128.

JEBRAM, D. (1973). The importance of different growth directions in the Phylactolaemata and Gymnolaemata for reconstructing the phylogeny of the Bryozoa. *In* "Living and Fossil Bryozoa" (G. P. Larwood, ed.), pp. 565–576. Academic Press, London.

KARAKASHIAN, S. and MILKMAN, R. (1967). Colony fusion compatibility types in *Botryllus schlosseri*. *Biol. Bull. mar. biol. Lab. Woods Hole* **133**, 473.

KNIGHT-JONES, E. W. and MOYSE, J. (1961). Intraspecific competition in sedentary marine animals. *Symp. Soc. exp. Biol.* **15**, 72–95.

LARWOOD, G. P. (1962). The morphology and systematics of some Cretaceous cribrimorph Polyzoa (Pelmatoporinae). *Bull. Br. Mus. nat. Hist.* (Geol.) **6**, 1–285.

LEVINSEN, G. M. R. (1902). Studies on Bryozoa. *Vidensk. Meddr dansk naturh. Foren.* **1902**, 1–31.

LEVINSEN, G. M. R. (1907). Sur la régénération totale des bryozoaires. *Overs. k. danske Vidensk. Selsk. Forh.* **1907**, 151–159.

LEVINSEN, G. M. R. (1909). "Morphological and Systematic Studies on the Cheilostomatous Bryozoa" Nationale Forfatterers Forlag, Copenhagen.

LUTAUD, G. (1961). Contribution a l'étude du bourgeonnement et de la croissance des colonies chez *Membranipora membranacea* (Linné), bryozoaire chilostome. *Annls Soc. r. zool. Belg.* **91**, 157–300.

LUTAUD, G. (1969). Le "plexus" pariétal de Hiller et la coloration du système nerveux par le bleu de méthylène chez quelques bryozoaires chilostomes. *Z. Zellforsch.* **99**, 302–314.

LUTAUD, G. (1973). The great tentacle sheath nerve as the path of an innervation of the frontal wall structures in the cheilostome *Electra pilosa* (Linné). *In* "Living and Fossil Bryozoa" (G. P. Larwood, ed.), pp. 317–326. Academic Press, London.

LUTAUD, G. (1974). Le plexus pariétal des cténostomes chez *Bowerbankia gracilis* Leydi [*sic*] (Vesicularines). *Cah. Biol. mar.* **15**, 403–408.

LUTAUD, G. (1977). The nervous system in bryozoans. *In* "The Biology of Bryozoans" R. M. Woollacott and R. L. Zimmer, eds), pp. 329–337. Academic Press, New York.

MACKIE, G. O. (1963). Siphonophores, bud colonies and superorganisms. *In* "The Lower Metazoa: Comparative Biology and Physiology" (E. C. Dougherty, ed.). University of California Press, Berkeley and Los Angeles.

MARCUS, E. (1926a). Beobachtungen und Versuche an lebenden Meeresbryozoen. *Zool. Jb.* (*Syst.*) **52**, 1–102.

MARCUS, E. (1926b). Beobachtungen und Versuche an lebenden Susswasserbryozoen. *Zool. Jb.* (*Syst.*) **52**, 279–350.

MARCUS, E. (1938). Bryozoarios marinhos brasileiros, II. *Bolm Fac. Filos. Ciênc. S. Paulo*, Zool. **2**, 1–196.

Marcus, E. and Marcus, E. (1962). On some lunulitiform Bryozoa. *Bolm Fac. Filos. Ciênc. S. Paulo*, Zool. **24**, 281–324.

Matricon, I. (1963). Dégénérescence du polypide femelle et formation d'une poche incubatrice chez *Alcyonidium polyoum* (Hassall) (bryozoaire cténostome). *Archs Zool. exp. gén.* **102**, Notes et revue 2: 79–93.

Moyano, G., H.I. (1967). Sobre la fusion de dos colonias de *Membranipora hyadesi* Jullien, 1888. *Notic. mens. Mus. nac. Hist. nat. Santiago* **11** (126), 1–14.

Nielsen, C. (1970). On metamorphosis and ancestrula formation in cyclostomatous bryozoans. *Ophelia* **7**, 217–256.

Norman, A. M. (1903). Notes on the natural history of east Finmark. Polyzoa. *Ann. Mag. nat. Hist.*, Ser. 7, **11**, 567–598; **12**, 87–128.

Numakunai, T. (1960). An observation on the budding of the stolon of a bryozoan, *Bugula neritina* Linné. *Bull. biol. Stn Asamushi* **10**, 99–101.

Ott, J. A. (1972). *Monobryozoon bulbosum* n. sp., a new solitary interstitial bryozoon from the west Atlantic coast. *Cah. Biol. mar.* **13**, 421–428.

Pohowsky, R. A. (1973). A Jurassic cheilostome from England. In "Living and fossil Bryozoa" (G. P. Larwood, ed.), pp. 447–461. Academic Press, London.

Powell, N. A. (1967). Polyzoa (Bryozoa)—Ascophora—from north New Zealand. '*Discovery*' *Rep.* **34**, 199–394.

Powell, N. A. and Cook, P. L. (1967). Notes on *Tremogasterina* Canu and *Tremogasterina robusta* (Hincks) (Polyzoa, Ascophora). *Cah. Biol. mar.* **8**, 7–20.

Remane, A. (1936). *Monobryozoon ambulans* n.g. n.sp., ein eigenartiges Bryozoon des Meeressandes. *Zool. Anz.* **113**, 161–167.

Remane, A. (1938). Ergänzende Mitteilungen über *Monobryozoon ambulans* Remane. *Kieler Meeresforsch.* **2**, 356–358.

Ryland, J. S. (1970). "Bryozoans". Hutchinson University Library, London.

Ryland, J. S. (1976). Physiology and ecology of marine bryozoans. *Adv. mar. Biol.* **14**, 285–443.

Ryland, J. S. (1977). Taxes and tropisms of bryozoans. In "The Biology of Bryozoans" (R. M. Woollacott and R. L. Zimmer, eds), pp. 411–436. Academic Press, New York and London.

Ryland, J. S. and Hayward, P. J. (1977). A synopsis of the British anascan bryozoans. *Syn. Br. Fauna* (N.S.) **10**, 1–188.

Schopf, T. J. M. (1973). Ergonomics of polymorphism: its relation to the colony as the unit of natural selection in species of the phylum Ectoprocta. In "Animal Colonies: development and function through time" (R. S. Boardman, A. H. Cheetham and W. A. Oliver, eds). Dowden, Hutchinson and Ross, Stroudsburg.

Silén, L. (1938). Zur Kenntnis des Polymorphismus der Bryozoen. Die Avicularien der Cheilostomata Anasca. *Zool. Bidr. Upps.* **17**, 149–366.

Silén, L. (1942). Origin and development of the cheilo-ctenostomatous stem of the Bryozoa. *Zool. Bidr. Upps.* **22**, 1–59.

Silén, L. (1944a). The main features of the development of the ovum, embryo and ooecium in the ooeciferous Bryozoa Gymnolaemata. *Ark. Zool.* **35a**, No. 17: 1–34.

Silén, L. (1944b). On the formation of the interzoidal communications of the Bryozoa *Zool. Bidr. Upps.* **22**, 433–488.

SILÉN, L. (1944c). On the anatomy of *Labiostomella gisleni* (Bryozoa Protocheilostomata), with special regard to the embryo chambers of the different groups of Bryozoa and to the origin and development of the bryozoan zoarium. *K. svenska VetenskAkad. Handl.* Ser. 3, **21**, No. 6, 1–111.

SILÉN, L. (1947). On the spines of *Flustrella* (Bryozoa). *Zool. Bidr. Upps.* **25**, 134–140.

SILÉN, L. (1950). On the mobility of entire zoids in Bryozoa. *Acta zool., Stockh.* **31**, 349–386.

SILÉN, L. (1972). Fertilization in the Bryozoa. *Ophelia* **10**, 27–34.

SILÉN, L. (1977). Polymorphism in marine bryozoans. *In* "The Biology of Bryozoans" (R. M. Woollacott and R. L. Zimmer, eds), pp. 184–227. Academic Press, New York and London.

SILÉN, L. and HARMELIN, J.-G. (1974). Observations on living Diastoporidae (Bryozoa Cyclostomata), with special regard to polymorphism. *Acta zool. Stockh.* **55**, 81–96.

SIMMA-KRIEG, B. (1969). On the variation and special reproduction habits of *Aetea sica* (Couch). *Cah. Biol. mar.* **10**, 129–137.

SMITT, F. A. (1868). Kritisk förteckning öfver Skandinaviens HafsBryozoer, III. *Öfvers K. VetenskAkad Förh.* **24**, 279–429.

STACH, L. W. (1938). Observations on *Carbasea indivisa* Busk (Bryozoa). *Proc. zool. Soc. Lond.* **108**, 389–399.

STEBBING, A. R. D. (1973). Observations on colony overgrowth and spatial competition. *In* "Living and Fossil Bryozoa" (G. P. Larwood, ed.), pp. 173–183. Academic Press, London.

TAVENER-SMITH, R. (1969). Skeletal structure and growth in the Fenestellidae (Bryozoa) *Palaeontology* **12**, 281–309.

THEODOR, J. L. (1976). Histo-incompatibility in a natural population of gorgonians. *Zool. J. Linn. Soc.* **58**, 173–176.

THORPE, J. P., SHELTON, G. A. B. and LAVERACK, M. S. (1975). Electrophysiology and co-ordinated responses in the colonial bryozoan *Membranipora membranacea* (L.). *J. exp. Biol.* **62**, 389–404.

WOOLLACOTT, R. F. and ZIMMER, R. L. (1972). Origin and structure of the brood chamber in *Bugula neritina* (Bryozoa). *Mar. Biol.* **16**, 165–170.

WOOLLACOTT, R. M. and ZIMMER, R. L. (1975). A simplified placenta-like system for the transport of extraembryonic nutrients during embryogenesis of *Bugula neritina* (Bryozoa). *J. Morph.* **147**, 355–378.

ZIRPOLO, G. (1924a). Sulla genesi delle colonie primaverili del *Zoobotryon pellucidum* Ehrbg. *Bol. Soc. Nat. Napoli* **35**, 113–128.

ZIRPOLO, G. (1924b). Le restituzione dei rami coloniali del *Zoobotryon pellucidum* Ehrbg. *Pubbl. Staz. zool. Napoli* **5**, 97–135.

13 | Coloniality in Vermetidae (Gastropoda)

ROGER N. HUGHES

Department of Zoology, University College of North Wales, Bangor, North Wales

Abstract: True coloniality is confined to those vermetids which form sheet-like aggregations of contiguous individuals, the rest form loose aggregations which cannot be regarded as true colonies. The adaptive significance and possible origins of true coloniality are explored by comparing the biology of three loosely aggregating species from South Africa and the Indo-Pacific with that of a colony-forming species from South Africa. The formation of loose aggregations seems to have arisen as a mechanism promoting successful fertilization by water-borne sperm, perhaps reinforced by a tendency of juveniles to use the presence of adults as an indication of suitable micro-habitats for settlement. This basic vermetid gregariousness has been developed further in certain species, notably within the genus *Dendropoma*, to form extensive sheet or reef-life colonies which allow the monopoly of space in stable, homogeneous habitats.

INTRODUCTION

Vermetids are completely sedentary prosobranch gastropods with loosely coiled or vermiform shells cemented to the substratum. Only the newly hatched young are motile and these have coiled snail-like shells similar to those of most other juvenile prosobranchs.

It is debatable whether vermetids can justifiably be regarded as colonial organisms. All vermetids form aggregations but only in certain species do the aggregations form entities which are more than the sum of their parts, an attribute which is perhaps the essence of coloniality. Rather than restricting discussion to the relatively few truly colonial vermetids, the general advantages and disadvantages of gregariousness in vermetids will be reviewed in the hope of

---

Systematics Association Special Volume No. 11. "Biology and Systematics of Colonial Organisms", edited by G. Larwood and B. R. Rosen, 1979, pp. 243–253. Academic Press, London and New York.

explaining why aggregations are formed at all and why in some species the aggregations become true colonies.

THE PENALTIES OF IMMOBILITY

Because post-larval vermetids are permanently fixed to the substratum they are subject to certain restrictions and dangers. Firstly, food supply is limited to particulate matter brought to the vermetids by environmental water currents. Although some, or perhaps all, vermetids are able to filter suspended food particles out of the inhalant current (Yonge, 1932) their main source of food particles is those trapped on a mucous net secreted by the large pedal gland and spun out into the water by the grooved, highly developed pedal tentacles (Fig. 1).

Secondly, the fixed adults are unable to copulate so that sperm must be shed into the sea water before being taken up by the inhalant current of females. Individuals must therefore settle sufficiently close to one another for fertilization to be successful. Hadfield (1969) reported the production of spermatophores by the densely aggregated *Petaloconchus montereyensis* (Dall) but no published information is available on the mode of transmission of spermatophores to the females or on spermatophore production in other species of vermetid.

Thirdly, the fixed post-larval stages are unable to move away should the local environmental conditions deteriorate. However, the orientation of the shell mouth can be altered by eroding the old shell with the radula and laying down new shell so that it grows in a different direction. This was observed, for example, by Hadfield *et al.* (1971) with *Vermetus allii* (Hadfield and Kay) kept in the labroatory. Alterations of shell orientation may happen in response to changes of current direction or to silting. Such alterations occur quite frequently and the robust central and lateral teeth on the radula may be designed specially for eroding the shell, leaving the more delicate, hooked marginals to the task of raking in the mucuous net while feeding (Hughes and Lewis, 1974).

Fourthly, the fixed post-larval stages are in danger of being smothered by other sedentary organisms competing for space on the substratum. Although a small area outside the mouth of the shell may be kept clear of fouling organisms by rasping with the radula, vermetids cannot extend far out of the shell to repel intruders.

CASE HISTORIES

1. Species Forming Loose Aggregations

All vermetid genera contain species which form loose aggregations, i.e. only a proportion of the individuals are contiguous; and this habit is probably wide-

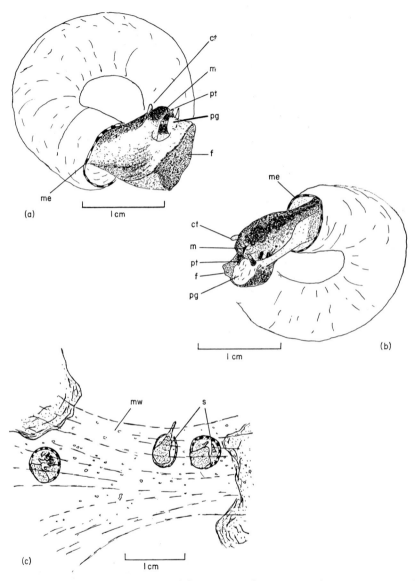

Fig. 1a. *Serpulorbis natalensis* positioned for secreting the mucous web. Mucus is secreted by the pedal gland and spun out by the pedal tentacles. ct—cephalic tentacle, f—foot, m—mouth, me—mantle edge, pg—pedal gland, pt—pedal tentacle.

Fig. 1b. *S. natalensis* positioned for hauling in the mucous web. The mouth is brought between the pedal tentacles and the mucous web is grasped by the jaws and radula. Labels as for Figure 1a.

Fig. 1c. Communal mucous web overlying three specimens of *S. natalensis*. mw—mucous web, s—*S. natalensis*.

spread within most genera. Examples are the South African *Serpulorbis* spp. studied by Hughes (1978a, b) and the Indo-Pacific *Dendropoma maximum* (Sowerby) (Hughes and Lewis, 1974).

FIG. 2. A loose colony of *Serpulorbis natalensis* on the underside of a stone. The stone is about 30 cm wide.

Serpulorbis natalensis (Mörch) is quite large, 8 mm wide, 12 cm long, and forms loose aggregations on the sides of boulders or under stones (Fig. 2) in sheltered waters which may carry considerable loads of detritus (Hughes, 1978a). *S. aureus* Hughes replaces *S. natalensis* in the warmer, more northerly waters of the eastern seabord (Hughes, 1978b). The type locality is a sheltered estuary where silting is heavy. In this habitat individual *S. aureus* form concentric whorls on the upper surfaces of boulders. A vertical "feeding-tube" raises the shell mouth away from accumulating silt. As with *S. natalensis*, the individuals are close to each other but frequently not contiguous (Fig. 3), and the aggregations never assume any well defined structure.

2. *Species Forming Colonies*

Certain vermetids form dense sheet-like aggregations (Fig. 4) on rock surfaces exposed to severe wave impact. The multitudes of small contiguous shells in such aggregations form a calcareous layer over the substratum. Individuals within the aggregations benefit from this coalescence because it spreads over and monopolizes the entire rock surface thus excluding potential competitors

13. Coloniality in Vermetidae

1cm

FIG. 3. *Serpulorbis aureus* growing in turbid estuarine water in the Transkei, South Africa. The mouths of the shells are elevated away from the silt-covered substratum and face the prevailing water currents.

for space. Vermetids can extend a little way out of their shells and keep a correspondingly small area free of fouling organisms by rasping with the radula. If individuals are closely packed, the areas kept clean by neighbours will overlap so that nearly the entire surface of the colony will be kept free of intruders. In this way the whole group develops a superior competitive ability which would be denied to solitary or non-contiguous individuals. The interstices between contiguous shells become filled by a *Lithothamnion*-type alga (Fig. 4) so that the whole surface of the colony is relatively smooth. The low, even profile of the colony is probably more resistant to wave action than would be individual shells with more projecting profiles. Because of these emergent group properties, the sheet-like aggregations may be regarded as true colonies. Such colony formation has been reported largely within the genus *Dendropoma*. For example, *Dendropoma petraeum* (Monterosato) forms reefs in the S.E. Mediterranean (Safriel, 1966) while *Dendropoma irregulare* (d'Orbigny) is an important component of patch reefs or "boilers" of Bermuda (Stephenson and Stephenson, 1972). The reef-forming species of *Dendropoma* usually occur at or below low tide level where there is considerable exposure to wave action.

The South African *Dendropoma corallinaceum* (Tomlin) occupies a zone of about 1 m in tidal height centred on mean low water springs on very exposed coastlines, but seeking rock surfaces sheltered from the main impact of breakers. On steep or vertical surfaces the colonies form a layer 1–2 cm thick (Fig. 4).

On flatter surfaces, especially where wave action is less severe, the colonies may grow to 6 cm thick. At this thickness the colonies become relatively fragile and susceptible to lateral erosion followed by sloughing of large patches of the colony. A vertical section through the colony shows that the individuals, which are about 2 mm in diameter, grow spirally upwards to a height of about 2 cm. The interstices of the contiguous coils become filled with several species of

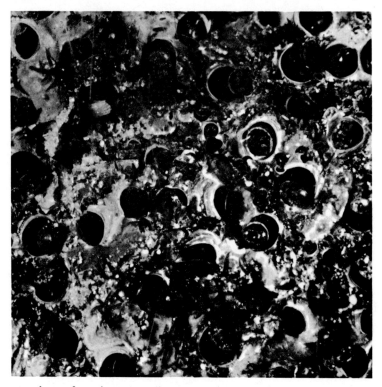

Fig. 4. A colony of *Dendropoma corallinaceum*. Light patches between the individuals are "*Lithothamnion*". Individuals are about 2 mm in diameter.

sponge and shelter numerous polychaetes and sipunculans. Each year, the newly crawling young of *Dendropoma* settle beside or on the adult shells. Gradually they add to the thickness of the colony but do not form well defined layers of new individuals upon old. Rather, there is patchy recruitment with simultaneous growth or death of adults.

Very often the *Dendropoma* colonies take on a pink hue due to the encrustation between individuals of a *Lithothamnion*-type calcareous alga. In addition to smoothing and so streamlining the colony, it is possible that the hard "*Lithothamnion*" also acts as a cement between the individual *Dendropoma*, thereby increasing the strength of the whole colony. Circumstantial evidence in the field suggests that the newly crawling young settle preferentially on the "*Lithothamnion*" associated with colonies. Small patches of bare rock among colonies are not settled on. After settlement *Dendropoma* erodes a groove in the substratum with its radula to accommodate the growing shell. When the young are forced to settle on hard surfaces such as glass they are unable to excavate the groove and are consequently readily detached from the substratum. "*Lithothamnion*" provides an easily excavated material to which the young can become firmly attached very quickly as demonstrated with laboratory animals. Post-larval *Dendropoma* rasp away "*Lithothamnion*" which is encroaching too quickly but both organisms coexist in a balanced fashion. It is not clear whether "*Lithothamnion*" could benefit from this association, thus making it a case of true mutualism.

THE SIGNIFICANCE OF COMMUNAL MUCOUS NETS

In most vermetid aggregations, whether of the loose or colonial type, individuals are so close to each other that their mucous nets overlap and coalesce to form a "communal net". Observations on *Serpulorbis natalensis*, *Dendropoma corallinaceum* and *Dendropoma maximum* (Hughes, 1978b; Hughes and Lewis, 1974), all with communal mucous nets, have shown that when an individual hauls in mucus, it may also drag in parts of the communal net secreted by its neighbours. Naturally occurring contiguous pairs of *Serpulorbis natalensis* and *Dendropoma maximum* showed synchronized feeding rhythms when compared with isolated controls. This is probably an adaptive mechanism which ensures that an individual does not lose most of its mucous net when the contiguous neighbour begins to haul. Synchronized feeding activities were not detected in *Dendropoma corallinaceum* and would be of doubtful value in colonies where the feeding area of each individual overlaps with those of several neighbours.

However, synchronized feeding of the occasional contiguous pairs in loose colonies poses the question of the adaptive significance, if any, of communal mucous nets in such species. If contiguous individuals synchronize feeding to avoid mutual robbing of the mucous net, why are individuals not spaced out to avoid the overlap of nets? Either the disadvantages of net overlap are outweighed by other benefits of settlement in that restricted area, or the net overlap itself is advantageous. Net overlap could be advantageous if the communal net catches more food per individual when averaged over a long time period than small, separate nets. The shape and orientation of the net depends on the water currents prevailing at the time of mucus secretion. If water currents change rapidly relative to the timing of net production, then individual nets might vary considerably in shape, orientation and hence food catching ability from time to time. If individuals are close enough for nets to overlap, the latter will stick together to form a communal net. While the mucus put out by a given individual at a certain time may not be trapping particles very efficiently, the mucus secreted by its neighbours may be more effective. When the individual hauls in its mucus, some of the neighbours' food-laden mucus will also be dragged in. This individual has thus increased its food intake by virtue of the communal net. For this scheme to be advantageous to an individual in the long term, the gains from an overall increased trapping efficiency of a communal net must outweigh the disadvantages of net-robbing by neighbours. Where robbing is likely to be very severe, as with contiguous neighbours, the communal net will be disadvantageous. Synchronized hauling of mucus may be developed by contiguous neighbours to offset this disadvantage. Individuals with contiguous shell mouths are rare in loosely colonial species and may be only temporary phases. Vermetids can alter the direction of shell growth and may diverge from one another to avoid feeding interference.

Several points argue against the adaptive communal net hypothesis. Firstly, a communal net could only be more efficient than single nets if the water currents varied unpredictably and rather frequently. *Dendropoma maximum* lives on the outer parts of coral reef-flats where the direction of water currents generated by onshore swells is very predictable, yet *Dendropoma maximum* has a communal net. Secondly, Hadfield (1970) reports that isolated individuals of *Serpulorbis squamigerus* (Carpenter) are often of much greater size than aggregated individuals, implying that the average ingestion rate of the latter is less than that of the former. Thirdly, communal nets are possessed by species with dense colonies of contiguous individuals. Here, the nets of neighbours overlap and, since each individual has several contiguous neighbours, the practicality of synchronized feeding is lessened. In these species the communal net seems to be merely a consequence of close packing, the advantages of which must exceed the disadvantages of net robbing by neighbours.

If overlapping mucous nets are not advantageous but merely an unfortunate consequence of close grouping, then one might expect the crawling young to space themselves out as do barnacle cyprids after settling in the general vicinity of adults (Crisp, 1961). Since such spacing out of settling young does not appear to happen, we must assume that the effect of net-robbing is of little importance, perhaps being reduced to a minimum by adjustments of orientation during growth of the shell.

DISCUSSION

True coloniality is found only in those vermetids, usually within the genus *Dendropoma* but occasionally in other genera, for example, *Petaloconchus montereyensis* (Hadfield, 1970), which form continuous sheet-like aggregations where the competitive ability of the group far exceeds the potentiality of non-contiguous individuals. The mode of colony formation in *D. corallinaceum* suggests that the space resource available to new recruits remains in constant limited supply for long periods after a rock face has been colonized and completely covered. Causes of death within the colony are unknown but sections through colonies show that most individuals reach adult size but are later overgrown by new recruits which have settled on the surface of the colony. Suitable settlement sites are limited to the spaces which cannot be reached by adults grazing the surface with their radulae. New spaces will appear when adults die, probably at a low constant rate. There is therefore a hint that colonial vermetids have rather stable populations kept close to the carrying capacity of the habitat. This would accord with their reproductive strategy of liberating relatively few crawling young over a very protracted breeding season (Hughes, 1978b, 1979). Over longer time periods there may be cyclical collapses in population density among local patches because old, thick colonies become unstable and slough off. However, coloniality would seem to be a modification of vermetid gregariousness, adapted by virtue of its competitive superiority, to the exploitation of a stable though somewhat severe habitat. Species which form loose aggregations, such as *Serpulorbis* spp. are often found attached to small boulders or stones in calmer water. These species cannot monopolize space over large areas and the fact that they have not evolved to do so may reflect a relative instability and spatial heterogeneity of their habitat. Small changes in current speed or direction or the growth of other sedentary organisms could alter microenvironmental conditions.

Vermetids shed sperm into the seawater so that the habit of forming breeding aggregations may have preceded the evolution of a fixed adult life. The promotion of successful fertilization seems to be the main function of loose aggregations, although a secondary benefit would be that juveniles settling near to adults are choosing a site already proven for its suitability. Such gregariousness paved the way for true coloniality with its superior competitive qualities in stable, homogeneous habitats.

Why vermetids should have evolved adult immobility in the first place is difficult to explain. The distantly related Turritellidae also employ a mucous net to catch suspended particles but some remain mobile as adults. Within this family the genus *Turritella* is restricted to soft substrates where individuals anchor themselves while feeding by burrowing into the sediment. It is necessary to remain stationary while operating the mucous net and whereas this can be successfully achieved by anchoring the shell in sediments, it would be quite difficult to achieve using the foot on hard surfaces subjected to strong currents or wave action. The capacity to cement the shell to rock surfaces enables vermetids to exploit a wide array of habitats denied to *Turritella*. Permanent attachment relieves the foot of its locomotory and adhesive functions so that in the vermetids there is an enormous development of the mucus-secreting pedal gland and pedal tentacles associated with the production of the mucous net, while the rest of the foot atrophies in the adult (Hughes, 1978b). Such extreme morphological commitments to mucous net production cannot be afforded in *Turritella* which requires a well developed foot for locomotion on unstable sediments. An intermediate stage of development is perhaps represented by the genus *Vermicularia* within the Turritellidae and by the family Siliquaridae.

SUMMARY

The individuals of all species of vermetid tend to form aggregations, solitary individuals being a minority and resulting from chance events during settlement.

Aggregations are necessary to ensure successful fertilization by sperm liberated into the seawater.

Certain species, particularly of the genus *Dendropoma*, form dense reef-like aggregations which, apart from facilitating fertilization, enable these species to monopolize space on the substratum in the face of competition from other sedentary organisms and produce a smooth profile less susceptible to wave damage.

Aggregations cause the overlap of individual mucous nets so that communal nets are formed. It is possible that under certain conditions of rapidly and unpredictably varying currents, the communal nets catch more food particles per individual over long periods than single nets. However, in many situations this cannot be so and the mucous nets seem to be mere consequences of aggregation. Contiguous neighbours within otherwise loosely aggregated colonies develop synchronized feeding rhythms to reduce mutual net robbing. Individuals within closely aggregated reef-like colonies do not develop synchronized feeding.

The fixture of adults to the substratum may have arisen as an adaptation enabling suspension feeding with mucous nets to be operated from rock surfaces subjected to strong current or wave action. It would then have allowed the extreme modification of the foot to produce large mucous nets. Gregariousness is necessary to ensure fertilization among the stationary adults but has been developed further in certain species to form true colonies which confer a superior competitive ability in stable, homogeneous habitats.

REFERENCES

CRISP, D. J. (1961). Territorial behaviour in barnacle settlement. *J. exp. Biol.* **38**, 429–441.
HADFIELD, M. G. (1969). Nurse eggs and giant sperm in the Vermetidae (Gastropoda). *Am. Zool.* **9**, 520.
HADFIELD, M. G. (1969). Observations on the anatomy and biology of two Californian vermetid gastropods. *Veliger* **12**, 301–309.
HADFIELD, M. G., KAY, E. A., GILLETTE, M. U. and LLOYD, M. C. (1971). The Vermetidae (Mollusca: Gastropoda) of the Hawaiian Islands. *Marine Biol.* **12**, 81–98.
HUGHES, R. N. (1978a). The biology of *Dendropoma corallinaceum* and *Serpulorbis natalensis*, two South African vermetid gastropods. *Zool. J. Linn Soc.* **64** (in press).
HUGHES, R. N. (1978b). A new species of *Serpulorbis* (Gastropoda: Vermetidae) from South Africa. *Veliger* **20**, 288–291.
HUGHES, R. N. (1979). Notes on the reproductive strategies of the South African vermetid gastropods *Dendropoma corallinaceum* and *Serpulorbis natalensis*. *Veliger* **21** (in press).
HUGHES, R. N. and LEWIS, A. H. (1974). On the spatial distribution, feeding and reproduction of the vermetid gastropod *Dendropoma maximum*. *J. Zool. Lond.* **172**, 531–547.
SAFRIEL, U. (1966). Recent vermetid formation on the Mediterranean coast of Israel. *Proc. malac. Soc. Lond.* **37**, 27–34.
STEPHENSON, T. A. and STEPHENSON, A. (1972). "Life Between Tidemarks on Rocky Shores" W. H. Freeman and Co., San Francisco.
YONGE, C. M. (1932). Notes on the feeding and digestion in *Pterocera* and *Vermetus* with a discussion on the occurrence of the crystalline style in the Gastropoda. *Scient. Rep. Gt Barrier Reef Exped.* **1**, 259–281.

14 | The Colonial Behaviour of *Modiolus modiolus* (L) (Bivalvia) and its Ecological Significance (Abstract)

C. D. ROBERTS

*Botany and Zoology Department, The Ulster Museum,
Botanic Gardens, Belfast, N. Ireland*

Abstract: The horse mussel *Modiolus modiolus* is a versatile epifaunal or infaunal species colonizing a wide variety of *stable* substrates ranging from bedrock to soft mud, colony form varying with substrate. It ranges from the littoral to over 200 m in depth sublittorally at both exposed and sheltered sites, often occurring in areas with fast currents.

A colony can modify the nature of its substrate by biodeposition which on bedrock and boulders will create an infaunal habitat and on gravel and mud will increase the fine fraction and organic content. The reduction in particle size will be partly offset by the influx of shell debris which will increase the stability of the substrate and enable an areal and numerical expansion of the colony.

Coloniality is of great reproductive advantage to free spawning invertebrates such as *Modiolus*. The limited information on spawning behaviour of *Modiolus* is conflicting. The larvae have a long pelagic development and are therefore widely dispersed. Despite the large number of eggs spawned spatfall is small. The variety of substrate which *Modiolus* colonizes indicates that the larvae do not discriminate between substrates, but there is some evidence that larvae are induced to settle by the presence of adults. Settling behaviour and recruitment are important aspects of *Modiolus* ecology probably being largely responsible for determining its distribution.

Colonial behaviour protects the mussels from predation. Large numbers greatly reduce the chance of any individual being eaten and young mussels which are especially vulnerable to predation are protected by the more resistant large mussels which dominate the population.

An adverse effect of coloniality is intense intra-specific competition. Several environmental, morphological and physiological factors contribute to the reduction of competition in *Modiolus*.

Systematics Association Special Volume No. 11, "Biology and Systematics of Colonial Organisms", edited by G. Larwood and B. R. Rosen, 1979, pp. 255–256. Academic Press, London and New York.

The colonies attract a characteristic rich and diverse fauna which generally shows a consistent speciation. *Modiolus* and its associated fauna form a successful and stable community with a large biomass and should be considered an important part of the benthos.

EDITORS' NOTE

At the Symposium, the author showed a short film to illustrate his talk. The author did not submit a manuscript for publication.

15 | Gregariousness and Proto-cooperation in Rudists (Bivalvia)

P. W. SKELTON

*Jane Herdman Laboratories of Geology, University of Liverpool,
Liverpool, England**

Abstract: Rudists were sessile epifaunal suspension-feeding bivalves that inhabited carbonate dominated shelf sea environments, during the latter half of the Mesozoic. Most were gregarious, and some closely aggregative. This paper attempts to analyse characteristics peculiar to the latter habit, and the benefits derived therefrom. Paradigms are constructed for individual epifaunal suspension-feeders, of bivalve affinities, in a representative range of substrate-based environments. The evolution of rudists is reviewed, to assess the degree to which they approached these paradigms, and to distinguish exclusively social characteristics. The closely aggregative habit arose amongst those forms that inhabited relatively calm, muddy environments, in which attachment surfaces were spatially and temporally restricted. The earliest of such clustered forms (monopleurids) approach the appropriate paradigm (a tall form with a broad, elevated feeding apparatus), but show no special modifications for aggregation *per se*. The most advanced types (hippuritids) come very close to the paradigm, though the aggregative forms (as well as some similar radiolitids) also show several additional features that are either irrelevant or deleterious to full attainment of individual paradigmatic form. Such features (engaging ornament, narrow cylindrical attached valve form, monospecific clustering, steeply inclined inner margin to the attached valve) appear to have been social, indicating positive selection for aggregation, which enabled these rudists to spring up rapidly together from temporarily exposed, localized shell carpets. The clusters spread laterally, by means of pseudo-stoloniferous growth, so as to establish broad bushes and platforms. Individuals nevertheless remained physiologically discrete, and are therefore described as having been "proto-cooperative".

Systematics Association Special Volume No. 11, "Biology and Systematics of Colonial Organisms", edited by G. Larwood and B. R. Rosen, 1979, pp. 257–279. Academic Press, London and New York.

* Present address: Department of Earth Sciences, The Open University, Milton Keynes, England.

INTRODUCTION

The rudists (superfamily Hippuritacea, of the subclass Heterodonta) are amongst the most bizarre of fossil Bivalvia. Some of the most specialized forms externally resemble certain corals, and are indeed barely recognizable as bivalves (Pl. I, 3). The group has a stratigraphical range from the Upper Jurassic (topmost Oxfordian) to the very top of the Cretaceous—except for one Paleocene form.

In life, they were sessile, epifaunal, and usually gregarious—sometimes densely so (Pl. I, 3). It is the purpose of this paper to investigate how and why the latter growth habit was adopted.

It is first necessary to define the degree of aggregation achieved. None of the closely aggregative rudists provides evidence (such as growth-line continuity) of any organic fusion between individuals; the shells are merely in intimate contact with one another. Nor is there any evidence of a co-operative division of labour induced by indirect organic communication between individuals. Hence, rudists cannot be described as "colonial" in the sense of Barrington (1967). Specific modifications for aggregative growth did, nevertheless, arise in later rudists. There must therefore have been some adaptive advantages in close-packing of the "critical spaces" required by individuals for growth, feeding, self-cleansing and so forth; such forms may be called "social", in the sense of Schäfer (1972).

By comparing the morphology of aggregative rudists with models of individual epifaunal suspension-feeders, ideally efficient in the environments in which rudists lived, it is possible to separate out those of their many strange features that were of exclusively social value. This paper attempts to demonstrate how these features merely enabled the aggregations to overcome, with greater efficiency, certain of the environmental problems also faced by solitary forms; they did not generate any "super-organisms", capable of exploiting new adaptive zones.

I shall use the word "gregarious", herein, merely to denote clumped life distribution, with no causal connotation (i.e. whether caused passively by a patchy distribution of suitable attachment surfaces, or actively by settling preference).

INVESTIGATION

1. Rudist Life Habits

The rudist shell is more or less inequivalve. In the majority of forms, the larger valve was umbonally attached to, or shallowly embedded in the substrate during life, and the smaller valve acted as a lid to it. A few species were, however, broadly recumbent—though unattached—upon the substrate. The juvenile shell invariably encrusted a hard surface, such as another shell or shell fragment (Pl. I, 1). Rudists were thus sessile and essentially epifaunal; the commissure was generally elevated well above the sediment surface (see Philip, 1972; Kauffman and Sohl, 1974).

In most forms, evidence of ligamentary atrophy and extreme shortening of the adductors indicates greatly restricted gaping (Skelton, 1974, 1976a, and Fig. 3).

In some of the more advanced forms, a superficial radial symmetry was achieved, with the axis passing through both umbones (Pl. I, 5 and Fig. 2—*Sauvagesia*).

Sessility, elevation of a restricted feeding margin from the substrate, and radial symmetry collectively suggest that the rudists were essentially epifaunal suspension feeders; such characteristics are incompatible with macrophagy or deposit feeding, at least. It has been proposed that some forms possessed symbiotic zooxanthellae (see Kauffman, 1969). How widespread or nutritionally important these might have been amongst rudists, if indeed they were present at all, has yet to be demonstrated; certainly, the most specialized rudists—hippuritids—exhibit an elaborate pore and canal system in the upper valve (Pl. III, 1), that is most readily interpretable in terms of a suspension feeding function (Skelton, 1976a).

Virtually all rudists were gregarious, living in a variety of shallow shelf–sea environments within the Tethyan Realm, and in certain contiguous areas (Kauffman, 1973). They were restricted to zones of minimal terrigenous input, and therefore their fossils usually occur in carbonate dominated rocks. In such facies, their distribution tends to be biostromal or biohermal, in matrices varying from biopelmicrite and shelly marls, of low current energy aspect, to biosparite and biomicrudite (particularly in the bioherms), of higher current energy character. Those rudist biostromes in which the individuals are largely in life position represent original carpet-like aggregations upon the flat sea floor. The bioherms represent rudist-inhabited carbonate banks and their flanking deposits. Some of these banks were apparently wave-resistant (Young, 1959; Bein, 1976), though most are thought not to have been (Kauffman and Sohl, 1974; Philip, 1972).

Of the closely aggregative rudists, all hippuritid and most radiolitid examples inhabited those environments normally subject to lower current energies (with little or no winnowing), although they generally exploited scattered (? sometimes storm-derived) shell debris carpets for initial attachment. In contrast, the consistently high current energy zones of large carbonate banks were typically inhabited only by big recumbent forms and a few particular species of aggregative radiolitids (Young, 1959; Philip, 1973; Kauffman and Sohl, 1974; Bein, 1976).

The range of substrate-based habitats originally occupied by rudists may therefore be represented by the three following (idealized) examples (Fig. 1):
(a) Substrates of marly or lime-mud, with some poorly sorted shell sand, and a few temporarily exposed, scattered shells or shell carpets; currents usually weak, and thus normally insufficient to winnow away mud, except during occasional stormy episodes (probably responsible for shell debris carpets); little or no preferred current orientation; permanently exposed hard surfaces therefore highly scattered in time and space. As stated above, most closely aggregative rudists lived in this kind of environment; it was the normal environment of hippuritids in South-East France, for instance (personal communication of J. Philip, 1974, and personal observation).
(b) Substrates of partially winnowed, marly or lime-muddy lime-sand, with many scattered shells and shell fragments; current energies commonly sufficient to winnow away mud; sustained periods of non-deposition, with moderate water agitation, sometimes permitting the development of grapestone, firmground, or hardground surfaces; permanent hard surfaces therefore common.
(c) Substrates of winnowed, loose shell sand, with scattered shells and shell fragments (as well as some interstitial fine material, presumably added later mainly as a result of bio-erosion and baffling); current energies usually high

---

PLATE I

1. Small bouquet of *Hippurites socialis* Douvillé (four right valves), from the Santonian of Piolenc (Vaucluse), South-East France. A young individual, showing the spirogyrate juvenile attachment stage of its right valve (j), in broken section, is attached to the front of the cluster.
2. Internal moulds of *Eodiceras perversum* (G. B. Sowerby), from the topmost Oxfordian of Mortagne-au-Perche (Orne), North-West France. A young individual encrusts the upper (right) valve of another specimen.
3. Bouquet of *Hippuritella toucasi* (d'Orbigny) right valves from the Santonian of La Cadière (Var), South-East France. At least fifteen individuals are visible, and many are heavily bored.
4. Two joined specimens of *Hippurites socialis* Douvillé (same provenance as 1). Relict concentric colour banding around the right valves is visible.
5. Outer surface of upper (left) valve of *Barrettia multilirata* Whitfield, described as coming from the Campanian of Jamaica (British Museum (Natural History), specimen No. LL. 30293). Radial lines of eroded pore-canals, opening onto infoldings of the right valve rim beneath, generate an approximate radial symmetry.
All white scale bars are 1 cm.

PLATE I. (Caption on facing page)

enough to winnow away and prevent settling of mud; and sometimes coarser particles moved as well; burial, or breakage and removal of shell fragments rendering spatially common hard surfaces impermanent.

Such substrates commonly capped rudist-inhabited carbonate banks.

2. *Construction of paradigms*

A series of variably important functional requirements—based on the given environmental conditions and the feeding type under investigation—can be postulated for individual organisms living in this range of habitats. Simple paradigms (ideally efficient models) of single organisms (of bivalve affinities) representing appropriately compromised, expected responses in design to these variables can then be constructed for the three exemplary habitats (Fig. 1); Paul (1975) has stressed the importance of considering all the vital activities necessary for survival, in the analysis of functional design by means of the paradigm method.

(*a*) *Means of exploiting diffuse, water-borne food and oxygen supply.* Efficient food and oxygen entrapment would entail evolution towards maximal areal spread and even exposure of the organs concerned (subject to the restrictions below), at the exposed upper surface of the animal; this would allow most direct contact with, and therefore rapid processing of the immediately surrounding water, with the minimum of anatomical elaboration. A one-way system of water flow over the entrapment organs would also assist efficient food gathering, particularly in calm water, and indeed exists in bivalves. Where water currents are weak and of no preferred orientation (Fig. 1a), these organs could be expected to tend towards radial symmetry, so as to be able to exploit inhalant water all around the exposed parts of the animal. In stronger current regimes, in which one current direction would probably be dominant (Figs 1b, 1c), the organs would best be exposed with a corresponding preference of orientation—either to the upstream (food-source) side of the animal, or alternatively perhaps to its leeward side, in regimes of violent current strength. Exhalant, processed water, being essentially clean, could be ejected anywhere from the animal such that it would be unlikely to be re-cycled.

So that food and energy are not wasted, economy of shell construction (economical growth geometry and utilization of porous fabrics) would of course be preferred.

(*b*) *Means of self-cleansing.* Faecal and pseudofaecal material would need to be ejected from the animal such that it would stand least chance of being returned to the feeding organs, either directly, or secondarily, as a result of precipitation and re-suspension next to the animal. It should be stressed that the essentially clean exhalant water of the feeding currents is irrelevant to the latter problem; the exhalant currents need not therefore entirely coincide with faecal rejection. In an environment of weak, variably directed currents (Fig. 1a), the best site for effective rejection of a stream of unwanted material would be directly overhead. From there the material could easily be carried far from the animal by the merest of open-water currents; the process may be likened to the operation of a factory chimney. In contrast, exhalant water, if not employed in impelling faeces, would best be channelled away sideways from the animal, so as minimally to inhibit the reception of inhalant water over its broadly exposed upper surface (as in (a)). In environments of elevated current energy (Fig. 1b, 1c), both unwanted particles and exhalant water would best be ejected on the normally downstream side of the animal. This would avoid fouling of the inhalant supply.

(c) *Effective utilization of a secure substrate for larval attachment and early growth support* (so as to avoid sinking, burial, or transport). Juvenile stability would be secured by an encrusting or byssate mode of attachment to any stable surface. Preference for specifically reliable surfaces would increase survivorship. Rapid upward growth from this surface would reduce the chances of subsequent burial —current induced toppling allowing.

(d) *Means of balanced stability in adult* (to prevent over-toppling). Centralization and lowering of the centre of gravity would ensure stability of the shell. Elevated elements of the shell would therefore have to be of low density; the body chamber, for instance, would best be located near the top of the shell. Tall forms could survive in calm environments (Fig. 1a), though forms living in rough environments (Fig. 1c), would need to be squatter.

In calm environments, permanently exposed hard surfaces are restricted in space and time (Fig. 1a). A balanced upright growth form would allow rapid establishment on the smallest of initial attachment surfaces. The free outer surfaces surrounding the point of attachment of the basal cone or platform of such a growth form (depending on the extent of surrounding sediment deposition), would thus play a supplementary supporting rôle. Such forms would be unlikely to be toppled by the weak currents of this type of environment. Xenomorphic growth over neighbouring upright surfaces would also assist stability. In the environment of medium current energies (Fig. 1b), permanently exposed hard surfaces are common and frequently expansive. Continued

Fig. 1. Paradigms for individual epifaunal suspension feeders (of bivalve affinity), in a variety of shallow, carbonate-rich, shelf sea environments. Their construction is explained in the text. All three are shown in radial section. Large blank arrows represent reject streams of unwanted particles. Small arrows represent inhalant/exhalant current flow.

(a) *"Elevator"*. Environment normally calm, with no preferred current direction: Predominantly marly to lime-muddy, bioturbated sediment, easily resuspended by the merest of localized currents, so as to produce a low-lying turbid water zone; scattered shells and lime-sand grains present, as well as a few shell fragment carpets generated by occasional episodes of high current energy.

The paradigmatic shell has a tall, upright, broadly conical to barrel-like form, with a small apical attachment site (the shell carpets being most apt as the initial substrate). Its outer layer is calcitic, and its bulky inner layers, of whatever composition, highly porous. Inhalation of relatively clear water for feeding and gas exchange takes place through an areally extensive, but linearly restricted set of apertures, over the entire elevated upper surface, supplying the radially spread entrapment organs beneath. Exhalant water is assumed to escape at the sides, away from the inhalant zone. The whole current system is as externalized and elevated as possible. Unwanted particulate matter is ejected directly overhead. The body chamber, being a low density zone, is situated near the top of the shell.

(b) *"Encruster"*. Currents often sufficient to winnow some mud, and possibly in a preferred direction: bioturbated lime-sand with variably copious matrix of lime-mud or marl–surface commonly stabilized as hardground, firm ground, or grapestone, such that overlying water is usually fairly clean; numerous scattered shells and shell fragments present; sedimentation spasmodic.

The paradigmatic shell has a broad, bun-like form, with a large encrusting lower surface. Its shell composition is as in (a). Inhalation of clean water for feeding and gas exchange takes place, via more or less restricted apertures, over most of the slightly raised upper surface. Exhalant water is assumed, however, to be released on the predominantly downstream side, along with unwanted particulate matter. The current system is again maximally externalized.

(c) *"Recumbent"*. Currents, usually with a preferred orientation, normally sufficient to winnow and prevent settling of mud: Bioturbated, mostly clean lime-sand or coarser shell debris (permitting later development of cement or secondary micritic matrix), perhaps with some mud trapped by baffling—surfaces rarely settled enough to become stabilized; overlying water clean; shells and their fragments common.

The paradigmatic shell has a low, broadly recumbent form, though with only a small site of initial attachment. Its outer layer may be calcitic, though more importantly, the whole shell is finely porous, as a protection against shattering. Inhalation takes places, via slightly raised, more or less restricted apertures, on the predominantly upstream side, whilst exhalation and rejection of unwanted particulate matter occurs on the downstream side.

encrusting growth—as from the juvenile—could readily provide a secure base for the animal. In the rough water environment (Fig. 1c), hard surfaces are again of unreliable permanence, hence a predominantly recumbent base is

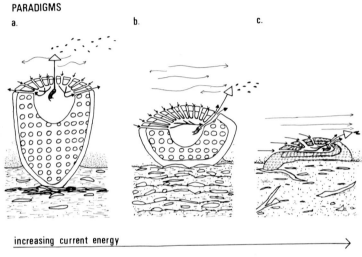

Fig. 1. (Caption on facing page)

again called for. This base must, however, be proportionally large, commensurate with the drag of the water currents; growth in a broadly recumbent, rather than upright form would be preferred.

(e) Means of avoiding burial by settling sediment and clogging of the food entrapment organs by (re-)suspended mud near the sediment surface. High sedimentary accretion rate did not pose a major problem in any of the carbonate dominated environments considered. In the highest energy environments (Fig. 1c), however, some lateral shifting of shell sand has to be taken into consideration; a degree of substrate-escaping growth may thus be shown by forms living in these conditions, as well as some elevation of the suspension feeding apparatus.

Far more important a consideration is the clogging effect of (largely resuspended) mud on the feeding apparatus. This would usually take the form of a turbid water zone immediately overlying muddy parts of the sea floor, and would therefore be a particular problem in the calmer regimes (Fig. 1a), where such loose mud is available for re-suspension. Rhoads (1970) has drawn attention to the means by which mud may easily be re-suspended in calm environments. Kulm *et al.* (1975) have described how the water immediately overlying offshore muddy areas, on the Oregon continental shelf, is commonly turbid, despite the relative calmness of the water there. In contrast, they found the water overlying elevated sandy areas to be normally clear. Lime-mud bottoms in calm waters of the Great Bahama Bank may be stirred by the very slightest of localized currents (personal communication of R. G. C. Bathurst, 1976). In such conditions there would therefore be a premium on elevation of the food entrapment system, away from the sediment surface, and from the bottom water zones of potentially greatest water turbidity. In Fig. 1a the animal is thus assumed to grow as tall as possible, with the suspension feeding apparatus presented at its crest. Such an extent of elevation is deemed to be neither necessary, nor indeed wise in the higher current energy environments (Figs 1b, 1c).

Additional particle screening (or removal) devices might be expected, for protection against swamping of the food entrapment organs themselves, in all environments. In the case of bivalves, a linearly extensive—even contorted—but narrowly restricted inhalant aperture, or set of apertures, that could be rapidly occluded, would serve this screening function. Such a feature would represent a limitation to the tendency for exposure of the food entrapment organs (see Section (a)), necessitating the maintenance of a channelled system of feeding currents to supply them (this already exists in bivalves). Placement of the entire feeding current circuit near the outer surface of the shell ("externalization") would also assist in the exclusion of unwanted particles from the inner parts of the main body/mantle cavity of the animal. Such features would be of particular value in muddy environments, in which sudden small fluxes of muddy water could readily occur.

(*f*) *Means of avoiding predatory attack and fouling by epibionts*. A strong covering shell is to be expected for the protection of the soft parts from predators (indeed it already exists in bivalves). Some kind of screened protection, involving a restricted gape, of the feeding and respiratory organs is again likely. A porous (oyster-like) shell structure would limit the tendency of the shell to crack on sudden impact (see Taylor and Layman, 1972).

Poisons and camouflage would also provide for defence against predators, and the former also against epibionts. Rugose ornament might also inhibit some epibionts (although epibionts themselves might afford some protection against predators).

(*g*) *Means of counteracting the depredations of boring organisms*. It is reasonable to restrict the models to a calcareous shell, in keeping with their bivalve nature. In the habitats here considered, the animals would thus be open to considerable attack by boring organisms (particularly algae). Bathurst (1971) has drawn attention to the importance of such attack in warm, shallow shelf-seas.

A thick shell—possibly porous for the sake of economy—would provide some protection against the weakening effects of boring. A calcitic outer shell layer, being less soluble than aragonite, and therefore probably less easily penetrated by chemical borers, might also be expected, or conversely, a thickened periostracum.

(*h*) *Means of counteracting physical abrasion and battering*. High current energies and their attendant problems need only be considered for Fig. 1c. A low, current-resistant form, with a strong, thick (porous) shell and a broad, stable base would be expected in such circumstances. Whereas any outer layer, for protection against borers (*g*), need not be porous in the environments of Figs. 1a and 1b, a finely porous texture would certainly be expected in that of Fig. 1c.

It will be useful in the following discussion to designate the three paradigms by names: "elevator" (Fig. 1a), "encruster" (Fig. 1b), and "recumbent" (Fig. 1c).

3. Comparison of the Functional Design of Rudists with the Paradigms, with Particular Reference to Closely Aggregative Forms

General features of rudist design will be reviewed briefly, in the light of the foregoing paradigmatic predictions; where appropriate, the latter will be referred to by their paragraph numbers, in parentheses.

A thick, usually porous shell, with a commonly rugose, calcitic outer shell layer (*f*, *g*, *h*) is found in all rudists (e.g. Pl. II, 4).

Examples may be found of disruptive colour banding (Pl. I, 4), which may have served as camouflage (*f*).

Kauffman and Sohl (1974) have suggested that the expanded mantle margin of some rudists could have secreted a "biochemical halo" for protection (*f*); they observed that epibionts, excepting other rudists, were mostly unsuccessful in colonizing the upper surfaces of most rudists.

All rudists possess an encrusting, spirogyrate juvenile stage (Skelton, 1976b, and see Pl. I, 1), satisfying paragraph (*c*).

The gape of most rudists was very narrow (*e*, *f*), as indicated by the extreme shortness of the adductors and atrophy of the ligament, in the majority of forms. This was particularly so in the commonly closely aggregative radiolitids and hippuritids (Skelton, 1974, 1976a).

The most spectacular evolutionary advances made by rudists, towards the paradigms of Fig. 1, involve the overall shell form. The earliest (late Oxfordian)

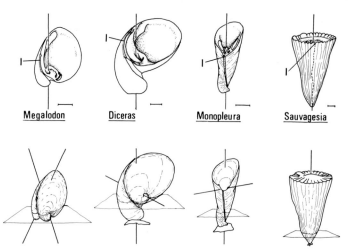

FIG. 2. Demonstration of the geometrical effects of progressive uncoiling in the evolution of the rudists from their ancestors. Top row: oblique apertural views of right valves, with their coiling axes shown vertically disposed, and their active ligamentary zones indicated (l). Scale bars are all 1 cm. Bottom row: bivalved specimens viewed in life position, with probable normal sediment surface level shown (though this could have varied greatly), and with coiling axes of each valve indicated.

In *Megalodon* (Devonian; of the stock ancestral to the rudists) and *Diceras* (Upper Jurassic), the ligament was fully external. In *Megalodon*, the bulky, projecting umbones served as anchors for the uncemented shell, and the commissural plane remained more or less vertical in life. In *Diceras*, the spirogyrate growth mode of the right valve, which became cemented to the substrate, was greatly exaggerated to permit growth elevation of the steeply tilted commissure. In *Monopleura* (Cretaceous) the ligament was shallowly invaginated during development, such that the spirogyrate mode of growth elevation gradually became unnecessary, being replaced by a sub-conical growth geometry in the right valve; marginal accretion could take place all around the commissure, without the need for a tangential component of growth to compensate for a lack of external accretion in the ligamentary zone. The coiling axis became more properly a focus of radial growth, and growth elevation of the commissure became much more economical as a result. In the radiolitid *Sauvagesia* (Middle and Upper Cretaceous), the ligament was deeply invaginated and uncoiling reached completion in the adult (both valves). The central axis was thus entirely a focus of radial growth. External shell form became simply bi-conical, and the elevated commissural plane more or less horizontal. A similar growth geometry was possessed by the hippuritids.

rudists (diceratids) comprised encrusters (Pl. I, 2), commonly overgrowing firm sand (Fig. 2). Ascending growth of an extended feeding and particle rejection margin (a, b, d and e) was effected by exaggerated spirogyrate growth, particularly of the attached (usually left) valve (Fig. 2). Such growth involved continuous anterior splitting, and posteriorward calcareous overgrowth of the split parts, of external (parivincular) ligament (Yonge, 1967). Since they had inherited this an external ligament from their ancestors, such a growth style—albeit elaborate—was indeed the only means by which the fixed animal could continuously raise its commissure from the substrate throughout life; it hardly represented an efficient deployment of shell material, though.

A novel growth geometry subsequently arose amongst the "inverse" rudists (those attaching by the right valve), the earliest members of which appeared during the Kimmeridgian (Favre and Richard, 1927). In forms such as *Monopleura* (appearing in the Valanginian), the ligament became shortened and invaginated (Pl. II, 3), such that outer marginal valve accretion could take place evenly all the way around the commissure, unrestricted by an external ligamentary zone. Such rudists were thus released from the earlier confinement to an uneconomical, upwardly spiralling growth geometry; growth elevation of the whole commissure (e) therefore became progressively more direct (d), and efficient (a, c), in inverse rudists (Pl. II, 5), particularly amongst the later Radiolitidae and Hippuritidae (Fig. 2), in which "uncoiling" is complete.

Early uncoiled forms such as *Monopleura* are commonly found in closely packed clusters (Pl. II, 2); their sub-conical form lends itself to such a habit. The constituent individuals show, however, no additional modifications for this clustering, although they do exhibit sustained lateral xenomorphism (d). It is probable, therefore, that such crowding merely reflects a paucity of suitable attachment surfaces; small clusters and individuals of *Monopleura* equally commonly exploited other rudist shells as attachment surfaces (Pl. II, 1). Preferential settling has yet to be demonstrated in this genus, though it still cannot be ruled out. In the Urgonian part of the Lower Cretaceous of South-East France, small bouquets of *Monopleura* and related forms are locally common in poorly bedded bioclastic pelmicrites. This suggests that the substrates they inhabited were mostly of the type depicted in Fig. 1a, in which permanently exposed hard surfaces are indeed expected to be rare.

The Middle and Late Cretaceous witnessed several broad rudist radiations, from monopleurid roots. Many forms evolved a secondarily recumbent habit, in keeping with their exploitation of rougher water environments like those depicted in Fig. 1c (d, h). Some members of the probably polyphyletic (Macgillavry, 1937) families Caprotinidae and Caprinidae provide the best examples of such rudists (see Philip, 1973; Kauffman and Sohl, 1974; Bein, 1976, for examples). Current controlled preferred commissural orientation (a, b) has been suggested in the case of *Titanosarcolites* (Kauffman and Sohl, 1974). Such forms were usually gregarious, but did not live in closely packed clusters.

It was amongst the Hippuritidae and Radiolitidae that almost perfect economical elevators arose. It was also in such forms that closely aggregative growth became common, often leading to the formation of bulky rudist bushes and platforms upon the sea-floor. The hippuritids, which tended to live in calm water environments, dominated by marl and lime-mud (Philip, 1973), most closely approached the elevator paradigm (Fig. 1a). Apart from efficient elevation, resulting from uncoiling, they also evolved an expansive system of pores and radiating canals in the upper (left) valve, which appears to have permitted the flow of feeding and respiratory currents without the need for active gaping (Skelton, 1976, and see Pl. III, 1, 3, and Fig. 3). They were thereby enabled to exploit water for feeding through a protective screen, from over the

---

PLATE II

1. Small bouquet of *Monopleura urgonensis* Matheron right valves (m), attached to the anterior face of a left valve of *Requienia ammonia* Goldfuss, from the Urgonian of Orgon (Bouches-du-Rhône), South-East France.
2. Small bouquet of silicified *Monopleura marcida* White, from the Albian of Texas, collected by W. J. Kennedy. Ledge-like myophores can be seen flanking the two large projecting teeth, on the inside of the displaced left valve (top).
3. Right valve of *Monopleura marcida* (provenance as in Fig. 2), showing the invaginated ligamentary posterior bourrelet (l)—the previovsly commarginal nymph having atrophied. The large single tooth and ledge-like myophores are also visible.
4. Radial section of right valve of *Vaccinites gosaviensis* (Douvillé) from the Senonian of Gosau, Austria (British Museum (Natural History) specimen No. 33972). The thick calcitic outer shell layer (c) is visible at the left (representing the posterior wall of the valve, and, in a thin slither, representing one of the infolded pillars), but has spalled off on the right. The economical construction of the aragonitic interior of the valve is demonstrated by the presence there of spaced tabulae (t), which even invade the anterior myophore, giving it a cellular appearance (a).
5. Left valve of a requieniid, *Toucasia texana* Roemer (r), itself attached to an upper (left) valve of *Monopleura marcida* (a), and serving as an attachment surface for the right valve of another specimen of *M. marcida* (m). Provenance as in 2. The aggregate nicely demonstrates the greater economical efficiency of upward growth in monopleurids, in which the ligament was invaginated, in comparison with requieniids, in which the ligament was external: at m the young monopleurid attached itself to the already well established requieniid, but it subsequently grew upwards, to point p, in advance of the latter, which therefore had to grow xenomorphically around it.

All white scale bars are 1 cm.

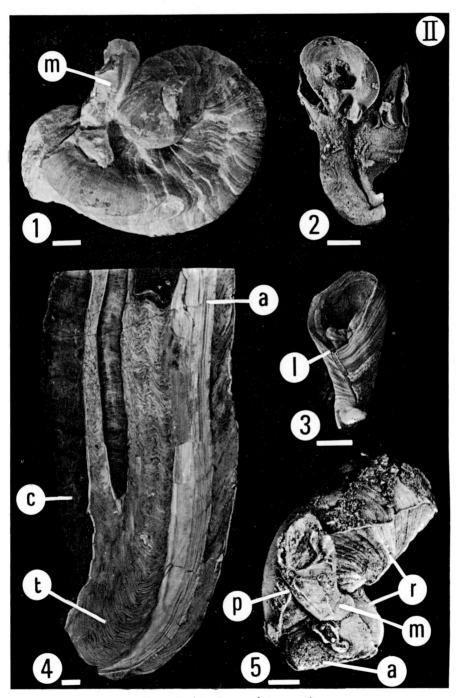

PLATE II. (Caption on facing page)

entire upper surface of the left valve, exhalant water being ejected laterally all around the commissure (Fig. 4b). Predictions *a, d, e* and *f*, relative to the elevator paradigm, were thus completely satisfied. Infolding of two zones at the posterior

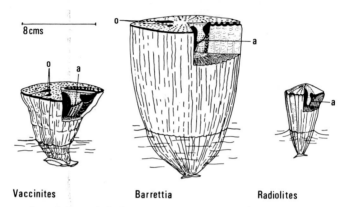

FIG. 3. Examples of elevated feeding apparati in various advanced rudists, all in dorsal aspect. The anterior marginal parts of each valve and the myophoral sites of insertion (a) of the very short anterior adductor muscle are shown in section. The level of the sediment surface during life is suggested (that for *Barrettia* after Kauffman and Sohl, 1974).

In the hippuritid *Vaccinites*, numerous pores on the surface of the left valve open into radial canals, which themselves open onto the broad inner margin of the right valve. Water probably entered through the pores, and left at the outer margin, food particles being trapped *en route* on the right mantle margin (Skelton, 1976a). Faeces and pseudofaeces are supposed to have passed out from the infolded oscules (o). The inner shell of the right valve is tabulate throughout. In the hippuritid *Barrettia*, the canals have atrophied, and the pores open almost directly onto numerous infoldings of the right valve—representing a considerable expansion of the food entrapment surface. Tabulate inner shell extends between the infolded secondary pillars of the right valve. The radiolitid *Radiolites* has no canal and pore system, but does have greatly expanded mantle margins, that of the right valve being supported by a cellular calcitic outer shell layer. The precise mode of food entrapment, although probably involving the expanded mantle margins, is as yet uncertain. There are no oscules in this form.

shell margins also created two pseudosiphonal oscules in the left valve, and a pair of occluding pillars beneath them in the right valve (Pl. III, 1, 3). Water flowing from the canals immediately surrounding these oscules probably caught up faecal and pseudofaecal material and carried it away directly above the shell (Skelton, 1976a), thus satisfying prediction *b*. It should be pointed out that in this reconstruction, most of the exhalant water was ejected separately from the faecal and pseudofaecal rejection streams; the elevator paradigm (Fig. 1a) allows for such a separation. As in other rudists, the hippuritids possessed a spirogyrate juvenile attachment valve (*c*), and a thick outer shell layer of calcite (*g*). Hippuritids were thus good elevators.

Radiolitids were less elaborately organized but nevertheless also possessed greatly expanded mantle margins (Fig. 3) and a restricted gape (Skelton, 1974). It would thus appear that they could, at least, have drawn feeding currents from all around the elevated commissure, except perhaps at two localized raised zones ("siphonal bands" of Douvillé, 1886), from which pseudofaeces and faeces may have been washed. They therefore approached predictions *a*, *b*, *d* and *e*, relative to the elevator paradigm. Many radiolitids, however, also exploited environments like those of Fig. 1b and 1c (see Kauffman and Sohl, 1974; Bein, 1976), with appropriate modifications of form. Many of the latter, though by no means all, were not closely aggregative.

The commonly aggregative hippuritids, in their near perfect achievement of the individual elevator paradigm (Fig. 1a), provide good material for the investigation of additional characteristics peculiar to the closely aggregative habit. Concentration of discussion will therefore be upon these, although most of the observations also apply to some closely packed radiolitids.

There are four remaining features that are outstanding in the morphology of closely aggregative hippuritids, but not predicted by the simple individual elevator paradigm. These are: exaggeratedly salient radial ornament of the right valve (Fig. 4a), narrowly elongate cylindrical form of the adult right valve (Pl. III, 2), a distinct tendency towards monospecific clustering (Pl. I, 3) and a steeply inclined inner margin to the right valve (Fig. 4b)—in contrast to that of such normally solitary forms as *Vaccinites*, in which this margin approaches the horizontal, in life position.

4. Effects of the Additional Modifications found in Closely Aggregative Rudists upon the Individuals Concerned, their Social Values and the Benefits thus Derived by the Aggregates

Pronounced radial ribbing exists in most hippuritids and many radiolitids, whether aggregative or not. In some, particularly aggregative, forms the ribs project sharply (Fig. 4a). Although it is difficult to assess the specific value of this kind of ornament to isolated individuals, it clearly expanded the area of mutual contact of closely packed forms and thus assisted the strength of their engagement to one another (Philip, 1972). In so far as solid clustering was advantageous —as will be demonstrated later—this contributory feature could be described as social.

As far as the isolated individual was concerned, an elongate cylindrical form would have been inimical to the maintenance of balanced stability (*d*), except where a similarly tall vertical surface was available for side-to-side attachment. Such a construction also limited the area of the porous upper surface of the left valve, representing an unparadigmatic restriction of the feeding zone (*a*). It did, however, allow the continuous and direct upward growth of closely packed adult individuals, without undue fanning (Philip, 1972), such that balanced stability was corporately preserved. Such social individuals were thus enabled to concentrate their energies on elevation of the feeding margin (*e*), relying upon each other for lateral support and protection. By this means, hippuritid clusters could grow up rapidly from the sea-floor, having exploited any temporarily exposed shell debris carpet as a base.

The elongate cylindrical right valve of aggregative hippuritids also played another socially valuable rôle. In many small hippuritid clusters, the extended right valve apices may commonly be seen to converge finely towards what originally must have been a small attachment surface, such as a single shell fragment. The adult shells, having thus initially radiated from this restricted site, subsequently curved upwards with further growth (Pl. III, 4 and Fig. 4c). In larger, bush-like clusters, however, many individuals around the periphery

---

PLATE III

1. View, from above, of upper (left) valve of *Vaccinites giganteus* (d'Hombres-Firmas), from the Senonian of Le Beausset (Var), South-East France (British Museum (Natural History) Specimen No. LL 27783). Both oscules are visible (o), opening on to infolded pillars from the posterio-dorsal margin of the right valve beneath. Most of the reticulately porous roofing (p) to the radial canal system has been eroded away, rendering the latter visible. Outwardly, the canals of the left valve open onto the broad, concave inner margin (r) of the right valve.
2. Right valve of *Hippuritella toucasi* (d'Orbigny), from the Santonian of La Cadière (Var), South-East France, with part of another attached at base. Note extremely narrow and elongate form.
3. Radial section through both valves of *Hippurites bioculatus* Lamarck, of unrecorded provenance (British Museum (Natural History) specimen No. 88936). The pore and canal system of the upper (left) valve is clearly visible in section (though the canal to the right passes out of the plane of section), as is also one of the oscules (at left), overlying a pillar head. The body cavity (b) is relatively small.
4. Basal view of bouquet of *Hippurites socialis* Douvillé, from the Santonian of Piolenc (Vaucluse), South-East France. The umbonal apices of the right valves converge tightly towards a small (now eroded) zone of initial attachment, from which the adults subsequently spread out in bush-like fashion.
All white scale bars are 1 cm.

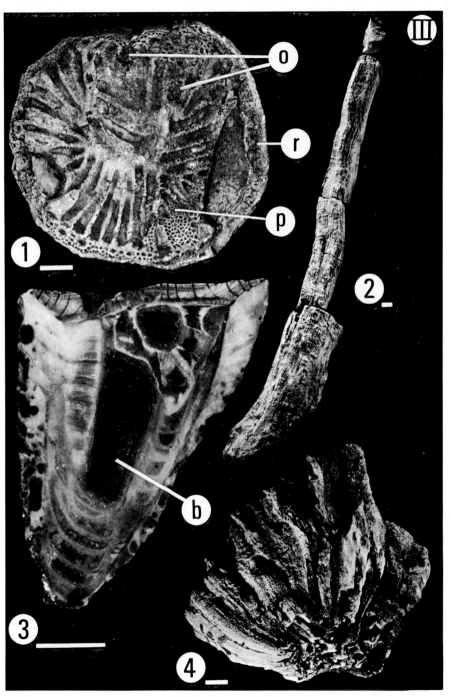

PLATE III. (Caption on facing page)

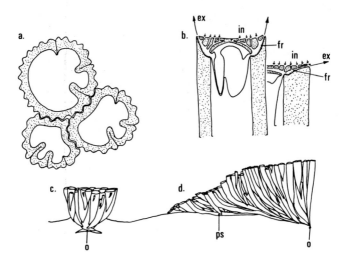

FIG. 4. Social features of Hippuritidae:
    (a) Engaging ornament, as in *Hippuritella toucasi* (D'Orbigny)—see Philip, 1972. The outer shell layers of three right valves are shown in concentric section.
    (b) Radial section of the left valve and part of the right valve in a closely aggregative hippuritid (left), and of a solitary hippuritid (right), showing the steeper inclination of the food entrapment rim (fr) in the former. The inhalant (in) and exhalant (ex) currents, associated with feeding, are indicated. The exhalant current of the aggregative form would not impinge on the inhalant area of a neighbour.
    (c) Small bouquet of hippuritids springing from a restricted attachment point (o). Their narrow adult cylindrical form enables them to grow rapidly upwards together, once established.
    (d) Section, of the sort commonly seen in outcrop, of part of a bush-like lenticular biostrome. From an initial attachment site (o), most individuals have grown outwards and upwards, as in (c) above. The lower-lying individuals, however, grew in pseudo-stoloniferous fashion (ps), and thus served to spread the attachment base of the whole aggregate (diagram based on biostromes of *H. toucasi* observed at La Cadière (Var), South-East France).

actually grew radially outwards for a considerable distance before curving upwards (Fig. 4d). The individual of Pl. III, 2 was found in such a life position, near the edge of a large hippuritid bush. This growth form not only allowed for exploitation of a small attachment site by a large number of individuals, but also served to spread the base of the hippuritid bush, creating new attachment sites for later settlers. The horizontally elongate individuals at the periphery thus effectively served a pseudo-stoloniferous function, and permitted continuous lateral spreading of the bush (Fig. 4d). In such a fashion, broad hippuritid biostromes could develop from a few scattered nuclear clusters; thus not only did their form allow such rudists to spring up rapidly from the sea-floor (as stated earlier), but it also assisted the development of broad platforms of individuals on the basis of spatially as well as temporally restricted attachment surfaces.

The tendency for monospecific clustering can have been of no value to the metabolism of the individual as such, for each appears to have been physiologically independent. In so far as it might signify preferentially clumped settlement of juveniles, particularly upon adults of the same species, it would clearly have been of value in the rapid establishment of rudist bushes, as described above.

The steep inclination of the right valve inner margin, in aggregative hippuritids, necessitated a greater deflection of feeding currents than did the nearly horizontal inner margin of essentially solitary hippuritids, such as *Vaccinites* (Fig. 4b). The latter system, involving more direct flow, was presumably more efficient (Skelton, 1976a). Furthermore, the upward deflection of the exhalant stream in aggregative individuals must have limited, to a certain extent, the overlying zone from which each could draw its inhalant water. Both these effects therefore reduced the feeding efficiency of each individual. They did, however, allow individuals to be packed closely together, such that the exhalant currents of each were ejected vertically, rather than sideways across the inhalant upper surfaces of neighbours. This was clearly an entirely social adaptation, in the sense of Schäfer (1972).

CONCLUSIONS AND DISCUSSION

Four features: salient radial ornament of the right valve, elongate cylindrical right valve form, preferred monospecific clumping and steep inclination of the right valve inner margin, appear to have been either irrelevant or even inhibitive to the full attainment of the individual elevator paradigmatic form of Fig. 1a by aggregative hippuritids. These characteristics were clearly of social value, however, permitting the rapid construction of large bush- or platform-like clusters upon the sea-floor. Members of these clusters, although remaining functionally discrete for feeding purposes, co-operatively achieved (1) rapid elevation of their feeding surfaces from the substrate, (2) broadly based, balanced stability, (3) maximally efficient exploitation of hard attachment surfaces (which tended to be spatially and temporally limited in the environments in which hippuritids normally lived), and (4) effective reduction of exposure of individuals to attack by borers, epibionts, and predators. It will nevertheless be observed that these socially derived advantages still only represent responses to some of the controlling parameters referred to in the construction of the individual elevator paradigm (Fig. 1a)—*viz*. predictions c, d, e, f and g; the aggregates therefore consisted only of co-operative individuals, showing no super-organismal division of labour for the attainment of new modes of feeding, etc.

It would thus appear that the aggregative growth mode in these rudists was principally of adaptive value to those forms that inhabited the lower current energy, marl to lime-mud rich environments.

Radiolitid form approached the individual elevator paradigm less perfectly than did that of the hippuritids. In contrast to the latter, however, they exploited all the habitats represented in Fig. 1; perhaps their less specialized design enabled them to be more opportunistic. However, many of the radiolitids were aggregative and such forms have at least some of the modifications observed in aggregative hippuritids (engaging ornament, narrow conical to cylindrical right valves and preferred monospecific clustering). Most of these closely packed radiolitids lived, like hippuritids, in calmer environments, thus it may be assumed that this growth habit served the same function as in the hippuritids. A few inevitable exceptions are provided, however, by certain species of *Biradiolites* (Kauffman and Sohl, 1974) and *Eoradiolites* (Bein, 1976). These latter forms successfully exploited high current energy environments, by means of the construction of highly stable clusters, of recumbent, or broadly fanned form.

In so far as the individuals of such clusters were never subjugated to the generation of a super-organism, entailing a division of labour, I feel it would be inappropriate to describe hippuritid and radiolitid aggregations as being fully "co-operative". I therefore borrow the ecological term "proto-cooperative" (although this strictly denotes an interspecific relationship; Odum, 1959) to describe the exact nature of their aggregation; individuals derived clear benefit from living thus, though this was not crucial to their functioning.

ACKNOWLEDGEMENTS

I would like to thank Dr Chris Paul for his useful criticism of the paper, and Dr Jean Philip for numerous enlightening discussions. Major blunders are entirely my own, of course.

REFERENCES

BARRINGTON, E. J. W. (1967). "Invertebrate Structure and Function." Nelson, London.
BATHURST, R. G. C. (1971). Carbonate sediments and their diagenesis. Developments in sedimentology, **12**, Elsevier, Amsterdam.
BEIN, A. (1976). Rudistid fringing reefs of Cretaceous shallow carbonate platform of Israel. *Bull. Am. Assoc. Petrol. Geol.* **60** (2), 258–272.

Douvillé, H. (1886). Essai sur la morphologie des rudistes. *Bull. Soc. Géol. Fr.* **XIV** (3), 389–404.

Favre, J. et Richard, A. (1927). Etude du Jurassique Supérieur de Pierre Châtel et de la cluse de la Balme (Jura méridional). *Abh. schweiz. paläont. Ges.* **46**.

Kauffman, E. G. (1969). Form, function and evolution. In L. R. Cox et al., "Bivalvia; Treatise on Invertebrate Palaeontology" (R. C. Moore, ed.) "(N) Mollusca 6 (1)." Lawrence, Kansas, pp. N.129–N.205.

Kauffman, E. G. (1973). Cretaceous Bivalvia. In "Atlas of Palaeobiogeography" (A. Hallam, ed.), pp. 353–383. Elsevier, Amsterdam.

Kauffman, E. G. and Sohl, N. F. (1974). Structure and evolution of Antillean Cretaceous rudist frameworks. *Verh. naturf. Ges. Basel* **84** (1), 399–467.

Kulm, L. D., Roush, R. C., Harlett, J. C., Neudeck, R. H., Chambers, D. M. and Runge, E. J. (1975). Oregon continental shelf sedimentation: interrelationships of facies distribution and sedimentary process. *J. Geol.* **83** (2), 145–175.

Macgillavry, H. J. (1937). "Geology of the Province of Camaguey, Cuba, with revisional studies in rudist palaeontology". Utrecht Rijks-Univ., Geogr. Geol. Meded., Phys., Geol. Reeks Diss. No. 14.

Odum, E. P. (1959). "Fundamentals of Ecology". 2nd edition. Saunders, Philadelphia.

Paul, C. R. C. (1975). A reappraisal of the paradigm method of functional analysis in fossils. *Lethaia* **7**, 15–21.

Philip, J. (1972). Paléoécologie des formations à Rudistes du Crétacé Supérieur: l'exemple du Sud-Est de la France. *Palaeogeogr., Palaeoclimatol., Palaeoecol.* **12**, 205–222.

Philip, J. (1973). (Ed.) *Groupe Français de Crétacé—Excursion en Provénce 5–7 Mai 1973*. Duplicated typescript, Univ. de Provence, Marseille.

Rhoads, D. C. (1970). Mass properties, stability, and ecology of marine muds relative to burrowing activity. In "Trace Fossils" (T. P. Crimes and J. C. Harper, eds.), pp. 391–406. Liverpool Geol. Soc., Liverpool.

Schäfer, W. (1972). "Ecology and Palaeoecology of Marine Environments" (transl. from German by I. Oertel, Ed., by G. Y. Craig), Oliver and Boyd, Edinburgh.

Skelton, P. W. (1974). Aragonitic shell structures in the rudist Biradiolites, and some palaeobiological inferences. *Géologie Méditerranéenne* **1** (2), 63–74.

Skelton, P. W. (1976a). Functional morphology of the Hippuritidae. *Lethaia* **9**, 83–100.

Skelton, P. W. (1976b). Investigations into the palaeobiology of rudists. Oxford University, D.Phil. Thesis.

Taylor, J. D. and Layman, M. (1972). The mechanical properties of bivalve (Mollusca) shell structures. *Palaeontology* **15** (1), 73–87.

Yonge, C. M. (1967). Form, habit and evolution in the Chamidae (Bivalvia) with reference to conditions in the rudists (Hippuritacea). *Phil. Trans. R. Soc.* (B) **252**, 49–105.

Young, K. (1959). Edwards fossils as depth indicators. In "Symposium on Edwards Limestone in Central Texas" (F. E. Lozo, H. F. Nelson, K. Young, O. B. Shelburne and J. R. Sandidge), pp. 97–104. Univ. of Texas, Austin.

16 | Different Causes of Mass Occurrence in Serpulids

H. A. TEN HOVE

Laboratory for Zoological Ecology and Taxonomy, State University of Utrecht, The Netherlands

Abstract: Colonies in the strict sense do not occur in serpulids. In view of the ambiguity of the word colony when applied to serpulids, it seems better to use the term aggregate for serpulid mass occurrence.

Asexual reproduction may reinforce aggregation in three genera, *Filograna*, *Josephella* and *Filogranula*. The reason why aggregation behaviour came to be developed is unclear. Due to the smallness of the branching tubes these aggregates are not spectacular. It is inferred that asexual reproduction by scissiparity as such is not the only cause of aggregation in these genera.

Next to this, there are mainly environmentally induced aggregations, shown by about 20 species out of the approximately 200 Recent Serpulinae, belonging to 8 out of 30 genera. All species occurring in aggregations may also occur singly. Despite the relative smallness of the tubes, aggregates may cover several m^2 with a layer up to 1 m thick.

Three rough groupings of aggregating genera may be recognized. Each group inhabits a major type of habitat, though individual species may occur in more than one. Each group consists of genera which are morphologically closer to each other than to the members of the other groups. The first group consists of belt-forming species from open coasts and mainly from cold-temperate waters. Factors causing this zonation will be discussed. The second group is mass-forming in quiet, lagoonal habitats where mass occurrence of several filter feeders is quite common. The third group occurs in brackish water and also lagoons.

In general, mass-formation occurs in unstable environments, by euryoecious serpulids only. It is likely that it is caused mainly by a complex of environmental factors such as competition for food and space, predation and physical factors as, for example, salinity. Attributes of the animal itself, length of larval stage, habitat selection, gregariousness of larvae, and conditioning of water by presence of congeners are all contributory factors to aggregation.

Systematics Association Special Volume No. 11, "Biology and Systematics of Colonial Organisms", edited by G. Larwood and B. R. Rosen, 1979, pp. 281–298. Academic Press, London and New York.

INTRODUCTION

Serpulid colonies have been known for more than a hundred years (Nardo, 1847; Mörch, 1863; Soulier, 1902; Hargitt, 1909). If, however, colonies are defined as groups of interdependent, organically connected zooids (sometimes with different functions), there are no colonies in the Serpulidae (Polychaeta). It is not yet entirely certain if colonies, defined as groups of organisms deriving some benefit from their living together, exist in serpulids. Colonies defined as groups of organisms originating by budding from one common ancestor are present in serpulids. In view of the ambiguity of the term colony when applied to serpulids, it seems better to use the term aggregate for all serpulid mass occurrences.

To my knowledge, the first mention of a Recent serpuline reef was made by Thomson (1877) from the Bermudan Challenger Bank. This, however, in reality was a *Lithothamnion*-bank with serpulids as a minor constituent only (compare Murray and Renard, 1891; Zaneveld, 1958). Serpulid reefs, atolls or "boilers" are recorded from the Caribbean area by Agassiz (1895) and Verrill (1903). These reefs are built up by coralline algae, vermetids, barnacles, serpulids and madreporarian corals (see also Ekman, 1967; Pérès, 1961). Similar reefs have been described by de Buisonjé and Zonneveld (1960) and Adey (1975). Serpuloid reefs are reported from Veracruz, Mexico by Heilprin (1890), quoted by Hedgpeth (1954). Nevertheless, in a recent study of the reefs of the same area, there is absolutely no mention of serpulids as reef builders (Kühlmann, 1975). Kühlmann mentions only raised rims, consisting of coral debris with a mainly algal overgrowth.

Yet reefs do exist where serpulids are the main constituent. Fossil reefs up to 2 m thick are recorded, for example by Desor and Cabot (1849, Miocene), Packard (1867, Miocene), Garwood (1931, Carboniferous), Schmidt (1955, Miocene), Regenhardt (1964, Jurassic) and Leeder (1973, Carboniferous). The partly vermetid origin of some of these has been demonstrated by Burchette and Riding (1977). Subfossil reefs up to 30×100 m, and 75 cm thick, are mentioned by Reed (1941), Hedgpeth (1953, 1954) and Andrews (1964) from Texas. Recent patch-reefs 3 m in diameter and up to 50 cm thick, are reported by Heldt (1944) and Vuillemin (1965) from the Lake of Tunis, fringing reefs by von Gaertner (1958) from the inner Norwegian skerries and by Bosence (1973) from a lagoon in the Irish Republic. Regenhardt (1964) states that the case of this mass occurrence is still a biological problem.

SCISSIPARITY AND LARVAL INCUBATION

A seemingly self-evident cause of aggregates is found in the genus *Filograna* (incl. *Salmacina*, see Zibrowius, 1973) which reproduces asexually, by means of transverse fission of the body near the middle of the abdomen (cf. Schroeder and Hermans, 1975). This "budding", or scissiparity, has been studied by many authors, e.g. Cresp (1953, 1956, 1964), Vannini and Ranzoli (1957, 1961) and Vannini (1975). It may be concluded that it is a very complicated process, which only takes place between a minimal and maximal abdominal length. Generally sexual maturity does not inhibit scissiparity; on the contrary, scissiparity causes a regression of sexuality. However, a high number of female segments may prevent scissiparity.

Cresp (1964) was unable to find internal (histochemical) or external causes of asexual reproduction. According to Faulkner (1929) budding can continue all the year round, the buds being somewhat more abundant during spring and early summer, whilst sexual reproduction occurs during summer and autumn. "In the summer, when budding and sexual reproduction are both possibilities, it is generally found that members of one colony tend to synchronise and pass through the same phase at the same time, though this is never strictly the case" (Faulkner, 1929). There are indications that scissiparity is regulated mainly by internal processes, and sexuality is partially correlated with climatic factors (see Dales 1970 and Schroeder and Hermans, 1975 for a general discussion of asexual reproduction in annelids).

Under laboratory conditions, successive fissions of one specimen gave rise to five individuals in 61 days (Cresp, 1964); a simple calculation shows that this rate of fission would give approximately 15 000 individuals in one year. In view of the size of the individuals (7×0.25 mm) it seems hardly possible that this rate of multiplication can explain the large clusters (30 cm across) of worms. Fouling experiments mention up to 4500 specimens on one panel per month (Simon-Papyn, 1965), so the scissiparity is not the only reason for aggregation in this genus. It is likely that larval brooding (e.g. Saint-Joseph, 1894; Gravier, 1923; Thorson, 1946; MacGinitie and MacGinitie, 1949) and maybe larval gregariousness as shown by other serpulid genera (see below), contribute to aggregation.

It is still not clear how the asexually formed individual, lying in the tube posterior to its parent, gains access to the exterior. Hanson (1948) and Anon. (1952) state that it bores a hole in the side of the tube. However, this statement is an inference from the fact that the tubes are branched (branching has been described by Benham, 1927; and Hanson, 1948). It appears that scissiparity may also lead to chains of individuals (Hanson, 1948).

TABLE I. Mode of occurrence of the main examples of aggregated serpulids, compiled from various sources. Some of the lagoonal worms occur in brackish environments; all brackish forms occur in lagoons too ("yes"). "Otherwise" refers to incidental mass occurrences, not belonging to any of the three major categories.

	Beltforming in mid- to upper sublittoral zone	Lagoons and similar environments	Brackish, sometimes beltforming	Otherwise
Galeolaria caespitosa Lamarck	Temperate Australia			
hystrix Mörch	New Zealand			
Pomatoceros triqueter (L.)	United Kingdom (Mediterranean)			Norway
lamarckii (Quatrefages)				Mediterranean
Pomatoleios krausii (Baird)	Tropical Australia South Africa			
Spirobranchus cariniferus (Gray)	New Zealand			
paumotanus (Chamberlin)	Pacific Islands			
polycerus (Schmarda)	Caribbean			
polytrema (Philippi)	Mediterranean			

Species	Distribution		Notes
Crucigera zygophora (Johnson)			Japan
Serpula massiliensis Zibrowius sp. nov.			Mediterranean
vermicularis L.	Caribbean		Caribbean
Hydroides elegans (Haswell)	Eire, Mediterranean		
	circummundane, tropical temperate	(yes)	fouling
dianthus (Verrill)	Eastern USA, Mediterranean	(yes)	
dirampha Mörch	Hawaii, Caribbean, Mediterranean	(yes)	
sanctaecrucis Mörch	Caribbean, tropical W. Africa		
Vermiliopsis s.str., sp. nov.	Caribbean		
Ficopomatus enigmaticus (Fauvel)		(yes)	circum-subtropical temperate
miamiensis (Treadwell)		(yes)	Neotropical
uschakovi (Pillai)		(yes)	Paleotropical
macrodon Southern		(yes)	India, Thailand
Marifugia cavatica Absolon and Hrabĕ			Yugoslavia, fresh water, caves

Asexual reproduction has also been reported from two other serpulid species *Filogranula gracilis* Langerhans (Clausade, 1969, as *Omphalopoma gracilis*; Zibrowius, 1973) and *Josephella marenzelleri* Caullery & Mesnil (Fauvel, 1927; Dew, 1959; Clausade, 1969; George, 1974). In the latter species it leads to a network of branching tubes (George, 1974). Scissiparity causes chains of individuals in *Filogranula gracilis* with the greater part of each tube growing along the substrate. However, its youngest part is generally free and erect, causing the mouth to lie at some distance from the substrate. Very thin tubes of new individuals bud at the mouths of established tubes and descend to the substrate, where they gradually attain the appearance and dimensions of mature tubes. Chains of only a few individuals are not infrequent (Zibrowius, personal communication). Owing to the smallness of the animals concerned (2×0.13 mm), these aggregates are never spectacular in size. Comparable chains of individuals are described from the ahermatypic scleractinian coral *Guynia annulata* Duncan, 1873, by Wells (1973). Evidence pertaining to the causes of the evolution of asexual reproduction within these three genera, and incubation of larvae in one of them, is difficult to find and interpretations are highly speculative. It is remarkable that all three genera have small size and fragility of the tube in common. To my knowledge however there are few other serpulids with these characters. As for *Filograna*, it is a very commonly occurring serpulid with possibly the widest distribution of all serpulid genera. *Josephella* has a very wide distribution also. It would be too much of a coincidence that the subfamily Spirorbinae, composed of tiny species with widely distributed genera, should also show incubation of larvae (Knight-Jones et al., 1972, 1975a). See the discussion of brooding on p. 291. It can be inferred that the success of these genera is at least partly due to the larval incubation, reducing the length of the planktonic larval stage and thus the period of greatest mortality. This agrees with the view that the success of any animal is not only due to its ability to maintain itself as an adult, but should also be regarded as the sum of the successes of the various stages of its life-history, and that the weakest link in this chain normally will be found during the breeding period and larval development (see e.g. Thorson, 1950). It is also in accordance with a general biological phenomenon that a low potential rate of reproduction is correlated with a longer period of incubation. The size of the eggs of *Filograna* and spirorbids is approximately 0·15–0·23 mm, the maximum female area is about 2 mm. It is inferred that only 10 to, at the most, 200 eggs can be produced simultaneously by each female. This is very few in comparison with larger species. MacGinitie and MacGinitie (1949) record 10–14 eggs for *Filograna*.

In conclusion, it seems that the loss of sexual reproductive fecundity, as a result of small size, is amply compensated for by incubation, and possibly by asexual reproduction in the case of *Filograna*.

ENVIRONMENTALLY INDUCED AGGREGATES

1. General

About 20 out of the 200 Recent serpuline species, belonging to 9 out of the 30 genera, are known to occur in aggregations (see Table 1). This number will certainly become greater with increasing knowledge of the subfamily. All species known to aggregate also occur singly. It should be noted that the four brackish species are now arranged in one genus, *Ficopomatus* (ten Hove and Weerdenburg, 1978). Notwithstanding the relative smallness of the tubes, 4–10 cm (exceptionally 20 cm) long and up to 5 mm wide, the aggregates may reach considerable sizes, as already mentioned above. In only three years, *Hydroides elegans* may form a layer 20 cm thick, encrusting all sides of a 2 m wide concrete tunnel (Behrens, 1968, as *H. norvegica*). In only three months *F. enigmaticus* may deposit 13 kg m^{-2} of lime (Rullier, 1946). *Marifugia cavatica* is known to cover every side of karst caves in Yugoslavia with a layer up to 1 m thick, and may even form stalactites (Absolon and Hrabě 1930; Stammer, 1935; Remy, 1937).

As can be seen in the table, we can discern three rough groupings of genera, each occupying a major type of habitat (though particular species may occur in more than one habitat), and consisting of genera which are morphologically closer to one another than to the members of the other groups (with the exception of *Vermiliopsis*).

The three major habitats are all to a certain extent unstable. Even the karst caves environment in Yugoslavia is unstable, for during several months of the year the caves are not submerged, and the serpulid *Marifugia* survives in only slightly damp conditions (Absolon and Hrabě, 1930; Remy, 1937).

Several other serpulids can occur aggregated, but never to the extent of the species mentioned above. Often these smaller aggregates are confined to microhabitats such as those provided by corals (*Spirobranchus giganteus* (Pallas) on reefs, *S. tetraceros* (Schmarda) in lagoons) and sponges (*Hydroides spongicola* Benedict and sometimes *Pseudovermilia multispinosa* (Monro)).

Thorson (1957) pointed out that knowledge of larval ecology is of paramount importance for the understanding of animal communities. As far as is known, most Serpulinae have a planktonic larval stage of between six days and two months (a partial survey is given by ten Hove, 1974). This period may vary depending upon season (Segrove, 1941), salinity (Hill, 1967), or food availability. It is well known that larvae of several species can delay settlement under unsuitable conditions (Hill, 1967). The delay, however, may cause a decreased discrimination during settling (see Knight-Jones, 1953). A larval stage of more than one week is long enough to produce a widespread dispersal of larvae, which can be compensated for by aggregation. Various mechanisms of aggregation are discussed by Knight-Jones and Moyse (1961).

2. Gregariousness

Gregariousness of Spirorbinae has been studied in particular by Knight-Jones and co-workers (e.g. Knight-Jones, 1951; Knight-Jones and Moyse, 1961; Knight-Jones *et al.*, 1971, 1975a, b). They supply evidence that gregariousness is due to the ability of the larvae to react to chemical stimuli, even to recognize their own species. Moreover, in a single spirorbid species, various strains may exist with a genetically determined preference for different algae (Knight-Jones *et al.*, 1971). Williams (1964) and Gee (1965) proved that certain algal extracts may stimulate gregarious settling of spirorbids. To date, active searching of larvae for their own species in other serpulids has been described only for *Ficopomatus uschakovi* (as *Mercierella enigmatica*: Straughan, 1972). Mechanical assistance during aggregation in the latter species is provided by eddy currents (Straughan, 1972).

There are indications that gregariousness in some serpulids is density dependent. Straughan (1972) concludes from fouling experiments with *Ficopomatus uschakovi* (as *M. enigmatica*) that settlement is generally randomly spaced on initially bare surfaces at low densities, while in denser populations (1 mm^{-2} and above) the larvae aggregate during settlement. To explain differences in settling peaks at lower and upper sides of experimental fouling panels, Sentz-Braconnot (1968) presumes that populations of *Pomatoceros triqueter* become attractive to conspecific larvae above a certain density only. This is corroborated by the experiments of Klöckner (1976a, b). Settlement of *Galeolaria caespitosa* larvae occurs in cultures only if they contain adult specimens as well (Andrews and Anderson, 1963). Fouling panels, with a population of *F. uschakovi*, placed near the upper limit of the vertical distribution of the species, had a denser spatfall than the controls at the same, abnormal height (Straughan, 1972). Moreover, the presence of adult congeners may promote the development of juveniles, according to experiments carried out by Srinivasagam (1966) with *Hydroides elegans* (as *H. norvegica*) and with balanids. Mussels showed the opposite reaction to the presence of adults.

The agent of attraction is still unknown. From experiments with *Ficopomatus uschakovi*, Straughan (1972) concludes that whole tubes encourage settlement of larvae in an area while living animals further stimulate settlement. Damaged tubes, however, seem to repel the larvae. The ridge formed by the serpulid tube is important in attracting settling larvae, but chemical composition, shape, height and colour seem not to be important in this respect. Nevertheless, the density dependence and the fact that living animals are more attractive than empty tubes suggest there is a chemical agent with increasingly wider effect as the number of generating individuals increases. Zottoli and Carriker (1974), having found that stationary tubicolous polychaetes (including *Hydroides dianthus*) produce external protease, hypothesized that this chemical helps to keep the internal surfaces of their tubes free of attached organisms. If there is a large number of excreting animals this might also help to prevent alien organisms from settling on the aggregate, while attracting conspecific larvae. In this case gregariousness would not only help the larvae to find suitable habitats and maintain or establish a stock sufficiently large to ensure breeding, but would also enable the species to monopolize an entire area (compare Jackson, this volume, for colonial strategies).

In these aggregates two growth forms can be found; one irregularly and sinuously encrusting and the other with parallel growth perpendicular to the substrate (e.g. Behrens, 1968; Klöckner, 1976a, b; for several literature references, Bosence, this volume). These growth-forms are probably not directly due to different environmental parameters, as assumed by Andrews (1964) and Leeder (1973), but mainly to the density of spatfall (Hartmann-Schröder, 1967). Density-dependent gregariousness will cause a further crowding of larvae in subsequent spatfalls, and thus cause parallel growth (Klöckner, 1976a, b).

3. Response to Light

Reviews of factors causing vertical zonation in the tidal zone are given by Doty (1957), Lewis (1964) and Barnes (1969) and are beyond the scope of the present paper. I shall try only to summarize a few facts about serpulids. One of the mechanisms, leading to aggregation in the tidal zone, would be a positive phototactility of the larvae, causing their crowding near to the surface of the water. This has been postulated for *Pomatoceros triqueter* by Thorson (1950, 1957) and Lewis (1964). However, conclusive proof has still to be furnished that serpulid larvae are either photopositive, photonegative, or successively photopositive then photonegative. It is a very well known feature that most serpulids are found on the lower side of experimental fouling slabs (see e.g.

Sentz-Braconnot, 1968; Relini and Sarà, 1971). Nevertheless, the seemingly logical conclusion that serpulid larvae are therefore photonegative is disputable since this type of settling behaviour may also have been caused by such factors as more variable temperature, direct insolation, denser growth of algae or balanids and greater sedimentation on the upper side of the plates. Sentz-Braconnot (1968) points out the probability that the reactions of the larvae to light are more complex, perhaps dependent in particular, upon the nature and intensity of the light. The final settling place of the larvae will certainly be a compromise between light-factors and several other biotic and abiotic factors, as already pointed out by Wilson (1952). This is corroborated by, for example, Dybern (1967) who found that the settling of *Pomatoceros triqueter* is a compromise between a stratified salinity gradient, oxygen content and light conditions. Thus, this species forms intertidal belts in Scotland, but occurs at 3–5 m depth in Norwegian fjords. In a literature survey Schroeder and Hermans (1975) reach the conclusion that "The (settlement) patterns obtained were the combined result of the timing of spawning, the length of larval life and hydrographic factors, in addition to the settling behaviour of the larvae". A different approach to this subject is taken by Klöckner (1976 a, b).

4. Biotic Factors

According to Dakin *et al.* (1948) tidal belts of serpulids may be 60 cm high and 20 cm thick on vertical surfaces, and may have the aspect of reefs on horizontal surfaces of sheltered coasts. In general, the lower limit of distribution of intertidal organisms is mainly determined by the action of biotic factors such as competition for space or predation. The upper limit is probably more often set by physical factors. Experiments by Mohammad (1975), for instance, indicate that the lower limit of the *Balanus*-zone is set by the faster growing (and faster aggregrating) *Pomatoleios*, and predation by crabs, while in its turn that of the *Pomatoleios*-zone is set by competition for space by ectoprocts, sponges and algae. For both *Balanus* and *Pomatoleios* the upper level is determined by physical factors (Mohammad, 1975). However, specific interactions may be important in determining both levels. In *Pomatoleios*, while exposure is important in limiting the upper level of the population, predation by crabs is also a significant factor in fixing an upper limit on distribution. No predation was recorded at lower intertidal levels where the population appeared to be limited entirely by seasonal competition for space with colonial ascidians (Straughan, 1969). Dew and Wood (1955) observed "a marked antagonism between *Balanus* and *Hydroides* so that if one colonizes an area the other does not". This might also be the case with the serpulid species under discussion here, since balanids generally occupy the zone immediately above the serpulid zone. Endean et al. (1956) observed that in the more sheltered areas where the ascidian association was replaced by an algal mat, the presence of this mat seemed to be the effective agent in determining the lower limits of the *Galeolaria* zone. The presence of algae was apparently inimical to the settlement of *Galeolaria* larvae, for the adults immediately adjacent to the algae always appeared to be quite healthy. Both observations, antagonism between some balanids and algae on the one hand, and certain serpulids on the other, are corroborated by Pyefinch's (1950) and Straughan's (1972) conclusions from fouling experiments, although local circumstances, as well as the occurrence of other species, may annul such a trend. Reish (1961) mentions crowded aggregates of *Hydroides elegans* (as *H. norvegica*) overgrowing and killing *Balanus*; *H. elegans* is also found growing on *Acetabularia* in a very quiet part of a lagoon in Curaçao (ten Hove, unpublished). Reimer (1976) states that a species of serpulid has its greatest abundance in the *Tetraclita stalactifera panamensis* (and other balanid-species) zone.

5. Brooding

Larval incubation, as discussed earlier, may contribute to mass occurrence of serpulids. However, there is only one single well documented case of larval incubation and this is in the serpuline species *Chitinopoma serrula* (Stimpson) which does not occur in aggregations; see for instance Thorson (1946, as *Miroserpula inflata*) or Hartmann-Schröder (1971). Saint-Joseph's (1894) mention of embryos in the coelomic cavity of *Pomatoceros triqueter* L. has never been confirmed in the extensive later literature on the species; the same holds for Fischer-Piette's (1937) remark on *Ficopomatus enigmaticus* (Fauvel) (as *Mercierella enigmatica*). Finally Augener (1914), in discussing the embryos of a commensal or parasitic isopod, may have raised misinterpretations of brooding in *Serpula vasifera* Haswell, as mentioned by Schroeder and Hermans (1975). Although biological data on Serpulinae are scarce, most of the species mentioned in Table I are known well enough to conclude that brooding does not contribute to mass occurrence in Serpulinae.

6. Other Physico-Chemical Factors

It is well known that the few species surviving in a more or less extreme environment often occur in very large numbers of individuals. An extreme example is provided by three serpulid species: *Pomatoceros triqueter*, *P. lamarckii* and

Hydroides elegans, which are found in enormous numbers in chlorine-injected cooling systems in the Mediterranean (Zibrowius and Bellan, 1969; Parenzan, 1965).

It is also well known that several of the serpulid species, mentioned above, are tolerant of fluctuating salinity (e.g. Heldt, 1944 and Soldatova and Turpaeva, 1960, *Ficopomatus enigmaticus* (as *Mercierella enigmatica*); Hill, 1967, *F. uschakovi* (as *M. enigmatica*) and *Hydroides* sp. (as *H. uncinata*); Dybern, 1967, *Pomatoceros triqueter*; Straughan, 1969, *Pomatoleios krausii*). A change in temperature from 23° to 17·7°C does not affect adult *F. uschakovi*, despite causing 100% mortality among the juveniles. Aggregates of *H. elegans* are known from waters with low dissolved oxygen values (below 4 mg^{-1} l; Reish, 1961, 1973 as *H. norvegica*, and *H. pacificus* respectively). It is therefore evident that these species are tolerant creatures.

In lagoons and estuaries the environmental factors are certainly less constant than in the open sea. Moreover, these habitats are often more eutrophic. Decrease in competition (and sometimes predation), caused by the less predictable environment, and increased food supply, are the main factors permitting mass occurrence of serpulids here. Larval retention, as discussed by Bosence (this volume) may be an additional factor.

7. *The Brackish Environment*

Species of the genus *Ficopomatus* can survive in conditions ranging from oligohaline to hyperhaline ("Venice system") but are mass-forming in mixohaline and hyperhaline environments only (c.f. Heldt, 1944; Hill, 1967 and Straughan, 1972). Knowledge of factors preventing their mass occurrence in oligohaline, mixoeuhaline and euhaline conditions may clarify the situation in other species too. These factors have been studied in *F. uschakovi* by Straughan (1972 as *M. enigmatica*) by moving populations from their normal habitat to the borders of their horizontal range. In the mixoeuhaline area, mass occurrence is prevented by a high incidence of predation by crabs and molluscs, combined with increased competition, mainly by balanids, for food and space. Hill (1967) states that the species does not mature at all in mixoeuhaline conditions. In oligohaline conditions *Ficopomatus uschakovi* is incapable of breeding, because of its inability to spawn, inactivity of sperm and the non-development of embryos.

8. *Temporary Mass Occurrence*

The mass occurrences discussed above are broadly permanent in nature. However, mass occurrence can also be of a temporary nature. In Japan, although the largest number of settling *Hydroides elegans* is found in July, the species is not very conspicuous on fouling panels. However, owing to the decrease of other settling organisms, the species occurs in masses during November (Kawahara, 1965). From various long-term fouling studies it is evident that serpulids belong to the pioneer species but do not generally form a conspicuous part of the climax community.

CONCLUSIONS

It is probable that scissiparity is not the only cause of aggregation in asexually reproducing serpulids. Gregariousness of larvae and the short duration of the free larval stage apparently also contribute to mass occurrence.

The remaining serpulids which occur in masses are euryoecious and consequently they occur in various habitats. At the periphery of their ranges, these species may occur in great numbers due to a decrease in competition, mainly for space, with other organisms. Density dependent gregariousness of larvae may contribute to aggregated occurrences, maybe even to the exclusion of other species.

Serpulids may occur in transient aggregates on substrates previously unoccupied by any other organisms.

REFERENCES

Absolon, K. and Hrabě, S. (1930). Über einen neuen Süsswasser-Polychaeten aus den Höhlengewässern der Herzegowina. *Zool. Anz. Leipzig* **88**, 249–264.

Adey, W. H. (1975). The algal ridges and coral reefs of St. Croix, their structure and Holocene development. *Atoll Res. Bull.* **187**.

Agassiz, A. (1895). A visit to the Bermudas in March, 1894. *Bull. Mus. comp. Zool. Harvard* **26**, 209–281.

Andrews, J. C. and Anderson, D. T. (1963). The development of the polychaete *Galeolaria caespitosa* Lamarck (Fam. Serpulidae). *Proc. Linn. Soc. NSW* **87**, 185–188.

Andrews, P. B. (1964). Serpulid reefs, Baffin Bay, Southeast Texas. *In* "Depositional environments South-Central Texas coast" (A. J. Scott, ed.), pp. 102–120. Gulf Coast Ass. Geol. Soc. 1964, Field Trip guidebook.

Anonymous (1952). Marine fouling and its prevention. *Contr. Woods Hole oceanogr. Inst.* **580**, 1–388.

Augener, H. (1914). Polychaeta, II, Sedentaria. *In* "Die Fauna Südwest-Australiens" vol. 5, (W. Michaelsen and R. Hartmeyer, eds). pp. 1–170. Verlag von Gustav Fischer, Jena.

Barnes, H. (1969). Some aspects of littoral ecology: the parameters of the environment, their measurement; competition, interaction and productivity. *Am. Zoologist* **9**, 271–277.

Behrens, E. W. (1968). Cyclic and current structures in a serpulid reef. *Contr. mar. Sci.* **13**, 21–27.

Benham, W. B., (1927). Polychaeta. British Antarctic ("Terra Nova") Expedition, 1910. Brit. Mus. (Nat. Hist.), *Nat. Hist. Rep. Zool.* **7**, 47–182.

Bosence, D. W. J. (1973). Recent serpulid reefs, Connemara, Eire. *Nature* **242**, 5392, 40–41.

Buisonjé, P. H. de and Zonneveld, J. I. S. (1960). De kustvormen van Curaçao, Aruba en Bonaire. *Nieuwe West-Ind. Gids* **40**, 121–144.

Burchette, T. P. and Riding, R. (1977). Attached vermiform gastropods in Carboniferous marginal marine stromatolites and biostromes. *Lethaia* **10**, 17–28.

Clausade, M. (1969). Peuplement animal sessile des petits substrats solides récoltés dans trois biocoenoses des fonds détritiques des parages de Marseille. *Tethys* **1** (3), 719–750.

Cresp, J. (1953). Régénération et bourgeonnement chez la salmacine (annélide polychète). *C.R. Hebd. Séanc. Mém. Soc. Biol. Paris* **147**, 844–846.

Cresp, J. (1956). Sur les processus normaux et anormaux du bourgeonnement chez le serpulide *Salmacina incrustans* (Clap.). *Bull. Soc. zool. France* **131**, 183–191.

Cresp, J. (1964). Études expérimentales et histologiques sur la régénération et le bourgeonnement chez les Serpulides *Hydroides norvegica* (Gunn.) et *Salmacina incrustans* (Clap.). *Bull. Biol. France Belgique* **98**, 3–152.

Dakin, W. J., Bennett, I. and Pope, E. (1948). A study of certain aspects of the ecology of the intertidal zone of the New South Wales coast. *Austr. J. Sci. Res. B* **1**, 176–230.

Dales, R. P. (1970) (1963). "Annelids" 2nd edition Hutchinson, London.

Desor, M. E. and Cabot, E. C. (1849). On the Tertiary and more recent deposits in the Island of Nantucket. *Proc. geol. Soc. Lond.* **5**, 340–344.

Dew, B. (1959). Serpulidae (Polychaeta) from Australia. *Rec. Aust. Mus.* **25** (2), 19–56.

Dew, B. and Wood, E. J. F. (1955). Observations on periodicity in marine invertebrates. *Aust. J. mar. freshw. Res.* **6**, 469–478.

Doty, M. S. (1957). Rocky intertidal surfaces. In "Treatise on Marine Ecology and Paleoecology" 1, Ecology. (J. W. Hedgpeth, ed.), *Mem. Geol. Soc. Am.* **67**, 535–585.

Dybern, B. I. (1967). Settlement of sessile animals on eternite slabs in two polls near Bergen. *Sarsia* **29**, 137–150.

Ekman, S. (1967) (1953). "Zoogeography of the Sea" 2nd edition Sidgwick and Jackson Ltd, London.

Endean, R., Kenny, R. and Stephenson, W. (1956). The ecology and distribution of intertidal organisms on the rocky shores of the Queensland mainland. *Aust. J. mar. freshw. Res.* **7**, 88–146.

Faulkner, G. H. (1929). The anatomy and the histology of bud-formation in the serpulid *Filograna implexa*, together with some cytological observations on the nuclei of the neoblasts. *J. Linn. Soc. Zool.* **37**, 109–190.

Fauvel, P. (1927). Polychètes sédentaires. Addenda aux errantes archiannélides, myzostomaires. In "Faune de France" Vol. 16, pp. 1–494. Paul Lechevalier, France.

Fischer-Piette, E. (1937). Sur la biologie du serpulien d'eau saumâtre *Mercierella enigmatica* Fauvel. *Bull. Soc. zool. France* **62**, 197–208.

Gaertner, H. R. von (1958). Vorkommen von Serpelriffen nördlich des Polarkreises an der norwegischen Küste. *Geol. Rdsch.* **47**, 72–73.

Garwood, E. J. (1931). The Tuedian beds of Northern Cumberland and Roxburghshire east of the Liddel Water. *Q. J. Geol. Soc. London* **87**, 97–159.

GEE, J. M. (1965). Chemical stimulation of settlement in larvae of *Spirorbis rupestris*. *Anim. Behav.* **13**, 181–186.
GEORGE, J. D. (1974). The marine fauna of Lundy. Polychaeta (marine bristleworms). *Rep. Lundy Fld Soc.* **25**, 33–48.
GRAVIER, C. (1923). La ponte et l'incubation chez Annélides Polychètes. *Ann. Sci. Nat. Paris. Zool.* (10), **6**, 153–248.
HANSON, J. (1948). Formation and breakdown of serpulid tubes. *Nature* **161**, 610–611.
HARGITT, C. W. (1909). Further observations on the behavior of tubicolous annelids. *J. Exp. Zool. Phila.* **7**, 157–187.
HARTMANN-SCHRÖDER, G. (1967). Zur Morphologie, Ökologie und Biologie von *Mercierella enigmatica* (Serpulidae, Polychaeta) und ihrer Röhre. *Zool. Anz.* **179**, 412–456.
HARTMANN-SCHRÖDER, G. (1971). Annelida, Borstenwürmer, Polychaeta. *Die Tierwelt Deutschlands und der angrenzenden Meeresteile nach ihren Merkmalen und nach ihrer Lebensweise* **58**, 1–594.
HEDGPETH, J. W. (1953). An introduction to the zoogeography of the North-western Gulf of Mexico with reference to the invertebrate fauna. *Publ. Inst. mar. Sci. Univ. Texas* **3**, 107–224.
HEDGPETH, J. W. (1954). Bottom communities of the Gulf of Mexico. In "Gulf of Mexico, its Origin, Waters, and Marine Life." (P. S. Galtsoff, ed.), *Fish. Bull. Fish Wildlife Serv.* **55**, 203–214.
HEILPRIN, A. (1890). The corals and coral reefs of the western waters of the Gulf of Mexico. *Proc. Acad. nat. Sci. Phila.* **1890**, 303–316.
HELDT, J. H. (1944). Sur la présence de *Mercierella enigmatica* Fauvel, serpulien d'eau saumâtre dans les eaux très salées du Lac de Tunis. *Notes stat. océanogr. Salammbo* **30**, 1–4.
HILL, M. B. (1967). The life cycles and salinity tolerance of the serpulids *Mercierella enigmatica* Fauvel and *Hydroides uncinata* (Philippi) at Lagos, Nigeria. *J. anim. Ecol.* **36**, 303–321.
HOVE, H. A. TEN (1974). Notes on *Hydroides elegans* (Haswell, 1883) and *Mercierella enigmatica* Fauvel, 1923, alien serpulid polychaetes introduced into the Netherlands. *Bull. Zool. Mus. A'dam* **4**, 45–51.
HOVE, H. A. TEN and WEERDENBURG, J. C. A. (1978). A generic revision of the brackish-water serpulid *Ficopomatus* Southern 1921 (Polychaeta: Serpulinae), including *Mercierella* Fauvel 1923, *Sphaeropomatus* Treadwell 1934, *Mercierellopsis* Rioja 1945 and *Neopomatus* Pillai 1960, *Biol. Bull.* **154**, 96–120.
KAWAHARA, T. (1965). Studies on the marine fouling communities. III. Seasonal changes in the initial development of test block communities. *Rep. Fac. Fish., Pref. Univ. Mie* **5**, 320–364.
KLÖCKNER, K. (1976a). Zur Ökologie von *Pomatoceros triqueter* (Linne 1758) (Serpulidae, Polychaeta). *Thesis Univ. Tübingen*, 1–168.
KLÖCKNER, K. (1976b). Zur Ökologie von *Pomatoceros triqueter* (Serpulidae, Polychaeta). I. Reproduktionsablauf, Substratwahl, Wachstum und Mortalität. *Helgoländer wiss. Meeresunters.* **28**, 352–400.
KNIGHT-JONES, E. W. (1951). Gregariousness and some other aspects of the setting behaviour of *Spirorbis. J. mar. biol. Ass. U.K.* **30**, 201–222.

KNIGHT-JONES, E. W. (1953). Decreased discrimination during setting after prolonged planktonic life in larvae of *Spirorbis borealis* (Serpulidae). *J. mar. biol. Ass. U.K.* **32**, 337–345.

KNIGHT-JONES, E. W. and MOYSE, J. (1961). Intraspecific competition in sedentary marine animals. *Symp. Soc. exp. Biol.* **15**, 72–95.

KNIGHT-JONES, E. W., BAILEY, J. H. and ISAAC, M. J. (1971). Choice of algae by larvae of *Spirorbis*, particularly of *Spirorbis spirorbis*. In "Fourth European Marine Biology Symposium", (J. D. Crisp, ed.). Cambridge University Press, Cambridge. pp. 89–104.

KNIGHT-JONES, E. W., KNIGHT-JONES, P. and VINE, P. J. (1972). Anchorage of embryos in Spirorbinae (Polychaeta). *Mar. Biol. Berlin* **12**, 289–294.

KNIGHT-JONES, E. W., KNIGHT-JONES, P. and AL-OGILY, S. M. (1975a). Ecological isolation in the Spirorbidae. In "Ninth European Marine Biology Symposium" (H. Barnes, ed.). Aberdeen University Press, Aberdeen. pp. 539–561.

KNIGHT-JONES, P., KNIGHT-JONES, E. W. and KAWAHARA, T. (1975b). A review of the genus *Janua*, including *Dexiospira* (Polychaeta: Spirorbinae). *Zool. J. Linn. Soc.* **56**, 91–129.

KÜHLMANN, D. H. H. (1975). Charakterisieurung der Korallenriffe vor Veracruz/Mexico *Int. Revue ges. Hydrobiol.* **60**, 495–521.

LEEDER, M. R. (1973). Lower Carboniferous serpulid patch reefs, bioherms and biostromes. *Nature*, **242**, 41–42.

LEWIS, J. R. (1964). "The Ecology of Rocky Shores" The English Universities Press Ltd., London.

MACGINITIE, G. E. and MACGINITIE, N. (1949). "Natural History of Marine Animals". McGraw-Hill Book Co.

MÖRCH, O. A. L. (1863). Revisio critica serpulidarum. Et bidrag til rørormenes naturhistorie. *Naturh. Tidsk. Henrik Krøyer*, København (3) **1**, 347–470.

MOHAMMAD, M.-B. M. (1975). Competitive relationship between *Balanus amphitrite amphitrite* and *Pomatoleios kraussii* with special reference to their larval settlement. *Hydrobiologia* **46**, 1–15.

MURRAY, J. and RENARD, A. F. (1891). Report on the deep-Sea deposits based on specimens collected during the voyage of H.M.S. Challenger in the years 1872 to 1876. *Rep. Sci. Res. Challenger, Deep-Sea Deposits*.

NARDO, G. D. (1847). "Prospetto della fauna marina volgare del Veneto estuario con cenni salle principali specie comestibili dell'adriatico, ecc." Venezia, 1847 (Annelidi p. 10–13).

PACKARD, A. S. JR. (1867). Observations on the glacial phenomena of Labrador and Maine, with a view of the recent invertebrate fauna of Labrador. *Mem. Boston Soc. natl hist.* **1**: 210–303.

PARENZAN, P. (1965). Eccezionale resistenza del polichete *Hydroides norvegica* Gunn. all'azione del Cl, nel Mar Piccolo di Taranto. *Rivista Chimico-Sanitaria* **3**, 3–5.

PÉRÈS, J.-M. (1961). "Océanographie biologique et Biologie marine" Tome premier. La vie benthique. Presses Universitaires de France, Paris.

PYEFINCH, K. A. (1950). Notes on the ecology of ship-fouling organisms. *J. anim. Ecol.* **19**, 29–35.

REED, C. T. (1941). Marine life in Texas waters. *Texas Acad. Publ. Nat. Hist.*, non-techn. ser.

REGENHARDT, H. (1964). "Wurm-" und Serpuliden-Röhren in Geschieben unter besonderer Berücksichtigung von "Riffbildungen". *Lauenb. Heimat* N.F. **45**, 57–62.

REIMER, A. A. (1976). Description of a *Tetraclita stalactifera panamensis* community on a rocky intertidal Pacific shore of Panama. *Mar. Biol.* **35**, 225–238.

REISH, D. J. (1961). The relationship of temperature and dissolved oxygen to the seasonal settlement of the polychaetous annelid *Hydroides norvegica* (Gunnerus). *Bull. S. Calif. Acad. Sci.* **60**, (1), 1–11.

REISH, D. J. (1973). Marine and estuarine pollution. *J. WPCF* **45**, 1310–1319.

RELINI, G. and SARÀ, M. (1971). Seasonal fluctuations and successions in benthic communities on asbestos panels immersed in the Ligurian Sea. *Thalassia Jugoslavica* **7** (1), 313–320.

REMY, P. (1937). Sur *Marifugia cavatica* Absolon et Hrabĕ, serpulide des eaux douces souterraines du karst adriatique. *Bull. Mus. Hist. nat. Paris* (2) **9**, 66–72.

RULLIER, F. (1946). Croissance du tube de *Mercierella enigmatica* Fauvel. *Bull. Lab. marit. Dinard* **27**, 11–15.

SAINT-JOSEPH, A. DE. (1894). Les annelides polychètes des Côtes de Dinard. III. *Ann. Sci. Nat.* (7) *Zool. Paris* **17**, 1–395.

SCHMIDT, W. J. (1955). Die tertiären Würmer Österreichs. *Denkschr. Österr. Akad. Wiss. math.-nat. Kl.* **109**, 1–121.

SCHROEDER, P. C. and HERMANS, C. O. (1975). Annelida: Polychaeta. In "Reproduction of Marine Invertebrates. 3. Annelids and Echiurans" (A. C. Giese and J. S. Pear, eds), pp. 1–213. Academic Press, New York and London.

SEGROVE, F. (1941). The development of the serpulid *Pomatoceros triqueter* L. *Quart. J. micr. Sci. London* **82**, 467–540, 25 figs.

SENTZ-BRACONNOT, E. (1968). Données ecologiques et biologiques sur la fixation des Serpulidae dans le Rade de Villefranche-sur Mer (Alpes Maritimes). *Vie Milieu* **19** (B, Océanogr.) **1**, 109–132.

SIMON-PAPYN, L. (1965). Installation expérimentale du benthos sessile des petits substrats durs de l'étage circalittoral en Méditerranée. *Rec. Trav. St. mar. Endoume* **55** (Bull. 39), 51–94.

SOLDATOVA, I. N. and TURPAEVA, E. P. (1960). O prodolzuteljnocti adaptatsii pri iz menii solenosti srecly u dvustvorchatogo molljieksa *Teredo navalis* L. i. mnogoshchetinkovogo chervja *Mercierella enigmatica* Fauvel. (On the duration of adaptation with change of salinity in the bivalve mollusc *Teredo navalis* L. and the polychaetous worm *Mercierella enigmatica* Fauvel). *Dokl. Akad. Nauk SSSR* **130**, 646–648.

SOULIER, A. (1902). Révision des annélides de la région de Cette, I. *Trav. Inst. Zool. Univ. Montpellier* (2), **10**, 1–55.

SRINIVASAGAM, R. T. (1966). Effect of biological conditioning of sea water on development of larvae of a sedentary polychaete. *Nature* **212** (5063), 742–743.

STAMMER, H. J. (1935). Untersuchungen über die Tierwelt der Karsthöhlengewässer. *Verh. int. Ver. theor. angew. Limn.* **7**, 92–99.

STRAUGHAN, D. (1969). Intertidal zone-formation in *Pomatoleios kraussii*. (Annelida: Polychaeta). *Biol. Bull.* **136** (3), 469–482.

STRAUGHAN, D. (1972). Ecological studies of *Mercierella enigmatica* Fauvel (Annelida: Polychaeta) in the Brisbane River. *J. anim. Ecol.* **41**, 93–136.

THOMSON, C. W. (1877). "The Voyage of the 'Challenger'. The Atlantic: a preliminary account of the general results of the exploring voyage of the H.M.S. 'Challenger' during the year 1873 and the early part of the year 1876" vol 1. Macmillan and Co., London.

THORSON, G. (1946). Reproduction and larval development of Danish marine bottom invertebrates, with special reference to the planktonic larvae in the Sound (Øresund). *Medd. Komm. Danm. Fisk. Hav. Plankton* **4**, 1–523.

THORSON, G. (1950). Reproductive and larval ecology of marine bottom invertebrates. *Biol. Rev. Cambr. Phil. Soc.* **25**, 1–45.

THORSON, G. (1957). Bottom communities (sublittoral or shallow shelf). *In* "Treatise on Marine Ecology and Paleoecology" vol. I, Ecology (J. W. Hedgpeth, ed.). (*Mem. geol. Soc. Am.*) **67** (1), 461–534.

VANNINI, E. (1975). Cicli reproduttivi nei policheti *Salmacina dysteri* e *Salmacina incrustans*. *Pubbl. Staz. zool. Napoli* **39**, Suppl.: 335–346.

VANNINI, E. and RANZOLI, F. (1957). Osservazioni sui fenomeni di toracizzazione negli schizozooidi di *Salmacina incrustans*. *Boll. Zool.* **24**, 145–151.

VANNINI, E. and RANZOLI, F. (1961). Correlazione del numero di metameri toracici fra schizonte e schizooide in *Salmacina dysteri*. *Atti. Acad. Naz. Linc. Rend.* (*Sci. fis. mat. nat.*) (8) **30**, 94–99.

VERRILL, A. E. (1903). The Bermuda Islands: their scenery, climate, productions, physiography, natural history, and geology; with sketches of their early history and the changes due to man. *Trans. Connect. Acad. Arts Sci.* **11**, 413–911.

VUILLEMIN, S. (1965). Contribution à l'étude écologique du Lac de Tunis. Biologie de *Mercierella enigmatica* Fauvel. *Thèse Univ. Paris* (*A*) 4622, 5469.

WELLS, J. W. (1973). *Guynia annulata* (Scleractinia) in Jamaica. *Bull. Mar. Sci.* **23**, 59–63.

WILLIAMS, G. B. (1964). The effect of extracts of *Fucus serratus* in promoting the settlement of larvae of *Spirorbis borealis*. *J. mar. biol. Ass. U.K.* **44**, 397–414.

WILSON, D. P. (1952). The influence of the nature of the substratum on the metamorphosis of the larvae of marine animals, especially the larvae of *Ophelia bicornis* Savigny. *Ann. Inst. Oceanogr.* **27**, 49–156.

ZANEVELD, J. S. (1958). A *Lithothamnion* bank at Bonaire (Netherlands Antilles). *Blumea*, Suppl. **4**, 206–219.

ZIBROWIUS, H. (1973). Serpulidae (Annelida Polychaeta) des côtes ouest de l'Afrique et des archipels voisins. *Ann. K. Mus. Midden-Afrika, Tervuren, België, Reeks in 8°, Zool. Wet.* **207**, 1–93.

ZIBROWIUS, H. & BELLAN, G. (1969). Sur un nouveau cas de salissures biologiques favorisées par le chlore. *Téthys* **1**, 375–381.

ZOTTOLI, R. A. and CARRIKER, M. R. (1974). External release of protease by stationary burrow-dwelling polychaetes. *J. mar. Res.* **32**, 331–342.

17 | The Factors Leading to Aggregation and Reef Formation in *Serpula vermicularis* L.

D. W. J. BOSENCE

University of London, Goldsmiths' College, London, England

Abstract: Ardbear Lough, Co. Galway, Eire is an enclosed brackish lagoon which contains reefs formed by the serpulid worm *Serpula vermicularis* L. The Lough is sheltered, has a restricted tidal range and on average exhanges about 10% of its water with each tide. The Lough waters are well mixed by inflowing spring tidal currents but during neap tides the salinities are lowered and the Lough waters stratified.

Most of the Lough is floored with mud and there are occasional rock outcrops. Serpulid reefs cover about 25% of the Lough and individually are about 2 m high and several hundred metres across.

Reefs develop initially from worms encrusting areas of hard substrate. When the original substrate is covered, larvae settle on adult tubes and eventually grow upwards to form bush-like aggregations of tubes. Parts of these aggregations fall off to provide further areas for reef growth. The reefs support a diverse and abundant fauna.

The reefs are restricted to depths between 2 and 20 m. The upper limit is thought to be due to low salinity waters and algal growth and the lower limit to suspended mud and low oxygen levels.

The factors leading to the aggregation of *S. vermicularis* are considered to be a combination of larval retention, limited substrate, slightly reduced competition for space and possible high primary production.

INTRODUCTION

The mass occurrence of serpulid worms is reviewed in this volume by ten Hove. Although aggregation by serpulids has a widespread distribution and is reported from a variety of environments the factors responsible for large aggregations of these worms are still not understood. In this paper aggregation and reef formation are described from a lagoonal environment which was first reported

---

Systematics Association Special Volume No. 11, "Biology and Systematics of Colonial Organisms", edited by G. Larwood and B. R. Rosen, 1979, pp. 299–318. Academic Press, London and New York.

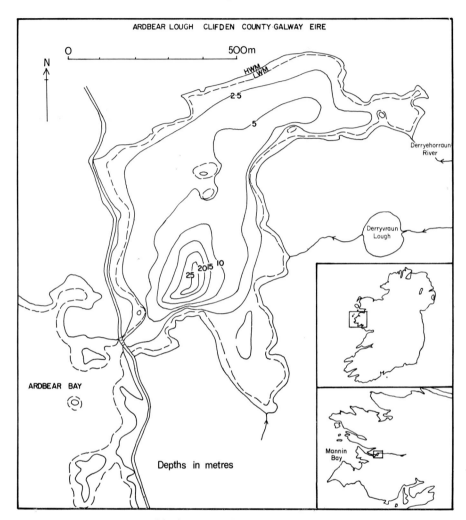

Fig. 1. Location and bathymetry of Ardbear Lough, Co. Galway, Eire.

by Bosence (1973). The main objects are to describe the reefs and to analyse the environmental conditions which are thought to have permitted the formation of these aggregations.

This preliminary investigation of the Lough and reefs was carried out during the months of July to September 1972 using SCUBA and boat.

LOCATION, BATHYMETRY AND HYDROLOGY

Ardbear Lough (Fig. 1) near Clifden, Co. Galway, Eire (Irish National Grid No. L 6649) is a glacially formed sea lough (marine lagoon) which has been eroded out of Connemara Schist. The schists are covered by thin glacial drift which is eroded by waves to form narrow pocket beaches. The bathymetry of the Lough can be seen in Fig. 1. Most of the Lough is shallow but in the south west depths of nearly 30 m are found.

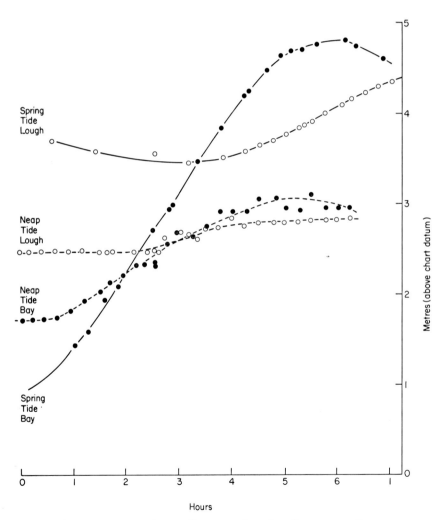

FIG. 2. Tide curves measured in Ardbear Lough and Ardbear Bay. All measurements have been reduced to chart datum, Clifden Bay.

The hydrology of the area has been studied by diving observations, Dentan CM2 direct reading current meter, measurement of tidal curves and through the interpretation of salinity distributions.

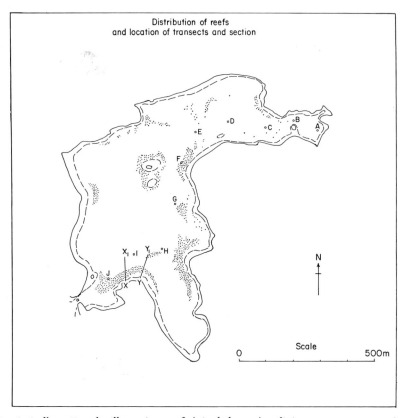

FIG. 3. Ardbear Lough, illustrating reefs (stippled areas), salinity measurement stations and transects $X-X_1$ and $Y-Y_1$.

Because of the restricted opening the tidal range within the Lough is considerably smaller than that in Ardbear Bay. Tidal ranges were measured over half a tidal cycle (Fig. 2) for neap and spring tides. The ranges are reduced by a factor of about three to give spring tidal ranges of 1·0 m within the Lough and neap tidal range of 0·4 m. The mid tide level within the Lough is much lower for neaps than it is for springs because the Lough progressively empties during neap tides and gradually fills during spring tides. This results in a fortnightly tidal cycle with a range of about 1·3 m.

The exchange of water over one tidal cycle can be calculated from the volume of the Lough (2 million m³) and the measured tidal ranges. The amount of water exchanged over a spring tidal cycle with a range of 1·0 m is about 300 000 m³. For neap tides with a range of 0·4 m the exchange is about 120 000 m³. This gives an average exchange of 210 000 m³ or about 10% per tidal cycle.

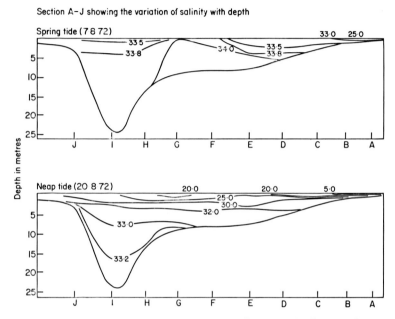

FIG. 4. Bathymetric sections from stations A–J in Ardbear Lough illustrating variations in salinity (in parts per thousand).

Current speeds measured during the flood phase of a spring tide (10.8.72) indicate currents up to 150 cm sec^{-1} and 40 cm sec^{-1} in a neap tide (4.7.72) within the narrow mouth to the Lough. Away from the Lough mouth tidal currents are too small to measure. However, winds blowing across the Lough produce waves and wave currents. The maximum observed wave height during July–September 1972 was about 0·5 m.

Information on the hydrology can be obtained from a study of the distribution of water salinities. These were measured at stations A–J (Figs. 3, 4) during a neap tide of 20.8.72 and over the flood phase of a spring tide on 7.8.72.

From the measured tidal cycles and salinity distribution the interpretation of the hydrology of the Lough is as follows.

During spring tides the Lough is progressively filled with sea water from Ardbear Bay. This raises the water level and results in dominantly marine conditions with salinities varying from 33–34‰ (cf. Mannin Bay (Fig. 1), summer salinities 34–35‰). At station A salinities are reduced to 23‰ where the less dense river water flows out over the Lough.

The high salinities found at G and H are thought to be due to sea water flooding in through the mouth of the Lough across the deep area to meet the north-south slope in between I and H. This would direct the water upwards to form the mass of higher salinity water, the fresher water being pushed back to the head of the Lough. The spring tides therefore produce well-mixed waters. This is in contrast to neap tides when the Lough is progressively drained. This results in a greater proportion of fresh water and the establishment of layering of the Lough waters. The top one or two metres of water have reduced salinities where freshwater flows out over denser sea water. Below this upper zone salinities gradually increase to about 33‰ at the bottom.

In conclusion, currents in the Lough are only measurable in the mouth where tidal currents race during flood and ebb. The environment is dominantly marine but the upper one to two metres are subjected to lower salinities over neap tides. As well as a diurnal tidal cycle the Lough fills during spring tides and empties during neap tides.

SUBSTRATES

The substrates of the Lough were investigated mainly by diving. In addition to this Van Veen grab samples were taken in the south western corner of the Lough (Fig. 7). There are five main substrates: mud, rock outcrops, terrigenous gravels and pebbles, serpulid reefs and carbonate gravels.

1. Mud

Mud is the most common substrate, covering an estimated 65% of the Lough. Surface waves prevent mud from being deposited where the water is less than 2–3 metres deep. Below this depth weak currents allow mud to settle over most of the Lough with the exception of the tidal currents sweeping over the platform near the Lough mouth.

2. Rock outcrops

These are common in the littoral zone but only occur over about 5% of the sublittoral. The main areas of sublittoral rock are firstly the steep slope in the south and secondly the island near the centre of the Lough.

3. Terrigenous pebbles and gravels

These are found intertidally and in the shallow subtidal areas. They are produced by waves breaking on the shoreline and eroding the thin cover of glacial drift.

4. Serpulid Reef

The term *serpulid reef* is here used to describe aggregates of intergrowing and encrusting calcareous tubes of *Serpula vermicularis* L. These aggregates are mainly in life position and cover areas up to hundreds of metres across and up to 2 metres in height (Fig. 3, Plate I: 1,2). Serpulid reefs occupy about 25% of the area of the Lough (Fig. 3). The reefs occur sublittorally between depths of 2–20 m and are mainly initiated on rocky substrates. The distribution of the reefs is discussed in detail below.

5. Carbonate gravels

These are composed mainly of eroded serpulid reef material, together with some shell debris from molluscs living within the reef (see below) and on rocky areas, especially from molluscs on the littoral and sublittoral rocks at the mouth of the Lough (Fig. 7).

REEF STRUCTURE

Brief descriptions and notes on *S. vermicularis* may be found in McIntosh (1923), Fauvel (1927), Nelson-Smith (1967) and Zibrowius (1973). However, no detailed morphological and ecological studies have yet been published.

S. vermicularis has a world wide distribution in marine waters and has been collected from depths down to 250 m (Zibrowius, 1973). The worm will encrust most hard substrates but is most commonly found on large bivalve shells. The tube is constructed of calcitic cone shaped lamellae with 5-10 mole % $MgCO_3$ (Bornhold and Milliman, 1973) secreted by glands in the anterior part of the thorax (Hedley, 1956, Neff, 1969). The tubes are white tinted with pink.

In Ardbear Lough young tubes are always encrusting, sinuous and commonly have three longitudinal ridges (Plate I: 3, 7). With age the tubes become smoother and lose the longitudinal ridges, the main ornament now being growth ridges around the tubes. Mature tubes have an outside mean diameter of 5·2 mm, thickness of 0·46 mm, average length of about 120 mm and a maximum observed length of 180 mm. This is slightly larger than the sizes recorded from worms in the open sea. After the original encrusting stage the worm grows away from the substrate in a sinuous fashion (cf. Fauvel, 1927; Nelson-Smith, 1967).

REEF DEVELOPMENT

Specimens have been sketched and photographed underwater, and collected for later dissection in order to provide the information for this section. All stages between single isolated tubes encrusting rocks and reefs 2 m high are found in Ardbear Lough. As mentioned above, young worms normally have an encrusting habit and later grow upwards and free from the substrate. The advantages to the worm would appear to be, firstly, that the water is clearer above the bottom and therefore better for suspension feeding, and secondly the worm is less likely to be buried by sediment. With further larval settlement the young tubes will cover the substrate (Fig. 5A and Plate I:3) and eventually encrust older tubes. The free growing tubes may gain strength by intertwining and encrusting their neighbours to form an aggregation of tubes growing upwards away from the substrate (Fig. 5B). These aggregations now provide further sites for larval settlement and the worms grow upwards and outwards resulting in a bush-like growth structure illustrated in Plate I: 1. This pattern of growth continues, to form aggregations up to 2 m in height and 1 m in diameter which have originated from small rocky outcrops. The construction of these upward growing masses of tubes appears to follow a simple pattern. The distance to nearest neighbour has been measured for adult tube apertures (Fig. 6). The distribution is bimodal with a distinct peak at 5 mm and another

---

PLATE I

FIG. 1. Portion of reef illustrating pattern of tube growth. Scale, × 1/10.
FIG. 2. Underwater photograph of reef surface showing new white tubes and older tubes with algae and debris. Scale from *Asterias rubens* (15 cm diameter) feeding on *S. vermicularis*.
FIG. 3. Detail of tubes with encrusting young *S. vermicularis* and an encrusting fauna of bryozoans, spirorbids and *Pomatoceras triqueter*. Scale, × 2.
FIG. 4. Cross section through old tube of *S. vermicularis* to show algal borings. Scale, × 50.
FIG. 5. Old broken and encrusted portion of reef. Scale, × 1.
FIG. 6. Portion of fallen reef segment showing reorientation of tubes. Scale × 1/3.
FIG. 7. Cross section through tube aggregation to show young *S. vermicularis* and encrusting bryozoans. Scale, × 25.

Figs. 1 and 3 reproduced, by permission of the Editor, from D. W. J. Bosence, Recent serpulid reefs, Connemara, Eire. *Nature, Lond.* **242** (5392), 40–41 (1973).

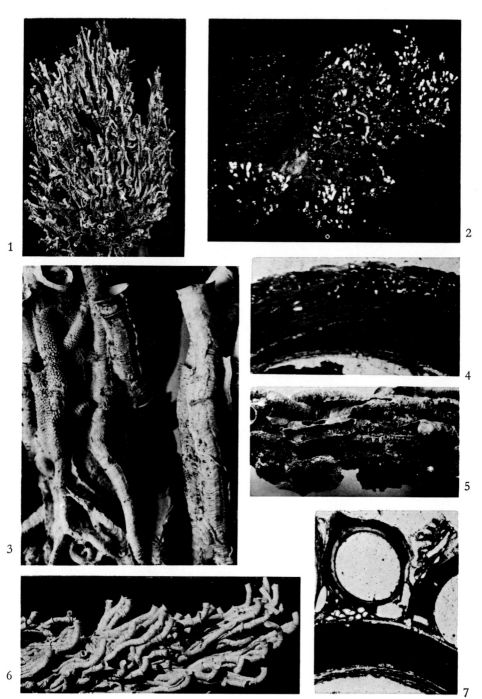

PLATE I. (Caption on facing page)

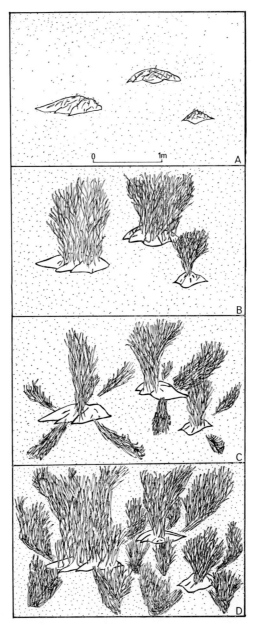

FIG. 5. Sketches to illustrate reef development. A, larval settlement restricted to hard rocky substrate. B, with increased settlement and restricted substrate larvae settle on old tubes and worms grow upwards and outwards to form bush-like aggregation. C, Bioerosion in the old parts of the reef causes fragments to fall off into the surrounding soft substrate areas. D, The fallen reef segments provide new areas for larval settlement and extend the original aggregations to form reefs up to 2 m high.

between 10 and 15 mm. The tubes at 5 mm apart are encrusting and the peak reflects the average tube diameter (5·2 mm). The second peak at 10–15 mm reflects an even (i.e. non random) spacing of non-encrusting tube apertures.

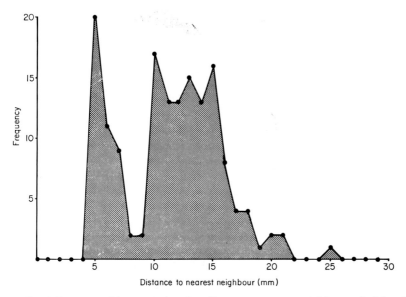

FIG. 6. Frequency histogram showing distance to nearest neighbour of adult tube apertures.

This is thought to be the result of the worms growing to a position where the expanded branchial crowns (diameter c. 15 mm) will not overlap or interfere. This spacing of the tubes gives the aggregations optimum strength from encrusting and optimum spacing for suspension feeding.

The shape of the aggregations (Plate I: 1, 6) is thought to be the result of the worms growing upwards and outwards to suitable suspension feeding sites.

As growth proceeds, the old base of the reef is weakened mainly as a result of biological erosion (Plate I: 4, 5) by boring sponges and algae, and biting by fish and echinoids (for details of fauna see below). This results in segments of the reef falling off and settling with the tubes orientated parallel to the substrate. Subsequent growth is reorientated vertically away from the substrate (Plate I: 6). These fallen reef segments provide large new areas for larval settlement and this is the main way in which a reef growing from an original rocky outcrop can expand to cover large areas of previous soft substrate (Fig. 5C and D) This process explains the extension of the reef downslope over an original mud bottom on transect $Y-Y_1$ (Fig. 8).

REEF BIOTA

The fauna of the reef is complex and has not been studied in detail. However, the following ecological and trophic groups have been recognized.

Encrusting on the *S. vermicularis* tubes are other hard substrate epifauna. The serpulid *Pomatoceras triqueter* is common and spirorbids (unidentified) are numerous (Plate I: 3). Calcified bryozoans are abundant and Dr P. Hayward has kindly identified the following species: *Berenicea patina*, *Escharella immisca*, *Escharoides coccineus*, *Microporella ciliata*, *Lichenopora hispida*, *Tubulipora phalangea* and *Callopora dunnerilii*. Non-calcified encrusting organisms including *Metridium* sp., *Clavelina lepadiformis* and *Halichondria* sp.

The sponge *Cliona celata* is the principal boring organism in the serpulid tubes but endolithic algal borings have been found in thin section.

Byssate attached bivalves are frequently found within the interstices of the reef. *Monia patelliformis* is common with very irregular shell forms conforming to the shapes of the serpulid tubes. The upper right valve may be subsequently encrusted by *S. vermicularis* and other encrusters. *Chlamys varia* is also common within the reef attached loosely by its byssus.

Predators and scavengers include the echinoids *Echinus esculentus* and *Psammechinus miliaris*. They have been observed feeding on serpulid tubes and later dissection shows the stomach full of tube debris. The rapid movement of *S. vermicularis* within its tube would suggest that it could only be caught if trapped in the end of its tube and it appears more likely that the tube is being eaten indiscriminately for the sake of its epifauna, flora and boring organisms. In addition the echinoids may be obtaining additional calcium for test growth. The asteroid *Asterias rubens* is frequently found feeding on the reef surface (Plate I: 2) with its stomach everted down the tubes of serpulids. *Buccinum undatum* is also common but has not been observed eating *S. vermicularis*. Adult specimens of *Cancer pagurus* are occasionally found, but they have not been observed feeding on serpulids either, and they may be using the reef just for protection. *Galathea squamifera* is abundant but it has not been seen feeding. It seems unlikely that it could break open serpulid tubes and it is probably using the reef for protection or feeding on something else. The wrasses *Ctenolabrus rupestris* and *Crenilabrus melops* are very common on and within the reef. They are frequently seen biting open serpulid tubes and extracting the worms. This is apparently a trial and error behaviour until a worm is seen and therefore generates a considerable amount of gravel sized serpulid debris.

In conclusion the serpulid reefs support a rich and diverse biota which is utilizing the aggregated worm tubes as a food source, for a substrate and for protection. The ecology of the fauna has parallels with the much more diverse

17. Serpula vermicularis: *Aggregation and Reef Formation*   311

and complex tropical coral reefs thus providing an example of ecological convergence at the community level.

DISTRIBUTION OF SERPULID REEFS

Figure 3 shows the distribution of serpulid reefs in Ardbear Lough. Outside the Lough *S. vermicularis* only occasionally occurs encrusting shells and pebbles. The serpulid reefs have a peripheral distribution and may be described as

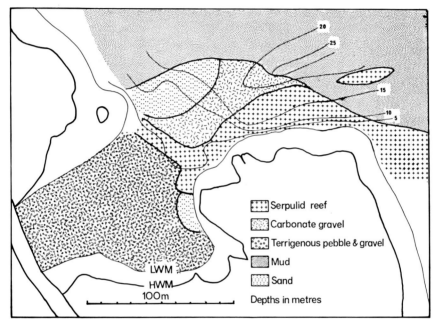

FIG. 7. Sediment distribution map for the south western corner of Ardbear Lough.

fringing reefs. To investigate the more detailed aspects of reef distribution grab sampling (stations about 20 m apart) and four transects (Fig. 3) along lines normal to the shoreline were undertaken. The results of the grab sampling are shown in Fig. 7. The four transects were lines labelled at 5 m intervals. Using a 50 cm × 50 cm quadrat positioned on the line the percentage abundance of the five substrate types and the algal cover was estimated and recorded on slates. The results of two transects are shown on Fig. 8.

The maps and transects show that the steep slope leading to depths of 30 m was originally rocky and this levels off to a muddy bottom on transects $X-X_1$, $Y-Y_1$. A rocky outcrop is found at the end of $Y-Y_1$ forming a ridge along the slope. This steep slope is in contrast to the shallow platform at the mouth of the Lough over which tidal currents sweep. On this platform terrigenous pebbles and gravels are found in equal proportions with carbonate gravels. Some of

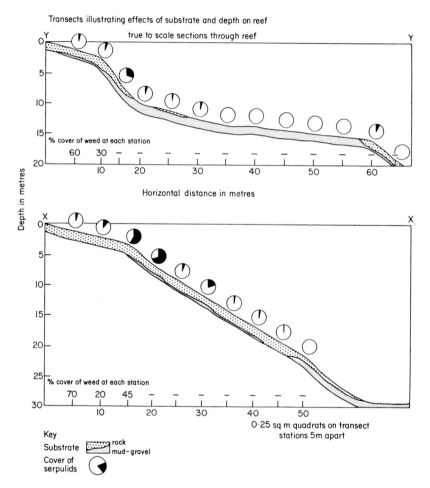

FIG. 8. Transects illustrating effects of substrate and depth on reef development. The percentage of cover for each substrate category is depicted in the band immediately beneath the transect profile. The thickness of the band represents 100%.

these carbonate grains are derived from occasional reefs and some from neighbouring littoral and sublittoral rock. The shoreline and shallow subtidal zone is either rocky or composed of terrigenous pebble and gravel. The transects also show the cover of serpulids, brown and green algae (*Codium tomentosum*

Ascophyllum nodosum, *Aspercoccus* sp. and *Chorda filum*) and epilithic corallines. The uncalcified algae are restricted to the upper 5 m of water whilst the corallines extend to 15 m. The reefs are seen to have a maximum development between 2 and 15 m but can extend to 19 m. The presence of the reefs is dependent on the original substrate. In general, where a soft substrate bottom is developed no reefs are found. This is due to the serpulid larvae requiring a hard substrate for settlement. This is well demonstrated in transect $Y-Y_1$. Pebbles and rocks support reefs to a depth of 7 m. The substrate then changes to a muddy gravel and then to a gentle muddy slope. Some patches of *S. vermicularis* are found below the rocky area. These are portions of the reef that have fallen downslope, so providing additional areas for larval settlement. The mud bottom continues to 17 m with no reef until a rock outcrop appears which supports serpulids. The serpulids do not, however, extend beyond 18 m despite the available substrate.

Besides substrate availability there appears to be a depth restriction on reef development. In shallow waters an inverse relationship exists between the growth of the brown and green algae and serpulids. All these organisms require a hard substrate and there may be competition for sites in this shallow water area. The algae may be restricted from growing at greater depths in the Lough due to light requirements but thrive in the shallow sheltered shores of the Lough. The algal community includes species which are known to inhabit quiet waters (Lewis, 1964).

At the lower limit of reef development and down to depths of 30 m the water contains a large proportion of mud. The mud is suspended in horizontal layers and the concentration increases downwards to pass into a very fluid muddy substrate. The mud is black and the presence of hydrogen sulphide in all the grab samples indicates a reducing environment. Additional evidence for the low oxygen levels comes from the absence of any macro-organisms. In this area all rocky outcrops are covered in layers of mud. The conditions suggest a very quiet environment with low oxygen levels and the water is probably only mixed during spring tides (Fig. 4) (cf. Lough Ine, Bassindale *et al.*, 1957).

Therefore from the transects the serpulids can be seen to have, in addition to a substrate restriction, a limited depth distribution. To investigate this further experiments were devised with settling plates and the transfer of reef sections.

1. Settling Plates

Plates of slate (*c*. 20 cm × *c*. 20 cm) were placed sub-horizontally at a range of depths along transect $X-X_1$ to investigate settling of the larvae. The depths and numbers of newly settled *S. vermicularis* on the plates, which were left for two weeks at the end of August 1972, are shown in Table I.

These results show that settlement was most abundant between the depths

TABLE I. Depths and numbers of newly settled *S. vermicularis* on plates left for two weeks of August 1972

Depth (m)	No. of young on upper surface	Average tube length (mm)	No. of young on lower surface	Average tube length (mm)
1·5	19+ algae	1·2	—	—
4·6	24+ algae	1·5	5	2
7·4	—	—	180	1·7
11	—	—	5	1·7
14	—	—	120	1·8
17	—	—	34	2·2
20	—	—	—	—
26	—	—	—	—

of 1·5 and 17 m. On the shallowest plates it is interesting to note that in addition to *S. vermicularis* there are algae settling on the upper surface. This is further evidence that the serpulids and algae are in competition in the shallow-water zone. The lower limit of larval settlement coincides with the mud-rich and possibly oxygen-poor water described above. These lowest settling plates were covered in a thin layer of mud which may well have prevented settlement (Straughan, 1972).

These results compare well with Dybern's (1967) work on larval settlement in Norwegian fjords. He found an upper low-salinity zone colonized mainly by algae but the serpulid *Pomatoceras triqueter* was found to be abundant in middle depths. However, below 6 m in the inner part of the fjord at oxygen levels of 2·2–6·2 ml O_2 l^{-1} no settlement was found. The greater settlement on the lower surface of the plates corresponds with previous work (e.g. Sentz-Braconnot, 1968). This may be attributable to a negative phototropism and positive geotropism which was demonstrated for the rock encrusting *Spirorbis tridentatus* by de Silva (1962). He concludes that the larvae are led away from the brightly lit areas with algae, to the bottom for settlement on rocky substrates. Similarly Dybern (1967) shows an increased settlement of *P. triqueter* on the shady side of vertically held plates. However, other factors may be involved such as temperature, algal growth etc. (ten Hove this volume). The settlement on the upper side of some of the shallow plates is not understood.

2. Transfer of reef segments

For this experiment portions of living reef were removed from the central area of maximum growth and placed at depths varying from 11 to 26 m. The results of the experiments are shown below. The condition of the transferred worms was assessed by (1) testing for a positive response to changes in light, (2) testing for a positive response to touch, (3) noting mud deposition over the tubes, and (4) finding dead serpulids in tubes. These tests indicate progressive stages of deterioration of the serpulids. The numbers in Table II show the results of the above assessments and therefore indicate how far the serpulids had deteriorated during the period of the experiments.

TABLE II. Deterioration of serpulids during reef segment transfer experiments

Depth (m)	20.7.72 (Date of transfer)	1.8.72	8.8.72	12.8.72
		(Dates of subsequent inspections)		
11	1	1	1	1
14	1	1	1	1
17	1	1	Serpulids not found	4
20	1	2	3	4
24	1	2	3	4
26	1	Serpulids not visible	4	Serpulids not visible

These experiments indicate a lethal environment to the transferred serpulids at depths greater than 20 m. This depth corresponds to the maximum depth limit of the reefs and of larval settlement. I interpret the results as reflecting the unfavourable effects of settling mud clouds, which would smother the worms, and of probable low oxygen levels.

In conclusion, the reefs have an overall horizontal distribution controlled by the occurrence of rock outcrops in sublittoral areas, and a vertical distribution of between 2 and 20 m. The upper limit of distribution is interpreted as being due to lowered salinity, higher light values and competition for space by algae. The lower limit is thought to be the result of suspended mud and low oxygen levels.

DISCUSSION

In this study the factors leading to the aggregation of *S. vermicularis* show some of the features of belt-forming species and some of the features associated with lagoonal mass occurrences (ten Hove, this volume). The vertical distribution of the reefs is shown to be limited to depths of 2–20 m, within a possible range provided by hard substrates from 0m (MLW) to 25 m depth. This restricted zone is the result partly of biological competition for space and partly of physical factors such as mud deposition, low oxygen levels and salinity. The limited availability of hard substrate in the Lough is also restricting settlement sites for *S. vermicularis*. Previous to reef growth about 5% of the Lough floor would have been hard substrate. This has now increased to about 25% by the development of reefs.

The importance of the restricted substrate for gregariousness is greatly increased by the enclosed nature of the Lough which is thought to prevent the dispersal of the planktonic larvae. Ardbear Lough exchanges about 10% of its water per tidal cycle. However, because of the narrow and enclosed nature of Ardbear Bay (Fig. 1) a large proportion of the outflowing water will be returned in the succeeding flood tide. In addition to this, water is only exchanged in the area near the Lough mouth during spring tides, whilst in neap tides only the uppermost low salinity water is lost (Fig. 4). The duration of the planktonic phase of *S. vermicularis* is not known but from a comparison with other serpulids (ten Hove, 1974) it would be expected to be between one and eight weeks. Ketchum (1952) found the loss of larvae in open estuaries to be much higher than in enclosed estuaries. This is well illustrated by Korringa's (1952) work on the nearly landlocked Oosterschelde. Here only 4% of oyster larvae are lost per tidal cycle due to dispersion. This low figure means that the greatest losses are incurred through predation which in this case results in a 14% loss per tidal cycle and between 2·5 to 10% of the larvae reaching the setting stage.

Similarly Bousfield (1955) studied the retention of barnacle larvae in the fairly open Miramachi estuary. In this case the larvae managed to remain in the estuary despite tidal currents because of vertical movement by the larvae in the water column. More recently Pearson (1970) has discussed the effect of larval retention in fjordic systems in relation to recruitment and colonization of benthic soft substrate communities. Finally Keegan (1974) has shown that abnormally high concentrations of *Paracentrotus lividus* on an unusual substrate (maerl) may be the result of larval retention right up to their critical period for settlement.

From the information found in Ardbear Lough it would appear that larval retention is an important contributory factor leading to the serpulid aggregations.

This is supported by my recent finding of similar *S. vermicularis* aggregations in the very enclosed Lhinne Mhurich, near Tayvallich, Argyllshire, Scotland. In this case, as in Ardbear Lough, there is limited hard substrate for larval settlement.

The biological factors controlling the formation and distribution of the reefs do not appear to be as important as the physical factors. Competition for space can only be demonstrated with respect to algae in the shallow sublittoral. The absence of barnacles (possibly due to low salinity) may be of importance in reducing competiton for shallow water sites in Ardbear Lough. Straughan (1972) describes the efficiency of *Balanus* in removing *Mercierella enigmatica* from hard substrates in the higher salinity areas of the Brisbane river. It seems unlikely that predation is much lower in the Lough than in the open sea as there are many predators on the reefs which in the open sea would be dispersed over a wide range of habitats. Primary production has not been measured but comparisons with estuarine environments suggest a slightly higher production than the open marine environment (Ketchum, 1967; Walne, 1971).

The dense settlements of larvae on adult tubes of *S. vermicularis* may indicate larval gregarity but, it must be emphasized, that there are few other hard substrates available for larval settlement in the Lough. *S. vermicularis* living in the open marine environment is not normally gregarious.

In conclusion there are many factors which may be responsible for the aggregations of *S. vermicularis* in Ardbear Lough. Without extended experiments it is not possible to point to any, one, overriding factor. The factors which are thought to be combining to cause these large serpulid aggregations are larval retention in the partially enclosed Lough, limited availability of hard substrates, slightly reduced competition for space and possible high primary productivity. Larval gregarity can not be ruled out as an additional contributory factor.

REFERENCES

Bassindale, R., Davenport, E., Ebling, F. J., Kitching, J. A., Sleigh, M. A. and Sloane, J. F., (1957). The ecology of the Lough Ine rapids with special reference to water currents. IV. Effects of the rapids on the hydrography of the south basin. *J. Ecol.* **45**, 879–900.

Bornhold, R. D. and Milliman, J. D. (1973). Generic and environmental control of carbonate mineralogy in serpulid tubes. *Journ. Geol.* **81**, 363–373.

Bosence, D. W. J. (1973). Recent serpulid reefs, Connemara, Eire. *Nature* **242**, 40–41.

Bousfield, E. L. (1955). The ecological control of the occurrence of barnacles in the Miramichi estuary. *Bull. natn. Mus. Canada*, no. 137, Biol. Ser. No. 46.

de Silva, P. H. D. H. (1962). Experiments on the choice of substrates by spirorbis larvae. *J. exp. Biol.* **39**, 483–490.

Dybern, E. I. (1967). The settlement of sessile animals on eternite slabs in two polls near Bergen. *Sarsia* **29**, 137–150.

Fauvel, P. (1927). Faune de France 16. Polychètes Sédentaires. Fed. Franc. Soc. Sci. Nat. Paris.

HEDLEY, R. H. (1956). Studies of serpulid tube formation. I. The secretion of the calcareous and organic components of the tube by *Pomatoceras triqueter*. *Quart. Journ. Micros. Sci.* **97**, 411–419.

HOVE, H. TEN (1974). Notes on *Hydroides elegans* (Haswell, 1883) and *Mercierella enigmatica* (Fauvel, 1923), alien serpulid polychaetes introduced into the Netherlands. *Bull. Zool. Mus. Univ. Amsterdam* **4**, 45–51.

KEEGAN, B. (1974). The macrofauna of maerl substrates on the west coast of Ireland. *Cahiers de Biol. Mar.* **XV**, 513-530.

KETCHUM, B. H. (1952). The relation between circulation and planktonic populations in estuaries. *Woods Hole Oceanographic Inst. Coll. Rept.* No. 598.

KETCHUM, B. H. (1967). Phytoplankton nutrients. *In* "Estuaries" (G. H. Lauff, ed.), American Assoc. Adv. Sci.

KORRINGA, P. (1952). Recent advances in oyster biology. *Q. Rev. Biol.* **27**, 266–308.

LEWIS, J. R. (1964). "The ecology of rocky shores" English Universities Press, London.

MCINTOSH, W. C. (1923). "A monograph of the British marine annelids. **4**. Polychaeta–Sabellidae to Serpulidae." pp. 251–538. Ray Society.

NEFF, J. (1969). Mineral regeneration by serpulid polychaete worms. *Biol. Bull.* **136**, 76–90.

NELSON-SMITH, A. (1967). "Catalogue of marine fouling organisms" Vol. 3, Serpulids. O.E.C.D. Paris.

PEARSON, T. (1970). The benthic ecology of Loch Linnhe and Loch Eil, a sea loch system on the west coast of Scotland. 1. The physical environment and distribution of the macrobenthic fauna. *Journ. Exp. mar. biol. ecol.* **5**, 1–34.

SENTZ-BRACONNOT, E. (1968). Données écologiques et biologiques sur la fixation des serpulidae dans la Rade de Villefranche-sur-Mer. *Vie Milieu*, **19** B (Oceanogr.), 109–132.

STRAUGHAN, D. (1972). Ecological studies on *Mercierella enigmatica* Fauvel in the Brisbane river. *J. anim. ecol.* **41**, 93–136.

WALNE, P. R. (1971). *In* "The Estuarine Environment" (R. S. K. Barnes and D. S. Green, eds), Applied Science Publishers.

ZIBROWIUS, H. (1973). Serpulidae des côtes ouest de l'Afrique et des archipels voisins. *Mus. Roy. de l'Afrique centrale. Tervuren, Belgique Annales.* Ser. 8. Sci. Zool., No. 207, 1–89.

18 | Dispersal and Re-aggregation in Sessile Marine Invertebrates, particularly Barnacles

D. J. CRISP

N.E.R.C. Unit of Marine Invertebrate Biology, Marine Science Laboratories, Menai Bridge, Gwynedd, U.K.

Abstract: Unlike some other sessile marine invertebrates that form aggregations—such as *Sabellaria*, *Pomatoleios* and *Crassostrea*—all barnacles are internally fertilized. Many are also obligatory cross-fertilizers, so that their survival depends upon the adults being in close proximity to one another. In the context of the planktotrophic larval stage, common to nearly all cirripedes, the evolution of some means of re-aggregation becomes essential to the sessile habit. One alternative is by gregarious behaviour at settlement. In barnacles this is achieved by specific recognition of the surface of their own species. An alternative strategy for re-aggregation is for the adults to be associated with specific and relatively rare substrata, such as the surfaces of other organisms. However, the separation into different epizoic or epiphytic habitats may eventually lead to sub-speciation. Excessive aggregation might cause too much competition, and therefore a spacing out mechanism at close range has been evolved.

We might well speculate why a less complex solution was not evolved, such as the suppression of the pelagic stage as in fresh water species. Evidently in the sea, larval dispersal, high fecundity and the resulting genetic panmixia enabling species to colonize unstable, isolated and varied habitats offer sufficient advantage to allow the pelagic larva to persist despite its obvious drawbacks.

AGGREGATIONS OF MARINE ORGANISMS

Clusters of individuals can be formed either by reproduction without dispersal, or by reproductive dissemination followed by re-aggregation. Examples of the former mechanism include asexually budding animals, such as sea anemones, ascidians and polychaetes such as *Filograna*, all of which differ only trivially from colonial forms where the individuals fail to sever their mutual connections. There are also sexually reproducing organisms whose offspring remain close

---

Systematics Association Special Volume No. 11, "Biology and Systematics of Colonial Organisms", edited by G. Larwood and B. R. Rosen, 1979, pp. 319–327. Academic Press, London and New York.

together, such as the viviparous *Littorinas*, some hydroids such as *Tubularia*, and many marine algae.

Re-aggregation after dispersal of a free swimming larval stage is remarkably common in the sea. In some cases aggregations may become so firmly cemented together as to produce reef-like structures. Good examples are *Sabellaria alveolata*, *Pomatoleios kraussi* and *Crassostrea virginica*.

Rock-living barnacles also re-aggregate, often crowding the intertidal zone to the exclusion of other species. When individuals are thus closely packed, their lateral plates interlock (Gutmann, 1960) so that their growth, largely confined to a direction at right angles to the substratum, results in the formation of tall, columnar individuals. Those at the edge of such clusters may be forced to slide outwards (Crisp, 1960) while those at the centre may become so severely restricted that they acquire a trumpet-like form and eventually may lose contact with the substratum so that the whole group breaks away (Barnes and Powell, 1950). Some species, such as *Balanus hameri* and *Balanus psittacus*, form tree-like growths as successive settlements attach to the sides of older individuals. The remarkable powers of re-aggregation at the cyprid stage have led distinguished observers, quite mistakenly (Crisp and Knight-Jones, 1953), to doubt the intervention of a free swimming stage (Thorson, 1946; Broch, 1924; Barnes, 1953).

ADVANTAGES OF AGGREGATION

No doubt in all cross-fertilizing organisms greater proximity tends to produce a higher level of reproductive success. Yet whereas in externally fertilizing animals a relatively isolated individual has some chance of breeding through water-borne sperm, a sessile animal which is internally cross-fertilizing can breed only if it lies sufficiently close to its neighbour for the intromittent organ to reach it. Barnacles, as a group, are internally fertilizing hermaphrodites, and the majority of those studied are obligatorily cross-fertilizing (Barnes and Crisp, 1956). Consequently there is a premium not only on settling in close proximity but in monopolizing the whole of the selected surface. For were a mixture of individuals of several different species scattered at random, some might be unpaired with their own species and therefore unable to reproduce.

A few barnacles, such as the predominantly deep-water *Scalpellum*, the Gorgonian epizoite *Conopea* and the boring *Acrothoracica*, all of which often exist as isolated individuals, retain sexuality by the development of dwarf or complemental males. These males reach the larger hermaphrodite or female individual independently after a pelagic larval stage, and thus maintain the same degree of genetic exchange as do barnacles which form massed populations.

Other possible advantages accruing to species which form large aggregates may be put forward. Competitors will be excluded by the monopoly of the habitat, while large reef-like masses of interlocked individuals may provide mutually beneficial defences both against predators and physical forces, especially for the smaller individuals. Moreover, the most recently settled individuals, which would be vulnerable to intraspecific competition, or to predation at the edge of the colony, will enjoy the fastest growth since they will tend to settle on the outside and benefit most from the surrounding food-bearing water currents. In *Vermetes* reefs, individuals may benefit co-operatively by sharing mucus nets. All such beneficial interactions will lead to the evolution of aggregation, but by far the most widespread and essential reason that dispersed larvae should settle close together, and the only reason where an individual's interaction is mainly competitive, must be the successful exchange of gametes.

RESPONSES LEADING TO RE-AGGREGATION

There are two principle methods by which re-aggregation is achieved. Both depend on contact recognition of particular species, after which the exploratory movements of the larvae consummate in settlement and metamorphosis. The differences between the methods by which aggregation is achieved depend on the organism to which the larva responds. When the larva settles in response to contact with its own species, A, it displays "gregarious settlement"; when in response to another species, B, it displays "associative settlement" (Crisp, 1974). It may not be immediately obvious why associative settlement should result in aggregation; however, since the species, B, to which the larva, A, responds presents a very small proportion of available surfaces, larvae must congregate on these limited areas. Examples of gregarious settlement include not only many barnacles, most notably *Balanus balanoides*, but also oysters (Cole and Knight-Jones, 1949) and polychaetes (Knight-Jones, 1951). Some gregariously settling species also cement themselves together forming reef-like structures, for example *Crassostrea virginica* (Crisp, 1967), *Sabellaria alveolata* (Wilson, 1968) and *Pomatoleios kraussi* (M. Crisp, 1977).

Examples of associative settlement are also afforded by barnacles: *Pyrgoma* species are epizoic on corals, *Conopea* (Say) on *Gorgonia*, *Platylepas* on turtles, *Anelasma* on sharks and *Coronula* on whales, while *Chelonibia patula*, *Octolasmis* sp. and *Sacculina carcini* attach to crabs with increasing dependence on the host tissues. Many polyzoa and spirorbinae associate with various species of algae (see Crisp, 1974) while invertebrates living in deposits may respond to bacteria or bacterial films present on the surface of the grains (Wilson, 1955;

Gray, 1966). In general, it is believed that associative settlement is not so much a response to food but rather to chemical clues indicating a suitable choice of habitat. However, there are examples of predatory nudibranchs which settle specifically on their prey species (Thompson, 1958; Tardy, 1962). In a few rare instances, habitat choice is made by selecting clean or scoured surfaces, a good example being the planula of the hydroid *Hydractinia echinata* which is epizoic on hermit crab shells which are constantly abraded as the crab drags the shell over the sand (Teitelbaum, 1966).

The mechanism of gregarious settlement has been studied most comprehensively in barnacles, but the results probably apply to oysters and polychaetes with which there are close parallels. The settling stimulus in *Balanus balanoides* (Knight-Jones, 1953) and *Crassostrea virginica* (Crisp, 1967) is the virtually insoluble, yet quite specific, tanned proteinaceous material covering the adult's shell, while in *Sabellaria* sp. (Wilson, 1968, 1970) it is the cement binding the sand tubes which is probably equally insoluble. In all three examples the larvae must make contact before a response is elicited. Nevertheless, it has been found that surfaces can be made attractive to barnacles by coating them with crude extracts of barnacles or other arthropod integumental proteins (Crisp and Meadows, 1962, 1963); this "arthropodin" stimulates only when applied to a surface. Nott and Foster (1969), impressed by the presence of what appeared to be chemosensory organs on the attachment disc of the cypris larvae of barnacles, and by their evident use of this organ to probe a surface, suggested that it might secrete proteases and specifically recognize the soluble hydrolysates thus liberated. Crisp (1975), noting the manner in which the cyprids tug at the substratum, suggests on the contrary that physical adhesion between the pilose surface of the attachment disc and the substratum might provide indirect evidence of the composition of the latter, and cited the analogy of the specific surface recognition in immunological phenomena. Further, the work by Gabbott and Larman (1971) and Larman and Gabbott (1975) have shown that settlement substances of quite different origin have similar physical and biochemical properties and are probably less specific than was at first thought. This finding is consistent with the adhesion theory, but more evidence of the nature of the sensory mechanisms involved in the response is clearly needed to make a convincing case. The similarity of settlement-inducing substances in unrelated species has probably been significant in the evolution of associative settlement. A small heritable change in the recognition mechanism might result in association with another organism rather than with its own species. Were the new habitat beneficial, the individuals carrying this trait would be isolated and successful independent evolution could proceed. Evidence of the

18. Sessile Marine Invertebrates: Dispersal and Re-aggregation

evolution of a similar change in habitat choice has been provided by Knight-Jones *et al.* (1971).

COMPETITION AND TERRITORIALITY

A gregarious or associative response at settlement may bring individuals so closely together that intraspecific competition becomes a limiting factor. In a study of the productivity of an experimental settlement of *Balanus balanoides* initially spaced at various population densities, I found that at distances of less than a centimetre apart, the decrease in tissue growth through competition almost exactly compensated for any increase in the biomass of the additional number of individuals accommodated. Thus the biomass produced remained constant although the settlement density increased (Crisp, 1964). Such high settlement densities therefore represent a wastage of larvae. Furthermore, the rate of egg production actually fell as population numbers rose, implying that fitness was even further depressed. Not surprisingly, therefore, mechanisms have been evolved which oppose excessive aggregation of individuals at close range. This phenomenon may be termed territoriality.

Knight-Jones and Moyse (1961) noted that the settling larvae space themselves out from one another and from adults, while Crisp (1961) showed that barnacle cyprids, and Wiseley (1960) that *Spirorbis* larvae, achieved this end as a result of a distinct behaviour pattern. Towards the end of the exploratory phase they moved in a diminishing spiral of small steps which ensured that settlement took place at a distance from any small raised obstacles on the surface. Where settlement space was limited, the larvae would tend to swim away, thus reducing wastage. Knight-Jones and Moyse believed that larvae not only settled gregariously, but also avoided their own species through chemical recognition, yet attempted to settle in the proximity of other species which they would hopefully stifle. Crisp and Wiseley, however, concluded that a non-specific tactile response would better account for territoriality. Recently Barnett and Crisp (in prep.) were able to show that when larvae of two species of barnacle were induced to settle simultaneously, they did not avoid their own species but showed gregarious behaviour throughout. Even when settling in very close proximity within small, and very favourable, pits the two species still tended to segregate.

Although territoriality is clearly advantageous, the actual distance over which the individuals seem to exert mutual repulsion is much smaller than would allow for optimal growth. Evidently, there is a balance between some degree of intraspecific competition and the possibility of leaving unoccupied space for alien competitors.

GENETIC AND SEXUAL CONSEQUENCES OF DISPERSAL AND RE-AGGREGATION

Where groups of sessile organisms are associated closely together the degree of inbreeding among them will obviously depend on their individual origins. Isolated associations arising from the vegetative propagation of zygotes, such as clones of anemones and corals, ascidian and bryozoan colonies, can breed only if they are hermaphrodite, and even then will have only a limited repertoire of genetic variation restricted to the re-distribution and re-arrangement of the two sets of parental chromosomes. Frequently in such animals, anti-self mechanisms exist so that intercolony breeding alone is possible (Sabbadin, 1978; Francis, 1973), but where associations of sessile organisms result from re-aggregation of individual larvae from widely separated populations, a degree of outbreeding comparable with that of fully mobile species becomes possible. In some densely packed aggregations of this type, separate sexes exist (e.g. *Crassostrea, Pomatoleios*) but hermaphroditism or an alternation of sex remains a common feature of sessile and slow moving species, even of those with larval dispersal mechanisms. Possibly the risk that pairs or triplets of the same sex may settle together with resulting sterility is not worth the taking (see Tomlinson, 1966). Mechanisms preventing or reducing the possibility of self-fertilization in hermaphrodites exist in barnacles (Barnes and Crisp, 1956) and probably also in other sessile animals (e.g. Spirorbinae, Gee and Williams, 1965).

Occasionally organisms which are apparently single and isolated may in fact be chimeras of different individuals with outbreeding possibilities. Fry (1971) cites the swarming of sponges and their subsequent consolidation into a single mass, while Jones (1956) reports similar co-operative associations in the red alga *Gracilaria*. Such cases, though analogous to the assemblages of invertebrates derived from dispersed larvae, usually originate from the same or from closely proximate parents, thus not greatly reducing the degree of inbreeding.

In species with larval transport, genetic panmixia is maintained over large areas whose size will depend on the free-floating life of the larva and the prevailing water movements (Crisp, 1978). After settlement, however, barnacles and other sessile and semi-sessile animals can breed only with those close to them. Consequently, if severe local selection operates on the post-larvae, the large or small enclaves of similar genetic constitution which remain will inbreed and so enhance the proportions of locally advantageous genes and probably of their homozygotes. In the next generation dispersal will tend to re-assert panmixia, but it may not succeed completely in re-distributing the effects of selection, especially over the larger enclaves. Furthermore, selection is likely to take place, not only in the post-larva but also in the free-swimming stage, while different genes may be selected at different seasons. Thus the resulting population might be an assemblage of essentially different populations, in each of which particular homozygotes had been selected in relation to the different conditions and habitats in which they arose. Such a multiple niche situation in space and time might be a factor giving rise to the excess of homozygotes which so frequently cause deviations from Hardy-Weinberg expectation to be observed in populations of marine invertebrates.

Larval dispersal and subsequent re-aggregation is clearly a wasteful process requiring high fecundity and the assignment of much of the available energy to reproduction; it must therefore confer a great and continuing advantage. One of these advantages is the ability to survive on unstable and isolated habitats from which larval transport offers a means of escape (Crisp, 1976). Furthermore, species whose genetic material is sexually re-assorted just before dissemination, will enjoy a large number and range of genotypes in their offspring. Thus the combination of high genetic variability (Wilkins, 1975; Crisp, 1978) with the possibility of the larvae choosing from a variety of sites, combine to enable such species, generation after generation, to fill a diverse range of isolated niches in the sea. As Williams (1975) points out, evolution of sex meets a recurrent need for the organism to adapt itself to the new conditions that might prevail in the next generation. It is not primarily designed to create once and for all an optimal genotype. In exactly the same way, larval dispersal serves less for the rare occasion when a species might transgress a geographical barrier and invade new territory, than for the continuing requirement for each generation to find different habitats and escape from locations which have become unsuitable.

REFERENCES

Barnes, H. (1953). Orientation and aggregation in *Balanus balanus* (L.) Da Costa. *J. anim. Ecol.* **22**, 141–148.

Barnes, H. and Crisp, D. J. (1956). Evidence of self fertilisation in certain species of barnacles. *J. mar. biol. Ass. U.K.* **35**, 631–639.

Barnes, H. and Powell, H. T. (1950). The development, general morphology and subsequent elimination of barnacle populations *Balanus crenatus* and *B. balanoides* after a heavy initial settlement. *J. anim. Ecol.* **19**, 175–179.

Broch, H. (1924). Cirripedia thoracica von Norwegen und dem norwegischen Nordmeere. Eine systematische und biologisch-tiergeographische Studie. *Skr. Vidensk Selsk. Christ. I. Mat. Naturv. Klasse No.* 17.

Cole, H. A. and Knight-Jones, E. W. (1949). The settling behaviour of larvae of the European oyster *Ostrea edulis* L., and its influence on methods of cultivation and spat collection. *Fishery Invest., Lond.* Ser. II, 17 (3), 1–39.

Crisp, D. J. (1960). Mobility of barnacles. *Nature, Lond.* **188**, 1208–1209.

Crisp, D. J. (1961). Territorial behaviour in barnacle settlement. *J. exp. Biol.* **38**, 429–446.

CRISP, D. J. (1964). An assessment of plankton grazing by barnacles. In "Grazing in Terrestrial and Marine Environments" 4th Symp. Brit. Ecol. Soc. (D. J. Crisp, ed.), pp. 251–264. Blackwell Scientific Publications, Oxford.
CRISP, D. J. (1967). Chemical factors inducing settlement in *Crassostrea virginica* Gmelin. *J. Anim. Ecol.* **36**, 329–335.
CRISP, D, J. (1974). Factors influencing the settlement of marine invertebrate larvae. In "Chemoreception in Marine Organisms" (P. T. Grant and A. N. Mackie, eds), pp. 177–265. Academic Press, New York.
CRISP, D. J. (1975). Surface Chemistry and life in the sea (The Sir Eric Rideal Lecture 1974). *Chemistry and Industry*, March 1975, pp. 187–193.
CRISP, D. J. (1976). The role of the pelagic larva. In "Perspectives in Experimental Biology" Vol. I, Zoology (P. Spencer Davies, ed.). pp. 145–155. Pergamon Press.
CRISP, D. J. (1978). Genetic consequences of different reproductive strategies. In "Marine organisms: Genetics, Ecology and Evolution" (B. Battagha and J. Beardmore, eds), pp. 257–273. Plenum Press, New York.
CRISP, D. J. and KNIGHT-JONES, E. W. (1953). The mechanism of aggregation in barnacle populations. A note on a recent contribution by Dr H. Barnes. *J. anim. Ecol.* **22**, 360.
CRISP, D. J. and MEADOWS, P. S. (1962). The chemical basis of gregariousness in cirripedes. *Proc. R. Soc.* B **156**, 500–520.
CRISP, D. J. and MEADOWS, P. S. (1963). Adsorbed layers: the stimulus to settlement in barnacles. *Proc. R. Soc.* B **158**, 364–387.
CRISP, M. (1977). The development of the serpulid *Pomatoleios kraussii* (Annelida, Polychaeta). *J. zool. Soc. Lond.* **183**, 147–160.
FRANCIS, L. (1973). Clone specific segregation in the sea anemone *Arthropleura elegantissima*. *Biol. Bull. mar. biol. Lab. Woods Hole* **144**, 64–72.
FRY, W. G. (1971). The biology of larvae of *Ophlitaspongia seriata* from two North Wales populations. In "Proc. Fourth European Marine Biology Symposium" (D. J. Crisp, ed.), pp. 155–178. Cambridge University Press.
GABBOTT, P. A. and LARMAN, V. N. (1971). Electrophoretic examination of partially purified extracts of *Balanus balanoides* containing a settlement inducing factor. In "Proc. IV European Marine Biology Symposium" (D. J. Crisp, ed.), pp. 143–153. Cambridge University Press.
GEE, J. M. and WILLIAMS, G. B. (1965). Self- and cross-fertilisation in *Spirorbis borealis* and *S. pagenstecheri. J. mar. biol. Ass. U.K.* **45**, 275–285.
GRAY, J. S. (1966). The attractive factor of intertidal sands to *Protodrilus symbioticus*. *J. mar. biol. Ass. U.K.* **46** (3), 627–645.
GUTMANN, W. F. (1960). Functionelle Morphologie von *Balanus balanoides*. *Abh. Senckenb. naturf. Ges.* **500**, 43 pp.
JONES, W. E. (1956). Effect of spore coalescence on the early development of *Gracilaria verrucosa* (Hudson) Papenfuss. *Nature, Lond.* **178**, 426–427.
KNIGHT-JONES, E. W. (1951). Gregariousness and some other aspects of the settling behaviour of *Spirorbis*. *J. mar. biol. Ass. U.K.* **30**, 201–2.
KNIGHT-JONES, E. W. (1953). Laboratory experiments on gregariousness during settling in *Balanus balanoides* and other barnacles. *J. exp. Biol.* **30**, 584–598.
KNIGHT-JONES, E. W. and MOYSE, J. (1961). Intraspecific competition in sedentary marine animals. *Symp. Soc. exp. Biol.* **15**, 72–95.

KNIGHT-JONES, E. W., BAILEY, J. H. and ISAAC, M. J. (1971). Choice of algae by larvae of *Spirorbis* particularly of *Spirorbis spirorbis*. *In* "Proc. Fourth European Marine Biology Symposium" (D. J. Crisp, ed.), pp. 89–104. Cambridge University Press.

LARMAN, V. N. and GABBOTT, P. A. (1975). Settlement of cyprid larvae of *Balanus balanoides* and *Elminius modestus* induced by extracts of adult barnacles and other marine animals. *J. mar. biol. Ass. U.K.* **55**, 183–190.

NOTT, J. A. and FOSTER, B. A. (1969). On the structure of the antennular attachment organ of the cypris larva of *Balanus balanoides* (L.). *Phil. Trans. R. Soc.* B **256**, 115–134.

SABBADIN, A. (1977). Genetics of the colonial ascidian *Botryllus schlosseri*. *In* "Marine Organisms: Genetics, Ecology and Evolution" (B. Battagha and J. Beardmore, eds), pp. 195–209. Plenum Press, New York.

TARDY, J. (1962). Observations et des expériences sur la métamorphose et la croissance de *Capellinia exigue*. *C.r. Hebd. Séanc. Acad. Sci. Paris* **254**, 2242–2244.

TEITELBAUM, M. (1966). Behaviour and settling mechanism of planulae of *Hydractinia echinata*. *Biol. Bull. mar. biol. Lab. Woods Hole* **131**, 410–414.

THOMPSON, T. E. (1958) The natural history, embryology, larval biology and post-larval development of *Adelaria proxima* (Alder and Hancock) (Gastropoda Opisthobranchia). *Phil. Trans. R. Soc.* B **242**, 1–58.

THORSON, G. (1946). Reproduction and larval development of Danish marine bottom invertebrates. *Medd. Komm. Havunderseg Kbh. (Fisk)* **4** (1) 523 pp.

TOMLINSON, J. (1966). The advantages of hermaphroditism and parthenogenesis. *J. Theoret. Biol.* **11**, 54–58.

WILKINS, N. P. (1975). Phosphoglucose isomerase in marine molluscs. *In* "Isozymes. IV. Genetics and Evolution" (C. L. Markert, ed.). pp. 931–943. Academic Press New York.

WILLIAMS, G. C. (1975). "Sex and Evolution", Princeton University Press.

WILSON, D. P. (1955). The role of microorganisms in the settlement of *Ophelia bicornis* Savigny. *J. mar. biol. Ass. U.K.* **34**, 531–543.

WILSON, D. P. (1968). The settlement behaviour of the larvae of *Sabellaria alveolata*. *J. mar. biol. Ass. U.K.* **48** 387–435.

WILSON, D. P. (1970). The larvae of *Sabellaria spinulosa* and their settlement behaviour. *J. mar. biol. Ass. U.K.* **50**, 33–52.

WISELY, B. (1960). Observations on the settling behaviour of larvae of the tubeworm *Spirorbis borealis* Daudin (Polychaeta). *Aust. J. mar. Freshwat. Res.* **11**, 55–72.

19 | The Social Insects as a Special Case of Coloniality

M. J. ORLOVE

Department of Biology, University of Sussex, England

Abstract: Social insect colonies, in contrast to coelenterate colonies, are derived from more than one zygote. As a result, their evolution is not always in the direction of the "good of the colony". Hamilton worked out a system to cope with the evolution of organisms in genetically heterogeneous colonies. As an approximation it ingeniously predicts events in the real world. As an analytic argument it seems to fail because of an inherent paradox. The paradox is pointed out and side-stepped here. To avoid algebra, a physical model of the population is employed. The model is made of plastic capsules and marbles.

Pre-Hamiltonian evolutionists said animals evolve so as to maximize the numbers of their descendants. Hamilton stated that it is genes identical by descent, not offspring, which are maximized. This article states that it is some other criterion of identity, not by descent, which is relevant. This criterion is based on a correlation or regression coefficient. However in most social insects where only one sex—the females—are involved in the division of labour, still another criterion of identity is required to determine what evolution is maximizing.

If we look at a colony of animals, such as a swarm of bees or a Portuguese Man of War, we notice a division of labour among its members. The most striking feature of this division of labour is that most of the colony members live out their lives playing a less reproductive, if not a non-reproductive role (Wilson, 1975). According to Darwin, refraining from reproducing is advantageous only when it helps the animal survive to reproduce more efficiently in the future. Darwin later resolved the difficulty, and others with him (Emerson, 1950; Wheeler, 1911; Wells *et al.*, 1931) by regarding the colony as a super-organism. As pointed out by Wilson (1971, 1975) the super-organism notion, although

---

Systematics Association Special Volume No. 11, "Biology and Systematics of Colonial Organisms", edited by G. Larwood and B. R. Rosen, 1979, pp. 329–344. Academic Press, London and New York.

not false, is oversimplistic and causes us to miss important quantitative relationships by overlooking them. Haldane (1955) was aware of these relationships but he did not state them too clearly. He was however once heard to say (Maynard Smith, personal communication): "If natural selection be true I would be willing to lay down my life to save one twin or two brothers or eight cousins". Although probably said half in jest it went unrecorded at the time and attracted no attention. In 1964 Hamilton independently came up with the notion by saying that an altruistic tendency would be selected for only if

$$K > 1/r,$$

where altruism is defined as decrementing your fitness in order to increase the fitness of someone else. K is the numerical value of the increment divided by the numerical value of the decrement, and r is the coefficient of relationship between altruist and beneficiary. In dying to save eight cousins, $K = 8$ and $r = 1/8$ (supposedly) so $K = 1/r$ i.e. $8 = 1/(1/8)$ and Haldane breaks even. If he died to save nine cousins he wins, if he died to save seven cousins he loses. In short, pre-Hamiltonian thinkers estimated an animal's fitness as the number of successful offspring it produced. Hamilton estimated an animal's fitness as the number of genes identical by descent to itself reared or rescued by it divided by the number of genes per genome. He called this estimate "inclusive fitness" to distinguish it from the older concept of fitness, and Maynard Smith (1964) called natural selection based on inclusive fitness "kin selection".

Now in a clone, such as the human body or a coelenterate colony, $r = 1$ between immortal (germ plasm, reproductives) and doomed (soma, workers) portions of the entity. Thus $K > 1$ is the sufficient condition for altruism to evolve within a clone. This is why parts of a clone work "for the good of the whole", i.e. members of a clone are expected to agree on the allocation of resources among the clone members.

Some physiologists and slime mould geneticists would like us to believe that genetically heterogeneous colonies also operate for the good of the whole because colonial living imposes *controls* on the individual members. However the work of Hamilton (1971, 1972), Trivers (1974) and Trivers and Hare (1976) shows that colonies and families are wrought with internal conflict wherever the K–r situation predicts it. This accounts for weaning conflicts, workers eating queen-laid eggs, and conflicts over the sex-ratio of the reproductive brood in ants. It also accounts for the more numerous independently evolved origins of sociality in the Hymenoptera. The details of these arguments have been very thoroughly covered by West Eberhard (1975) and Trivers and Hare (1976), and I do not need to repeat them here.

19. Social Insects: A Special Case of Coloniality   331

The problem I will consider now is this: If the "$K > 1/r$" rule is very important in predicting the direction of selection, it is important to be able to measure r. Now r is a well-known concept and has been used by animal breeders to indicate the degree of inbreeding in their stock, or to evaluate the egg producing potential of a rooster or milk producing potential of a bull by comparing each to their female relatives. Hamilton was the first to use r in understanding the evolution of wild animals. Trivers and Hare (1976) extended the approach in predicting sex ratios (or bounds on them) with r. But here is the clinch. Most if not all of the formulae for r assume that there has been no selection, or at least only stabilizing selection in the population. It is assumed that no gene frequencies have changed, and no gene was "selected *for*". This is to make calculations easier, but "$K > 1/r$" means selection *for* altruism. But until recently, r was unknown when there was selection for something, because suitable formulae had not been derived. Hamilton (1964) tried to resolve this by stating that r probably changes very little from its non-selection values (1/2, 3/4 (for ants, bees and wasps), 1/8, 1/16, etc., which depend on positions on a pedigree) so long as the selection is slow. He also pointed out that the regression-coefficient may be used in place of r in uncharted ground (Hamilton, 1971, 1972, 1975).

Recently, the author (Orlove, 1975) tried a different approach, that is, to simulate a population and say that its value of r must be whatever value of $1/K$ which, if suddenly allowed to replace the current value of $1/K$, would cause selection to stop instantly. Even if the new value of $1/K$ were not implemented and the population continued along its old path, some value of $1/K$ could have been used to stop selection. That value of $1/K$, whether it was used or not, could be regarded as r at that instant. But this will generate difficulties, as we shall see. Since a population has only one possible past we cannot turn back the clock and try various histories on a real population. But we can play such tricks on time with a simulated one. The procedure for simulating a population undergoing kin selection was spelt out mathematically in Orlove (1975). Here I will spell it out in non-mathematical terms.

Imagine a plastic capsule as in Fig. 1, of the kind toy rings come in (in vending machines). In it are two marbles, both red or both blue, or a red one and a blue one. The marbles represent genes. The two colours represent alleles. In population genetics as laid down in the works by Fisher, Haldane and Wright, each capsule would be a zygote. Each would produce gametes (marbles) according to Mendel's laws, and the fecundity and survivorship of each zygote would be controlled by its genotype.

For example, all survive equally well; red–red makes seven gametes, red–blue makes ten, blue–blue makes nine. The new gametes (marbles) are put in a barrel (the gene pool) and shaken to randomize the mixture (random mating). The new zygotes can then be simulated by blindfold choice of pairs of marbles and putting them in capsules. This represents pre-Hamiltonian natural selection.

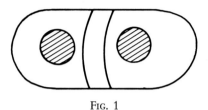

Fig. 1

Now let us simulate kin selection. Our colonies will be in a stage of evolution where the workers are free to leave and reproduce instead of staying and helping with the colony. In many species of social insect the workers can lay eggs. Some insects at the so-called semi-social stage of evolution (Michener 1969), the famous *Polistes* wasps for example (West Eberhard, 1969), found colonies in groups. It is very nebulously defined as to which wasps will work themselves to death as helpers without reproducing, which wasps will stay as

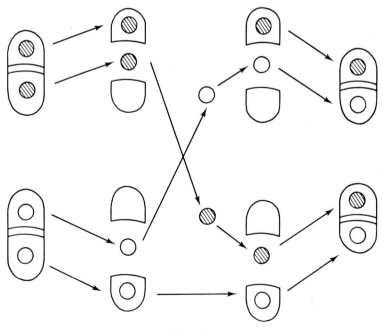

Fig. 2

egg-layers, and which wasps will leave the group and found nests alone. Since the members of the foundress association are sisters, the simulation involves a simplification in which pairs of sisters represent the foundress association.

Again we use the plastic capsules with pairs of marbles in them, but now each capsule is not a zygote. Instead it represents a pair of gametes from the same individual which upon fertilization will be each other's siblings. The capsules are put in the barrel and shaken. Pairs of capsules are next drawn at random (Fig. 2). Each capsule is transformed into a zygote by exchanging one of its marbles with its partner.

The zygotes in each pair are each other's siblings. The one at the bottom left is designated "worker", the one at the top left is designated "queen".

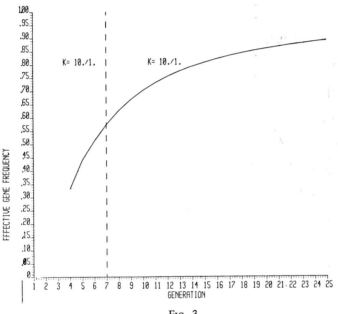

Fig. 3

Both the worker and queen put their gametes (if they produce any) in pairs in capsules. The gametes from each individual are shaken in a box to randomize them before being placed in their capsules. The number of gametes produced by the worker and the queen is controlled solely by the worker's genotype. This constraint ensures that the model is a pure kin selection model and that the results are unbiased by arbitrary selection of the pre-Hamilton kind. If the worker joins the colony the queen has more offspring and the worker has none. If the worker leaves, both produce but the queen produces fewer than an aided queen. K is kept constant throughout the population. To avoid random sampling errors (genetic drift) an infinite population is used. By an infinite population we mean the limit approached as larger and larger populations are considered. Deviations from the numbers we will get will be smaller the larger the population. An infinite population is achieved with a finite number of capsules by saying that exactly half of the gametes of a red–blue heterozygote will be red and exactly half will be blue. No shaking in boxes is allowed. All randomization is done by using exactly ideal ratios. A computer is used to simulate the operation. Now let us simulate a population. Figure 3 shows the effective gene frequency at each generation.

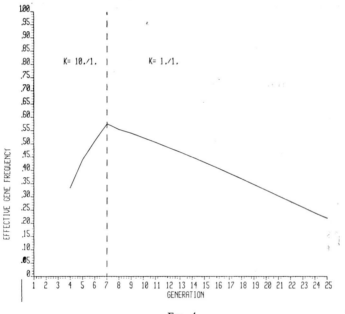

Fig. 4

Effective gene frequency is an estimate that takes both sexes into consideration, and in the ants, bees and wasps it equals

$$\frac{(2 \times \text{gene frequency in females} + \text{gene frequency in males})}{3}$$

(see also Appendix 1). Each worker who joins the colony gains ten nieces and nephews per offspring it loses. Because the slope is positive, $K > 1/r$, and r must be $> 1/10$. What is r in generation 7 for example? See Fig. 4.

After generation 7, the slope is negative and therefore $K < 1/r$ so we know

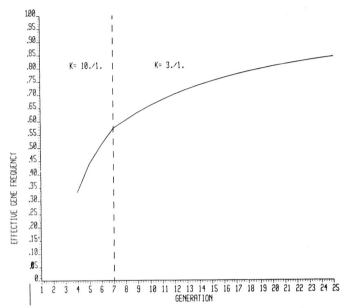

Fig. 5. After generation 7, $K < 1/r$ and $r > 1/3$.

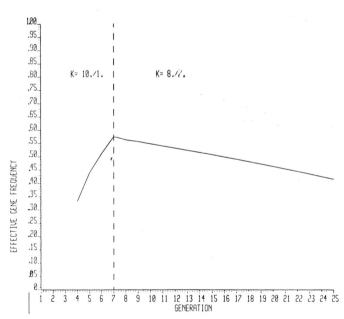

Fig. 6. After generation 7, $K < 1/r$ and $r < 7/8$.

$r < 1$. We can simulate the population over and over again starting with the same initial conditions and with $K = 10$ until generation 7, thereafter using different values of K, as in the examples given in Figs 5 and 6.

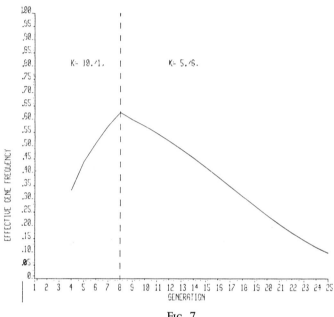

FIG. 7

If we make our next guess at K the average between the last guess in which $K < 1/r$ and the last guess in which $K > 1/r$, we will get an approximation of $1/r$. Each time we repeat this averaging procedure our guesses get closer and closer to $1/r$. We can estimate r as accurately as we please depending on how long we repeat the averaging procedure. Once we have estimated r in generation 7 we can estimate it in generation 8 the same way (Fig. 7). This is called a searching algorithm. It is inefficient but illustrates the philosophical assumptions involved in defining r as

$$\frac{1}{\text{equilibrium } K}.$$

This is the same idea as trying different values of X to find X in $3 + X = 7$; until we find an X value for which the equation is true. We could instead do algebra on the equation and find that: $X = 7 - 3$; by analogous algebraic manipulations we can make a formula for r. But life does not work out so simply. Even in our simplified model with two animals per colony and two

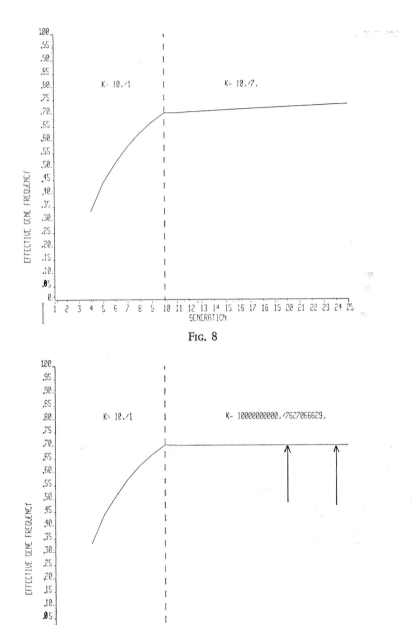

Fig. 8

Fig. 9. Arrows point to two generations with the same gene frequency. The searching algorithm was used to find the relevant *K* value. The same is true in subsequent figures.

genes per animal, we do not simply find trajectories as free from oscillations as they appear to be in Figs 6, 8 and 9.

We do find trajectories however as in Figs 10, 11 and 12.

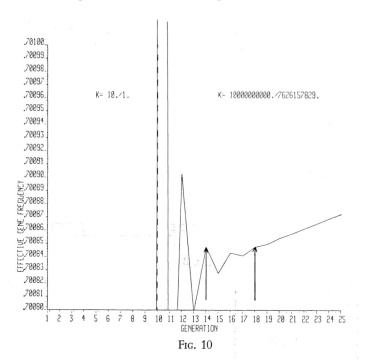

Fig. 10

The question then arises, which two generations should be compared in order to ascertain whether selection is for or against altruism?

If the comparison occurs after the oscillations appear to have died down (as in Fig. 13), an algebraic solution would be very unwieldy. Also the oscillations never completely disappear, but their effect will be large if the comparison is made only after a small number of generations. We are also faced with the fact that comparing two peaks (as arrowed in Fig. 10) gives a different result from comparing two troughs as arrowed in Fig. 11, or a peak and a trough as arrowed in Fig. 14, or a trough and a peak as in Fig. 15.

As odd as it may seem I believe in making the comparison between the beginning and end of the same generation—the generation in which K is first changed as in Fig. 16.

This implies that if we are to keep the gene frequency constant we must update K each generation. The idea of manipulating K to control the gene frequency of a simulated population appears contrary to the intuition of many

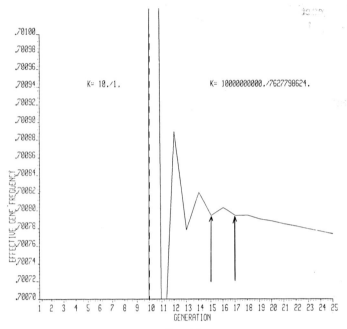

Fig. 11

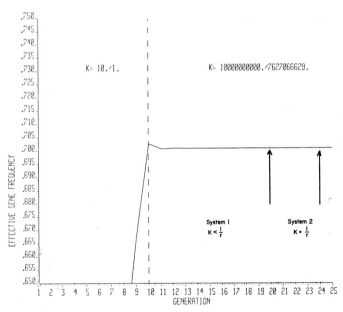

Fig. 12. In system 1, $K < 1/r$, in system 2, $K = 1/r$; for generation 10.

population geneticists who like to think of K as fixed and the gene frequency as changing. Furthermore, since we have two ways to define direction and two ways to define r, a pair of definitions (a direction definition paired with its

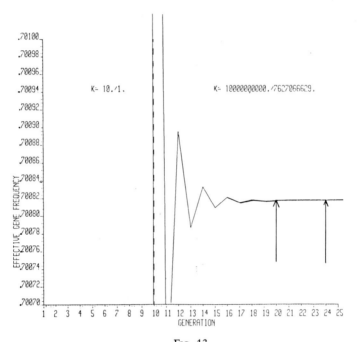

FIG. 13

corresponding r definition) is completely self consistent and preserves the K–r rule in its own terms as in Figs 12 and 15.

"System 1" means "update each generation" and "System 2" means "wait till the oscillations die down".

I prefer System 1 also because if the altruism gene has a frequency close to zero, or close to one, or if both genes are equally dominant (i.e. the heterozygote uses half of its resources on her own nest and half on the queen's), then the regression coefficient of Hamilton (1971, 1972) comes out equal to the r defined in System 1. This also happens if both males and females can participate in the altruism to male and female beneficiaries, as in termites, who have male and female workers, male and female soldiers and male and female reproductives (kings and queens). In the Hymenoptera (ants, bees, wasps) where the workers are all females, there exists no formula beyond that of System 1. This takes us up to Orlove (1975). What next? Well, as a double check I ran a simulation with

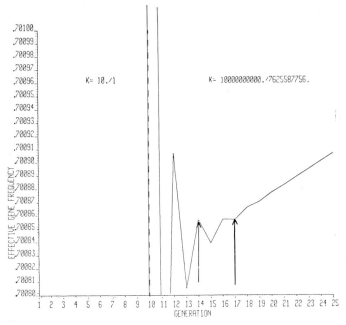

Fig. 14

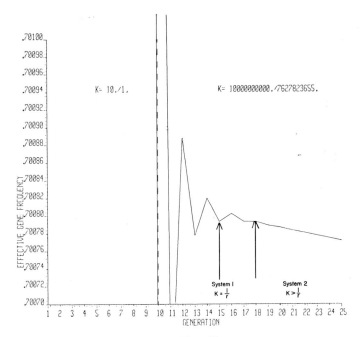

Fig. 15

three alleles. Two of them were altruism alleles, behaving identically but one was "marked" so the investigator could observe gene flow among altruism alleles much as smoke can be used to observe air flow. The third allele was a selfishness allele.

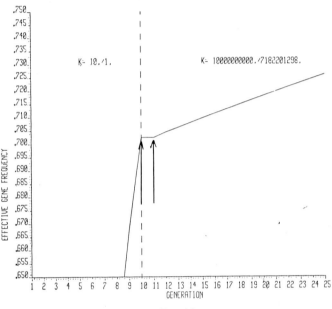

Fig. 16

The model thus functions as a 2-allele model in which some alleles are tagged. The tagged gene is introduced initially as a very rare gene. The coefficient of relationship is then the conditional probability that the queen has a tagged gene if the worker has one too. This is only true if the frequency of the tagged gene is so small that it is practically zero and there is no inbreeding. Because both the tagged and untagged altruism gene are being similarly selected (since they are phenotypically identical), and because the tagged gene is still rare, we can measure r when altruism is at intermediate frequencies.

The r defined by conditional probability proves to be the same r predicted by a pedigree so long as the frequency is constant; however with dominance of one gene or the other the regression coefficient and the K-up-dated r are not equal to the pedigree-defined r, nor is the oscillations-dying-down-definition equal to the pedigree-defined r. One thing is clear, however, and that is that r as a fraction of genes identical by descent, and r as a correlation or regression coefficient, are two different concepts which happen to coincide under certain simple conditions. How lucky is the coelenterate biologist who works with clones. No intra-colonial strife need trouble his mind—or need it? Manning (1975) indicates that clones too may have these problems, the selfish entities which leave the colony being cancer viruses capable of living as genes in a chromosome of a host for generations, or travelling between hosts to set up camp in some new host's chromosomes. And since the virus endangers its old host in so doing, it seems to weigh K against $1/r$ (r to its replicas in the germ cells of the old host) as it prepares to embark, but this is still another story.

APPENDICES

1. The Hymenoptera, ants, bees, wasps and allies, have haploid males arising from unfertilized eggs. As a result all genes are sex linked. In most other organisms, including termites, this is not the case and the capsule model described (Fig. 2) applies. For Hymenoptera, the model must be slightly modified as follows: after picking a pair of capsules at random, designate the one at the bottom as sperm and the one at the top as eggs. If the two sperm are not identical, replace the sperm at the top by a replica of the sperm at the bottom. Which capsule ends up top or bottom, or which marble ends up top or bottom, is random. The graphs shown here are from such a simulation.

2. The K-update-r between a worker and queen, called ρ turns out to be:

$$\frac{\text{regression of worker's genotype on queen's phenotype}}{\text{regression of worker's genotype on worker's phenotype}}$$

when we analytically solve for ρ between a worker and a queen. The algebra is shown in Orlove and Wood (1978).

REFERENCES

Emerson, A. E. (1950). The supraorganismic aspects of the society. *Colloques int. Cent. natn. Res. scient.* **34**, 333–353.

Haldane, J. B. S. (1955). Population genetics. *New Biol.* **18**, 34–51.

Hamilton, W. D. (1964). The genetical evolution of social behaviour I, II. *J. theor. Biol.* **7**, 1–16, 17–52.

Hamilton, W. D. (1971). Selection of selfish and altruistic behaviour in some extreme models. In "Man and Beast: Comparative Social Behaviour" (J. F. Eisenberg and W. S. Dillon eds.), pp. 57–91. Smithsonian Press, Washington D.C.

Hamilton, W. D. (1972). Altruism and related phenomena, mainly in the social insects. *Ann. Rev. Ecol. Syst.* **3**, 193–232.

Hamilton, W. D. (1975). Innate social aptitudes of man; an approach from evolutionary genetics. In "Biosocial anthropology" (R. Fox, ed.), pp. 133–155. Malaby Press.

MANNING, J. T. (1975). Sexual reproduction and parent-offspring conflict in the RNA "tumour" virus/host relationship: implications for vertebrate oncogene evolution. *J. theor. Biol.* **55**, 397–413.

MAYNARD SMITH, J. (1964). Group selection and kin selection: a rejoinder. *Nature, Lond.* **201**, 1145–1147.

MICHENER, C. D. (1969). Comparative social behaviour of bees. *Ann. Rev. Entomol.* **14**, 299–342.

ORLOVE, M. J. (1975). A model of kin selection not invoking coefficients of relationship. *J. theor. Biol.* **49**, 289–310.

ORLOVE, M. J. and WOOD, C. L. (1978). Coefficients of relationship and coefficients of relatedness in kin selection: a covariance form for the Rho Formula. *J. theor. Biol.* **73**, 679–686.

TRIVERS, R. L. (1974). Parent–offspring conflict. *Am. Zool.* **14**, 249–264.

TRIVERS, R. L. and HARE, H. (1976). Haplodiploidy and the evolution of the social insects. *Science, N.Y.* **191**, 249–263.

WELLS, H. G., WELLS, G. P. and HUXLEY, J. (1931). "The science of life". 1st Edition, pp. 1184–1185. Doubleday Doran & Co., London, New York.

WEST EBERHARD, M. J. (1975). The evolution of social behavior by kin selection. *Q. Rev. Biol.* **50**, 1–33.

WEST EBERHARD, M. J. (1969). The social biology of polistine wasps. *Misc. Publ. of University of Michigan, Museum Publ.* **140**, 1–101.

WHEELER, W. M. (1911). The ant colony as an organism. *J. Morphol.* **22**, 307–325.

WILSON, E. O. (1971). "The insect societies" Belknap Press, Cambridge, Mass.

WILSON, E. O. (1975). "Sociobiology: the new synthesis" Belknap Press, Cambridge, Mass.

20 | The Uniqueness of Insect Societies: Aspects of Defence and Integration

P. E. HOWSE

Chemical Entomology Unit
Department of Biology, The University of Southampton, England

Abstract: Differences between species of social insect are reflected in nest structure and in systems of chemical communication and defence. The nest structure is adapted to the defensive strategies of the species and to requirements for homeostasis, and may sometimes differ markedly in a given species according to local environmental conditions. Recent work has shown that volatile chemical secretions are of immense importance in the integration of social life in insects, but the use of differences in pheromones and defensive secretions as taxonomic characters will be of only limited value until the adaptive significance of such differences is understood.

THE EVOLUTION OF SOCIALITY

Sociality in insects is confined to the bees, wasps, ants and termites. The social insects show co-operation between individuals, extending to altruism. The colony is, with some exceptions, a single family group living in a common shelter, and has a large proportion of non-reproductive forms. Inter-individual co-ordination is developed to such a high degree that ancient writers such as Aristotle and Pliny saw clear evidence of purpose in the behaviour of individuals. Later writers, among them Maeterlinck ("The Soul of the Bee") and Marais ("The Soul of the White Ant"), regard the insect society as a *super-organism* analogous, for example, to a human body or a foraminiferan. This concept had distinct appeal for systematists such as Emerson (e.g. 1956) who considered that evolutionary trends (in nest structure for example) were thrown into relief by considering a colony *in toto* as a "supra-organism". Nevertheless,

---

Systematics Association Special Volume No. 11, "Biology and Systematics of Colonial Organisms", edited by G. Larwood and B. R. Rosen, 1979, pp. 345-374. Academic Press, London and New York.

analogies that were drawn tended to be superficial and easily break down when pressed beyond a certain point.

It would, perhaps, be more instructive to compare an insect colony with a metazoan showing division of labour among similar constituent parts, such as a hydroid or an echinoid. The pedicellariae and tube feet of an echinoid are semi-autonomous and show specializations for defence, feeding and hygiene that are paralleled in many insect societies, and a study of the similarities might well shed light on evolutionary trends in coloniality.

However, in general terms, such analogies have limited usefulness, partly because detailed research has revealed very complex behavioural mechanisms in social insects which are greatly different from those known in non-social organisms, partly because the cells or organs of a multicellular organism have a similar genetic constitution while different castes of social insect colonies often have marked differences in their gene complements. In the "Origin of Species", Darwin dwelled upon the difficulties implicit in explaining the evolution of worker forms in social insects. Being sterile, these could not, apparently, pass on their genes which contributed to increased efficiency in the colony. To explain this, he supposed that selection acted upon the family group, rather than on the individual.

Sterile castes tend to show altruistic behaviour, in that they contribute to the welfare of one another and of the reproductives. This altruism commonly extends to self-sacrifice in defence of the colony, as in honeybee workers which die after loss of the sting, and certain termites like *Globitermes* in which the soldiers dehisce under extreme provocation, releasing a sticky defensive secretion. Hamilton (1964) showed that such behaviour could be expected of social Hymenoptera in which the males are haploid and the females diploid. If a female is fertilized only once, the coefficient of relationship (a measure of the likelihood that genes will be identical) between mother and daughter is 0·5. Between daughters it is half as much again (0·75) but between males and their sisters it is only 0·25. From this it can be seen that worker-like behaviour in males is unlikely to be perpetuated if it occurs, while the converse is true among females. Altruism that increases the overall fitness of the colony will be strongly selected among female workers, irrespective of their sterility, and altruism between new queens and old queens will also be favoured. Colonies with multiple queens are not common, however, except in primitive hymenopteran societies. In the first place it is of greater benefit for sisters to put their effort into rearing more sisters rather than more queens, and in the second, foraging, nest-building and other activities necessary in an incipient colony soon become more important than egg-laying. The possession of many queens by a social wasp colony may be an adaptation to rapid build-up of a colony, as Richards and Richards (1951) have suggested, and to rapid re-establishment of colonies that are subject to heavy predation by ants.

The coefficients of relationship in a hymenopteran society also indicate that sisters contribute more to the perpetuation of their own genes by rearing more sisters rather than by having their own offspring. It is to be expected, in fact, from the coefficients of relationship that the effort put into rearing sisters would be three times that put into rearing males. Trivers and Hare (1976) looked for such a relationship in the sexual brood of a number of social Hymenoptera. They found a good fit to this ratio in 21 species of monogynous ants. In solitary bees and wasps as well as in slave-making ants in which the slaves have no genetic relationship with the queen, a different situation pertains because a premium is on the queen's fitness alone, and she is most likely to perpetuate her own gene complement if she produces equal numbers of males and females. Trivers and Hare found in these insects that the ratios of alate males to females were close to 1 : 1.

At this point we should stop to consider what is meant by true sociality in insects. According to Lin and Michener (1972) a colony can be regarded as the interacting individuals of a species in one nest, while a communal colony consists of individuals of a single generation which show no division of labour. In the "fully social" or *eusocial* colony there is division of labour occasioned by the mother surviving until the maturation of her offspring which assist her in the further care of the brood. Lin and Michener argue that there are two routes to the eusocial condition, one through a *subsocial* stage, in which the mother protects and feeds her own offspring, another through a *semisocial* stage in which individuals of a single generation associate together and show division of labour in caring for the brood.

In some halictine bees such as *Augochloropsis*, which nest in aggregations, a nest is founded by a single bee but others then join with her and take part in the rearing of the brood. Some of the foundresses appear to change from queen-like behaviour to worker-like behaviour accompanied by regression of their ovaries, but the data on this, as Wilson (1971) points out, is not unequivocal. Pleometrosis (founding of colonies by more than one queen) is akin to semi-sociality. Some *Polistes* wasps and some bees are in this sense temporarily semisocial, with auxiliary queens dying or disappearing when the worker brood becomes adult.

Lin and Michener argue that the transition to eusociality along the semisocial route involves both the development of increased longevity in the queen and the ability to produce females under sub-optimal conditions that will join with her. They quote a number of examples in which it is known that transfer of individuals occurs between colonies, a process that is sometimes extended to the continuity of a large number of colonies to form a "supercolony" as in *Iridomyrmex humilis*, the Argentine ant. It is, incidently, difficult to see the adaptive significance of such fusion in "highly eusocial" insects: there is a great deal of evidence suggesting that socialization engenders individuality of colonies so that outsiders are all treated as enemies (Haskins, 1970; Howse, 1975). It is also difficult to explain selection for semisocial behaviour using the kinship theory, because the coefficient of relationship between joining sisters and nieces is 0·375 in Hymenoptera, while mothers living alone have a coefficient of 0·5 with their offspring.

Another route to eusociality is through a stage of subsociality, in which a mother protects and feeds her own offspring. In both bees and wasps, examples can be taken showing possible evolutionary trends. In some wasps the female first shows mass provisioning, providing the egg with sufficient food for the developing larva. The next step is progressive provisioning, in which the female reopens the nest from time to time and brings in more food according to the demands of the offspring. An important step in the development of eusociality in wasps according to Evans and Eberhard (1973) is maceration of the prey which makes it possible for females to feed the larvae individually in accordance with stimuli they produce. Once this stage has been reached kinship factors will begin to play a role in the evolution of queen and worker castes.

However, Richards (1971) has disputed that there are any truly primitive eusocial wasps. Wasps such as *Polistes* and *Belonogaster* have simple nests of uncovered single combs. They usually arise from several foundress queens and the worker caste is distinguished only by behaviour and typically includes some of the foundress queens. Behavioural differentiation (polyethism) depends upon a dominance hierarchy established among foundresses in which the alpha female behaves as a queen, developing the largest ovaries and potentiating her own offspring by eating the eggs of other queens. She rarely forages and then only for wood pulp. The organization of the society is certainly far from simple, and Richards (1971) has listed the features which may be associated with dominance and subordination in "primitive" wasps (Table I).

In vespine wasps, Montagner (1967) showed that dominance is an important factor in social organization, and is shown by aggressivity in nest-mates soliciting for food. The organization of the *Vespula* society breaks down in Autumn apparently as a result of the perturbation in the hierarchy brought about by the presence of large numbers of males and new queens which do not respond to solicitations for food.

The evolution of sociality in primitive ants has been surveyed in an excellent article by Haskins (1970). In the primitive myrmeciine ants of Australia and in many ponerine ants there is little morphological differentiation between queens

TABLE I. Characters associated with dominance and subordination in primitive social wasps. (Adapted from Richards, 1971)

Dominant wasp	Subordinate wasp
Larger	Often smaller
Raised posture	Depressed posture
Palpates subordinate with antennae	Antennae depressed
Receives wood pulp, food and nectar from others	Donates wood pulp etc.
Aggressive towards others	Flees or shows submission
Abdomen virbrated on comb	No such display
Occupies central position on comb	Rests on back of comb or near by
Rarely forages, brings back only wood pulp	Forages often, brings back food, nectar and wood pulp
Larger ovaries	Smaller ovaries
Eats more eggs, never her own	Eats few eggs

and workers. In the higher ants, nests are commonly formed by a queen who then metabolizes her large wing muscles for rearing the first brood in a closed cell. In *Myrmecia* and various ponerines the queen makes little use of her wing muscles but breaks out of the nest cell from time to time to forage for food. The larvae are also sufficiently independent when they are a few days old to crawl to fragments of arthropod prey and feed on them. Adults of the primitive ponerine *Amblyopone* are also able to break their way out of their cocoons unaided, and those of *Myrmecia* with very little or no help, while those of higher ants are quite incapable of emerging from the cocoon or pupal casing without aid from the other workers. Primitive ants thus have much of the behavioural repertoire that one would expect of wingless solitary wasps. Their survival in competition with higher ants that have a more advanced social organization is testimony to their often elaborate raiding behaviour. Many ponerine species are specialized feeders; e.g. *Megaponera foetens* and *Termitopone* are obligate termite feeders, and some *Leptogenys* species feed only on woodlice. Scout ants detect the prey and lead workers to it: observations show that the leader is indispensible to the raiding party (Fletcher, 1973).

There is no elegant explanation for the evolution of altruism in termites, which do not have the haploid–diploid method of sex determination, and special factors must be brought into consideration to explain their evolution. One of these is undoubtedly the dependence of the wood-dwelling, so-called primitive, species on symbiotic protozoa to break down cellulose. Similar intestinal symbionts are also found in the semisocial cockroach *Cryptocercus punctulatus*, Division of labour depends upon polymorphism which is not sex-linked in primitive forms such as *Kalotermes* and *Zootermopsis* but is in higher termites (Lüscher, 1975). Insects that live in wood or subterranean galleries are able to do so because they are apterous or have the wings folded and encased (Coleoptera). Apart from the primary founding pair, termite colonies consist entirely of juvenile forms, which are apterous. In primitive damp-wood termites such as *Zootermopsis* and *Archotermopsis* a true worker caste does not exist, this function being undertaken by the larval stages (without wing pads) and nymphs (with wing pads). It is conceivable that societies such as these evolved under the influence of juvabiones which are known to be present in some soft-woods (Slama and Williams, 1965). Juvenile hormone has been shown to be important in inducing soldier development and inhibiting the production of replacement reproductives in primitive termites (Lüscher, 1975), and is also present in large quantities in the anal fluid and haemolymph of the queen of the mound-building termite *Macrotermes subhyalinus* (Meyer and Lüscher, 1973) and in the fungus gardens of a number of Macrotermitinae (Sannasi *et al.*, 1972).

If polymorphism arose in termites as an adaptation to substances with juvenile hormone (JH) action, the problem is to explain how they escape this action in order to produce replacement reproductives and winged adults. Lüscher (1975) gives a possible clue to this: soldiers of primitive termites develop under high dosages of JH and, when they are present, stimulate the production of reproductives, possibly by eliminating JH. The production of a caste which is at the same time a defensive element and a sink for JH analogues could have been an elegant solution for early termites.

DIVERGENCE OF EVOLUTIONARY PATHWAYS

Figure 1 summarizes some of the major factors leading to the distinctiveness of different groups of social insects. The retention of wings and an aerial existence has had manifold effects on the social life of bees and wasps. The wings can be used for ventilating the nest and cooling it. Metabolic heat can be generated from the wing muscles to raise the nest temperature when necessary. Because of the efficient climatic control that is possible, nests can be very densely populated and the brood close-packed. Foraging tends to be dominated by visual route-following, which requires an excellent memory (as in vespine wasps)

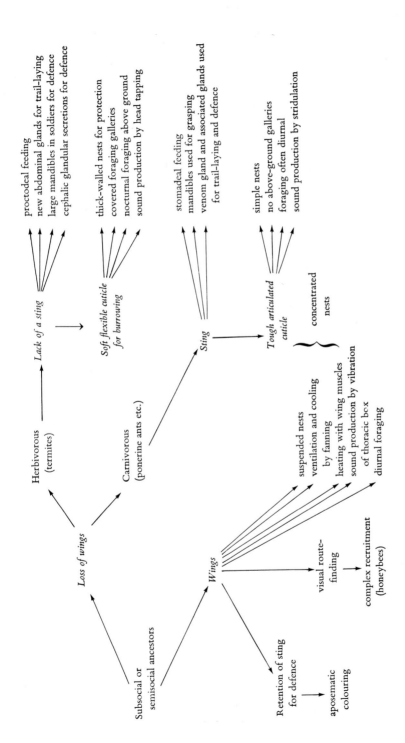

Fig. 1. A comparison of the factors in the evolution of termites, primitive ants and social bees and wasps.

or a complex system of information transfer (as in the "dance language" of honeybees). Foraging is thus diurnal and the sting, where it is not used in prey capture, is retained as a valuable defence against vertebrate predators, and its presence may be advertised by coloration of the wings and body.

Primitive carnivorous ants, such as the ponerines and myrmeciines also rely heavily upon the sting for prey capture and defence. The sting is also an ideal dispenser for pheromones used in recruitment. For its operation, it requires a tough, well-articulated cuticle. This also makes rapid locomotion possible, is a useful protection against predators and resists desiccation well, if it supports a good wax layer. These factors make possible diurnal foraging above ground. Sound production by stridulation is also possible with tough mobile body segments; a stridulatory organ is found between the postpetiole and the gaster in many ant species.

The lack of a sting in termites has had a number of consequences which bear interesting comparison with ants. Without a sting, proctodeal feeding can occur while stomodeal feeding of nest mates and brood is the rule in ants. Sternal glands in the mid-abdominal region have developed for trail laying. Defence has been taken up by massive jaws in the lower termites, superseded in the higher termites by means of applying chemical defensive secretions from cephalic glands. The lack of a sting has opened the pathway for evolution of a soft flexible cuticle, ideal for burrowing but offering little protection against predators and desiccation. Termites typically (with the exception of grass-harvesting species) forage above ground in covered galleries, usually at night, and many species build thick-walled nests. With a soft cuticle, stridulation by friction between body segments does not occur, but sound is produced by the workers and soldiers tapping their heads on the substrate as a warning signal.

THE ROLE OF THE NEST IN DEFENCE

1. Passive Defence

A simple nest such as an underground burrow or a system of galleries in wood fulfils a number of requirements: it is insulated from climatic changes and if it is sealed it may be difficult to find and break open. Some species of *Bembix* and *Sphex* construct short blind-ending burrows around the true nest entrance, which are thought to be means of misleading parasites (Evans, 1966). Social insects characteristically defend the nest entrance; some termite species which live in hard wood (e.g. *Cryptotermes*) and some ants (*Camponotus truncatus*) defend the nest entrance by plugging it with their rounded head capsules. This behaviour is known as *phragmosis*.

Passive means of defence include siting of nests in places that are not easily accessible and building them in association with another species of animal, or in a well-protected plant.

(a) *Nests "out of reach"*. There are many tree-nesting ants and termites, but while ants such as the weaver ants *Oecophylla* and *Polyrhachis* are very aggressive and make a living in the canopies, termites of the genus *Nasutitermes* evidently nest in trees to avoid predation and descend to the ground to feed on the leaf-litter. They are able to resist predatory ants with the aid of elaborate chemical defences and by enclosing their routes down the tree trunk in covered galleries. Colonies of South American wasps are believed to be subject to heavy predation pressure by ants and therefore tend to site their nests high in trees, on cliff overhangs and other inaccessible places (Richards and Richards, 1951).

(b) *Nest associations*. Some species of *Polybia* and *Protopolybia* nest in trees infested with dolichoderine ants, such as *Azteca*, which have well developed chemical defences. Birds, especially orioles, tend to nest alongside, no doubt benefiting from the hymenopteran arsenal available against other intruders. In S. America, W. D. Hamilton (personal communication) found a nest of the wasp *Chartergus chartarius* surrounded by nests of caciques, all on one branch. In the same tree he found four further species of social wasp nesting alongside *Azteca* ants and yet two more species nesting inside the *Azteca* nest.

(c) *Nests in plants*. A number of plants provide extra-floral nectaries or "food bodies" which are used by ants; they also have growths that are used by the ants as nest chambers. In S.E. Asia, *Macaranga* species harbour ants in this way and the plants are kept free of caterpillars (Ridley, 1910). Acacias, which appear to have developed spines in Africa and America as a protection against browsing mammals (Brown, 1960) also commonly harbour ants. In Mexico, Janzen (1966) found that removing *Azteca* ants from acacias resulted in heavy attacks by phytophagous insects. The African thorn tree, *Acacia drepanolobium*, has numerous gall-like growths that are inhabited by *Crematogaster* ants. These ants deter giraffes which are otherwise capable of browsing on the trees (Foster and Dagg, 1972). Such associations are thus mutually advantageous: the ants gain shelter, food and some protection, while the plants are provided with a mobile scavenging and defensive army.

2. Chemical Defences

Writing on social wasps, Richards (1971) suggests that large colonies are at a greater risk where predation is intense, compared to many smaller colonies. Thus the genera with the greatest number of species in South America are *Polistes* and *Mischocyttarus* in which a colony rarely has more than 100 individuals. These are genera, along with *Belonogaster* and *Ropalidia* which are found outside S. America, in which a single comb is suspended from a petiole. The petiole is constructed from a mandibular gland secretion and is slender with a smooth hard surface. Such wasps also possess a secretory organ on the terminal abdominal sternite (Van der Vecht's organ), which Jeanne (1970) showed is used by *Mischocyttarus drewseni* to apply an ant repellent to the petiole. It may be that this is the function of Van der Vecht's organ wherever it is found.

3. Behavioural Defences

In some social insects threat or aggressive behaviour alone is used to defend the exposed brood. The wasp, *Apoica pallida*, builds a single comb without a petiole. During the day, the wasps rest in a regular array around the edge of the comb, their large whitish abdomens providing a visual display. There is one wasp on top of the comb, apparently as a look-out. If the nest is disturbed the peripheral ring of wasps climbs on to the top of the comb and shows increased alertness.

The wasp *Synoeca surinama* also constructs a sessile nest, but covers this with a domed envelope. When disturbed, the insects drum on the nest envelope and eventually swarm on the outside of the nest and raise and lower their wings in synchrony with the sound (Evans and Eberhard, 1973).

FIG. 2. Adaptive radiation of social wasp nests (not drawn to scale). One line of evolution from an unknown ancestral type leads to combs suspended from a secreted petiole, as in *Polistes* (a) and *Mischocyttarus* (b), which are protected by chemical repellents. The petiole has been lost in *Apoica* (c). The combs are protected by simple envelopes in *Protopolybia* (d), which also retains petioles, and in *Synoeca* (e). Support for multiple combs of large nests is achieved by strengthening the envelope, as in *Chartergus* (f), and by constructing the nest about a vertical support, as in *Polybia dimidiata* (g). In vespine wasps (h) the petiole is made of plant fibre, combs are supported also by secondary petioles, and a multi-layered envelope provides thermal insulation. Secondary loss of the envelope has occurred in *Stelopolybia testacea* (i) which builds multiple combs stacked vertically in hollow trees, and *S. angulata* (j) which builds a subterranean nest with cylindrical combs.

(Based on Jeanne, 1975b, modified after various authors).

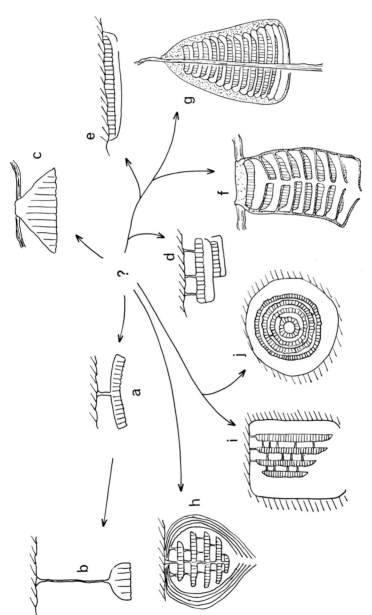

Fig. 2. (Caption on facing page)

The most adaptable kind of nest is undoubtedly that of the doryline ants, which is formed by the bodies of the workers themselves. In forming a bivouac, *Eciton* species make "ant ropes", attaching to one another with the aid of their large tarsal hooks. The ropes then pull together to form a continuous sheet. The large workers, which are more formidable and more resistant to desiccation are found on the outside of the nest (Schneirla, 1971).

4. The Role of Nest Walls in Defence

In wasps, the main alternative to a chemical barrier on the petiole is a protective envelope that restricts the entrance to a small hole that can be actively defended (Fig. 2). Jeanne (1975) has argued that this has led to the secondary loss of the petiole (as in *Synoeca*) when the comb is built against a flat surface. In other species the envelope is thickened by the wasps so it is sufficiently strong to support tiers of combs, as in *Chartergus*. An interesting side-line of development is found in *Polybia dimidiata*, where the combs are built around a central axis formed by a small branch. An alternative line of evolution involves multiplication of short petioles constructed mainly of plant fibres, not of mandibular gland secretion, which are used to support combs stacked either horizontally or vertically, according to the species. Such developments lead to large colonies, in which the queen and attendants cannot flee to found replacement nests, as they can in *Polistes*, for example, and with a consequent premium upon defence which involves focused attacks upon intruders. Such behaviour is, of course, characteristic of temperate region vespines and also of *Apis* species. Richards (1971) notes that tropical polybiine wasps which form large nests have a barbed sting, like that of *Apis*, which when torn out has an autonomously contractile poison sac.

Papier-maché nests demand intensive labour for building. One way of reducing this demand is by making nests of earthen material. This occurs in some polybiines which make clay nests, but is very common among termites. The most striking nests of all are those constructed by the Macrotermitinae. Members of this subfamily have evolved a symbiosis with fungi (*Termitomyces*) which are grown on fungus gardens of triturated vegetable material. This is made possible in the first place by siting the nests below ground where conditions of temperature and humidity are sufficiently stable. As the nest grows in size, problems of control of the internal climate arise because of the metabolic products of both the fungi and the insects. In the genus *Macrotermes* the nest proper with fungus gardens (the hive) almost reaches ground level, retaining a thick outer wall in which air ducts develop (see below and Fig. 3). In *Macrotermes subhyalinus* this wall may be several feet thick. In *M. bellicosus* it is usually somewhat thinner, but composed of very hard clay. The foraging galleries of the insects are below ground and when they come to the surface the insects forage under covered galleries. Predatory ants are therefore unlikely to penetrate into the nest unless mammalian predators first break it open. Specialized predators are an exception: in Nigeria the ponerine ant *Megaponera foetens* preys upon *Macrotermes* that are feeding under covered galleries at the surface.

Fungus-growing termites can also develop large nests by dispersing the fungus gardens in separate chambers. This leads to extremely large nests: up to 30 m in diameter in *Macrotermes falciger* in the Katanga region of Zaire (Grassé and Noirot, 1961). Nests of a somewhat similar kind are built by some leaf-cutting (attine) ants in the neotropics, which have also evolved a fungal symbiosis.

REGULATION OF INTERNAL CLIMATE

Doryline ants often nest in tropical forest where humidity and temperature fluctuate relatively little compared with scrub or savanna. According to Schneirla (1971) they are nevertheless capable of regulating the internal conditions of the nest. In *Eciton hamatum* the ants on the outside of the bivouac are the largest and most resistant to desiccation. In cool conditions the ants form a solid outer curtain in which the strands of ants are pulled together so that the bivouac tapers downwards providing a rain-shedding device, but as the temperature rises during the day they dissociate, leaving more gaps in the curtain.

Mound-building termites such as *Thoracotermes* and certain *Cubitermes* species habitually build nests in tropical forest. The mounds do not have a thick outer wall and consist of numerous interconnected chambers of carton (a mixture of vegetable and earthen material). The temperature fluctuations in such nests are far lower than in similar nests of species like *Amitermes evuncifer* which are found in forest fringe regions (Lüscher, 1961).

Procubitermes, which builds against tree trunks, constructs chevron-shaped ridges above the nest which serve to divert rain water away from it. *Cubitermes intercalatus* builds a sheaf of umbrella-like projections over the nest in a similar situation.

Bees and wasps have a particular advantage over ants and termites: they are able to use their wings as fans to ventilate the comb. *Polistes* wasps are capable of lowering the temperature of their open comb by up to 12°C: on hot days they bring water droplets and deposit them on the comb and speed their evaporation by fanning (Himmer, 1932). To help with water collecting, many *Polistes* species are able to float on the surface of the water. European *Polistes* are not able to raise the temperature of their combs but usually place their nests facing southwards.

Some of the social bees and wasps which have enclosed nests are known to be capable of raising the temperature of the brood and are able to maintain the nest at a very stable temperature. The apotheosis of this is found in the honeybee, *Apis mellifera*, in which the temperature of the comb is maintained at $34.8 \pm 0.55°C$ when brood is present: development stops below 32°C and 36°C is the upper lethal temperature (Himmer, 1932). A crucial factor in this thermoregulation is the ability of the workers to raise their body temperature by muscular activity to a temperature which may be on average 12·4°C above that of the ambient temperature. This ability is also well developed in the bumble bee *Bombus vosnesenskii* in which the queen sits astride her brood and raises her thoracic temperature to 34·5–37·5°C and her abdominal temperature to 31–36°C (Heinrich, 1974). By this means, the queen is able to lift the temperature of the brood by 20°C when the external temperature is only 5°C. Heinrich calculated that at 5°C the bee consumes about 150 calories of carbohydrate per hour, which is equivalent to the nectar content of 100 of the flowers which the insect commonly visits.

The incubation behaviour of *B. vosnesenskii* is apparently a response to a species-specific pheromone which she herself deposits on the brood clump. A similar phenomenon is found in the Vespinae: Ishay (1972) found that wasps or hornets offered a comb containing brood of their own species entered the vacant cells adjacent to pupae and warmed them by abdominal pumping movements to a temperature of 30–32°C when the temperature outside the comb was 20–22°C. The thermoregulatory behaviour can be elicited from filter paper soaked in alcohol extracts of pupae. Pupae which were not kept warm usually developed into malformed adults.

The key element in the maintenance of stable temperature conditions and, of course, relative humidity, lies in behavioural responses towards water, but in tropical rain forests there is a danger of accumulation of excess water within the nest. Workers of *Polybia scutellaris* regularly remove drops of water from the nest (Evans and Eberhard, 1973). Many wasps and bees with enclosed nests indulge in fanning, which moves air through the nest and is also used to evaporate water from the combs, thereby cooling them. The mechanism is best understood in honeybees from the work of Lindauer (1954). As the temperature within the hive rises, the number of bees that are engaged in fanning increases; some of them spread water with their mouthparts in a thin film over the surface of the comb, others speed its evaporation by fanning. A further group devotes itself to collecting water outside the hive. The total number of visits to water, however, is closely geared to the temperature in the hive, and when the requirement for water is very great the water collectors perform a recruitment dance on return to the hive. Most of the time, though, the water collectors do not enter the hive but transfer their water to bees at the hive entrance. Lindauer showed that time taken by the hive bees to take up the water loads is a measure of the requirements of the hive. Collecting activity is reduced if the load takes over two minutes to be accepted; it is accelerated with the aid of some dancing if accepted within 100 seconds, and there is a great deal of dancing to recruit more water collectors if it is accepted in under one minute. A sudden temperature rise cannot be compensated for immediately; there is a time lag of 15–20 minutes before the thermostatic behaviour just described begins to operate. In such an emergency situation the bees are able to use the contents of their honey stomachs, which have 30–70% water, as an immediate short-term remedy.

A high degree of homeostasis is achieved in the nests of a number of termite species. Some of the mound-building termites have overcome the problems presented by possessing thick-walled clay mounds, which provide good insulation and defence, but which restrict air exchange. The so-called "compass mounds" built by several *Amitermes* in Australia are wedge-shaped with the long axis in a north–south line. Grigg (1973) confirmed earlier speculations that the shape and orientation were adaptations to thermoregulation. In a mound of *A. laurensis* the temperature was found to reach a plateau at 33–35°C during the greater part of the daylight hours, although it dipped to 13–15°C at night. When the mound was sawn off at its base and rotated through 90° the temperature rose to a maximum of 40–42°C.

Nests with fungus gardens pose special problems; climatic control is needed to cope with the results of metabolism of both the insects and the fungus. Some fungus-growing termite species have nests with funnel-shaped openings above ground. On the Serengeti plains, *Macrotermes subhyalinus* builds low mounds with several openings to the exterior which channel considerable volumes of air through the mound (up to 1113 l min^{-1} in a large mound in late afternoon). The mound is cooled overnight, reheats rapidly in the morning sun, and is cooled by evaporation engendered by air flow in the afternoon (Weir, 1972). In the Ivory Coast, *Macrotermes bellicosus* (= *natalensis*) has a closed nest and the air is exchanged with the aid of a thermosiphon effect (Lüscher, 1955, 1961). Warm moist air, rich in carbon dioxide, ascends in the centre of the mound by convection into a space above the hive (Fig. 3). From there, it passes into ducts marked by prominent ribs, more or less vertically orientated, on the side of the mound. The air in these ducts is exchanged and cooled, so that cool air enriched again with oxygen descends to the base of the mound. *M. bellicosus* mounds vary considerably in form (see below) and the way in which air is exchanged no doubt also shows great variation. In Ugandan mounds, the ribs are absent and air enters through holes at the base and leaves through thin-walled cavities at the top of the mound (Lüscher, 1955). In mounds in western Zaire the thermosiphon effect is weak and changes in the direction of air-flow occur, particularly under the influence of wind gusts (Ruelle, 1964).

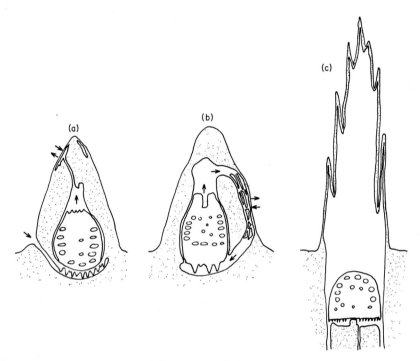

FIG. 3. Sections of *Macrotermes bellicosus* mounds from Uganda (a), Ivory Coast (b) and northern Nigeria (c). a and b after Lüscher, 1955; the arrows indicate the pathways of air circulation and exchange. c, from Howse and Imevbore (unpublished).

THE UNIQUENESS OF THE INSECT SOCIETY

1. The Nest

Nests are crystallized forms of behaviour and it is therefore tempting to use them in taxonomy. This immediately raises the question of whether behavioural and morphological features are equally conservative. Undoubtedly they are not, for behaviour must be extremely sensitive to environmental exigencies.

This implies that small environmental changes can have major effects upon nest structure.

Richards (1971) noted considerable variability in nest structure in species of wasp which on morphological grounds clearly belonged to a single genus. For example, *Stelopolybia testacea* builds flat parallel combs in a hollow tree, while the closely-related species *S. angulata* builds an underground nest of concentric cylindrical combs. An undescribed species of *Stelopolybia* builds double combs (back to back) in a hollow tree. *S. multipicta* and *S. flavipennis* are almost identical morphologically, but the former builds irregular parallel combs in hollow trees and the latter builds a nest consisting of an umbrella-shaped structure on the branch of a tree, with one comb below and one above. *S. areata* of Mexico also builds tree nests of concentric spheres of combs connected by ramps, while *S. myrmecophila* builds inside *Azteca* nests. (Jeanne, 1975). Conversely, very different species can build very similar nests.

The change from one nest type to another may depend upon very subtle factors. The oriental hornet (*Vespa orientalis*) starts its nest as a single comb supported from a vertical pedicel and covered with a paper envelope. Ishay (1975) showed that fertile queens with amputated wings built sessile combs but if they were cooled for several days or treated with queen pheromone they built normal combs with a pedicel. If one wing was amputated they built combs with the pedicel horizontal. It thus appears that the wings are used as measuring devices; wing length as well as primer pheromones can influence nest structure markedly. Ishay and Sadeh (1975) also showed that hornets learn the direction of gravitational force during the first two days of adult life and build new combs accordingly. Juvenile hornets kept in a centrifuge built cells in the direction of the resultant force, starting from the side walls of the container.

Among termites, the effects of climate, soil conditions and vegetation can have a pronounced effect upon nest structure. *Nasutitermes ephratae* which normally builds carton nests in trees, builds mounds in tree-less areas of Trinidad (Harris, 1971). In the fungus-growing Macrotermitinae regional variations in nest structure are so marked that there has been lively controversy over the correctness of published descriptions. In western Uganda, *Macrotermes subhyalinus* builds mounds with very thick walls and no openings to the exterior. On the Serengeti plains where the soil is volcanic ash, it builds very low mounds with many pit-like openings through which air circulates (Weir, 1973), while in semi-arid regions of eastern Africa the same species constructs steeple-shaped mounds with a narrow base, around a central open chimney which may reach 9 m in height.

Similar variability is seen in *Macrotermes bellicosus*. Grassé and Noirot (1961) described three nest types; a "cathedral" type with a spire and subsidiary spires coming to point, a dome nest, with flanks of subcylindrical towers, and a

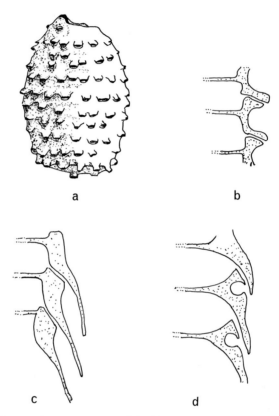

FIG. 4. (a) Small nest of *Apicotermes gurgulifex* (diameter 5·5 cm) with (b) section of the nest wall. (c) nest wall section of *A. lamani* nest, and (d) section of *A. desneuxi* nest. (a, b and d, original, c, modified from Schmidt, 1955).

rounded mound nest with neither towers nor spikes. Around Enugu in Nigeria on deep, coarse, sandy soils, *M. bellicosus* does not build mounds and the only evidence of the nest is a funnel-shaped opening in the ground (W. A. Sands, personal communication). This contrasts strongly with the mounds around Mokwa, 450 km to the NW, which are made of clay and reach heights of over 10 m (Fig. 3). They are steep-sided and strongly fluted: another peculiarity is that the hive has its base well below ground level; it rests on a central pillar and has a remarkable vane-like spiral thickening of the base (Howse and Imevbore, unpublished). This contrasts with the normal base-plate which has downwardly-projecting cones providing a support. Differences of this kind are clearly an invitation for further taxonomic studies. On the basis of the nest structure of *Apicotermes*, Emerson (1956) developed the concept of the "ethospecies". *Apicotermes* build subterranean nests mainly in the sandy soils of the Congo basin. The nests are among the most remarkable in the animal world: about the size of a small melon, they consist of many horizontal floors connected by spiral ramps. At each floor level the outer wall is pierced by a regular series of pores or slits (Fig. 4). In *A. lamani* these are extended into long tubes on the outside of the nest; in *A. gurgulifex* the openings are short hollow nipples and in *A. desneuxi* the openings are in downwardly projecting flaps. On the basis of the openings, Emerson (1956) distinguished between the two latter species in the absence of any clear morphological characters and Schmidt (1955) constructed an evolutionary tree for the genus. Machado (1959) later found morphological differences in the soldiers of *A. gurgulifex* and *A. desneuxi* and criticized the use of nest structures in deducing phylogenetic relationships on the basis that different species could build similar nests according to the nature of soils and climate. This points to the logical difficulties involved in basing systematic judgements on structures of unknown function. The writer's observations suggest that the openings of the nest prevent waterlogging by providing an airlock to water percolating through the soil, but this may be only one of their functions.

It must be concluded that nest structures can provide valuable systematic material, but should not be considered apart from morphological differences.

2. Chemical Communication

The past decade has seen a rapidly growing interest in pheromones and other chemical secretions of social insects. Those which have proved easiest to study are alarm pheromones. Trail pheromones are secreted in such minute quantities that the elucidation of their chemical structure presents great problems. It has been tacitly assumed by most workers in the field that a pheromone is a secretion with one major component that controls behaviour, differing effects depending upon the concentration. Where sufficiently detailed studies have been made they suggest that this is an oversimplified general picture. First, behaviour can depend very much upon context. For example, the main components of the honeybee queen substance are 9-hydroxydec-*trans*-2-enoic acid and 9-oxodec-*trans*-2-enoic acid, but neither of these attracts workers or drones within the hive. The oxodecenoic acid acts as an attractant for workers when released less than 4 m above ground, but the hydroxy acid must also be present for a stable swarm cluster to form (Butler and Simpson, 1967). The oxodecenoic acid is also a sex attractant for drones, but only when presented at a height of 4–25 m above ground, the height depending on the wind speed (Butler and Fairey, 1964). Secondly, pheromones have commonly proved to be complex mixtures containing not only synergists but minor components that control ancillary behaviour patterns.

(a) *Trail pheromones*. One of the few trail pheromones to have been isolated and synthesized is methyl-4-methylpyrrole-2-carboxylate, from the poison gland of the leaf-cutting ant, *Atta texana* (Tumlinson et al., 1972). Although this is

TABLE II. The cross-specificity of poison gland trails of leaf-cutting ants: the number of ants following trails when given a choice of two (from Robinson et al., 1974)

Follower	Donor			
	Ao	Ac	As	Ac
Acromyrmex octospinosus	42	$19\,(P < 0\cdot01)$	—	—
Atta cephalotes	40	$98\,(P < 0\cdot001)$	73	$62\,(P > 0\cdot3)$
Atta sexdens	—	—	93	$48\,(P < 0\cdot001)$

only one component of the trail pheromone, its potency is amazing: an estimate being that 0·33 mg would theoretically be sufficient to draw a detectable trail around the world. Robinson et al. (1974) investigated the species-specificity of this compound and found that a number of non-attines tested did not follow an artificial trail made with it, but workers of 12 out of 13 attine species did so (Table II). A more revealing assay was a test of the cross-attractivity of trails made from the crushed poison glands of three species. Ants given a choice of trails tended to follow those made by poison glands of their own species.

Species-specificity of trails is known to exist from field studies, and the results of these experiments suggest that minor components are present in the trail pheromone which are responsible. A detailed investigation of trail following in the harvester ants of the genus *Pogonomyrmex* has shown that they have two kinds of trail (Hölldobler and Wilson, 1970). A poison gland secretion is very attractive to recruited foragers but is very volatile. The trail is stabilized by a secretion of Dufour's gland which is less volatile and appears to be used for marking trunk routes. Different species can distinguish their own trails and the separation of trails is greatly enhanced by intra- and interspecific aggression between neighbouring colonies (Hölldobler, 1974). Hölldobler also showed that visual landmarks are of considerable importance in stabilizing trunk trails and in controlling fidelity of workers to a given trail. Chemical differences may be important in colony- or species-specificity of trails. Regnier et al. (1973) showed that the Dufour's gland secretions of *P. rugosus* and *P. barbatus* share the same mixture of hydrocarbons but differences are found in the trace components.

In attines and *Pogonomyrmex*, therefore, it appears that species-specificity of trails depends upon subtle chemical differences, but then only partly: the behaviour of the insects in their natural environment includes other mechanisms which are most important in specificity.

(b) Alarm pheromones. Alarm pheromones need to be rapidly acting but non-persistent. Bossert and Wilson (1963) predicted that they would generally prove to be of low molecular weight (100–200) to give high volatility and, because of the relatively limited choice available to insects, therefore generally lacking in species specificity. The lack of specificity was confirmed by Maschwitz (1964) in a study of various social insects, including 21 ant species representing 6 genera. Alarm was sometimes communicated between ants of different genera, and the wasp *Vespula vulgaris* reacted more strongly to *Paravespula germanica* alarm pheromone than to that of its own species.

While many volatile chemicals can generate alarm behaviour in social insects, it may be for one of a number of reasons such as detection of predators, or toxic food materials and may depend upon volatiles they give off. Detailed chemical studies of the components of alarm pheromones are therefore again necessary, coupled with sensitive behavioural bioassays. Where this has been done it has sometimes shed light on phylogenetic relationships (Blum, 1969). As an example, we can take the work of Crewe and Blum (1972) on leaf-cutting ants. An analysis of the mandibular gland secretions tended to confirm the separation of N. American *Trachymyrmex* into two separate species, and the derivation of the genera *Acromyrmex* and *Atta* from *Trachymyrmex* (Table III). *Atta* species produce 2-heptanone which sets them apart from the other three genera considered. Consistent differences were also found in the relative titres of 3-octanone and 3-octanol present in some of the species. Work of this kind, however, is still at a preliminary stage: the variability of pheromone mixtures in ants of the same species from different geographical areas needs to be established and the significance of species differences to behaviour needs to be made clear. Apart from this, components thought to be absent in any one species may later be shown to be present as techniques improve.

Bradshaw, Baker and Howse (1975, and in prep.) found a complex system for the chemical communication of alarm in the African weaver ant, *Oecophylla longinoda*. The main control of alarm behaviour comes under the mandibular

TABLE III (a) Major components of the mandibular gland secretions of various attine ants, showing possible phylogenetic relationships. (b) Ratios of two major components. (Modified after Crewe and Blum, 1972)

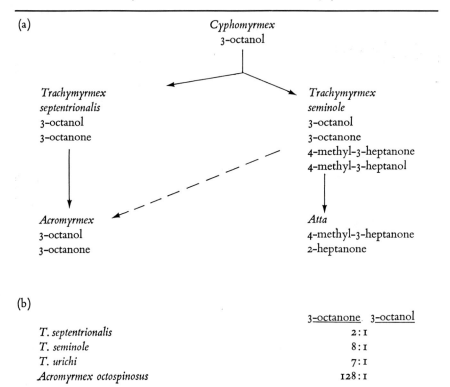

gland, but abdominal gland secretions play an important role in high alarm. In major workers, the mandibular gland secretion has over 30 components. Of these, hexanol has an alerting function, 1-hexanol, which is less volatile is an attractant and 2-butyl-2-octenal and 3-undecanone, both of low volatility, elicit localized biting. Some other components may act as synergists. The secretion of the minor workers is markedly different, the monoterpenes nerol and geraniol being present among the main components. These compounds have a relatively low volatility and are suited for communication of alarm in the confined atmosphere of the leaf nest where the minor workers tend to remain.

Quantitative differences were found in the mandibular gland constituents

TABLE IV. Variations in the quantities of some mandibular gland components of major workers of *Oecophylla longinoda* from two colonies located 230 km apart in Nigeria (Bradshaw, Baker and Howse, unpublished data). The figures are means expressed as percentages of the major component, 1-hexanol.

Component	Ibadan	Benin	
Hexanal	20·3	12·3	$P < 0.2$
Octanal	1·5	19·0	$P < 0.1$
3-Undecanone	2·3	0·1	$P < 0.1$
1-Nonanol	1·1	2·6	$P < 0.1$
2-Butyl-2-octenal	6·0	1·9	$P < 0.2$

of major workers from Ibadan and Mokwa (about 250 km to the NW) and Benin (230 km to the SE) (see Table IV). In Mokwa colonies it was possible to correlate these with behaviour: synergists such as 1-hexanol were necessary to elicit a biting response to 2-butyl-2-octenal, while in Ibadan colonies in which the latter chemical is present in greater quantities, they were not. The significance of such differences is not clear. One suggestion is that individuals of a given colony may adapt to their own local "brew", which in turn depends upon twists given to biosynthesis by raw materials in locally available foods. Learning mechanisms resembling imprinting do exist in ants. Wallis (1964) noted that the collective odour of a colony is imprinted on callows in *Formica* species. Jaisson (1975) has shown that *F. polyctena* can become imprinted on the cocoons of other species during the first 15 days of adult life, and then reject their own. The use in taxonomy of *minor* chemical differences in alarm pheromone chemistry is therefore likely to be inadmissible for a long time to come.

3. Defensive Secretions

Defensive secretions can be repellent, toxic, or glutinant. They can be applied by injection through a sting, as in honeybees; secretion into a wound cut by the mandibles, as in many termites and ants; direct application without a lesion; spraying or squirting, as in *Formica* and nasute termites, and diffusion from an exposed source, as in dolichoderine ants.

Maschwitz and Kloft (1971) have pointed out that among ants, sting fighters are less efficient than mandibular fighters, because the former have to spend time finding a soft area of cuticle through which their sting can penetrate. Stings are thus confined to the primitive highly active and predatory species of ants such as the bulldog ants (Myrmeciinae) of Australia and the ponerines. The primitive nature of the bulldog ants is emphasized by the presence of histamine in their venoms which relates them to the social bees and wasps, more than to other more advanced ant genera (Cavill and Clark, 1971).

Predation and interspecific competition are two important factors likely to influence the nature of defensive secretions and the behaviour associated with their use (see Howse, 1975). Furthermore, the manufacture of a powerful toxin must run parallel with mechanisms that make the insect concerned resistant to its own toxin. On these grounds, one would expect defensive secretions to show considerably more species-specificity than do alarm pheromones. The dangers of such a generalization are, however, revealed by studies of Cavill and Hintenberger (1960) on two dolichoderine ants in which a volatile defensive secretion is given off from anal glands. *Dolichoderus scabridus* has two colour morphs in New South Wales. The black-legged variety has a defensive secretion containing the lactone dolichodial. One colony of the red-legged variety produced predominantly iridodial, but colonies from a different location produced isoiridomyrmecin. Different colonies of another species, *Iridomyrmex rufonigra*, produced either iridodial or dolichodial; in one case nests of the different types were only 50 m apart.

Termites are heavily predated by ants, especially dorylines and ponerines. Some of the latter, including *Megaponera foetens* in Africa and *Termitopone* in South America, are obligate termite feeders. It is not surprising that termite soldiers have developed powerful and diverse defensive secretions. The manner in which the secretions are applied shows great variety. In *Cubitermes* and *Amitermes* soldiers a secretion of the frontal gland in the head runs down a groove on to the mandibles, or whatever they are grasping. In *Schedorhinotermes*, the secretion runs forward on to the long forwardly-projecting labrum, which is tipped by a brush of fine cuticular projections (Quennedey, 1973). The insect applies the secretion, which is toxic to many ants, on to the opponent with the aid of this "brush". Soldiers of most of the subfamily Nasutitermitinae have the head capsule extended into a nozzle through which the frontal gland secretion is ejected under pressure. The terpenaceous secretion solidifies rapidly in air forming a sticky thread which immobilizes other insects and may also have toxic properties. The soldiers of *Tenuirostritermes tenuirostris* often oscillate to and fro while "firing", which produces a series of loops in the thread (Nutting *et al.*, 1974). An aberrant form of defence is found in the southern African species *Skatitermes*, in which the workers deposit, with remarkable accuracy, a droplet of anal fluid on the head of a predatory ant, effectively immobilizing it (Coaton, 1971).

The defensive secretion may kill the insect that produces it. Some *Macrotermes* soldiers produce a secretion containing quinones only in states of high alarm, and may die from the effects. A more spectacular sacrifice is seen in the amitermitid *Globitermes sulphureus* in which the soldier exudes a sticky yellowish secretion, sometimes rupturing its body wall in the process. The soldier-less termite *Astalotermes quietus* has well-developed circular and longitudinal muscle bands in the abdomen, and on even a relatively small provocation, such as grasping the extremity of a leg, the muscles contract causing abdominal dehiscence (Sands, 1972). The gut and haemolymph are sticky and effectively block the narrow subterranean galleries in which the insect lives, as well as immobilizing predators.

Table V shows the major components of defensive secretions of a number of Amitermitinae and Macrotermitinae. *Macrotermes* species that have been studied make use of quinones, which have fairly general toxicity and irritant properties. They are found very commonly in defensive secretions of invertebrates (Eisner, 1970) and in the Macrotermitinae are sometimes produced with sticky proteinaceous secretions which harden on drying.

The conservative nature of *Macrotermes* and *Odontotermes* secretions may reflect the fact that they have similar habits, tending to build large, relatively impregnable mounds. *Ancistrotermes cavithorax*, which forage in soil, do not build mounds and their defensive secretion is markedly different from those of other species and may well be an adaptation to deal with different kinds of predators.

In comparison with the Macrotermitinae, the Amitermitinae of Australia make great use of monoterpenes, species differing in the mix of compounds they produce and the relative amounts of each (Moore, 1968). The African species show a sharp contrast; *A. evuncifer* has a secretion consisting mainly of a novel oxygenated sesquiterpene (Wadhams *et al.*, 1974) which is toxic to some ants. The closely-related species *A. unidentatus* and *A. lonnbergianus* have completely different chemicals (Prestwich, 1975). Further studies of the defensive secretion chemistry of termites are likely to be an aid to the taxonomy of this difficult order and studies of the toxicity of the secretions are likely to shed light on the specificity of predator–prey interactions.

TABLE V(a). Defensive secretions of Amitermitinae

Australian species	Major components
Amitermes herbertensis[a]	Terpinolene, α-phellandrene
laurensis[a]	α-pinene, limonene, α-phellandrene
vitiosus[a]	α-pinene, limonene, terpinolene α-phellandrene
Drepanotermes ruficeps[a]	Limonene, terpinolene, α-phellandrene
African species	
A. evuncifer[b]	4, 11-epoxy-*cis*-eudesmane
A. messinae[c]	4, 11-epoxy-*cis*-eudesmane, limonene
A. unidentatus[d]	2-tridecanone, 2-pentadecanone
A. lonnbergianus[c]	None of the above.

[a] From Moore, 1968; [b] From Baker et al., 1974; [c] Prestwich, unpublished; [d] From Prestwich, 1975.

TABLE V(b). Defensive secretions of Macrotermitinae

S.E. Asian species	Major components	
Macrotermes carbonarius[a]	Benzoquinone, toluquinone	
M. gilvus[b]	No quinones	
Odontotermes redemanni[b]	Toluquinone	
O. praevalens[b]	Toluquinone	
O. horni[b]	Terpenes? No quinones	+ glutinant secretion
Hypotermes obscuriceps[b]	Benzoquinone	
Microtermes globicola[b]	Toluquinone	
African species		
Macrotermes bellicosus[c]	Toluquinone, naphthalene	
M. subhyalinus[d]	C_{23}–C_{29} olefins and alkanes, naphthalene	
Odontotermes badius[e]	Benzoquinone, protein	
O. stercorivorus[e]	Benzoquinone, protein	
Ancistrotermes cavithorax[f]	Furanoid sesquiterpenes	

[a] From Maschwitz et al. 1972; [b] From Maschwitz and Tho, 1974; [c] Baker, Briner, Evans and Howse, unpublished; [d] Prestwich et al., 1977; [e] From Wood et al., 1975; [f] Evans et al., 1977.

CONCLUSIONS

Over 25 years ago, Schneirla (1952) in a paper similar in scope to this one, emphasized the importance of bonds provided by mutual feeding and grooming (trophallaxis) in the integration of insect societies, controlling the responses of queen to brood, workers to the queen and worker to worker. In the intervening period the trend has been towards the study of behavioural mechanisms and we have learnt much about the sensory mechanisms involved in foraging in the honeybee (von Frisch, 1967). During the same period there has been a rapid growth of interest in chemical factors controlling insect social behaviour, generated largely by the development of new techniques in chromatographic and spectroscopic chemistry. Because of this, it is possible to see that trophallaxis is only one facet of communication and not the central pillar as Schneirla believed. Volatile chemical secretions play a pervasive role in the integration of the society, not only in communication, but also in defence where their use may be linked to a greater or lesser extent with nest structure.

The taxonomy of some social insect groups is notoriously difficult. Differences and affinities between species can sometimes be seen much more clearly in nest structures, pheromone chemistry and chemicals used in defence, than in morphology. However, there are obvious pitfalls here unless the adaptive features of such characters are fully understood and they are at best used to provide an overlay that highlights existing morphological differences.

ACKNOWLEDGEMENTS

I am grateful to Dr R. Baker for his help and Dr W. D. Hamilton for information on nests of social wasps.

REFERENCES

BLUM, M. S. (1969). Alarm pheromones. *A. Rev. Ent.* **14**, 57–79.
BOSSERT, W. H. and WILSON, E. O. (1963). The analysis of olfactory communication among animals. *J. theor. Biol.* **5**, 443–469.
BRADSHAW, J. W., BAKER, R. and HOWSE, P. E. (1975). Multicomponent alarm pheromones of the weaver ant. *Nature, Lond.* **258**, 230–231.
BROWN, W. L. (1960). Ants, acacias and browsing animals. *Ecology* **41**, 587–592.
BUTLER, C. G. and FAIREY, E. M. (1964). Pheromones of the honeybee: biological studies on the mandibular gland secretion of the queen. *J. Apicult. Res.* **3**, 65–76.
CAVILL, G. W. K. and CLARK, D. V. (1971). Ant secretions and cantharidin. *In* "Naturally occurring insecticides" (M. Jacobson and D. G. Crosby, eds), pp. 271–305. Marcel Dekker Inc., New York.
CAVILL, G. W. K. and HINTENBERGER, H. (1960). The chemistry of ants. IV. Terpenoid constituents of some *Dolichoderus* and *Iridomyrmex* species. *Aust. J. Chem.* **13**, 514–522.
COATON, W. G. H. (1970). Five new termite genera from South West Africa (Isoptera: Termitidae). *Cimbebasia* (A) **2**, 1–34.
CREWE, R. M. and BLUM, M. S. (1972). Alarm pheromones of the Attini and their phylogenetic significance. *J. Insect. Physiol.* **18**, 31–42.
EISNER, T. (1970). Chemical defense against predation in arthropods. *In* "Chemical Ecology" (E. Sondheimer and J. B. Simeone, eds), pp. 157–218. Academic Press, New York, London.

EMERSON, A. E. (1956). Ethospecies, ethotypes, taxonomy, and evolution of *Apicotermes* and *Allognathotermes* (Isoptera, Termitidae). *Am. Mus. Novit.* **2236**, 1–46.

EVANS, D. A., BAKER, R., BRINER, P. H. and MCDOWELL, P. G. (1977). Defensive secretions of some African termites. *Proc. 8th Int. Cong. I.U.S.S.I.*, Wageningen, Pudoc., 46–47.

EVANS, H. E. (1966). "The comparative ethology and evolution of the sand wasps" Harvard U.P., Cambridge, Mass.

EVANS, H. E. and EBERHARD, M. J. W. (1973). "The Wasps" David and Charles, Newton Abbot, U.K.

FLETCHER, D. J. C. (1973). "Army ant" behaviour in the Ponerinae: a re-assessment. *Proc. VIIth Int. Congr. IUSSI*, London, 116–121.

FRISCH, K. VON (1967). "The dance language and orientation of bees". Belknap Press, Cambridge, Mass.

FOSTER, J. B. and DAGG, A. I. (1972). Notes on the biology of the Giraffe. *E. Afr. Wildl. J.* **10**, 1–16.

GRASSÉ, P.-P. and NOIROT, C. (1961). Nouvelles recherches sur la systématique et l'éthologie des termites champignonnistes du genre *Bellicositermes* Emerson. *Insectes soc.* **8**, 311–359.

GRIGG, G. C. (1973). Some concequences of the shape and orientation of "magnetic" termite mounds. *Aust. J. Zool.* **21**, 231–237.

HAMILTON, W. D. (1964). The genetical evolution of social behaviour I and II. *J. theor. Biol.* **7**, 1–52.

HARRIS, W. V. (1971). "Termites, their recognition and control". Longman Group, London.

HASKINS, C. P. (1970). Researches in the biology and social behaviour of primitive ants. *In* "Development and evolution of behaviour." (C. L. R. Aronson, E. Tobach, D. S. Lehrman and J. S. Rosenblatt, eds), pp. 355–388. W. H. Freeman and Co., San Francisco.

HEINRICH, B. (1974). Thermoregulation in bumble bees. 1. Brood incubation by *Bombus vosnesenskii* queens. *J. comp. Physiol.* **88**, 129–140.

HIMMER, A. (1932). Die Temperaturverhältnisse bei den sozialen Hymenopteren. *Biol. Rev.* **7**, 224–253.

HÖLLDOBLER, B. (1974). Home range orientation and territoriality in harvesting ants. *Proc. natn. Acad. Sci. U.S.A.* **71**, 3274–3277.

HÖLLDOBLER, B. and WILSON, E. O. (1970). Recruitment trails in the harvester ant *Pogonomyrmex badius*. *Psyche* **77**, 385–399.

HOWSE, P. E. (1975). Chemical defences of ants, termites and other insects: some outstanding questions. *In* "Pheromones and defensive secretions in social insects" (C. Noirot, P. E. Howse and G. Le Masne, eds), pp. 23–40. French Section of the International Union for the Study of Social Insects at the University of Dijon.

ISHAY, J. (1972). Thermoregulatory pheromones in wasps. *Experientia* **28**, 1185–1187.

ISHAY, J. (1975). Hornet nest architecture. *Nature, Lond.* **253**, 41–42.

ISHAY, I. and SADEH, D. (1975). Direction finding by hornets under gravitational and centrifugal forces. *Science, N.Y.* **180**, 802–804.

JAISSON, P. (1975). L'imprégnation dans l'ontogénèse des comportements de soins aux cocons chez la jeune fourmi rousse (*Formica polyctena* Forst). *Behaviour* **52**, 1–37.

JANZEN, D. H. (1966). Coevolution of mutualism between ants and acacias in Central America. *Evolution, Lancaster, Pa.* **20**, 249–275.

JEANNE, R. L. (1970). Chemical defense of the brood by a social wasp. *Science, N.Y.* **168**, 1465–1466.

JEANNE, R. L. (1975a). The adaptiveness of social wasp nest architecture. *Q. Rev. Biol.* **50**, 267–287.

JEANNE, R. L. (1975b). Social biology of *Stelopolybia areata* (Say) in Mexico. *Insectes soc.* **22**, 27–34.

LIN, N. and MICHENER, C. D. (1972). Evolution of sociality in insects. *Q. Rev. Biol.* **47**, 131–159.

LINDAUER, M. (1954). Temperaturregulierung und Wasserhaushalt im Bienenstaat. *Z. vergl. Physiol.* **36**, 391–342.

LÜSCHER, M. (1955). Der Sauerstoffverbrauch bei Termiten und die Ventilation des Nestes bei *Macrotermes natalensis* (Haviland). *Acta trop.* **12**, 289–307.

LÜSCHER, M. (1961). Air-conditioned termite nests. *Scient. Am.* **205**, 138–145.

LÜSCHER, M. (1975). Pheromones and polymorphism in bees and termites. *In* "Pheromones and defensive secretions in social insects" (C. Noirot, P. E. Howse and G. Le Masne, eds), pp. 123–189. French Section of the International Union for the Study of Insects at the University of Dijon.

MACHADO, A. DE B. (1959). Le concept d'espèce éthologique et son application prématurée à la systématique des termites *Apicotermes*. *Proc. 15th Intern. Congr. Zool., London* **1958**, 205–207.

MASCHWITZ, U. (1964). Gefahrenalarmstoffe und Gefahrenalarmierung bei sozialen Hymenopteren. *Z. Vergl. Physiol.* **47**, 596–655.

MASCHWITZ, U. and KLOFT, W. (1971). Morphology and function of the venom apparatus of insects—bees, wasps, ants and caterpillars. *In* "Venomous animals and their venoms'" (W. Bücherl, E. E. Buckley and V. Deulofev, eds), Vol. 3, pp. 1–60. Academic Press, London and New York.

MASCHWITZ, U. and THO, Y. P. (1974). Chinone als Wehrsubstanzen bei einigen orientalische Macrotermitinen. *Insectes soc.* **21**, 231–234.

MASCHWITZ, U., JANDER, R. and BURKHARDT, D. (1972). Wehrsubstanzen und wehrverhalten der termite *Macrotermes carbonarius*. *J. Insect. Physiol.* **18**, 1715–1720.

MEYER, D. and LÜSCHER, M. (1973). Juvenile hormone activity of the haemolymph and the anal secretion of the queen of *Macrotermes subhyalinus* (Rambur) (Isoptera, Termitidae). *Proc. VIIth Int. Congr. IUSSI, London*, 268–273.

MONTAGNER, H. (1967). Le mécanisme et les conséquences des comportements trophallactiques chez les guêpes du genre *Vespa*. *Bull. biol. Fr. Belg.*, **100**, 189–323.

MOORE, B. P. (1968). Studies of the chemical composition and function of the cephalic gland secretion in Australian termites. *J. Insect Physiol.* **14**, 33–39.

NUTTING, W. L., BLUM, M. S. and FALES, H. M. (1974). Behaviour of the North American termite *Tenuirostritermes tenuirostris* with special reference to the soldier frontal gland secretion, its chemical composition, and use in defense. *Psyche, Cambridge* **81**, 167–177.

PRESTWICH, G. D. (1975). Chemical analysis of soldier defensive secretions in several species of East African termites. *In* "Pheromones and defensive secretions in social insects" (C. Noirot, P. E. Howse and G. le Masne, eds). French Section of the International Union for the Study of Social Insects at the University of Dijon.

PRESTWICH, G. D., BIERL, E. D., DIVILBISS and CHAUDHURY, M. F. B. (1977). Soldier frontal glands of the termite *Macrotermes subhyalinus*: morphology, chemical composition and use in defence. *J. Chem. Ecol.* **3**, 579–590.

QUENNEDEY, A., BRULÉ, G., RIGAUD, R., DUBOIS, P. and BROSSUT, R. (1973). Le glande frontale des soldats de *Schedorhinotermes putorius* (Isoptera): analyse chimique et fonctionnement. *Insect. Biochem.* **3**, 67–74.

REGNIER, F. E., NIEH, M. and HÖLLDOBLER, B. (1973). The volatile Dufour's gland components of the harvester ants *Pogonomyrmex rugosus* and *P. barbatus*. *J. Insect. Physiol.* **19**, 981–992.

RICHARDS, O. W. (1971). The biology of the social wasps (Hymenoptera, Vespidae). *Biol. Rev.* **46**, 483–528.

RICHARDS, O. W. and RICHARDS, M. J. (1951). Observations on the social wasps of South America (Hymenoptera, Vespidae). *Trans. R. ent. Soc. Lond.* **102**, 1–170.

RIDLEY, H. N. (1910). Symbiosis of ants and plants. *Ann. Bot.* **24**, 457–483.

ROBINSON, S. W., MOSER, S. L., BLUM, M. S. and AMANTE, E. (1975). Laboratory investigations of the trail-following response of four species of leaf-cutting ants with notes on the specificity of a trail pheromone of *Atta texana* (Buckley). *Insectes soc.* **21**, 87–94.

RUELLE, J. E. (1964). L'architecture du nid de *Macrotermes natalensis* et son sens fonctionnel. In "Études sur les termites Africains" (A. Bouillon, ed.) pp. 55–68. Masson, Paris.

SANDS, W. A. (1972). The soldier-less termites of Africa (Isoptera: Termitidae). *Bull. Brit. Mus.* (Nat. Hist.) Entomology Suppl. **18**, 1–244.

SANNASI, A., SEN-SARMA, P. K., GEORGE, C. J. and BASALINGAPPA, S. (1972). Juvenile hormone activity for various sources of termite castes and their fungus gardens. *Insectes soc.* **19**, 81–88.

SCHMIDT, R. S. (1955). Termite (*Apicotermes*) nests—important ethological material. *Behaviour* **8**, 344–356.

SCHNEIRLA, T. C. (1952). Basic correlations and coordinations in insect societies with special reference to ants. *Coll. int. Centr. natn. Res. Scient.*, Paris (34), 247–269.

SCHNEIRLA, T. C. (1971). "Army ants: a study in social organisation" W. H. Freeman and Co., San Francisco.

SLAMA, K. and WILLIAMS, C. M. (1965). Juvenile hormone activity for the bug *Pyrrhocoris*. *Proc. natn. Acad. Sci. U.S.A.* **54**, 411–414.

TRIVERS, R. L. and HARE, H. (1976). Haplodiploidy and the evolution of the social insects. *Science, N.Y.* **191**, 249–263.

TUMLINSON, J. H., MOSER, J. C., SILVERSTEIN, R. M., BROWNLEE, R. G. and RUTH, J. M. (1971). Identification of the trail pheromone of a leaf-cutting ant, *Atta texana*. *Nature, Lond.* **234**, 348–349.

WADHAMS, L. J., BAKER, R. and HOWSE, P. E. (1974). 4, 11-epoxy-*cis*-eudesmane, a novel oxygenated sesquiterpene in the frontal gland of the termite *Amitermes evuncifer* Silvestri. *Tet. Letters* **18**, 1697–1700.

WALLIS, D. I. (1964). Aggression in social insects. In "The natural history of aggression" (J. D. Carthy and F. J. Ebling, eds), pp. 15–22. Academic Press, New York and London.

WHEELER, J. W., EVANS, S. L., BLUM, M. S. and TORGERSON, R. L. (1975). Cyclopentyl ketones: identification and function in *Azteca* ants. *Science, N.Y.* **187**, 254–245.

WEIR, J. S. (1973). Air flow, evaporation and mineral accumulation in mounds of *Macrotermes subhyalinus* (Rambur). *J. anim. Ecol.* **42**, 509–530.

WILSON, E. O. (1971). "The Insect Societies." Belknap Press, Cambridge, Mass.

WOOD, W. F., TRUCKENBRODT, W. and MEINWALD, J. (1975). Chemistry of the defensive secretion from the African termite *Odontotermes badius*. *Ann. ent. Soc. Am.* **68**, 359–360.

21 | Aggregation in Echinoderms

G. F. WARNER

Department of Zoology, University of Reading, England

Abstract: Dense aggregation is not uncommon amongst echinoderms. It has been recorded in all major subdivisions of the Phylum. Aggregation may be the result of responses to the general environment or the result of social behaviour. The latter has been demonstrated in certain brittle-stars and probably occurs in other echinoderms. One advantage of aggregation is that it is likely to promote reproductive success, this advantage being most important in sedentary species and in rare, nomadic species. Reproductive social behaviour has been reported in several groups.

Many dense aggregations are permanent and sedentary, requiring a high delivery rate of food. It is proposed that suspension feeding in currents is the only method that can support such aggregation. In line with this, a possible suspension feeding method is suggested for certain densely aggregating sand dollars. Dispersion within aggregations is often clumped and may appear crowded. Such distribution suggests the existence of special advantages associated with crowding. In crowded aggregations of suspension-feeding crinoids and brittle-stars the forest of arms extended into the current may create a baffle effect thus promoting particle capture. In some brittle-stars crowding promotes stability in currents. In contrast to crowding, spacing out occurs within groups of certain brittle-stars; arm-loop feeding on small organisms is suggested to account for this pattern.

Starfish aggregations are mobile and predatory. Feeding smells appear to draw individuals together. In *Asterias*, aggregation serves the rapid exploitation of a localized superabundance of food. In *Acanthaster*, aggregation in small groups may be a sound reproductive strategy while starfish plagues may indicate food shortage.

Apart from suspension feeding parallels, echinoderm aggregations have little in common with colonial organisms.

INTRODUCTION

Echinoderm aggregations are less impressive than herds of caribou, flocks of starlings and shoals of mackerel only in the sense that they tend to remain in one place, or if they do move, to move slowly. They are often more impressive

---

Systematics Association Special Volume No. 11, "Biology and Systematics of Colonial Organisms", edited by G. Larwood and B. R. Rosen, 1979, pp. 375–396. Academic Press, London and New York.

in terms of total numbers and population density (Table I) and, when one considers the diffuse nervous systems and simple ways of life of most echinoderms, the very existence of dense aggregations becomes impressive as a worthy

TABLE I. Population densities recorded within dense echinoderm aggregations

Species	Density m^{-2}	Authority
Antedon bifida	1200	Könnecker and Keegan, 1973
Ophiothrix fragilis	2196	Brun, 1969
Amphiura filiformis	2200	Könnecker and Keegan, 1973
Pseudocucumis mixta	200	Könnecker and Keegan, 1973
Paracentrotus lividus	1600	Könnecker and Keegan, 1973
Mellita quinquiesperforata	821	Salsman and Tolbert, 1965
Asterias rubens	97	Brun, 1968

enigma. How do they do it? Why do they do it? Reese (1966) found it difficult to credit echinoderms with social behaviour and preferred individual responses to the physical environment as a means of achieving aggregation. This interpretation was justifiable on the evidence available but even then Reese noted as "exceptions which prove the rule" records of spawning aggregations in echinoids and asteroids, "copulation" in asteroids and various other reproductive interactions between conspecifics. Since then more evidence for the involvement of social behaviour in the formation and maintenance of aggregations has emerged. In addition, the involvement of larvae in the selection of the adult habitat by responses to the presence of adult conspecifics has been shown to be a widespread phenomenon in marine invertebrates. There is some evidence that this occurs in brittle-stars, and in other species the possibility of larval social behaviour helping to maintain aggregations cannot be ignored.

On the question of why they do it Reese (1966) suggested reproduction and feeding. In the former case the advantages to animals with external fertilization of aggregation and synchronized spawning are obvious. But these advantages are not equally important in all species. Permanent aggregation is of great value to sedentary species since isolated individuals have little chance of successful reproduction. It should also be of value to rare, nomadic species since chance meetings during a restricted breeding season are likely to be infrequent. Small, permanent aggregations might here represent the best reproductive strategy except in cases where the food resource is widely dispersed. Common, nomadic species have little need for permanent, reproductive aggregation, but a mechanism to promote clustering during spawning would still have a distinct survival value. One cannot, of course, divide echinoderms into groups labelled sedentary and nomadic, or rare and common. Crinoids, for instance, are normally sedentary but can clamber slowly about and may even swim. The importance of reproductive aggregation must, therefore, be judged separately for each species. It is inevitable, however, that permanent aggregations, whether or not they have other functions, must promote fertilization success. Temporary spawning clusters probably have no other function than this.

On feeding, Reese (1966) noted: "...one would expect grazing, detrital-suspension feeding and deposit feeding species to reach the highest densities". Here the renewal of food resources by plant growth, or by suspended particles delivered in a current, or settling onto the sea bed, proceed relatively rapidly and continuously within particular areas. Aggregations can accumulate in these favourable places up to the limit set by the renewal rate of the resource. These renewal rates must vary according both to the resource and to the environment. One might expect, however, that in most cases the delivery rate of suspended particles in currents should be a more rapid method of resource renewal than either settling particles or plant growth. Consequently one expects the densest aggregations to occur amongst suspension feeders. In addition, aggregation should be expected more frequently in suspension feeders since, being comparatively sedentary, the reproductive advantage is particularly important to them.

Echinoderm aggregations, however, present features which lead one to doubt whether this represents the whole story. Aggregations are often crowded, in the sense that physical contact occurs frequently between individuals. Here one might expect competition between crowded conspecifics for food. Further, such crowds are often discontinuously distributed over an apparently uniform substratum, with bare patches occurring between crowds. This crowding and clumping is understandable as a temporary reproductive phenomenon but when it persists as a permanent feature of the population other reasons must be sought for its occurrence.

The following account is not intended as an exhaustive review and most of my references are to fairly recent work. The earlier work is very well reviewed by Reese (1966). I have attempted, however, to elaborate on the ideas introduced above by reference to specific examples which are either well covered in the literature or of which I have first hand experience through underwater observation. Throughout I shall be attempting to answer both how and why they aggregate. As a conclusion I have attempted to draw together these ideas to see whether there are any parallels between the biology of echinoderm aggregations and the biology of colonial animals.

CRINOIDEA

Crinoids are suspension feeders. They cling by means of cirri to hard substrata and stretch their long, pinnate arms out into the water. The arms and pinnules may adopt various orientations depending on the character of the prevailing water movement (Meyer, 1973). A minority of crinoid species become densely aggregated but it appears to be a common phenomenon in *Antedon bifida* (Pennant) (Keegan, 1974). Aggregations occur in areas subject to currents, the latter providing a continuous supply of food. Keegan's populations on the west coast of Ireland were densely crowded at more than 1200 m^{-2} but within these aggregations distribution was even. I have observed similar aggregations of *Antedon* in Torbay, Devon, in current-washed environments but here populations displayed a clumped distribution, dense aggregations occurring as discrete patches on rocky outcrops. In the large Caribbean crinoid *Nemaster grandis* Clark, Meyer (1973) noted that although isolated individuals were common, "two or more individuals often occur side by side" and at one site "very large aggregations ... were observed on coral mounds". This species, like *Antedon*, favours environments which are exposed to currents. In unidirectional currents its arms are held up to form a filtration fan orientated perpendicular to the current. The 40 long arms overlap somewhat such that the fan may be several layers thick. According to Meyer such filtration fans act as baffles to through-flowing water thus increasing the chances of particle capture by the tube feet. This effect is probably the result of micro-turbulence on the leeward side of each arm, slowing and mixing the water stream. Meyer suggested that the baffle effect might explain the existence of aggregations of *N. grandis*, promoting the capture of particles by the individuals. The same explanation can be used to account for crowding in *Antedon* since, although this species does not form a filtration fan, the forest of feathery arms must slow down the current close to the sea bed, promoting micro-turbulence and deposition.

One may ask why more species of crinoids, being fairly sedentary suspension feeders, do not aggregate; some reproductive advantage should apply to all. The answer is probably that all do not feed in quite the same way. To take three contrasting Caribbean examples (Meyer, 1973): *Tropiometra carinata* (Lamarck) attaches its cirri inside crevices in relatively shallow water where wave oscillation prevails. The arms are extended from the crevice and sway backwards and forwards as the waves pass. The crevice is an essential anchor point and although several crinoids may crowd together in a crevice (personal observation in Trinidad) the relative scarcity of crevices limits the formation of dense aggregations. *Nemaster rubiginosa* (Pourtalès) inhabits coral reefs. It is adapted to feed from slow moving, multidirectional currents such as occur amongst the corals rather than above them where *N. grandis* feeds. No baffle effect would be produced in this case by the addition of extra individuals. *Analcidometra caribbea* (A. H. Clark) is a small crinoid which clings to gorgonians and black corals. Thus supported in the current it spreads its arms in a filtration fan. I have observed in Trinidad as many as four individuals of this species spread along the margin of a fan-shaped black coral colony, but the narrow perch leaves no room for a dense aggregation to develop.

Whether the behaviour leading to aggregation in crinoids is social or the result of responses to the physical environment is not known. The patchiness observed in *Antedon* in Torbay, however, and the frequent occurrence of small groups in an otherwise dispersed population of *N. grandis* observed by Meyer (1973), suggest responses to conspecifics rather than simply to physical factors. The dense aggregations of *Antedon* in Torbay have been present since first observed ten years ago suggesting that these populations are permanently maintained and not simply the result of a freak settlement. Social behaviour by larvae or dispersal stages may account for such maintenance.

OPHIUROIDEA

The variety of feeding methods open to brittle-stars is very large. In just one species, *Ophiocomina nigra* (Abildgaard), Fontaine (1965) described suspension feeding and cited deposit feeding, herbivorous browsing and carnivorous behaviour as additional methods. Brittle-stars which form crowded aggregations, however, seem mostly to be specialized suspension feeders. The best known of these is *Ophiothrix fragilis* (Abildgaard) (Brun, 1969; Warner, 1970, 1971; Allain, 1974; Broom, 1975; Warner and Woodley, 1975). This species lives in densely crowded aggregations (Table I) below the normal depth of penetration of wave oscillation in areas subject to tidal currents. Distribution within a population is markedly clumped, individuals occurring in discrete patches of one to many square metres in area. Being subject to sometimes quite strong currents the substratum is normally rocky or gravelly but the brittle-stars are not restricted to one or other type of bottom and, unlike crinoids, cannot cling on but simply rest on the sea bed. During feeding an aggregation extends a forest of arms into the current (Plate IA). Each individual has its disc raised slightly above the substratum balanced on the bent, proximal parts of two or three arms, the remaining arms and the distal parts of the support arms are stretched into the current. The densely crowded nature of the aggregation allows individuals to economize on support arms since each brittle-star is partly propped up by its neighbours. As in crinoids it is likely that the forest of arms acts as a baffle, promoting deposition in the area of the aggregation and increasing the chances of capture of individual particles. Another, and in my opinion overriding, advantage associated with crowding in this species is that it promotes stability in currents. The current from which the brittle-stars feed is variable in strength according to the time of the tide. As current speed increases beyond about 20 cm sec^{-1}, individuals crouch down and cling to their neighbours (Warner, 1971; Warner and Woodley, 1975). This effect is first noticeable at the edges of patches but with increasing current speed it spreads inwards to affect all individuals. The interlocking brittle-stars thus form a mat which has much greater stability in the current than scattered individuals would have. The occurrence of crouching only at the edges of patches in intermediate currents is interesting and may be the result both of the firm support by neighbours in the centre of a patch and of the baffle effect producing slower currents away from the windward edge. Individuals in the centres of patches would appear to be favoured on this basis and indeed Allaine (1974) found disc diameters to be greater in individuals from the centres of dense beds than in individuals from peripheral sites.

Beds of *Ophiothrix fragilis* appear to be maintained by social behaviour on the part of both larvae and adults. The regular occurrence of recently settled young on the bodies of adults, their apparent clustering on particular adults and their virtual absence from aggregations of medium sized individuals (which cluster separately amongst the sessile epifauna on rocky outcrops) suggest fairly specific larval settlement preferences (Warner, 1971). The adults, for their part, show behavioural responses which tend to maintain aggregations and which provide accidentally isolated individuals with a means of re-locating an aggregation. Broom (1975) found that an individual which was removed from a patch and placed nearby on a bare area would pause for several seconds and then set off along the sea bed, pausing and changing direction from time to time. About 80% of such walks were across current but showed no orientation to the nearest aggregation of brittle-stars. Isolated individuals continued walking until they encountered other brittle-stars, at which point they would climb in amongst them, stop walking and start feeding. Walking was not terminated, however, by contact with other objects, both living and non-living, which might have served as feeding anchors. If, as Broom's work suggests, contact with conspecifics inhibits walking, then this behaviour constitutes a social mechanism for maintaining crowded aggregations. The fact that isolated individuals normally walk across currents is a sound strategy for re-locating an aggregation since these, and the bare areas between them, tend to be arranged in strips parallel to the prevailing current.

PLATE I

A, Crowded aggregation in the brittle-star *Ophiothrix fragilis*; the bare area suggests clumping. B, Food sharing in *Asterias rubens* occurring in an aggregation on a bed of the bivalve *Spisula*.

Other species of the genus *Ophiothrix* may form dense aggregations (see Warner and Woodley, 1975) but not all do so, and there are many other genera of spiny armed, epifaunal brittle-stars which probably suspension feed but which do not form crowded aggregations. Many of these are cryptic, inhabiting crevices or living amongst sessile epifauna such as sponges and corals. Certain burrowing brittle-stars, however, do form crowded aggregations; an example is *Amphiura filiformis* (O. F. Müller) (Table I), a rheophilic species inhabiting silty sand. Its feeding method was studied by Buchanan (1964) and turns out to be very similar to that practised by *O. fragilis*, with the proviso that here the individuals are anchored against the current by their bodies being embedded in the substratum. The arms are extended into the current in the same way, however, and one may therefore expect that in this species too a baffle effect is present which may provide some advantage to crowded individuals. One intuitively expects *A. filiformis*, being a burrowing species, to be more sedentary than are epifaunal brittle-stars; the reproductive advantages of aggregation may, therefore, be greater in this species than in other supposedly more mobile forms.

Buchanan (1964) compared the feeding of *A. filiformis* with that of another burrowing species *Amphiura chiajei* Forbes. He found the latter to be a deposit feeder favouring areas which were less subject to currents. It is noteworthy that Buchanan found *A. chiajei* at much lower densities than those reached by *A. filiformis*: 10–11 m^{-2} compared with 300–400 m^{-2}. *A. chiajei* was also found to be a relatively slow-growing species with a low metabolic rate. These points give some indication of the relative delivery rates of food to suspension and deposit feeders and provide fuel for the theory that crowded populations of sedentary animals can only flourish by suspension feeding.

Certain brittle-stars form aggregations in which the individuals are spaced out. This distribution pattern is found in *Ophiocomina nigra* which may occur in discrete patches in which individuals with arms spread horizontally do not quite touch each other (personal observation in Torbay, see Warner, 1971; Wilson *et al.*, 1977). The patches which I observed were located in narrow gullies between rocky outcrops and comprised 15–20 individuals; Wilson *et al.*, however, described patches up to 300 m across, containing thousands of individuals, located on extensive offshore plains of rippled gravel. Spacing out in this species has been confirmed by laboratory experiments (Wilson *et al.*, 1977); the behaviour involved may be antagonistic and results in a distribution which is significantly more dispersed than random. The explanation of spacing out may lie in the feeding methods used by this species. Fontaine (1965) described how *Ophiocomina*, when suspension feeding, swings the arms from side to side "fishing an area roughly equal to the diameter of the animal". This behaviour clearly requires spacing out at arms length, but it seems likely that the carnivorous behaviour of *Ophiocomina* is also relevant here. The species is capable of catching small organisms in an arm loop (Vevers, 1956), a response probably controlled by both tactile and chemical receptors. Mechanical interference by conspecifics would be likely to reduce the efficiency of any such predatory response.

Euryalous brittle-stars (other than basket-stars) may also feed by capturing small organisms in an arm loop. These forms live entwined amongst the branches of gorgonian or black coral colonies and hold their arms out into the current when the light intensity is low (personal observation in Trinidad). Their arms are not spiny and when they are "fishing" their tube feet are not extended. This suggests that simple suspension feeding by providing a large, sticky surface as practised by *O. fragilis* is not involved. On some of the larger gorgonian fans several individuals of one species (*Asteroschema* sp.) may occur together but they are always spaced out at arms length (Fig. 1). I have observed what were probably feeding movements in this species in which the arm tips were moving sinuously from side to side as if searching the surrounding area. These movements were taking place on the leeward side of the gorgonian fan, the brittle-star possibly taking advantage of a baffle effect. If such movements are concerned with arm-loop feeding, mechanical interference with adjacent brittle-stars must be avoided and spacing out is expected.

It is not known whether the distribution of these euryalous brittle-stars is clumped, in the sense of spaced aggregations occurring on some fans only, but *Ophiocomina* clearly occurs in patches implying the involvement of social behaviour. Under rather different circumstances Keegan (1974) reported a clumped distribution in *Ophiocomina*. Here the clumps, in which population density reached 90 m^{-2} (sparse compared with *O. fragilis*), occurred within dense aggregations of *Antedon* and were associated with areas where the population density of this crinoid was particularly high! That *Ophiocomina* is capable of social behaviour has been shown by Gorzula (1974) who described for this species a curious sexual behaviour in which small, male "riders" rode on the backs of larger "carriers" of both sexes. Copulation also occurred in which male riders embraced female carriers in a mouth-to-mouth position. The advantages to *Ophiocomina* of aggregation seem likely to be reproductive rather than concerned with feeding.

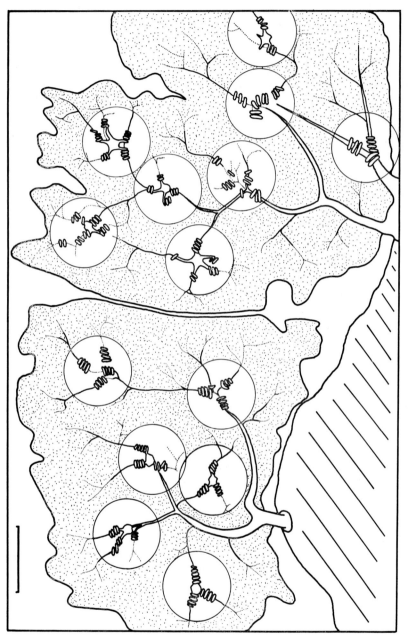

FIG. 1. Euryalous brittle-stars (*Asteroschema* sp.) spaced out on gorgonian fans. The long arms of each individual (circled) are coiled around the adjacent gorgonian branches. The discs of some individuals only are on the observer's side of the fans. Based on a photograph taken at 30 m off Trinidad. Scale: 10 cm.

HOLOTHUROIDEA

Most species of sea cucumber which occur at high population density are burrowing suspension feeders. Fish (1967) found *Cucumaria elongata* Duben and Koren at a density of more than 20 m^{-2} off the Northumberland coast but these figures have been greatly exceeded by the findings of Könnecker and Keegan (1973). They found *C. elongata* at more than 50 m^{-2} and, most spectacular of all, *Pseudocucumis mixta* Östergren at more than 200 m^{-2}. At this density the spread of the oral feeding tentacles of individuals is restricted and they project upwards rather than radially. Contact between adjacent animals is apparently minimal and therefore, presumably, avoided. Konnecker and Keegan suggested a combination of favourable environmental factors as being among the more important reasons for the occurrence of the *P. mixta* aggregation. While this is certainly true, it is possible that such a dense population would produce some baffle effect which might bring extra food to the individuals. Also in such a situation successful reproduction must be assured.

ECHINOIDEA

Regular sea urchins are typically grazing animals and may be herbivorous, carnivorous or omnivorous. They frequently form dense, but not crowded populations extending over large areas of suitable substratum. A good example is the large urchin *Echinus esculentus* L. which is often very common in kelp forests. Krumbein and Pers (1974) have described a vast population of 20–100 million *Echinus* feeding largely on the boring worm *Polydora ciliata* (Johnston) and producing as a result an estimated 30 000 tons of rock detritus per year! Population density up to 7 m^{-2} was found but does not represent crowding even in this large species. Such large populations occur as a result of the presence of favourable environments. Small scale clumping in the breeding season to promote fertilization success has been described in the Caribbean urchin *Tripneustes esculentus* Leske by Lewis (1958) and probably occurs in many such species. Even without reproductive clumping, synchronized gonad maturation, and spawning in response to the spawn of conspecifics—epidemic spawning—(e.g. Fox, 1923, 1924) greatly increases the chances of fertilization success in these large populations.

The homeostatic effect on the environment of large urchin populations has been noted many times (e.g. Jones and Kain, 1967; Schuhmacher, 1974). The random distribution of grazing individuals and the constant renewal of their food (usually plant material) must be responsible for this conservation effect. Reese (1966), however, described a different situation in Californian kelp forests in which urchins grazed in herds (Leighton, personal communication to Reese). Here the food consisted of the massive holdfasts of the giant kelp. This is an irregularly distributed and slowly regenerating resource and contrasts in these respects with the resources of most grazing urchins. An urchin population distributed at random in such a kelp forest and feeding on the holdfasts might weaken many holdfasts at once, leading to the destruction of the environment. Feeding in herds and concentrating on one holdfast at a time may be a better strategy for conservation. Such behaviour requires either that urchins behave socially, or that they are attracted by the smell of damaged kelp. These grazing herds of urchins appear not unlike the starfish aggregations discussed below.

Paracentrotus lividus (Lamarck), an otherwise typical grazing urchin (Kitching and Ebling, 1961), forms very dense, "layered" aggregations within deposits of branched, coralline algae or maerl (Table I) (Keegan, 1974). It occurs here as part of what Keegan called a "sub-surface epifauna" referring to the fact that there is relatively free penetration of water through the branched maerl. These dense populations are only found in certain beds of maerl. Keegan (1974) gave no details on the food of these urchins, but unless the maerl is considerably more productive than are most algae it seems unlikely that by itself it could support such dense populations. It is possible that the open structure of the maerl provides a baffle effect and that the urchins are feeding largely on detritus.

Crowded aggregation in irregular urchins is somewhat better known and here the sand dollars (Clypeastroida) take pride of place (Table I). *Mellita quinquiesperforata* (Leske) occurs in very large numbers in narrow belts many hundreds of metres long and orientated parallel to the coast just off sandy beaches. I have observed this species off the north coast of Trinidad in a sandy bay which is exposed to wave action throughout the year. Here the belt of sand dollars is 3–4 m wide and the density in the middle often exceeds 100 m^{-2}, the sand dollars occupying more than two thirds of the available space. The belt is situated where the largest waves begin to break as they approach the shore. This zone is characterized by quite strong to and fro surge on the bottom but there is relatively little turbulence of the type that occurs in shallower water under the breaking waves. The environment is therefore fairly stable for these flattened animals although the strongest surges do flip the occasional one over onto its back. Sand dollars are slightly domed on the upper surface and much less stable when on their backs so that individuals which flip over are soon turned the right way up again. I observed that the population remained in this environment over periods of changing sea state by migrating inshore or offshore as the swell became weaker or stronger respectively. On calm days the belt of sand dollars could be found in 1·5 m of water while on rough days they were 3–4 m deep. This Trinidad population was not as large as that described by Salsman and Tolbert (1965). Their belt was wider, denser (Table I) and peak population density occurred deeper in 7 m of water. Their sand dollars were, however, far smaller than the Trinidad ones (mean of 43 mm test diameter compared with about 90 mm in Trinidad). As in Trinidad, Salsman and Tolbert's population lived just beyond the surf zone.

Sand dollars are thought to be deposit feeders. In *Mellita sexiesperforata* (Leske), Goodbody (1960) showed that as the urchin burrowed along just under the sand, small particles would drop between the club shaped spines of its upper surface to be led by ciliary currents through the lunules and over the margins of the test to the mouth. *M. quinquiesperforata*, being very similar in general morphology, probably feeds in a similar way. Living just beyond the breaking waves, it may receive much potential food in the form of small particles disturbed from closer inshore or brought in on the waves. Such material might be sucked out close to the substratum by an undertow to accumulate close to the bottom just beyond the break. Would this material, however, be deposited in such a mobile environment or would it simply waft back and forth across the sand dollar belt? If the latter occurs, then in addition to deposit feeding, some suspension feeding involvement might be expected. In Trinidad I observed that individual *M. quinquiesperforata* were predominantly orientated with their long axes parallel to the shore. The large, posterior interambulacral lunule was usually open and its aperture raised above the general surface by the elevation of a row of flattened spines around its margin; the other lunules were generally closed by marginal spines folded over them. The large lunule was therefore orientated with its long axis across the direction of the wave surge (Fig. 2). It is likely that here a similar effect obtains to that described by Vogel *et al.* (1973) which ventilates the burrows of prairie-dogs, namely that the comparatively swift current over the raised aperture will induce water to flow out of it. This water would have to be replaced via the other lunules and from around the margins of the test. These areas might therefore be expected to accumulate finely divided material by filtering it from suspension. Further work must decide whether such a flow exists and if so whether it can be considered a suspension feeding mechanism. Since the induced currents would, in part, flow in the same direction as the ciliary currents described by Goodbody (1960) this possibility does not appear too improbable.

Dense sand dollar belts off sandy beaches probably occur in response to favourable environmental factors which produce a rich supply of food. Distribution within the belts is not normally clumped but Salsman and Tolbert

(1965) noted the formation of heaps of individuals in July. In August these heaps contained a proportion of dead individuals. It is possible that heaping represents spawning aggregation and that after spawning the animals die *in situ*.

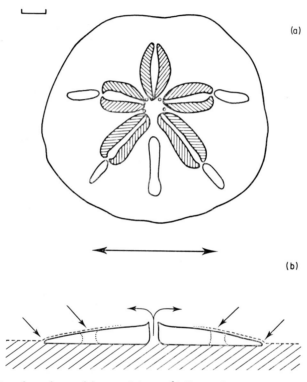

Fig. 2. (a) View from above of the test of the sand dollar *Mellita quinquiesperforata* showing the usual orientation to wave surge (indicated by double-headed arrow). Scale: 10 mm. (b) Diagrammatic transverse section through the posterior interambulacral lunule of a sand dollar in life position. The arrows show the directions of a possible suspension feeding current induced by surge.

ASTEROIDEA

The most spectacular aggregations of starfish that have been reported concern the common starfish *Asterias rubens* L. (Table I) and the coral-eating Crown of Thorns, *Acanthaster planci* (L.). Both can be regarded as grazers since they move as herds over a sedentary and closely packed food source: bivalve beds in *A. rubens* and coral reefs in *A. planci*. *A. rubens* individuals may be crowded together, often occurring in layers several starfish deep. Brun (1968) observed a herd in the middle of which the average density was 97 m^{-2}; this mass of starfish was feeding on Iceland Scallop. Seed (1969) reported similar aggregations on mussel beds. I have observed dense aggregations of *A. rubens* off the Dorset coast feeding on mussels, and off Devon feeding on dense populations (300–500 m^{-2}) of the burrowing bivalve *Spisula*. In the former case, as in Brun's observations, the starfish appeared to be moving as a herd over the beds leaving only dead shells behind them. On the *Spisula* beds, however, the starfish population was more dispersed although up to five individuals could often be counted in a 1/16 m^2 quadrat. Isolated individuals were found digging for bivalves. This appears to be a slow process in which the central part of the starfish sinks into the substrate while excavated material, presumably removed by the tube feet, piles up around the circumference. Feeding individuals were quite common but more than half of the feeding starfish were sharing their meal with one or more other starfish (Plate 1B); the maximum number found sharing a single bivalve was five. The relative difficulty, compared with mussels or scallops, of reaching these burrowing bivalves probably explains the dispersed nature of the aggregation on *Spisula*. In the case of the scallops and mussels the resource is destroyed and the aggregation is presumably temporary. The aggregation on *Spisula*, however, has persisted for at least six years, gradually reducing in density as the *Spisula* population declines.

The Crowns of Thorns starfish came to prominence recently when it was found that vast populations were devastating huge areas of coral reef in the Pacific. Chesher (1969), for instance, reported the destruction of 90% of living coral along 38 km of coastline in Guam from low water mark down to 65 m deep! Normal population densities of this starfish are in the region of 0·01 m^{-2}, whereas in "plagues" or "infestations" hundreds of individuals may be found in quite small areas (Dana *et al.*, 1972). Such plague populations are rare but relatively small aggregations are less uncommon. There appears, indeed, to be no clear separation between normal and plague densities since all intermediates occur (Dana *et al.*, 1972). Ormond *et al.* (1973) in the Red Sea found that in slightly denser than normal populations of *A. planci* clumping into groups of up to 15 starfish occurred. Such groups could remain intact for three months. Observations suggested some attraction between individuals during feeding which might help to maintain such groups. Stomach eversion is triggered in *A. planci* by coral extracts (Brauer and Jordan, 1970; Collins, 1974) and experiments in Y mazes have demonstrated that other feeding individuals are more attractive than live coral (Ormond *et al.*, 1973). Beach *et al.* (1975) have further shown that spawning *A. planci* are attractive to others, providing a mechanism for the formation of breeding aggregations. Since this starfish is not normally particularly common it may be that a population divided into small groups, each kept intact by feeding smells and aggregating more densely during spawning, represents a sound reproductive strategy.

Feeding smells are probably responsible for bringing *Asterias rubens* populations together into crowded aggregations on bivalve beds. Experiments on olfactory responses in several species of *Asterias* (Castilla and Crisp, 1970; Zafiriou et al., 1972; Heeb, 1973) show that these starfish are attracted by living bivalves and by bivalve extracts. Tissue extracts can also cause "humping", a characteristic feeding posture, and stomach eversion. Curiously Castilla and Crisp (1970) found that starfish avoided injured bivalves whereas Zafiriou et al. (1972) found just the opposite. According to Mauzey et al. (1968) starfish feed reluctantly in captivity and display great variability in their responses to food. My own observations on food sharing in *A. rubens* (Plate 1B) suggest the attractiveness of other feeding individuals, and Zafiriou et al. (1972) noted that "most individuals fed only after a few attacking sea stars had succeeded in opening a gap in the prey's valves". They suggested that tissue juices diffusing from a torn or partially digested bivalve provide a feeding stimulus to non-attacking starfish. An alternative possibility is that feeding starfish themselves release some chemical attractant. Brauer and Jordan (1970) with *Acanthaster planci* found that extracts of starfish gastric folds squirted onto the oral region did not elicit stomach eversion but that extracts of pyloric caeca produced strong and enduring eversion. Starfish digestive juices thus constitute a possible attractant.

In both *Asterias* and *Acanthaster*, therefore, feeding smells may help to promote the formation and maintenance of aggregations. In both cases feeding aggregations, if sufficiently large, can lead to the destruction of the food source. In *Asterias* dispersal can follow such destruction since its habitat is not limited in extent. In *Acanthaster*, however, the coral reef habitat is often limited and the food resource, unlike that of grazing urchins, is slow growing. Thus once a section of reef is destroyed the starfish can only move in one direction—along the reef. Such behaviour would tend to maintain aggregations and, indeed, by a "snowball effect", to increase their size since resident populations would be collected into the herd as it moves along the reef. Dana et al. (1972) have suggested this mechanism to account for the reported starfish plagues and further suggested that the original destruction may not be caused by starfish at all but by typhoons. *Acanthaster* are most frequent on leeward reefs which are particularly susceptible to typhoon damage and the resident population of a storm-damaged reef, according to this theory, forms a potential plague nucleus. Whether or not a plague occurs may then depend on the extent of suitable habitat within reach of the starfish; a narrow ribbon of reef would be most susceptible.

Plagues of *Acanthaster*, therefore, may occur through localized food shortage whereas in *Asterias* aggregations develop as a result of localized food superabundance. Similar behavioural mechanisms may be acting in both cases. As suggested above, the advantage of the mechanism to *Acanthaster* may be that it fosters an efficient reproductive strategy. In *Asterias* it promotes the rapid exploitation of a localized food source; a valuable function since such food sources are by no means rare and there are other predators to compete with. Is there any advantage to the crowded nature of the feeding herds? Food sharing may serve a useful function in preventing possible wastage and, to the sharers, provides food for free since the effort of attacking, perhaps digging up, and opening the bivalve is restricted to a single individual. In addition it is possible that the upper layers of the heaps of starfish feeding on mussels may help to prevent the escape of diffusable semi-digested material. It is not known whether the attraction operating between feeding starfish constitutes social behaviour but starfish are certainly capable of such behaviour in a reproductive context (Reese, 1966) and there is even a record of intraspecific agonistic behaviour in asteroids (Wobber, 1975).

CONCLUSIONS

1. Comparisons Between Groups

Aggregation is not a particularly precise term. The geographic range of an animal is, in a sense, an aggregation, while within this range aggregations occur in suitable environments. Here, highly favourable locations may occur in which yet more aggregations are found. Within these local aggregations individuals may be spaced out or clumped. I have been discussing the last two levels of this series: local aggregation and clumping. It is important to distinguish between them since local aggregation may simply be the result of the responses of individuals to the physical environment or to an abundance of food, while clumping suggests social behaviour. Further, where clumps are permanently crowded some advantage to the individuals beyond that concerned with reproductive success must be suspected.

Temporary clumps concerned with spawning have been observed in several species (Reese, 1966); they have not been considered in detail here. Permanent clumps have been shown to be often crowded and concerned with suspension feeding. It is probable that suspended food in currents is the only type with a sufficiently high renewal rate to sustain sedentary, crowded populations. However, there are other advantages which apply to crowding in suspension feeders. The baffle effect of a forest of feeding arms extended into a current is the most widespread advantage and probably applies to crowded crinoids, brittle-stars and sea cucumbers. A similar effect may make possible the subsurface aggregations of the sea-urchin *Paracentrotus lividus*. Two other advantages apply specifically to certain epifaunal brittle-stars: support during feeding and, in stronger currents, stability by crouching down and interlocking arms to form a mat. Some clumped brittle-stars are spaced out rather than being crowded. Arm-loop capture of small organisms has been suggested in these cases as a feeding method which requires spacing out at arms length. The occurrence of crowded but not clumped aggregations in sand dollars indicates a super-abundance of food but suggests no advantage to individuals through physical contact with neighbours. The dense populations and some underwater observations are put forward as evidence for the existence of suspension feeding in these urchins.

Crowded aggregations in starfish are quite different since they are mobile and therefore not limited by the local renewal rate of their food. Both of the well known aggregating starfish species appear to be attracted to conspecifics by feeding smells. This allows the rapid and efficient exploitation of localized food in *Asterias* whereas in the comparatively rare *Acanthaster* it may help to keep small breeding aggregations together. Plagues of *Acanthaster* may occur as a result of food limitation. Herds of sea-urchins feeding on kelp holdfasts are similar in some ways to the herds of predatory starfish since they completely destroy the food source and then move on. They contrast, however, with the many species of grazing urchins which occur in huge populations of randomly distributed individuals and tend to conserve their environments. The difference may relate to the rapidity with which the resource can be renewed: slowly in the case of predatory starfish and holdfast eating urchins, rapidly in the case of grazing urchins.

The possibility of social behaviour being involved in the formation and maintenance of echinoderm aggregations has been suggested throughout. Any departure from random dispersion in a uniform environment in which resources are randomly arranged provides evidence for the involvement of social behaviour. Maintenance of aggregation by the social behaviour of larvae probably occurs in *Ophiothrix fragilis* and in this species social behaviour of the adults also promotes aggregation. Social behaviour which fosters aggregation is not at all unlikely in other species, particularly in crinoids, brittle-stars and starfish in which non-random dispersion appears commonest. In the latter case, if the feeding smells which keep aggregations together consist simply of prey juices then no social component appears necessary, but starfish may release an additional, specific odour. The not uncommon occurrence of reproductive social behaviour shows that echinoderms are perfectly capable of social interactions.

2. Disadvantages of Aggregation

The very denseness of some echinoderm aggregations suggests the term "crowded" to describe them. Surely, one feels, there must be inter-individual competition for food. I hope that I have shown above that this may not normally be the case and that crowding provides more food rather than less. Where competition is a risk then spacing out occurs. Sand dollars, however, do not appear to benefit from crowding and it is interesting in this context to note that the individuals in Salsman and Tolbert's (1965) very dense populations were much smaller than those in the sparser Trinidad aggregations. Perhaps competition between sand dollars can occur.

Were any aggregating echinoderms palatable one might expect predation to be a considerable threat to the habit. That it does not appear to be so must indicate that these species do not make good eating. In Torbay I found that *Ophiothrix fragilis* was eaten by many common predators, but considering its abundance it was clearly not popular (Warner, 1971). Similarly in the Pacific, workers on *Acanthaster* have found that a number of predators will eat the starfish but none of them appears suitable for the biological control of plagues.

A final possible disadvantage of aggregation lies in the potential for the spread of disease and of parasites. Nothing is known about this aspect of aggregation biology.

3. Comparisons with Colonial Organisms

If one considers a colonial organism to be an entire, usually structured body formed of a clone of connected individuals, then the vast majority can be broadly described as suspension feeders. Being a composite organism, a colony has limited powers of co-ordination and is almost always sedentary, sessile or planktonic, moving at the whim of its environment. Food must be delivered to it and suspended food is most suitable. The delivery rate of suspended food is greatest in currents and in current-washed benthic environments sessile colonies often abound: hydroids, octocorals, bryozoans, etc. In this respect many colonial animals follow the same strategy as sedentary echinoderm aggregations—they suspension feed. This similarity leads to several possible comparisons. Colonies which feed passively in currents often have a skeleton which supports the constituent individuals and, being anchored to the substratum, maintains the location of the colony. The skeleton is a property of the entire colony. Equally, in *Ophiothrix fragilis* support and maintenance is a property of the group. The baffle effect, probably of considerable importance in crowded echinoderm aggregations, may be catered for in colonies by their shapes. In several black corals (Antipatharia), for instance, the food catching polyps occur mainly on the leeward side of the colony in a probable zone of micro-turbulence (personal observation in Trinidad). On an individual basis the orientation of pinnules or tube feet on the arms of suspension feeding echinoderms compare very well with the shapes of gorgonian and hydroid colonies in similar environments: planar or fan shaped orientations are found in unidirectional currents and radial or bushy shapes occur in turbulent water (Meyer, 1973; Warner and Woodley, 1975; Wainwright *et al.*, 1976). Most of these points arise since suspension feeding presents similar problems whatever the nature of the feeder.

Another point of comparison arises from the physical contact between individuals in crowded aggregations and in colonies. In *O. fragilis* aggregations, individuals may respond to a disturbance occurring some distance away from them, the stimulus being transmitted mechanically from individual to individual. Stirring a 2 cm radius circle with a finger in the middle of a dense patch produces lowering of arms within a 20 cm radius (Warner, 1971). Such mechanical transmission of stimuli may confer some protection from predation. Both nervous and mechanical transmission occurs between individuals in colonies and may serve similar functions.

Apart from these parallels, echinoderm aggregations have little in common with colonial organisms. The reproductive advantage is not present in a colony since all individuals are of the same genotype. To achieve this advantage would require a secondary aggregation of colonies; such a situation has been reported for the hydroid *Nemertesia antennina* (L.) by Hughes (1975 and personal communication). Starfish aggregations appear to have nothing in common with colonial organisms and can be more easily compared with vultures, elephants or locusts.

ACKNOWLEDGEMENTS

I thank Dr D. M. Broom for helpful discussion and criticism of the manuscript.

REFERENCES

ALLAIN, J.-Y. (1974). Ecologie des bancs d'*Ophiothrix fragilis* (Abildgaard) (Echinodermata, Ophiuroidea) dans le golfe normanno-breton. *Cah. Biol. mar.* **15**, 255–273.
BEACH, D. H., HANSCOMB, N. J. and ORMOND, R. F. G. (1975). Spawning pheromone in crown-of-thorns starfish. *Nature, Lond.* **254**, 135–136.

Brauer, R. W. and Jordan, M. R. (1970). Triggering of the stomach eversion reflex of *Acanthaster planci* by coral extracts. *Nature, Lond.* **228**, 344–346.

Broom, D. M. (1975). Aggregation behaviour of the brittle-star *Ophiothrix fragilis*. *J. mar. biol. Ass. U.K.* **55**, 191–197.

Brun, E. (1968). Extreme population density of the starfish *Asterias rubens* L. on a bed of Iceland Scallop, *Chlamys islandica* (O. F. Müller). *Astarte No.* 32, 1–4.

Brun, E. (1969). Aggregation of *Ophiothrix fragilis* (Abildgaard) (Echinodermata: Ophiuroidea). *Nytt Mag Zool.* **17**, 153–160

Buchanan, J. B. (1964) A comparative study of some features of the biology of *Amphiura filiformis* and *Amphiura chiajei* (Ophiuroidea) considered in relation to their distribution. *J. mar. biol. Ass. U.K.* **44**, 565–576.

Castilla, J. C. and Crisp, D. J. (1970). Responses of *Asterias rubens* to olfactory stimuli. *J. mar. biol. Ass. U.K.* **50**, 829–847.

Chesher, R. H. (1969). Destruction of Pacific corals by the sea star *Acanthaster planci*. *Science, N.Y.* **165**, 280–283.

Collins, A. R. S. (1974). Biochemical investigation of two responses involved in the feeding behaviour of *Acanthaster planci* (L.). I. Assay methods and preliminary results. *J. exp. mar. Biol. Ecol.* **15**, 173–184.

Dana, T. F., Newman, W. A. and Fager, E. W. (1972). *Acanthaster* aggregations: interpreted as primarily responses to natural phenomena. *Pacif. Sci.* **26**, 355–372.

Fish, J. D. (1967). The biology of *Cucumaria elongata* (Echinodermata: Holothuroidea). *J. mar. biol. Ass. U.K.* **47**, 129–144.

Fontaine, A. R. (1965). The feeding mechanism of the ophiuroid *Ophiocomina nigra*. *J. mar. biol. Ass. U.K.* **45**, 373–385.

Fox, H. M. (1923). Lunar periodicity in reproduction. *Proc. R. Soc.* B **95**, 523–550.

Fox, H. M. (1924). The spawning of echinoids. *Proc. Camb. phil. Soc. biol. Sci.* **1**, 71–74.

Goodbody, I. (1960). The feeding mechanism in the sand dollar *Mellita sexiesperforata* (Leske). *Biol. Bull. mar. biol. Lab., Woods Hole*, **119**, 80–86.

Gorzula, S. J. (1974). Sexual behaviour in the brittle-star *Ophiocomina nigra*. *Proc. Challenger Soc.* **4**, 237–238.

Heeb, M. A. (1973). Large molecules and chemical control of feeding behaviour in the starfish *Asterias forbesi*. *Helgoländer wiss. Meeresunters.* **24**, 425–435.

Hughes, R. G. (1975). The distribution of epizoites on the hydroid *Nemertesia antennina* (L.). *J. mar. biol. Ass. U.K.* **55**, 275–294.

Jones, N. S. and Kain, J. M. (1967). Subtidal algal colonisation following the removal of *Echinus*. *Helgoländer wiss. Meeresunters.* **15**, 460–466.

Keegan, B. F. (1974). The macrofauna of maerl substrates on the west coast of Ireland. *Cah. Biol. mar.* **15**, 513–530.

Kitching, J. A. and Ebling, F. J. (1961). The ecology of Lough Ine. XI. The control of algae by *Paracentrotus lividus* (Echinoidea). *J. anim. Ecol.* **30**, 373–383.

Könnecker, G. and Keegan, B. F. (1973). *In situ* behavioural studies on echinoderm aggregations. Part I. *Pseudocucumis mixta*. *Helgoländer wiss. Meeresunters.* **24**, 157–162.

Krumbein, W. E. and Pers, J. N. C. (1974). Diving investigations on biodeterioration by sea urchins in the rocky sub-littoral of Helgoland. *Helgoländer wiss. Meeresunters.* **26**, 1–17.

Lewis, J. B. (1958). The biology of the tropical sea urchin *Tripneustes esculentus* Leske in Barbados, British West Indies. *Can. J. Zool.* **42**, 549–557.

Mauzey, K. P., Birkeland, C. and Dayton, P. K. (1968). Feeding behaviour of asteroids and escape responses of their prey in the Puget Sound region. *Ecology* **49**, 603–619.

Meyer, D. L. (1973). Feeding behaviour and ecology of shallow-water unstalked crinoids (Echinodermata) in the Caribbean Sea. *Mar. Biol.* **22**, 105–129.

Ormond, R. F. G., Campbell, A. C., Head, S. M., Moore, R. J., Rainbow, P. R. and Saunders, A. P. (1973). The formation and breakdown of aggregations of the crown-of-thorns starfish, *Acanthaster planci*. *Nature, Lond.* **246**, 167–169.

Reese, E. S. (1966). The complex behaviour of echinoderms. *In* "Physiology of Echinodermata" (R. A. Boolootian, ed.), pp. 157–218. John Wiley and Sons, New York.

Salsman, G. G. and Tolbert, W. H. (1965). Observations on the sand dollar, *Mellita quinquiesperforata*. *Limnol. Oceanogr.* **10**, 152–155.

Schuhmacher, H. (1974). On the conditions accompanying the first settlement of corals on artificial reefs with special reference to the influence of grazing sea urchins (Eilat, Red Sea). *In* "Proceedings of the Second International Coral Reef Symposium" Vol .1, pp. 257–267. Great Barrier Reef Committee, Brisbane.

Seed, R. (1969). The ecology of *Mytilus edulis* L. (Lamellibranchiata) on exposed rocky shores. II. Growth and mortality. *Oecologia*, **3**, 317–350.

Vevers, H. G. (1956). Observations on feeding mechanisms in some ophiuroids. *Proc. zool. Soc. Lond.* **126**, 484–485.

Vogel, S., Ellington, C. P. Jr. and Kilgore, D. L. Jr (1973). Wind-induced ventilation of the burrow of the Prairie-Dog, *Cynomys ludovicianus*. *J. comp. Physiol.* **85**, 1–14.

Wainwright, S. A., Biggs, W. D., Currey, J. D. and Gosline, J. M. (1976). "Mechanical design in organisms." Edward Arnold, London.

Warner, G. F. (1970). Brittle-star beds in Torbay, Devon. *Underwater Ass. Rep.* **1969**, 81–85.

Warner, G. F. (1971). On the ecology of a dense bed of the brittle-star *Ophiothrix fragilis*. *J. mar. biol. Ass. U.K.* **51**, 267–282.

Warner, G. F. and Woodley, J. D. (1975). Suspension-feeding in the brittle-star *Ophiothrix fragilis*. *J. mar. biol. Ass. U.K.* **55**, 199–210.

Wilson, J. B., Holme, N. A. and Barrett, R. L. (1977). Population dispersal in the brittle-star *Ophiocomina nigra* (Abildgaard) (Echinodermata: Ophiuroidea). *J. mar. biol. Ass. U.K..* **57**, 405–439.

Wobber, D. R. (1975). Agonism in asteroids. *Biol. Bull. mar. biol. Lab., Woods Hole,* **148**, 483–496.

Zafiriou, O., Whittle, K. J. and Blumer, M. (1972). Response of *Asterias vulgaris* to bivalves and bivalve tissue extracts. *Mar. Biol.* **13**, 137–145.

22 | New Observations on the Mode of Life, Evolution and Ultrastructure of Graptolites

R. B. RICKARDS and P. R. CROWTHER

Department of Geology, University of Cambridge, England

Abstract: From the general morphology of dimorphograptids and observed preferred orientations of isolated monograptids in dense fluids, it is concluded that the widely-held assumption that graptolites were vertically orientated in sea water is broadly correct. It is thought that the most likely orientation was with the sicular aperture downwards in most graptoloids, and that the successive appearances and evolution of the Dendroid, Anisograptid, Dichograptid, Diplograptid and Monograptid Faunas reflect responses to certain disadvantages in the otherwise superbly pre-adapted genotype of the first planktonic dendroids. Evidence presented here suggests that the graptolite zooids were very like some hemichordate zooids in their ability to leave the thecal tubes and secrete a cortex composed of short bandages, each consisting of probable collagen fibrils parallel to the length of the bandages. This conclusion is in complete contrast to all the modern opinions of Kozlowski (1949), Bohlin (1950), Bulman (1955, 1970), Kirk (1972), Rickards (1975) and Urbanek (1976) which require a more or less continuous sheet of extrathecal tissue in permanent (secreting) contact with the cortex.

THE UPRIGHT RHABDOSOME

Most workers have assumed that the graptolite rhabdosome had an essentially vertical orientation in the sea water, even though they may debate (e.g. Kirk, 1969) the question of whether the nema was pointed upwards or downwards. This assumption has grown in graptolite studies presumably because it seems unnatural that a long monograptid should float like a canoe or be submerged like a submarine. However, only Rickards (1975) has suggested positive evidence for this verticality through a consideration of *Dimorphograptus* where the proximal uniserial portion, however short, is always curved toward the (lost) plane of the median septum (Rickards, 1975, Text fig. 73). Such a curvature

---

Systematics Association Special Volume No. 11, "Biology and Systematics of Colonial Organisms', edited by G. Larwood and B. R. Rosen, 1979, pp. 397–410. Academic Press, London and New York.

would be a reasonable hydrodynamic response of a vertically oriented biserial species whose immediate ancestors suddenly acquired a proximal uniserial part: the rhabdosome would thus be able to remain vertical rather than tilting to one side. The same argument would apply whether the Kirk (1969) or Rickards (1975) orientation, with nemata respectively downwards and upwards, were adopted.

More recently, whilst isolating monograptids chemically, we have observed a curious response of the rhabdosomes to the treatment undergone. The *usual* response of the colonies, upon isolation, is either to float with the assistance of tiny gas bubbles amongst the scum of carbonaceous matter and oily substances, or to sink and lie on the bottom of the beaker or bowl. In both cases their orientation is more or less horizontal. Of the hundreds of isolations successfully achieved we had until recently seen no exceptions to this, and discussions with colleagues elsewhere revealed similar experiences. However, some isolations involving hundreds of monograptid specimens resulted in almost all the graptolites remaining suspended with a vertical orientation (Pls I, IIA) in a particular density-layer of the resultant fluid (mostly calcium chloride, unspent HCl and water). These, and other graptolites preserved horizontally in glycerine for some years, were carefully washed in distilled water and were placed in columns of natural sea water (North Sea water) where the vast majority rapidly assumed a vertical orientation whilst sinking slowly. Work is still in progress but the preliminary conclusion can reasonably be drawn that the straight monograptid rhabdosome was very stable when orientated vertically in normal sea water. It would seem that the several chemical isolations, which were done in rather small beakers and which resulted in vertically oriented rhabdosomes, accidentally gave a residual fluid with the correct density, after standing overnight. Clear chemical/density layering was visible in the fluid each morning.

The amount of gas (incorporated in vacuolated tissue) needed to float a graptolite rhabdosome has been alluded to elsewhere (Rickards in press) but it can be said that only relatively tiny amounts would be needed perhaps distributed in the manner suggested by Rickards (1975), namely along the nema, or nemal modifications, in most planktonic graptolites.

Of the successive faunas outlined by Bulman (1958, 1970) the Anisograptid and Dichograptid Faunas had, of course, few vertically oriented rhabdosomes. Mostly they were slender, ramose, horizontally-spread stipes often with accompanying web structures. These, and their vertically oriented successors in the Diplograptid and Monograptid Faunas, are discussed below in what is taken to be their evolutionary setting.

DENDROIDS IN A NEW HABITAT

The vast majority of dendroid graptolites were benthonic throughout the range of the group, middle Cambrian to Carboniferous, and were attached to the substrate by means of holdfasts usually consisting of basal discs or fibrous root-like processes homologous to divided nemata. However, undoubted planktonic species existed from the Tremadoc, such as *Dictyonema flabelliforme* (Eichwald), and these had a nema, divided nematose bundles, or vane-like bodies almost certainly connected with flotation of the colony. The vane-like structures have recently been described by Bulman and Størmer (1971) from the Tremadoc of the Oslo region; but they certainly occur in the early Ludlow of Middle Asia (Koren' personal communication) in material examined by one of us (R.B.R.). It has been firmly established that the first planktonic dendroids gave rise to the Anisograptid Fauna, but from the point of view of the present discussion the most important factor is that whereas the ancestors of *D. flabelliforme* were benthonic and of very restricted geographic occurrence, the planktonic form was both abundant and cosmopolitan: the "new" habitat entered by the graptolites in the Tremadoc comprised the vast surface layers of the earth's seas.

Certain morphological changes accompanied this dramatic and sudden change of environment. In the first place the thick stem of benthonic dendroids, often occluding many proximal thecae, was lost, as was much of the thick cortex, so that the colony became somewhat lighter. There is no evidence whatsoever that benthonic dendroids, attached in early astogeny, became free of the substrate in later astogeny by resorption or breakage of the stem: all well-preserved planktonic forms have a nema (or nematose bundle) or vanes; and resorption or breakage would be clearly detectable if it had occurred. The disc or root-like attachment was replaced most commonly by a single nema, a straight, fine, black rod which Rickards (1975) considered to have been uppermost with the rhabdosome pendant.

DISADVANTAGES AND GENETIC RESPONSES

The *Dictyonema flabelliforme* genotype was superbly pre-adapted to accomplish with such success the change from a benthonic, inshore, shell-rich environment to that of oceanic plankton. However, subsequent evolution of the planktonic dendroids leading to anisograptids and then to graptoloids suggests that the *Dictyonema* rhabdosome had certain inherent disadvantages in the new habitat. Despite loss of some cortex the conical rhabdosome was still relatively heavy and rigidly held together by a combination of regular branching and dissepiments. The rapid appearance of vane-like flotation structures in numbers of specimens, and of nematose bundles in others, was probably an early adaptation improving buoyancy of the rhabdosome.

The Anisograptid Fauna arose almost immediately after *D. flabelliforme* in the Tremadoc and reached a peak of species diversity in the late Tremadoc–early Arenig. The derivation of anisograptids from *D. flabelliforme* is at present unquestioned and the most striking morphological changes which took place strongly suggest the acquisition of a more buoyant rhabdosome, i.e. longer, fewer, more slender branches with a horizontal spread, and the development of proximal web structures (Rickards, 1975, e.g. Text-figs. 59–67). The last structure would not only confer buoyancy, or retard sinking, but would afford a degree of protection to the vital proximal regions of the colony which, in an essentially horizontal colony, would be forced upwards into the uppermost surface layers of the water; at least, relative to the position occupied by the sicula in a pendant *Dictyonema*.

In the succeeding Dichograptid Fauna the tendency continued to reduce the number of stipes and to make the rhabdosome lighter. In addition there was an increasing tendency towards scandency, that is with the stipes changing from the horizontal or pendant position, to pointing upwards with the result that the proximal regions became further from the surface layers: as reclined colonies were achieved, so web structures were largely lost, and the penultimate stage was reached in the Diplograptid Fauna with its back-to-back stipes and a nema (or virgula) enclosed proximally and usually free distally (or upwards in our present model).

One immediate consequence of biserial scandency is that the rhabdosome, although of less mass than its ancestors, is relatively less buoyant, so that the fauna saw the *immediate* development of nemal vanes, extensive rhabdosomal spinosity, and further peridermal reduction (as in "*Retiolites*"), all of which changes would either confer buoyancy or retard sinking. However, the final stage, the development of the uniserial, scandent rhabdosome, (involving as

---

Plate I

A, *Monograptus* sp., Llandovery, Silurian, during process isolation from impure limestone using 10% HCl: increase in room temperature during the day has caused a circulation of fluid and a slow up and down movement of the vertically oriented rhabdosomes. Specimens at the bottom of the beaker (which has about twice the width illustrated) are coming free of the limestone chip, others at the top are caught up in bubbles of oily scum released at the same time from the chips. × 1.

B, As A, but before circulation, showing a concentration of rhabdosomes suspended vertically, towards the bottom of the beaker. × 1.

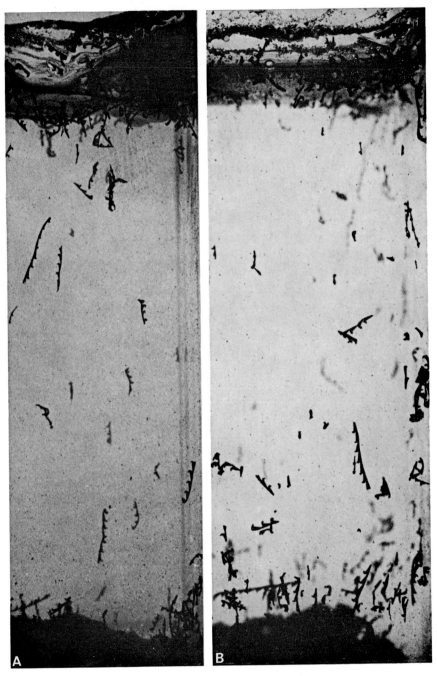

PLATE I. (Caption on facing page)

yet uncertain complex proximal end developmental changes) in many cases with long slender cladia or of long, flexuose rhabdosomes, resulted in colonies whose vital proximal regions were downwards, away from the most turbulent zone, and which had stipes essentially designed to retard sinking. As a result of this success the nemal vanes, extensive rhabdosomal spinosity and peridermal reduction became things of the past, and the graptoloids embarked on what was one of their most spectacular evolutionary expansions, namely that in the Silurian and early Devonian.

At this stage, therefore, the major disadvantages of the dendroids and anisograptids were overcome: the anisograptids survived only until the early Caradoc, whilst the planktonic dendroids are seen in sporadic occurrence at least until the low Ludlow. The bulk of the Diplograptid fauna became extinct in the middle Llandovery, although those with a most reduced periderm and thecal spinosity (*Holoretiolites, Spinograptus*) survived rarely into the middle Ludlow. Thecal spinosity in the Monograptid Fauna is usually seen in rather stiff, straight monograptids (*M. priodon, M. sedgwickii, M. flemingii*) or in rather robust spiral monograptids and robust cyrtograptids; and proximal webs are known in only one species of robust *Cyrtograptus*.

THE WANDERING ZOOID?

The nature of the graptolite zooid, essentially a hypothetical concept, is important in the understanding of any mode of life or evolutionary story. If it was similar to the hemichordate zooid and secreted a skeleton in similar fashion, then a common ancestor for the early hemichordates (Tremadoc) and early graptolites (middle Cambrian) would not be unreasonable. Similarly if the zooid had extensive soft parts in the form of extrathecal tissue it would have strong implications for the hydrodynamics of the colony as a whole and for any model based upon automobility of colonies. The method of periderm secretion itself is of paramount importance for any attempt to reconstruct the

---

PLATE II

A, Same stage as Pl. IB, enlarged to ×2, showing the concentration of rhabdosomes in one layer, and the orientation of thecae and thecal spines in rhabdosomes (top right and bottom left particularly).

B, *Climacograptus typicalis* (Hall), SM A97507, Viola Limestone. SEM photograph. The region of a single thecal excavation and aperture is shown, ×120, with clear bandaging of the cortex covering the whole of the rhabdosome surface: length, curvature and constant width of the bandages is clearly seen in relationship to the size of the thecal aperture.

PLATE II. (Caption on facing page)

graptolite zooid. Such a reconstruction must rely heavily on ultrastructural data due to the lack of any reliable evidence for soft part morphology from the fossil record.

There has been a growing body of opinion, following Kozlowski (1949) and Bohlin (1950), that the graptolite skeleton was broadly speaking, internal to an extrathecal layer of tissue which was at least responsible for secretion of the cortex. Some of the problems involved in connecting this layer to the zooids themselves have been discussed recently by Rickards (1975) and Urbanek (1976). Bohlin (1950) required that both cortex and fusellar layers of the graptolite periderm were secreted by an epithelial membrane whereas Kozlowski (1949) had a probable cephalic disc secreting the fusellar layer; and distal extensions of the zooids, connecting each other, responsible for secretion of the cortex. Beklemishev (1951) thought that the cortex was secreted by each zooid leaving its tube and creeping over the outer surface. Bulman (1955, 1970) essentially supported Kozlowski but found it difficult to envisage an extrathecal layer which allowed access to the outside for the proboscis of the zooid.

Kirk's (1972) ideas were very similar to those of Bohlin and she pictured secretion taking place by means of a double layered epithelial evagination of a mantle-like structure. Rickards *et al.* (1971) and Rickards (1975) supported the general principle of an extrathecal layer, in the latter case suggesting a combination of bifid pre-oral lobe and evagination to overcome the objections of Bulman while at the same time explaining the zig-zag arrangement of fuselli. The most recent paper on this difficult subject is that of Urbanek (1976) who bases his conclusions upon the ultrastructure of cortical and fusellar tissue, the similarity of fibrillar fabric in each and the occasional passage of fibrils of fusellar fabric into cortical fabric (as in *Didymograptus* sp.). He considered that the Kozlowski (1949) dual system (fuselli by the zooid; cortex by an almost independent extrathecal layer) is untenable and that a bryozoan-like secretion was more likely in which a perithecal membrane wrapped over and was in permanent contact with the growing edge of the peridermal wall; a similar idea to that of Kirk (1972). A shift in the secretory activity of the cells accounts for secretion of the fusellar and cortical layers. Evidence for this model relies largely on transmission EM studies of Urbanek and Towe (1974, 1975) which revealed

---

Plate III

(A) Same specimen as Pl. IIB, × 250, showing relationship of bandages one to the other, the ends of several (top right), and the small scarp-like edges of the bandages.
(B) Enlargement of the area shown in Pl. IIIA, showing fibril content of the bandages and traces of the enclosing membrane covering the fibrils, × 1200.

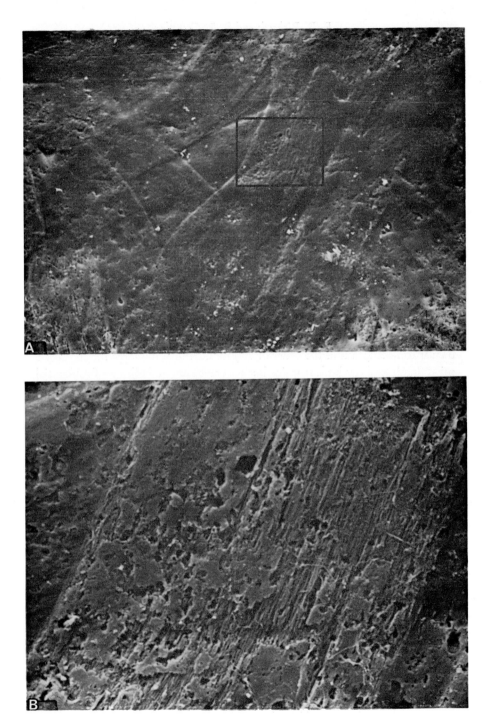

Plate III. (Caption on facing page)

a meshwork of fibrils within each fusellus and appeared to show that cortical tissue was laid down in layers of close-packed parallel fibrils, their orientation changing from layer to layer. There is no doubt that some fusellar outer lamellae substantially overlap each other proximally to form "cortical" layers, at least in some cases, e.g. *Didymograptus* sp. but the sections cover too small an area to show if all such layers are so formed.

The dangers involved in relying so heavily on two-dimensional sections has recently been revealed by scanning EM studies carried out by one of us (P.R.C.). The examples on Pl. IIB, III of *Climacograptus typicalis* show that what appear to be cortical "layers" on TEM micrographs are formed from well defined cortical bandages overlapping each other in an apparently random fashion. Each bandage consists of tightly packed fibrils directed parallel to the long edge and protected by a bounding membrane (see Pl. IIIB). This random arrangement of bandages would strengthen the skeleton and also produce the characteristic layered appearance of cortical tissue seen in ultrathin section. The change of fibril orientation from "layer" to "layer" is caused by the angle of section relative to successive bandages. The minimum width for a bandage cut perpendicular to its length is of the order of 70 μm, i.e. much too large for any single micrograph to show both margins but montages can be built up of substantial lengths of periderm which enable bandages to be recognized deep within the cortex. In *Climacograptus typicalis* there is a gradual increase in bandage size distally along the rhabdosome although at any particular point they are of constant width and length (Pl. IIIA). There is thus an overall size increase following the distal increase in thecal volume. This leads to the conclusion that there is a relationship between bandage size and zooid size as would be expected if they were plastered on directly by a secretory organ of the zooid.

TEM micrographs explain why the bandaging is so clear on *Climacograptus typicalis* at low magnification (see Pl. IV). Each bandage is several fibrils thick and at the long edge comes to an abrupt end with the bounding membrane outlining a stepped margin. Similar obvious cortical bandaging has been recog-

---

PLATE IV

TEM photograph. Transverse ultramicrotome section, ×6000, of the spongy fusellar fabric (to right) and denser, "layered", cortical fabric (to left), of *C. typicalis* (Hall) from the Viola Limestone. Each cortical "layer" is a section through one bandage. The change in fibril orientation from "layer" to "layer" is a reflection of the varied orientation of the superposed bandages which construct the cortex. The scarp-like edge of one bandage is visible (arrowed) where it can be seen that the next cortical bandage is underlain by a little spongy fusellar fabric at the foot of the scarp slope before the denser part is secreted.

PLATE IV. (Caption on facing page)

nized on *Climacograptus innotatus*, "*Climacograptus*" *inuiti* Cox and *Orthograptus gracilis*. However, several other genera show features indicative of a similar mode of cortical secretion, for example: *Dendrograptus*; *Dictyonema*; *Didymograptus*; *Monograptus*; *Cyrtograptus*. In these cases the bandaging is rarely clear at low magnifications but evidence for its existence comes from unconformities in fibre orientations occasionally visible when pressed through the bounding membrane or when the membrane has flaked off. The lack of a stepped bandage margin would explain the difficulty in recognizing bandages at low magnification, the step being replaced instead by a more tapered edge producing the overall effect of a smooth outer surface. Future TEM work should provide the answer.

The genera mentioned above span almost the whole classification of graptolites and it can be said without hesitation that evidence for bandaging is always visible whenever the peridermal wall is sufficiently well preserved. The structure was first observed under the light microscope (Holland et al. 1969) although it was impossible at that time to interpret such low magnification results.

This important new feature of cortical tissue appears to be of general occurrence throughout graptolite evolution and therefore requires a single universally applicable method for mode of secretion. In our opinion the simplest reasonable interpretation is that the zooid crawled out of the zooidal chamber, affixed the pre-oral lobe and moved forwards secreting the collagen-like fibrils by means of secretory cells in the pad of the pre-oral lobe. This interpretation is supported by:

(1) The width of the bandages, not unlike the width of a hemichordate pre-oral lobe, and their constant appearance at any point of the rhabdosome (Pl. IIB, IIIA).

(2) The slight regular increase in bandage size distally along the rhabdosome accompanying the overall increase in thecal volume and therefore zooid size.

(3) Parallelism of the cortical fibres to each other and to the long edge of each bandage (Pl. IIIB), combined with the sinuous shape of some bandages (Pl. IIIA), pointing to a "painted-on" form of application to the surface of older cortex.

This model requires that there be no perithecal membrane or any other mantle-like covering to the skeleton and an ability on the part of the zooid to extend itself outside the theca or at least to extend its secretory organ. This contradicts the recent conclusions of Kirk (1975) and Urbanek (1976) while supporting Beklemishev's (1951) original idea. Freedom of movement for the secreting zooid is an acceptable hypothesis for graptolites with simple thecae and large apertures such as dichograptids and most diplograptids but it is more difficult to justify for the many complex and constricted apertures seen particularly in some monograptids. A further problem involves Bohlin's (1950) argument for insisting that the graptolite skeleton was internal i.e. the rigid morphological control exerted on the growth and development of rhabdosomal form (see Urbanek, 1960). Such precision control is easier to envisage occurring within soft tissue rather than having to rely on the secretory "skill" of each zooid. Contrast the extreme regularity of thecal form in *Climacograptus typicalis* with the apparently haphazard deposition of cortical bandages (Pl. IIB) superimposed on them. It also seems most unlikely, as pointed out by Rickards (1975), that the nema or nemal vanes could be secreted by a thecal zooid, such is the high degree of their regularity and symmetry. It is far more likely that soft secreting tissue was extruded from the tip of the nema (or vane, holdfast or nematose bundle) and was also responsible for holding together rhabdosomes in synrhabdosomal associations. Also it is difficult to explain the occasional passage of fusellar into cortical fabric as observed by Urbanek (e.g. 1976) in some graptolites (see also Pl. IV), although if the same parts were carrying out the secretion such a passage would not be unreasonable.

Our provisional conclusions on this question are that the graptolite zooid was probably not unlike the *Cephalodiscus* zooid with a capability of considerable wandering in order to construct the main body of the cortex. The continuing nature of our research into this problem and the difficulties outlined above mean that this can only be a tentative conclusion at present. It is clear that future electron microscope work will have to examine larger portions of the periderm than has hitherto been the case, using both TEM and SEM techniques in order to interpret ultrastructure in terms of broader structure.

EDITOR'S NOTE

Revised manuscript received August 1976.

ACKNOWLEDGEMENTS

We are indebted firstly to Miss Lorraine Norman who prepared the "vertical" graptolites; and to Miss Shirley Newman and Mr D. Newling for electron microscope work and photography. Dr T. N. Koren' kindly gave us specimens and unpublished information; N.E.R.C. has supported the work of both authors.

REFERENCES

BEKLEMISHEV, V. N. (1951). K postroeniyu sistemy zhivotnykh. Vtorichnorotye (Deuterostomia), ikh proizkhozhdenie i sostav. *Uspekhi Sovremernoi Biologii* **32**, 256–270.

BOHLIN, B. (1950). The affinities of the graptolites. *Bull. Geol. Inst. Uppsala* **34**, 107–113.

BULMAN, O. M. B. (1955, 1970). Graptolithina with sections on Enteropneusta and Pterobranchia. *In* "Treatise on Invertebrate Paleontology". (R. C. Moore and C. Teichert, eds), (1st and 2nd editions) Part V. Geol. Soc. Am. and Univ. Kansas Press. New York and Kansas.

BULMAN, O. M. B. (1958). The sequence of graptolite faunas. *Palaeontology* **1**, 159–173.

BULMAN, O. M. B. and STØRMER, L. (1971). Buoyancy structures in rhabdosomes of *Dictyonema flabelliforme* (Eichwald). *Norsk geol. Tidsskr.* **51**, 25–31.

HOLLAND, C. H., RICKARDS, R. B. and WARREN, P. T. (1969). The Wenlock graptolites of the Ludlow District, Shropshire, and their stratigraphical significance. *Palaeontology* **12**, 663–683.

KIRK, N. H. (1969). Some thoughts on the ecology, mode of life and evolution of the Graptolithina. *Proc. geol. Soc. Lond.* no. **1659**, 273–292.

KIRK, N. H. (1972). Some thoughts on the construction of the rhabdosome in the Graptolithina, with special reference to extrathecal tissue and its bearing on the theory of automobility. *Univ. Coll. Wales, Dept. Geol. Publ.* **1**, 1–21.

KIRK, N. H. (1975). More thoughts on the construction and functioning of the rhabdosome in the Graptoloidea in the light of their ultrastructure. *Univ. Coll. Wales, Dept. Geol. Publ.* **7**, 1–24.

KOZLOWSKI, R. (1949). Les graptolithes et quelques nouveaux groupes d'animaux du Tremadoc de la Pologne. *Palaeont. pol.* **3**, 1–235.

RICKARDS, R. B. (1975). Palaeoecology of the Graptolithina, an extinct Class of the Phylum Hemichordata. *Biol. Rev.* **50**, 397–436.

RICKARDS, R. B. (in press). Graptolithina. *In* "Encyclopedia of Paleontology" (R. W. Fairbridge, and D. Jablonski, eds). Dowden, Hutchinson and Ross, Stroudsburg, Pa.

RICKARDS, R. B., HYDE, P. J. W. and KRINSLEY, D. H. (1971). Periderm ultrastructure of a species of *Monograptus* (Phylum Hemichordata). *Proc. R. Soc. B.* **178**, 347–356.

URBANEK, A. (1960). An attempt at biological interpretation of evolutionary changes in graptolite colonies. *Acta Palaeont. pol.* **5**, 127–234.

URBANEK, A. (1976). The problem of graptolite affinities in the light of ultrastructural studies on peridermal derivatives in pterobranchs. *Acta palaeont. pol.* **21**, 3–36.

URBANEK, A. and TOWE, K. M. (1974). Ultrastructural studies on graptolites, 1: The periderm and its derivatives in the Dendroidea and in *Mastigograptus*. *Smithsonian Contributions to Paleobiology*, **20**, 1–48.

URBANEK, A. and TOWE, K. M. (1975). Ultrastructural studies on graptolites, 2: The periderm and its derivatives in the Graptoloidea. *Smithsonian Contributions to Paleobiology*, **22**, 1–48.

23 | Thoughts on Coloniality in the Graptolithina

N. H. KIRK

Department of Geology, University College of Wales, Aberystwyth, Wales

Abstract: Graptolites were colonial animals which secreted their periderm in increments between the epithelium collectively covering the zooids and the epithelium lining the mantle collectively evaginated from them.

With this simple constructional equipment the graptoloids successfully exchanged a sessile habit on the seafloor for automobile suspension in the plankton. This change occurred with minimal modification of the constructional process, and generally without specialization of any of the zooids. The colonies appear to have attained and maintained their positions at the most favourable overall levels in the sea solely by co-ordinated feeding in association with a variety of rhabdosomal designs.

In earlier papers, which are cited in the bibliography, I have evolved ideas about the construction and mode of life of graptolites by a detailed discussion and reappraisal of structures previously described in the literature. In this essay I shall abandon the dialectical approach and present those ideas in a simple narrative as though they were established facts, leaving readers who are interested to refer to the evidence presented elsewhere. This will allow space to discuss the special aspects of coloniality in the graptolites, as they evolved from sessile to automobile forms, and from relatively unco-ordinated colonies to highly specialized super-individuals.

The class Graptolithina, as summarized by Bulman (1970), includes marine, colonial organisms in which the zooids were budded from a stolon system, and which secreted a sclerotized exoskeleton or rhabdosome showing characteristic growth lines. Of the six orders included by Bulman in this class I shall discuss only the Crustoidea, Dendroidea and Graptoloidea, which were characterized

---

Systematics Association Special Volume No. 11. "Biology and Systematics of Colonial Organisms", edited by G. Larwood and B. R. Rosen, 1979, pp. 411–432. Academic Press, London and New York.

by sympodial budding and a skeleton secreted by contiguous epithelium. These orders are represented by fossils found in sediments of marine origin ranging in age from the middle Cambrian to the Carboniferous.

The remaining orders, the Tuboidea, Camaroidea and Stoloneidea, were probably monopodially budded and may have mortared their "skeleton" or coenoecium by means of a mobile organ. They were sessile animals showing interesting adaptations parallel to those developed by the crustoids and dendroids. Their possible relationship to the Pterobranchia on the one hand, and to the sympodial, secreting graptolites on the other, has been discussed by Kirk (1974a) and will not be considered further here since the final answer to the problem awaits a description of their ultrastructure.

The colonies of the crustoids, dendroids and graptoloids were formed by budding from the internal stolon of each terminal, autothecal zooid in turn (Kozlowski, 1949 and 1962; Bulman, 1970). This process, which may have originated as a form of asexual reproduction with complete separation of the daughter zooid from the parent, resulted in the formation of rows of interconnected zooids enclosed in a continuous sheath of periderm. An aperture in the autotheca surrounding each zooid is presumed to have afforded access to the mouth, anus and feeding organ (Fig. 1).

During the last few years the ultrastructure of the periderm has been described, most fully by Urbanek and Towe (1974 and 1975), and has been interpreted by Kirk (1974b and 1975). The autothecal wall in longitudinal section is seen to be composed of superimposed arches, each arch corresponding to an increment of secretion. The bulk of each arch is made of a fusellar fabric of loosely anastomosing fibrils, but the outermost part of each increment is made of similar fibrils closely packed and with a parallel orientation; this is the lamella. Each lamella normally extends back further on the outerside of the thecal wall than on the inner, so that their overlap gives rise to a thicker, layered cortex on the outside, and to a thin cortical lining on the inside.

Because the parallel orientation of the fibrils changes from one lamella to another, the resulting layered cortex would have been extremely resilient and resistant to tensile stress. The mechanical efficiency of the graptolites was undoubtedly related to this mode of construction.

The macroscopic, microscopic and ultrastructural features of the graptolites suggest that the autothecal periderm was secreted beneath a mantle evaginated from the "shoulder" of each zooid peripheral to the presumed mouth, anus and origin of the feeding organ (Kirk, 1972b, 1974a, 1974b, 1975). The increments were presumably secreted by the epithelium covering the zooid and extending into the lining of the mantle. The arch-like form of each increment corresponded to the form of the epithelium in the "arm-pit" of the mantle evagination.

Each increment of fusellar fabric would seem to have been the result of

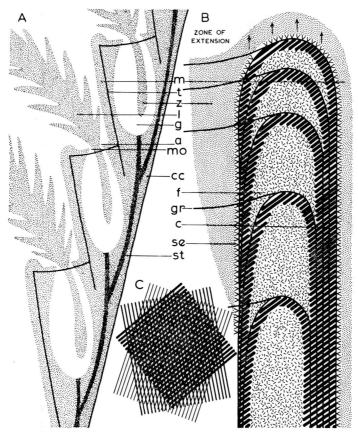

FIG. 1. Graptolite construction. A, Ideogram of branch of graptolite colony, approx. ×15. B, Ideogrammatic longitudinal section of apertural part of a graptolite theca and associated living tissues, approx. ×5000. C, Fragment of cortex (schematic) made of 3 layers with differently orientated, parallel fibrils. a, anus; c, cortex; cc, common canal; f, fusellar material; g, gut; gr, growth lines; l, lophophore; m, mantle; mo, mouth; se, secretory epithelium, st, stolon; t, theca; z, zooid.

rapid secretion following a period of rapid cell-division in the mantle "arm-pit". Its function would have been rapidly to extend the thecal sheath round the growing zooid. Each lamella of closely packed fibrils would have corresponded to secretion during the slowing down of the growth impulse, its function having been to strengthen the new structural addition.

Each theca was formed by a succession of such growth and secretory impulses, the characteristic growth lines being the trace of the closures of the incremental arches. As a theca neared completion, the growth lines became closer as the

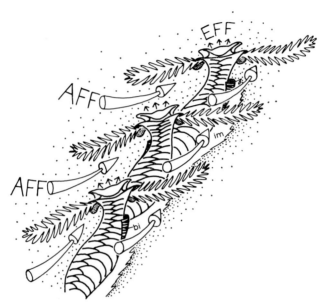

FIG. 2. Crustoid colony. Ideogrammatic reconstruction of a fragment of a crustoid colony showing three autothecae with expanded lophophores producing afferent and efferent currents. Bithecae (bi) open adjacent to the autothecae, the colony being attached by an encrusting, basal, interthecal membrane (im). Reproduced from Kirk (1974a) based on Kozlowski (1962), approx. × 50.

amount of the fusellar component decreased and the lamellae overlapped further causing the thecal aperture to develop a strong rim and the more proximal part of the theca to be thickened. This thickening occurred mainly on the outside of the thecae. On the inside the cortical lining tended to remain thin, because it would otherwise have reduced the living space of the zooids which filled the peridermal sheath.

The Crustoidea (Kozlowski, 1962; Bulman, 1970; Kirk, 1974a), were encrusting graptolites which have been found as fossils in marine sedimentary rocks of Ordovician age. In life they were probably attached to objects on the sea floor (Fig. 2). The fragments which have been found show that the zooids were regularly budded, and that their uplifted, lobed, thecal apertures were evenly spaced and orientated, suggesting some co-ordination of ciliary currents to expanded lophophores. A small bitheca opening adjacent to the aperture of each large autotheca may have housed a cleaning individual. The autothecae were about 1 mm in length, so that there would have been a distinct advantage in coloniality both for the generation of feeding currents and for the establishment and maintenance of an area of attachment.

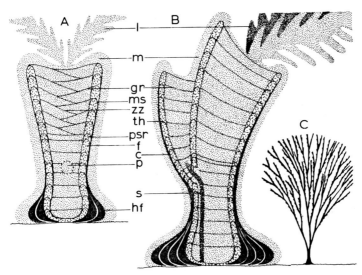

FIG. 3. Dendroid juvenile and adult colonies. A, Ideogram of dendroid metasicular individual, approx. × 50. B, Ideogram of young dendroid colony, approx. × 50. C, Tree-like dendroid colony, Schematic, × 1. c, cortex; f, fusellar fabric; gr, growth lines; hf, holdfast; l, lophophore, m, mantle; ms, metasicula; p, porus; psr, prosicular rim; s, stolon; th, daughter theca; zz, zigzag suture.

The thecae were thin-walled, the attachment of the colony to the substratum being strengthened by thick-walled, sclerotized stolons fused with the encrusting, lower thecal walls. The upper walls of the autothecae show the increments of secretion to have overlapped alternately from the two sides to give the median zigzag suture which is a characteristic feature of the graptolites.

The initial stages of development as well as the final form of the crustoid colony are unknown, but in the remaining two orders, the Dendroidea and the Graptoloidea, fossil evidence has provided remarkably detailed information. In both orders the youngest growth stage found preserved is a small tapering cup, the prosicula (Fig. 3). Its ultrastructure (Urbanek and Towe, 1975) shows it to have been formed of arch-shaped increments, beginning at the narrower end and finishing with a slight thickening of its apertural rim. Presumably it was secreted by the same contiguous epithelia as produced the other thecae (Kirk, 1975). The slowing down in growth coincident with the strengthening of the prosicular rim may have coincided with a metamorphosis of the zooid, for it always occurred when the prosicula had reached an overall length of about 0·5 mm. The metamorphosis may have involved a change from a circular or spiral lophophore to a two-branched structure, because the subsequent increments of the metasicula were added on the two sides, overlapping in dorsal and ventral zigzag sutures. The budding of the daughter zooid, which commenced at about this time in the dendroids and early graptoloids, could have been more easily accommodated to a two-branched lophophore, and the doubling of the overlapping lamellae along the sutures would have strengthened the branches of the colony from their inception.

In the dendroid or tree-like graptoloids the prosicula was undoubtedly attached by the narrower end to the sea floor. During further growth of the colony the attachment was extended by the secretion of increments with orientated fibrils to form a holdfast. At the same time the metasicula extended round the growing siculozooid and a succession of buds began to be produced. The thecae of these buds were secreted in continuity beneath a common mantle, and their outer lamellae forming the cortex would have been continuous with the increments of the holdfast.

Secretion by the lining of the all-enveloping mantle allowed the possibility of secretion anywhere over the surface of the graptolite colony at any time during its life, and in its early stages of development secretion may have occurred simultaneously over the entire surface. In later stages, however, secretion probably became restricted to areas of thecal extension and to those areas requiring extra strength or the addition of special colonial structures. Co-ordination of secretion must have been maintained via the communal mantle as well as via the interconnected bodies of the zooids.

While most dendroid colonies were tree-like, conical colonies evolved in the Tremadocian in which the many delicate branches were rather regularly forked (Fig. 4). In an upright position, with the cone widening upwards, these branches would have been prevented from falling apart by the numerous little cross-bars or dissepiments which connected them. The thecae were of two kinds: regularly spaced, large autothecae opening into the cone, and small bithecae opening at the side of the autothecae. This arrangement suggests that the conical form, varying from almost cylindrical to almost discoidal, was an adaptation to a sessile habit on the sea floor, and that the large zooids drew a feeding current down into the cone by the co-ordinated ciliary action of their lophophores. The small zooids could have been cleaning individuals, as in the Crustoidea, manoeuvring unwanted fragments through the meshes of the cone on the outgoing efferent current.

Very young conical colonies of *Dictyonema flabelliforme* have been found which suggest attachment by a fine stalk or nema. This may have been hollow, and could perhaps be regarded as an elongated tubular holdfast, elongating

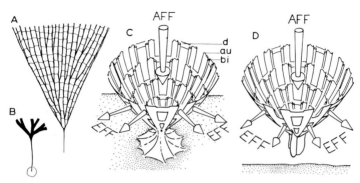

FIG. 4. Conical dendroids. A, *Dictyonema flabelliforme*, conical rhabdosome in side view, approx. ×0·5. (Schematic, after Bulman, 1970, fig. 13.) B, *D. flabelliforme*, juvenile form with possible attachment by tissue at distal end of tubular holdfast or nema, approx. ×2. (Based on Ruedemann, 1904–8.) C, Ideogrammatic representation of a sessile form of *Dictyonema*, showing regular spacing of autothecae (au) bithecae (bi) and supposed afferent and efferent currents. d, dissepiment. D, Ideogrammatic representation of *D. flabelliforme* Eichwald, balanced by 3 vanes extending from the nema and proximal branches, and held in suspension by the ciliary activity responsible for the afferent and efferent currents.

to hold the growing cone well above the sea floor. The young colony would have been supported on its delicate stalk, partly by its buoyancy in water, and partly by the drag exerted on it by the ciliary activity producing the feeding currents. The large adult colonies, which reached a length of 16 cm and are said (Urbanek, 1973) to have borne up to 30 000 zooids, could hardly have been supported in this way, and their worldwide occurrence as fossils in marine sediments accumulated at different depths suggests that they lived in the plankton. It seems probable that the colonies became liberated at an early stage in their development and were then maintained in free suspension by the drag exerted by the ciliary feeding activity. In some cases three vertical vanes, inclined to one another at 120°, were secreted at the base of the cone to serve, presumably, as a stabilizer and to prevent rotation.

It seems likely that the automobile colonies of the graptoloids evolved from conical dendroids of this kind.

In the early graptoloid *Clonograptus tenellus* (Hutt, 1974) the prosicula had a more strongly differentiated conical portion (conus) and a proximal stalk-like portion (cauda or nema-prosiculae), the tip of the cauda being resorbed preparatory to the secretion of a hollow stalk or nema (Fig. 5). In the absence of ultrastructural data it has to be supposed that the growth of the nema, to

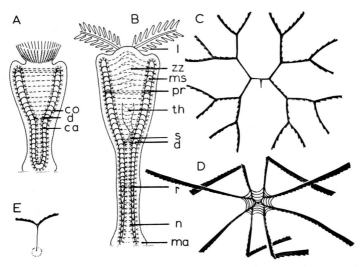

FIG. 5. Early graptoloids with numerous horizontal branches. A, Ideogram of *Clonograptus tenellus* prosicular stage of development with inferred ultrastructure and attachment by living tissue, approx. ×50. (Based on Hutt, 1974.) ca, cauda; co, conus; d, diaphragm. B, Ideogram of same at metasicular stage showing resorption followed by development of tubular nema. d, diaphragm; l, lophophore; ma, attachment by mantle; ms, metasicula; n, nema; pr, prosicular rim; r, resorption; s, stolon; th, daughter theca; zz, zigzag suture. C, Adult planktonic rhabdosome of *C. tenellus*, approx. ×1. (Based on Elles and Wood, 1902.) D, Adult planktonic rhabdosome of *Dichograptus octobrachiatus* showing proximal web and progressive elongation of the more distal thecae, approx. ×0·3. (Based on Elles and Wood, 1902.) E, Young rhabdosome of *Adelograptus lapworthi*, suggesting attachment by nema, approx. ×2. (Based on Ruedemann, 1904-8.)

a length of about 10 mm, would have been by increments added in the opposite direction by mantle covering the stalk.

In these early graptoloids the symmetrically arranged branches had become reduced in number to 64 and 32, 16 and 8, and in some horizontally branched forms they were connected proximally by a web. It seems possible that in their early growth stages these colonies were attached by the mantle enveloping the elongating nema, the branches being supported by the drag exerted by the ciliary feeding and by the web when one was present. The nema was too short to have supported the adult colonies, however, because the delicate branches, several cm long and sometimes less than 0·5 mm wide, extended far beyond the circumference of any proximal web. These adult colonies must at some stage have become liberated from their attachment to the substratum, possibly by fracture of the mantle, and have subsequently lived in free suspension, probably rising slightly when vigorously feeding and sinking slightly when less active. Balance could have been maintained by the symmetry of the branches, and by the nema functioning as a keel.

With the change from sessility to a free planktonic mode of life in the adult stage, there was no longer a need to form a large, rather dense colony overtopping other sessile organisms competing for living space on the sea floor. The new mode of life, on the other hand, favoured a more sparse colony, and evolution under the new controls produced colonies with fewer, but highly symmetrical branches supporting less than a tenth of the population carried by *D. flabelliforme*. Dissepiments were lost as there was no longer any need to hold up the branches from a crowded sea floor, and this in turn made the cleaning individuals superfluous so that the bithecae also disappeared.

Possibly associated with the increasing distance between adjacent branches in these more sparse, radiating colonies, was the development of a thecal size gradient in the graptoloids. This feature is not found in *D. flabelliforme* where the sicula and evenly spaced autothecae all had a length of about 1·5 mm. It seems logical to suppose that the increasingly long, distal thecae, reaching 3–4 mm in *Dichograptus*, accommodated larger zooids able to create stronger feeding currents and exploit the food in the wider spaces between the distal branches.

The combination, in the graptoloids, of a precision-built, symmetrically branched colony with a finely graded succession of thecae along the branches, is unusual among colonial animals and may be explicable in terms of their automobile existence in the plankton.

With the evolution of 4- and 2-branched colonies the inclination of the proximal thecae forming the branches sometimes changed rapidly from upright to drooping (Fig. 6). This suggests that the young colonies of such forms must have become detached quite early, for otherwise the drooping branches would have impinged upon the sea floor. (In the Corynoididae the sicula and one or two adnate thecae attained a length of 12 mm, though the prosicula was of normal length (0·5 mm) and the nema quite short. This suggests that these colonies may never have become automobile, but remained attached to the sea floor to which their upright, abnormally elongated, conical form was adapted. The last formed theca was minute and situated near the prosicula where it may have housed a cleaning or protective individual concerned with the attachment to the substratum.) Once liberated the drooping attitude would have greatly favoured a rise during active feeding and the outwardly facing zooids would have been more advantageously placed for producing ciliary currents and obtaining food from the surrounding water.

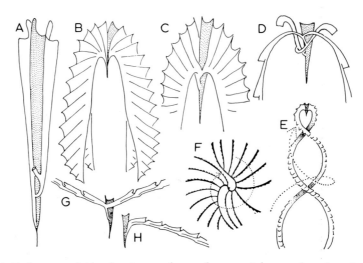

FIG 6. Early graptoloids showing a change from upright to drooping branches. A, *Corynoides*, a possibly attached form with long sicula and adnate thecae and a third ?bitheca, approx. × 10. (After Bulman, 1970, fig. 86.) B and C, *Tetragraptus phyllograptoides* and *Isograptus gibberulus*, approx. × 5. (After Bulman, 1970, figs. 82 and 83.) These show a gradual change from upright to drooping thecae possibly associated with a change from a sessile to a planktonic mode of life. D, *Dicellograptus*, thecal diagram, approx. × 12. (Based on Bulman, 1970, fig. 49.) This shows the change from upright to drooping within the first four thecae. E and F, *Dicellograptus caduceus* and *Nemagraptus gracilis*, approx. × 1. (After Bulman, 1970, figs. 72 and 87.) These spiral rhabdosomes suggest a spiral feeding rise (dotted arrows). G and H, *Holmograptus lentus* and *Nicholsonograptus fasciculatus*, approx. × 10. (Based on Skevington, 1967). The single-branched form may have been a spiralling mutant.

It was in the two-branched drooping colonies that rotation about the sicular axis may have originated as the result of uneven ciliary pull on the two sides of the branches. Rotation would have benefited such colonies, allowing the zooids to sweep a broad cylinder of water on their feeding rise. The spiral form of the rhabdosome probably evolved because it offered less resistance to spiral movement. In *Nemagraptus gracilis* the many side-branches conformed to the general curvature. This delicate rhabdosome could only have rotated very slowly, but its worldwide distribution and preservation provide remarkable evidence of the resilience of graptolite periderm.

A mutation in the two-branched *Holmograptus lentus* may have led to the development of the highly asymmetrical one-branched *Nicholsonograptus fasciculatus* (Skevington, 1966) which was markedly more successful than its symmetrical ancestor. It can only be supposed that its asymmetry resulted in spiralling which would have enabled its zooids to exploit the food in a broad cylinder of water.

The evolution of colonies such as *Apiograptus* (Cooper and McLaurin, 1974), with the two branches back-to-back, was the natural result of the change in inclination consequent upon the change to an automobile habit in the plankton. The biserial "scandent" colony, however, evolved in a different way—by means of a complicated budding procedure at the proximal end of the rhabdosome (Fig. 7).

This budding, combined with incremental secretion beneath a common mantle, resulted in maximum sharing of thecal walls and in a strong and compact skeletal structure. The sicula, although conservative in its conical construction and orientation, became an integral part of the colonial structure and the nema was also secreted in direct continuity with the more distal thecae beneath the common mantle (Berry, 1974).

The initial parts of the first thecae still grew in the same direction as the sicula and the young colony may still have been sessile and upward-facing at this stage of its development. However the later formed parts of the first thecae grew round to face in the opposite direction, suggesting that the colony had by then become detached. In these biserial graptoloids the cauda was much shorter, about one quarter of the total prosicular length as compared with two thirds in *Clonograptus tenellus* (Hutt, 1974), perhaps because it was only required to support the colony for a shorter period. The proximal end of the sicula in these forms also commonly shows signs of breakage, presumably the result of detachment, the nema being regenerated from the prosicular remnant or even from the metasicula.

After detachment the young colony was probably kept in suspension at first by the ciliary action of the siculozooid alone, though this would soon have been supplemented by the action of the laterally extended lophophore branches of the daughter zooids.

During the early growth stages of the colony the proportion of immature individuals to mature feeding zooids would have been relatively high and spines were frequently developed in the proximal part of the rhabdosome presumably to help to keep it in balanced suspension. A ventral apertural spine (virgella) was invariably developed on the sicula and this was sometimes accompanied by two dorsal spines. Apertural or subapertural spines were also frequently developed on the first two thecae, even when subsequent thecae were without

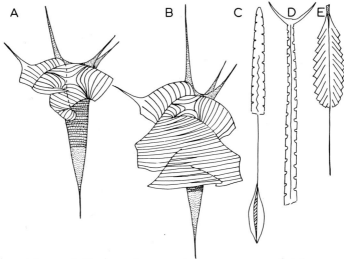

Fig. 7. Biserial graptoloids. A and B, *Amplexograptus*, young rhabdosomes, showing upward-growing initial parts of the first two thecae, alternating later thecae which omit dorsal walls, and balancing spines on sicula and first theca, approx. × 30. (Schematic, based on Bulman, 1970, figs 57 and 58.) C, *Climacograptus* with 3-vaned stabilizer on virgula, approx. × 1·5. (Schematic, after Bulman, 1970, fig. 70.) D, *Climacograptus* with strongly developed proximal spines, probably counter-rotational, approx. × 1. (Somewhat schematic, after Bulman 1970, fig. 91.) E, *Petalograptus* with shorter distal thecae, approx. × 4. (Somewhat schematic, after Bulman, 1970, fig. 91.)

them. Different combinations of sicular and thecal spines evolved, each producing a balanced, spinose, young colony with a well developed nema (virgula) acting as a keel. Where the two series of zooids were closely apposed, as in some climacograptids, rotation about the sicular axis would have conferred no advantage in feeding and three vertical vanes inclined at 120° were frequently secreted as extensions from the virgula. It seems probable that the further, exaggerated growth of the proximal spines in some old climacograptid colonies (Riva, 1974) also served to damp-down any tendency to rotate.

With the development of special proximal spines and distal virgular vanes the biserial "scandent" graptoloid had become much more than the sum of its constituent zooids. The simple size gradient was of little advantage in the biserial colony, and had become somewhat less important, or was even reversed—

as in *Petalograptus* (Urbanek, 1973). In its place there had evolved a highly complex discontinuous variation in thecal structure which developed under strict genetic control, so that each *colony* was essentially a replica of the parent. The biserial "scandent" graptoloid was approaching the condition of a many-mouthed super-individual with bipolarity and approximate bilateral symmetry. Yet this status was attained without the loss by any zooid of its full feeding powers or, in all probability, of its sexuality. More remarkably it was attained without any important alteration in the form or function of the siculozooid as it made the transition from an initially sessile state at the commencement of the colony's development, to the mobile state at the "head" of the detached colony. It was also attained without any zooid's incurring new duties, no-one was required to do anything but feed. The whole transformation of ancestral sessile colony into mobile super-individual had been achieved by a simple rearrangement of the zooids made possible by the budding mechanism. The corresponding rearrangement of the rhabdosome was made possible by the unique secretory process.

The problem of the automobile graptoloid had been one of reconciling the need to present its zooids most advantageously for the production of feeding currents, with the need to construct the lightest rhabdosome capable of withstanding the stress imposed by the co-ordinated ciliary action. The biserial "scandent" form offered the greatest possibility of sharing both ventral and dorsal thecal walls, and even of dispensing with dorsal walls altogether by the adoption of an alternating mode of budding throughout the colony. The reduction in thecal support was compensated by the incorporation of the virgula into the rhabdosome. This could occur without adding to the overall weight and without any reduction in its effectiveness as a stabilizer.

In the Lasiograptidae and Dicaulograptidae a further economy in the use of skeletal material was made by a general attenuation of the periderm, the rhabdosome being supported and strengthened by a framework of thickened lists (Fig. 8). (A rather similar but short-lived evolutionary experiment, in reduction of the periderm and its support by a framework of lists, had been made by some of the Ordovician, "reclined", two-branched graptoloids, the Abrograptidae.) Many lasiograptids were also highly spinose. The combination of reduced periderm and exaggerated spinosity probably resulted in an almost stationary colony, with little tendency to sink in periods of reduced ciliary activity and therefore with the necessity of producing only a small rise when vigorously feeding. The feeding currents to these colonies could therefore have been dominantly lateral, which would have been greatly to the advantage of zooids arranged in two vertical rows.

In the Retiolitidae a complete covering of periderm is not usually found, suggesting that perhaps it was not reinforced by a sufficient number of cortical lamellae to be preserved. The network of lists supporting the rhabdosome

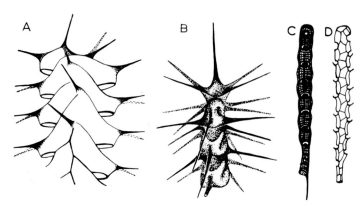

FIG. 8. Biserial graptoloids with reduced periderm. A, *Lasiograptus harknessi* showing strongly developed lists and spines, approx. ×20. (Schematic, after Rickards and Bulman, 1965.) B, *Dicaulograptus hystrix*, looking at one series of thecae. With strongly developed sicular and thecal spines. Approx. ×30. (Ideogrammatic, based on illustrations by Holm and Liljevall, reproduced in *Arkiv for Zoologi*, 24A, no. 9.) C, *Gothograptus nassa*, schematic, looking at one series of thecae. Periderm represented by a close network of lists. Approx. ×5. D, *Holoretiolites balticus* (schematic). The network of lists has become simpler and the appendix is smaller, possibly representing two reduced thecae. Approx. ×5.

followed a variety of patterns, some resembling the thecal framework of the lasiograptids, others forming a close reticulum. In some retiolitids it seems possible that even the siculozooid may have been planktonic since a sclerotized sicula seems not to have been secreted.

In the genera *Holoretiolites* and *Gothograptus* the more distal zooids may not have developed feeding organs since their thecae appear to have been without apertures. Their more proximal parts covering the "common canal" formed a tubular appendix. As in the Corynoididae (p. 419, and Fig. 6 above), specialization of the last-formed zooids resulted in the formation of a finite colony, but in the retiolitids it was adapted to a planktonic habit. It seems possible that the inflated proximal end of the rhabdosome was relatively light, the narrow distal appendix relatively heavy, so that the colony would have rapidly returned to verticality following deflection from it. *Gothograptus* and *Holoretiolites* lived in lower Ludlow times, long after the extinction of normal biserial graptoloids in the Llandovery. Their relative success may have been related to their small size and highly specialized construction which enabled them to live in the relatively turbulent but food-rich upper layers of the sea.

The gradual disappearance of normal biserial graptoloids in the Llandoverian was probably related to the appearance and increase of uniserial forms. In these there was less opportunity for skeletal economy and their relative success can perhaps be attributed to the fact that there was less interference between their feeding currents. In the biserial forms the two series of closely packed zooids were likely to have competed uneconomically in the production of currents and in the extraction of food from them.

In the uniserial graptoloids budding was from the developing aperture of the sicula, and the first theca grew directly downwards towards the sicular apex (Fig. 9). This suggests that the juvenile colony became detached at an even earlier stage than in the biserial graptoloids and the nema-prosiculae was correspondingly reduced to one twelfth of the total prosicular length. It was frequently damaged, presumably at the time of detachment, and the nema was regenerated from the prosicular remnant (Urbanek, 1958).

The liberated colony must, at first, have been maintained in suspension by the feeding current of the siculozooid alone. When this was supplemented by the currents to the laterally-spreading lophophore branches of the daughter zooids, it seems likely that the young colony rose spirally owing to the slight ventral or dorsal curvature of the rhabdosome. This would have been definitely to the advantage of zooids arranged in a single series. The slightly curved colonies probably rose in a narrow spiral with the fine, sicular end uppermost, but other forms became more strongly curved or coiled and these would have slowly swept a broad cylinder of water on their feeding rise. The strong curvature or coiling of the rhabdosome probably arose by the natural selection of forms offering the least resistance to spiral movement.

In the uniserial graptoloids the thecal gradient in size and form was probably associated with the spiral movement, because where a smooth curve was replaced by a marked bend in the rhabdosome, as in *Monograptus limatulus*, the accompanying change of form of the thecae was also abrupt. The thecal gradient undoubtedly reflected a gradient in size, attitude and ciliary activity of the zooids which would have been necessary to produce a spiral feeding rise by the colony. In some late Silurian monograptids of the genus *Cucullograptus* (Urbanek, 1966), the thecal apertures became strongly asymmetrical, probably reflecting an asymmetrical presentation of the lophophores associated with their asymmetrical ciliary action.

The thecal gradient, varying from one of size, shape and inclination, to the graded, asymmetrical details of the apertural lobes, was reproduced with

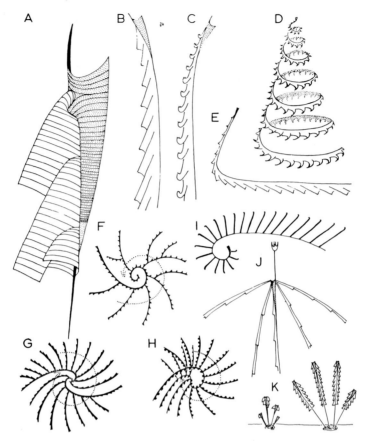

Fig. 9. Uniserial graptoloids. A, *Monograptus*, young rhabdosome, schematic, showing the first theca growing down from a notch in the immature sicular aperture, and the third theca growing in continuity with the virgula. Approx. × 40. (Schematic, based on Walker, 1953.) B and C, Monograptids with ventral and dorsal curvature suggestive of a narrow spiral feeding rise, approx. × 4. D, *Monograptus turriculatus*, suggestive of a broad spiral feeding rise. Approx. × 2. (Schematic, after Bulman, 1970, fig. 99.) E, *Monograptus limatulus*, showing the abrupt thecal change at the bend in the rhabdosome. Approx. × 2. (Schematic, after Elles and Wood, 1911.) F, G, H, *Cyrtograptus* (× 0·5), *Averianowograptus* (× 0·3) and *Sinodiversograptus* (× 1). Compound spiral rhabdosomes in which the thrust from the cladia probably contributed to a spiral feeding rise as suggested by the dotted arrow. (Schematic, after Bulman, 1970, figs. 102 and 103.) I, *Rastrites*, in which the elongation and isolation of the thecae probably led to the near immobilization of the spiral rhabdosomes in the plankton. Approx. × 1. (Schematic, based on Bulman, 1970, fig. 99.) J, *Linograptus posthumus*, with cladia radiating from the sicular aperture and a possibly buoyant virgellarium. Approx. × 8. (Based on Urbanek, 1963.) K, Synrhabdosomes of a biserial graptoloid shown as if attached to the sea floor by a tangle of possibly stolonal threads. Approx. × 3. (Based on Ruedemann, 1904–8.)

exactitude in every colony within a given species and so must have been the result of secretion under strict genetic control. Presumably this control was exerted on the secretory epithelium through the all-enveloping mantle as well as through the interconnected zooids filling the common canal.

In the cladia-bearing spiral Cyrtograptidae it has been observed (Thorsteinsson, 1955) that thecae secreted simultaneously on different parts of the colony were identical. This suggests that the internal environment, in which the secretory epithelium functioned, was changing with time. The incremental mode of secretion could have given precision to such a time control (Kirk, 1975).

During life, the co-ordination of growth and secretion must have been combined with co-ordination of the ciliary activity of the zooids. This is particularly obvious in the compound colonies of *Cyrtograptus*, *Averianowograptus* and *Sinodiversograptus*, where the inward thrust of the cladia presumably contributed to the spiral movement of the procladium.

A large colony of *Cyrtograptus linarssoni*, measuring more than a metre across and figured by Rickards (1975), has long, straight, radiating cladia. I suggest that this colony could only have moved spirally in the young stage represented by the initial curve of the procladium, and that it later became practically stationary held in suspension by its co-ordinated ciliary activity in the calm, lower-layers of the photic zone.

A different manner of achieving "sessility" in the plankton is found in *Rastrites*, which probably evolved from actively spiralling ancestors by increasing the friction between the surrounding water and its greatly elongated and widely spaced thecae.

Another kind of radiating colony was produced by budding from the sicular aperture of *Linograptus posthumus* (Urbanek, 1963). The virgellarium, a unique modification of the ventral virgellar spine on the sicula, may have been buoyant, keeping the sicular aperture uppermost in spite of the extra thickening which it underwent for the support of the cladia arising from it. These radiating colonies must also have been practically stationary, and are somewhat reminiscent of the many-branched colonies which evolved early in the history of the graptoloids. They may represent an adaptation to deeper, calm water with a more sparse food supply and, like the nebulous and spinose biserial lasiograptids, could be regarded as showing a return to a form of "near-sessility" within the planktonic environment.

In both biserial and uniserial graptoloids a different kind of sessility was achieved by the development of synrhabdosomes. In these, a number of colonies appears to have been budded, perhaps from a stolon, and to have remained attached by the growing distal ends of their virgulae until they had each developed a considerable number of thecae. Synrhabdosomes were therefore, colonies of colonies. Their overall form, radiating from a small centre, somewhat resembles that of *Linograptus posthumus*, but in the latter the cladia added thecae away from the centre, whereas in synrhabdosomes thecae were added towards the centre—the virgulae constantly elongating to accommodate them. The available evidence indicates that synrhabdosomes were sessile upon the sea floor, the constituent colonies radiating upwards and outwards from the small, possibly stolonal attachment. Upon detachment these colonies would have risen separately as a reaction to their feeding activity and would have completed their development in the plankton.

These were only interesting diversions from the main evolutionary stream. In the uniserial graptoloids, as in the biserial graptoloids and early, branched anisograptids and dichograptids, the general tendency was towards the evolution of smaller and simpler colonies. There were probably several interdependent reasons for this. Large colonies would have been more susceptible to damage in the food-rich but more turbulent, upper layers of the sea, and the spreading, branched forms would have been more liable to entanglement on their feeding rides. In the lower, calm layers of sea the more sparse food supply might have prevented their co-existence in the close proximity needed to ensure cross-fertilization between the gametes liberated from neighbouring colonies.

Small simple colonies, on the other hand, could have co-existed in vast numbers in the upper layers of the sea without exhausting the more abundant food supply and without much risk of entangling, or damage due to turbulence. Rapidly reaching maturity, their liberated eggs and sperm would have had a good chance of cross-fertilization with those from neighbouring colonies and the larvae could achieve successful settlement and development on the offshore sea bottom.

The broad evolutionary trend of the graptoloids is therefore explicable in terms of the transformation of the large colony, adapted to sessility on the crowded sea floor, to a small, streamlined, super-individual habit adapted to a fairly mobile existence in the plankton.

Why then did the graptoloids, after evolving for nearly 200 million years towards efficient simplicity, become extinct upon attaining it? In Britain the graptoloids disappeared from the fossil record in the middle Ludlow, the last found colonies being small, infrequent monograptids. However, elsewhere in the world, small simple monograptids and complex linograptids continued to occur in abundance in the early Devonian. Their disappearance in Britain was therefore clearly due to some factor other than the dying out of the race and might be attributed to unsuitable conditions either in the plankton, or on the sea floor where the colonies may still have commenced their development.

The disappearance of the specialized, complex colonies of *Linograptus* and *Abiesgraptus* occurred somewhat later. Their disappearance could have been associated with a change to shallower and rather more turbulent conditions to which they would probably have been ill adapted. The small, slightly curved monograptids with 20–40 thecae persisted, however, with little change and in considerable abundance until the late Emsian, when they too disappeared and are assumed to have become extinct.

In 1969 I naïvely attributed the extinction of the graptoloids to predation by the fish which undoubtedly evolved, multiplied and diversified after the graptoloids had attained automobility; I still maintain that these active predators would almost certainly have prevented the evolution of comparable, sluggishly automobile colonies in post-Palaeozoic times.

Professor Michael House, in a personal communication, has drawn attention to the way in which, during the Devonian, there appears to have been a successive dominance of planktonic graptoloids, planktonic coniconchines and planktonic ostracods (House, 1975) and has suggested that this represented sequential successful competition in the planktonic environment, and that extinction of the graptoloids was due to failure in this.

However, it seems doubtful to me if extinction due to competition occurs in the plankton and it is noteworthy that the increase in the coniconchines coincided with a change to a shallower and more calcareous type of sedimentation in those areas where the graptoloids had previously occurred in abundance, a change which would have been unfavourable to the latter's continued local existence and preservation.

It is remarkable that in both Alaska and Australia the late lower Devonian monograptids are found in association with vascular land plants, in grey or black shales. These shales are interbedded with high energy, calcareous sediments from which both graptolites and land plants are missing (Jaeger, 1970).

It seems likely that, after death, graptoloid colonies would have slowly sunk into lower layers of the sea. With the onset of decay and the production of gas bubbles, many colonies may later have risen to the surface where they could have been swept together into sheaves by wind and waves and surface currents, and perhaps mixed with the floating remains of land plants washed or blown far out to sea. With the further decay of the tissues the bubbles would have been released and the graptoloid colonies and vegetable trash would have gradually sunk to the sea floor. Under low energy conditions the organic remains might have been preserved, but under high energy conditions decay could have resulted in the total disappearance of both the graptolite and vegetable material.

Since the disappearance of the graptoloids appears to have approximately coincided with a change to high energy conditions in the areas where Emsian graptoloids had been preserved, it remains doubtful whether they actually became extinct everywhere at about that time. It also becomes desirable to try to extrapolate from the course of graptoloid evolution, so as to try to foresee the kind of graptoloid we might expect to find in suitable sedimentary rocks accumulated in a low energy environment.

Coloniality in the graptolites probably arose in the first place as a result of the non-separation of asexually-budded individuals. The colonial organization would have offered a better chance of occupying and holding a space on the sea floor, and the evolution of tree-like and conical colonies probably followed under the pressure of competition with other sessile organisms. The evolution of a tubular, elongating holdfast led to automobility, but the initially liberated colonies had been adapted to sessility, with radial symmetry and many branches, and needed to be drastically pruned to fit the new environment and mode of life.

The pruning proceeded for nearly 200 million years, leading towards smaller, approximately bilaterally symmetrical, bipolar super-individuals. In the lower Devonian this resulted in a very uniform but cosmopolitan graptoloid fauna of small colonies of only 20–40 zooids, with little potential for morphological evolution other than minor variations in apertural structure and a further reduction in zooid numbers. Can we be sure that such reduction did not occur, leading not to the extinction of the graptoloids but to the disappearance of the graptoloid *colony*? If the siculozooid had been able to produce eggs and sperm there would have been no obvious advantage to be gained from budding a chain of autothecal zooids. In the plankton they were superfluous, the colony merely an inconvenient hangover from the graptoloids' sessile past.

With the achievement of this ultimate reduction, the sclerotized periderm could also have become superfluous, since it would no longer have been desirable to secrete the overlapping lamellae which had been necessary to bind the thecae and virgula into a single resilient colony.

So instead of looking for the descendants of the graptoloids among the colonial animals of the sea floor, it would perhaps be more logical to seek them among the lightly clad individuals of the plankton. In the middle Devonian they might still have carried some traces of the conventional sicular structure, though this too could soon have disappeared as it did in some retiolitids which also may have omitted the initially sessile stage from their development.

REFERENCES

BERRY, W. B. N. (1974). Virgula structure and function in a monograptid and an orthograptid. In "Graptolite studies in honour of O. M. B. Bulman" (R. B. Rickards, D. E. Jackson and C. P. Hughes, eds), *Spec. Pap. Palaeont.* **13**, 131–140.

BULMAN, O. M. B. (1970). Graptolithina. In "Treatise on invertebrate paleontology" (C. Teichert, ed.), pt. V, Geol. Soc. America and Univ. Kansas Press, New York; Kansas.

BULMAN, O. M. B. and STØRMER, L. (1971). Buoyancy structures in rhabdosomes of *Dictyonema flabelliforme* (Eichwald). *Norsk Geol. Tidskr.* **51**, 25–31.

COOPER, R. A. and McLAURIN, A. N. (1974). *Apiograptus* gen. nov. and the origin of the biserial graptoloid rhabdosome. In "Graptolite studies in honour of O. M. B. Bulman." (R. B. Rickards, D. E. Jackson and C. P. Hughes, eds), *Spec. Pap. Palaeont.* **13**, 75–85.

HUTT, J. E. (1974). The development of *Clonograptus tenellus* and *Adelograptus hunnebergensis*. *Lethaia* **7**, 79–92.

JAEGER, H. (1970). Remarks on the stratigraphy and morphology of Praguian and probably younger monograptids. *Lethaia* **3**, 173–182.

KIRK, N. H. (1969). Some thoughts on the ecology, mode of life, and evolution of the Graptolithina. *Proc. geol. Soc. Lond.*, **1659**, 273–292.

KIRK, N. H. (1972a). More thoughts on the automobility of the graptolites. *Jl. geol. Soc. Lond.* **128**, 127–133.

KIRK, N. H. (1972b). Some thoughts on the construction of the rhabdosome in the Graptolithina, with special reference to extrathecal tissue and its bearing on the theory of automobility. *Geol. Dept. Publ., U.C.W., Aberystwyth*, **1**, pp. 1–21.

KIRK, N. H. (1974a). Some thoughts on convergence and divergence in the Graptolithina. *Geol. Dept. Publ., U.C.W., Aberystwyth*, **5**, pp. 1–29.

KIRK, N. H. (1974b). More thoughts on the construction of the rhabdosome in the Dendroidea, in the light of the ultrastructure of the Dendroidea and of *Mastigograptus*. *Geol. Dept. Publ., U.C.W., Aberystwyth*, **6**, pp. 1–11.

KIRK, N. H. (1975). More thoughts on the construction and functioning of the rhabdosome in the Graptoloidea in the light of their ultrastructure. *Geol. Dept. Publ., U.C.W., Aberystwyth*, **7**, pp. 1–24.

KOZLOWSKI, R. (1949). Les graptolithes et quelques nouveaux groupes d'animaux du Tremadoc de la Pologne. *Palaeont. Pol.* **3**, 1–235.

KOZLOWSKI, R. (1962). Crustoidea, nouveau groupe de graptolites. *Acta palaeont. Polo.* **7**, 3–52.

RICKARDS, R. B. (1975). Palaeoecology of the Graptolithina, an extinct class of the Phylum Hemichordata. *Biol. Rev.* **50**, 397–436.

RIVA, J. (1974). Late Ordovician spinose climacograptids from the Pacific and Atlantic faunal provinces. In "Graptolite studies in honour of O. M. B. Bulman." (R. B. Rickards, D. E. Jackson and C. P. Hughes, eds), *Spec. Pap. Palaeont.* **13**, 107–126.

RUEDEMANN, R. (1904–1908). "Graptolites of New York" Pts. I and II. *Mem. N.Y. St. Mus., nat. Hist.* **7**, 457–803. **11**, 1–583.

SKEVINGTON, D. (1966). The morphology and systematics of *"Didymograptus" fasciculatus* Nicholson, 1869. *Geol. Mag.* **103**, 487–497.

THORSTEINSSON, R. (1955). The mode of cladial generation in *Cyrtograptus*. *Geol. Mag.* **92**, 37–49.

URBANEK, A. (1958). Monograptidae from erratic boulders of Poland. *Palaeont. Pol.* **9**, (iv) + 105.

URBANEK, A. (1963). On generation and regeneration of cladia in some Upper Silurian monograptids. *Acta palaeont. Pol.* **8**, 135–254.

URBANEK, A. (1966). On the morphology and evolution of the Cucullograptinae (Monograptidae, Graptolithina). *Acta palaeont. Pol.* **11**, 291–544.

URBANEK, A. (1973). Organization and Evolution of Graptolite Colonies. *In* "Animal Colonies" (R. S. Boardman, A. H. Cheetham and W. A. Oliver, eds), pp. 441–514. Dowden, Hutchinson and Ross, Inc., Stroudsburg, Pennsylvania.

URBANEK, A. and TOWE, K. M. (1974). Ultrastructral studies on graptolites. 1. The periderm and its derivatives in the Dendroidea and in Mastigograptus. *Smithsonian Contributions to Paleobiology*, **20**, pp. 1–48.

URBANEK, A. and TOWE, K. M. (1975). Ultrastructural studies on graptolites. 2. The periderm and its derivatives in the Graptoloidea. *Smithsonian Contributions to Paleobiology*, **22**, pp. 1–48.

24 | Colonial Structure and Genetic Patterns in Ascidians

ARMANDO SABBADIN

Istituto di Biologia Animale, Università di Padova, Italy

Abstract: There are various modes of colony formation in ascidians. This causes many problems of interpretation at both the phylogenetic and taxonomic level. The extent to which the components of the colony are interrelated is highly variable from species to species. In the family Styelidae it reaches a maximum in the genus *Botryllus*, where it results in a complete integration of the zooids to form an individual unit. A synchronous development of the zooids of each new generation in strict correlation with the development of pre-existing generations is attained, budding being confined to a definite stage of the parental zooids and the individuals being connected to the vascular system of the common tunic. The colonial matrix, the persistent element in the continuous succession of blastogenic generations, acquires autonomy which expresses itself in the capacity to reconstruct the colony by vascular budding and to re-cycle the germ cells to successive generations. A highly adaptive value is conferred on the colony by its ability to regulate the number of individuals per generation and the number of co-existing generations depending on environmental conditions.

The highly integrated colonial pattern of *Botryllus* allows a genetic analysis of some prominent problems confronting sessile animals: the preservation of individuality by a mechanism of nonself recognition which prevents fusion of contiguous colonies; self-fertilization and inbreeding; the significance and adaptive value of certain polymorphisms.

COLONIAL ASCIDIANS AND BUDDING

Asexual reproduction leading to coloniality occurs in both ascidian orders Enterogona and Pleurogona. The first of these orders embraces the two sub-orders Aplousobranchiata and Phlebobranchiata, while the second consists of the single suborder Stolidobranchiata. Pleurogonids are considered to be derived

---

Systematics Association Special Volume No. 11, "Biology and Systematics of Colonial Organisms", edited by G. Larwood and B. R. Rosen, 1979, pp. 433–444. Academic Press, London and New York.

from an enterogonid ancestor. Coloniality is universal in the Aplousobranchiata, whereas in the Phlebobranchiata it is limited to the families Diazonidae and Perophoridae, and in Stolidobranchiata confined to the subfamily Botryllinae of the family Styelidae (Table I).

TABLE I. The main ascidian taxa

Orders	Suborders	Families
ENTEROGONA	Aplousobranchiata	Clavelinidae
		Polyclinidae
		Didemnidae
	Phlebobranchiata	Cionidae
		Diazonidae
		Perophoridae
		Ascidiidae
		Corellidae
		Agnesiidae
PLEUROGONA	Stolidobranchiata	Styelidae
		Pyuridae
		Molgulidae

The buds may have one of two principal origins: they may arise as epidermal outgrowths in different regions, enclosing totipotent tissues of various types (epicardial epithelium, atrial epithelium, mesenchyme, undifferentiated haemocytes); or they arise as epidermal constrictions in the abdominal or postabdominal regions—these constrictions may subdivide the epicardium only, or epicardium and other organs also. In the Botryllinae, at least in some genera, two types of budding occur: normal atrial budding and spontaneous or experimentally inducible vascular budding (Oka and Watanabe, 1957a). Vascular budding arises from the circulatory system of the tunic, this system being homologous to the stolonal apparatus from which buds form in some Enterogona. In vascular buds the morphogenetic tissue is a cluster of haemocytes while in stolonal buds it is a cluster of mesenchymal cells. Freeman (1964) has shown that undifferentiated haemocytes injected into irradiated stolons of *Perophora* can be responsible for the entire bud formation, the mesenchyme having been destroyed by the irradiation. This might well indicate the primitive nature of the colonial habit in ascidians (Millar, 1966).

The development of buds having only one kind of morphogenetic tissue corresponds to a complete somatic embryogenesis, starting from a double-layered vesicle (Berrill, 1950, 1951). The role of the epidermal layer is not merely a passive one. At least in stolonal and vascular buds, in which the morphogenetic material consists of free or loosely associated cells, the epidermis seems to act as an inductor for the aggregation of the cells into an organized tissue (Berrill, 1961). In *Botryllus schlosseri*, the bilateral asymmetry of developing zooids can be experimentally reversed and this reversed asymmetry is transmitted to the palleal buds, as well as to the vascular buds generated by the vessels of the tunic after removal of the zooids (Sabbadin *et al.*, 1975). This shows that both the epidermal wall of the zooids and the epidermal wall of the vessels arising from the zooids can impart information to the enveloped morphogenetic tissue and thus determine the kind of bilateral asymmetry that will develop.

COLONIAL STRUCTURE AND ADAPTIVE VALUE OF COLONIALITY

In ascidian colonies the zooids are integrated to different degrees depending on the species. They may be simply contiguous; or they may share a common tunic in which they are embedded independently or arranged in systems with the individual atrial siphons converging into common cloacal cavities; or they may be interconnected by a general vascular apparatus so that the clone forms a true colony—a morphophysiological unit.

These steps of progressive integration are found within the Botryllinae. In most species a sequential production of buds by the same zooid results in an intermingling of individuals at different developmental and sexual stages. It is only in the genera *Botryllus* and *Botrylloides* that a sharp distinction between the successive blastogenic generations occurs. In the colony, three generations co-exist, the stages of which are so correlated that the disappearance of the oldest generation coincides with the maturation of the next one and the initiation of a new generation (Berrill, 1941). Thus the zooids of each generation are at the same stage and behave as a single individual. This combines the advantages of both colonial and solitary modes of life. For illustration, reference will be made to *Botryllus schlosseri*, the species with which I am most familiar.

The colony is founded by a tadpole larva which metamorphoses into an oozooid. This bears a bud, the first blastogenic generation (Plate I, 1), which in turn produces dextral and sinistral bud primordia, the second generation (Plate I, 2). When the oozooid regresses and is resorbed, generation 1 reaches functional maturity replacing the oozooid and the buds of generation 2 give off bud primordia of generation 3 (Plate I, 3), and so on (Plate I, 4). The colonial cycle consists of a long series of such blastogenic generations, which at a temperature of 18°C succeed each other at weekly intervals. As the number of zooids increases, they arrange themselves in star-shaped systems (Plate I, 5). In the colony all the zooids are interconnected by a vascular network which traverses the tunic and opens at the periphery into a collecting marginal vessel from which numerous ampullae branch out (Plate I, 4).

The two elements of the colony, namely the zooids which are continually renewed and the persistent colonial matrix with its vascular system, collaborate closely in the dynamics of the colonial cycle. The colonial matrix, by way of the vascular connections between the zooids, controls the timing of asexual reproduction. If the vascular systems of zooids are separated, a difference between their stages later appears. If the connections are re-established, a re-phasing of the stages will follow.

Owing to the contraction of ampullae, which maintain an autonomous circulation, the colonial matrix can survive for weeks after removal of the zooids, and eventually reconstruct the colony by the process of vascular budding (Milkman, 1967; Sabbadin et al., 1975). Less serious crises, such as reduced food supply, are overcome by a different strategy. The adults and some buds regress, their resorption products being re-cycled through the general circulation to the benefit of the buds that survive. Normally two or more buds per zooid are produced, but when the colony is under stress only one is retained, the other being resorbed (Plate I, 6). Thus there is competition between contemporary buds and there may also be competition between co-existing generations. Under particularly unfavourable conditions the three generations may be reduced to two, both the resorption of the adults and the maturation of the following generation being so advanced in time as to precede the budding stage of the third generation (Sabbadin, 1958).

There is no evident reduction of the blastogenic potential in time, such as

---

Plate I

1–6: Development and growth of the colony of *Botryllus schlosseri*.
1, Larva metamorphosing into the oozooid carrying the bud primordium (bp) of the first blastogenic generation ($\times 10$). **2**, Adult oozooid with grown bud (b) of the first and bud primordia (bp) of the second generation ($\times 12$). **3**, Adult blastozooid of the first generation with grown buds (b) of the second and bud primordia (bp) of the third generation ($\times 12$). **4**, Regressing zooid (rz) of the first generation; zooids of the second generation with the buds of the third generation. At the periphery is the marginal vessel (mv) from which branch out the vascular ampullae (va) ($\times 10$). **5**, Zooids arranged in star-shaped systems. **6**, Stressed zooids each with only one bud ($\times 12$).

7–9: Self-, nonself-recognition in *Botryllus schlosseri*. **7**, Vascular ampullae of two facing colonies interdigitating ($\times 18$). **8**, Facing colonies fused ($\times 12$). **9**, Facing colonies repelling each other ($\times 14$).

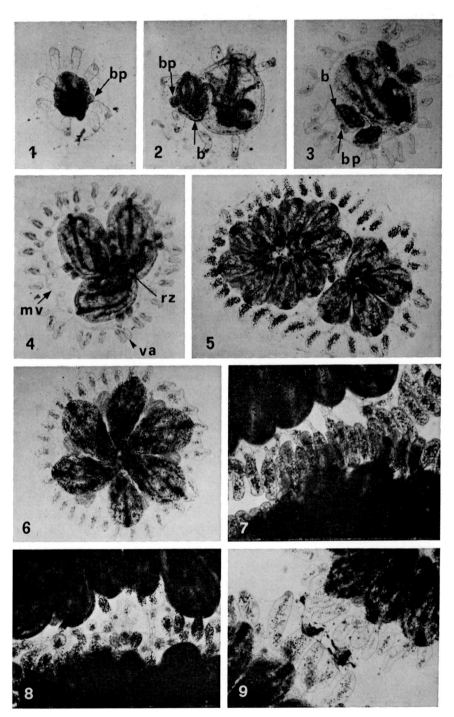

PLATE I. (Caption on facing page)

would reflect ageing in the colony. We have been culturing single colonies for several years, that is for hundreds of generations, without any apparent loss of vigour.

This capacity of the *Botryllus* colony, to reconstruct itself by vascular budding, to adjust its growth rate continuously, to rejuvenate through successive generations, is an excellent manifestation of the adaptive value of the colonial mode of life.

CONIALITY: GENETIC PATTERNS AND POPULATION DYNAMICS

In most ascidians the single colony, as a clone, would behave as a small intensely-inbreeding community unless some mechanism of sexual self-incompatibility had been evolved. Indeed hermaphroditism is the rule, so the difference between the sexual stages of the component zooids would tend to frustrate such simple devices as protogyny or protandry.

In the grown colonies of *Botryllus schlosseri* ovaries and testes form in the zooids of every succeeding generation. The eggs develop to full maturity in the buds and are fertilized as soon as the opening of siphons allows sperm to enter. Sperm are discharged from the same zooids somewhat later (Milkman, 1967). Therefore, despite first appearances, protogyny is effective in preventing self-fertilization in this species, for all the zooids of any given generation are at the same sexual stage.

There is thus no self-incompatibility. However self-fertilization, if induced, entails a remarkable inbreeding depression. This has been demonstrated by the following experiments (Sabbadin, 1971):

(1) Crosses were performed between colonies differing in their sexual stages, both colonies of each couple being homozygous for the dominant and the recessive allele of two Mendelian loci, respectively, with the alleles at the same locus being opposite in the paired colonies (AAbb × aaBB). Self- and cross-fetilization were inferred from the offspring phenotypes. The results (Table II) showed that self-fertilization was very unusual.

(2) Each colony was cut into two pieces. A difference in the sexual stages of the pieces from the same colony was induced by exposure at different temperatures; then they were placed together with another colony at one or the other stage. Either only self-fertilization, or only cross-fertilization, or a combination of the two in equal proportions, took place, depending on which mate the colony supplying the eggs happened to encounter (Table II).

(3) The progeny by self- and cross-fertilization were compared with respect to their ability to pass through the various developmental phases to form new

colonies. No difference was found in the percentages of eggs that developed but highly significant differences were observed in the percentages of normal and metamorphosing larvae, and in the survival and growth of the newly founded colonies (Table III).

TABLE II. Self- and cross-fertilization in Botryllus schlosseri

Genotypes of pairing colonies	No. of crosses	Offspring phenotypes	
		aB (self-fertilization)	AB (cross-fertilization)
aaBB × AAbb	15	1	471
aaBB ×(AAbb★ × *AAbb★*)	6	0	53
(*aaBB★* × AAbb) × aaBB★	9	49	0
aaBB★ ×(aaBB★ × AAbb★)	11	134	136

The colonies which supplied the eggs are in italics. Colonies in parentheses are at the same sexual stage. The *asterisks* mark pieces of the same colony.

Apart from self-fertilization, some degree of inbreeding between related individuals is to be expected among sessile organisms. This is particularly true in colonial ascidians since viviparity, and the capacity to choose a substrate, will favour larval settlement on and around the parental colony. Aggregation of larvae, sometimes so intimate as to form spurious colonies, is also rather common in ascidians. In *Botryllus* this kind of inbreeding is inhibited, and outbreeding is favoured, by the capacity for self/non-self recognition. Colonies which come into contact by means of their vascular ampullae (Plate I, 7) will fuse (Plate I, 8) or repel each other (Plate I, 9). The genetic control of this phenomenon is exerted both in *B. primigenus* (Oka and Watanabe, 1957b) and in *B. schlosseri* (Sabbadin, 1962) by a series of alleles at one locus. Fusion occurs when the contiguous colonies share at least one of these alleles. Each colony can fuse with either of the parental colonies, while the offspring of two non-fusing colonies ($F^1 F^2 \times F^3 F^4$) segregate into four groups ($F^1 F^3$, $F^1 F^4$, $F^2 F^3$, $F^2 F^4$), each of which recognizes two other groups as self. The fusion consists of an anastomosis of the vascular systems of the partners; this will result in a phasing of their stages (Watanabe, 1962) so that protogyny will prevent the fused colonies from intercrossing. In *B. primigenus* the process seems to have developed further than in *B. schlosseri*. It has been claimed (Oka, 1970) that in the first species self-recognition is associated with a haplo-diploid type of sexual self-incompatibility. The sperm cannot penetrate the egg envelopes if the recognition allele that they carry is shared by the diploid follicular cells. Thus for many related colonies intercrossing is prevented, even if they do not come into contact and do not fuse.

TABLE III. Development, survival and growth in the offspring of self- and cross-fertilization in *B. schlosseri*

	Self-fertilization (%)	(No.)	Cross-fertilization (%)	(No.)
Developed eggs	95·64	(1171)	96·49	(1027)
Normal larvae	76·89	(1991)	98·36	(1529)
Metamorphosed larvae	46·85	(572)	87·27	(314)
Surviving colonies	50·24	(205)	88·72	(266)
Mean no. of zooids per colony	2·74	(103)	7·54	(236)

The numbers of colonies from which the percentages or the means were calculated are in parentheses. The survival percentage and the number of zooids per colony were scored at the 5th blastogenic generation.

The composite colony resulting from the fusion of two or more colonies will preserve the genetic complements of the components. This would be true even if some of them later are resorbed or subsequently separate themselves. The memory of the previous colonial admixture is retained for a fairly long time. This has been demonstrated experimentally (Sabbadin *et al.*, 1970). Colonies genotypically *AAbb* and *aaBB* were allowed to remain fused for some time, then were separated and each was crossed with the double recessive genotype *aabb*. From the crosses *AAbb* × *aabb* the phenotype *Ab* and from the crosses *aaBB* × *aabb* the phenotype *aB* were obtained. However, from the first type of cross the phenotype *aB* and from the second the phenotype *Ab* also continued to appear in the offspring of many successive blastogenic generations. This suggests that, during fusion, germ cells had been exchanged between the colonies, re-cycled through the common circulation and transmitted to, and matured by, the zooids of successive generations.

The adaptive value of these devices should not be overlooked. They would serve to maintain genetic variability within communities even of reduced size. Drastic reductions in the size of sessile populations, either occasional or cyclically repetitive, are to be expected. For example, Goodbody (1963) has shown that communities of the solitary *Ascidia nigra*, which he had studied from settlement until the sixth week of life, had from 0 to 17·5% survival, the high mortality having been mainly due to predation and competition for food and space. The competition for space between *Botryllus* colonies would not result in such

TABLE IV. Frequencies of colour morphs of *B. schlosseri* in two areas and two biotopes from the Venetian lagoon. (Orange pigment and double band, each of them with the two phenotypes i.e. presence and absence, are controlled by independent loci. Figures are given for one phenotype per locus. The data of the same year from similar stations (piles) have been pooled.)

Area	Stations	Year	No. of colonies	Orange pigment without pigment (%)	Double band with band (%)
Venice	Piles	1963	1919	9·32	6·72
		1964	1418	11·70	3·66
	Piles	1963	1567	15·69	4·91
Chioggia		1964	883	21·17	4·53
	Zostera	1964	286	87·76	0·69

a high mortality, but rather in an inhibition of their growth and in the fusion of related colonies which come into contact.

Differentiation between various communities of sessile species will depend on the extent of gene flow allowed by larval dispersal and on local selection pressures. The genetic structure of sessile populations is little known for most species (for review, see Gooch, 1975).

In *B. schlosseri* there is a striking polymorphism in cells and pigmentation patterns. There are 48 recognizable colour morphs resulting from different combinations of alleles at five loci (Sabbadin and Graziani, 1967a). The distribution of the colour morphs related to two loci has been studied in two areas of the Venetian Lagoon, near Venice and near Chioggia, some 20 km apart (Sabbadin and Graziani, 1976b). Seasonal samples were collected during 1963 and 1964 at pairs of stations, separated by about 1300 m, in the two areas. These represent the two more typical biotopes, i.e. piles marking the canals, and beds of *Zostera*. A substantial homogeneity of samples both for each station and between similar stations was observed in each of the two years, with fluctuations from one year to the other (Table IV). There must have been an extensive gene flow to account for this homogeneity. A highly significant difference was found between the samples from the piles and those from *Zostera* in the area of Chioggia, much higher than the difference between the samples from piles taken at Venice and Chioggia. This might reflect different selection pressures in the two biotopes. The conclusion was drawn that the population of the Venetian Lagoon is subdivided into micro-geographical and ecological sub-populations significantly different with respect to frequencies of genetically controlled characters.

Deeper insight into spatial–temporal dynamics, and hence into evolution of the populations of this and other colonial species, will be achieved by genetic analysis of biochemical characters which, among other advantages, will largely bypass the necessity for controlled breeding. The work of specialists in the biology and ecology of the different groups will still be needed in order to weigh the contributions of those factors, such as substrate preferences, growth and survival trends and rates, sexual behaviour and larval biology, which are basic to evolution.

SUMMARY

There are various modes of colony formation in Ascidians. This causes some problems of interpretation at the phylogenetic level.

The extent to which the zooids of the colony are interrelated is highly variable from species to species. It reaches a maximum in the genus *Botryllus* of the family Styelidae, where it results in a complete integration of the clonal zooids to form an individual unit. Members of each new blastogenic generation develop synchronously, budding being confined to a definite stage of the parental zooids; synchroneity of the individual developmental phases is assured by the vascular connections between the zooids of the colony.

A high adaptive value is conferred on the colony of *Botryllus schlosseri* by its ability to regulate the number of individuals per generation and to regulate the number of co-existing generations depending on environmental conditions.

In *B. schlosseri* self-fertilization is prevented by protogyny. A self/non-self recognition mechanism controlled by many alleles at one locus is at work. Self-recognition enables those colonies coming into contact and sharing a recognition allele to fuse. The fusion consists of an anastomosis of the vascular system of the contiguous colonies which results in their sexual stages coming into phase with each other. As a result intercrossing between the fused colonies comes to be prevented by protogyny. At the same time an interchange of germ cells between the fused colonies occurs, the composite colony thus preserving the genetic complement of the components, even if some of them later are resorbed or break contact. It is suggested that these strategies, which limit inbreeding and ensure genetic variability even within very small communities, are of paramount importance in the population dynamics of the species.

A study of the frequency distribution of genetically controlled colour morphs has shown that, in the Venetian Lagoon, the population of *B. schlosseri* is subdivided into microgeographical and ecological subpopulations which differ significantly from each other.

ACKNOWLEDGEMENTS

The experimental work on *Botryllus schlosseri* has been supported by C.N.R. grants through the Istituto di Biologia del Mare, Venice. I am indebted to W. J. Canzonier for correction of the English manuscript.

REFERENCES

BERRILL, N. J. (1941). The development of the bud in *Botryllus*. *Biol. Bull.* **80**, 169–184.
BERRILL, N. J. (1950). "The Tunicata." Ray Society, London.
BERRILL, N. J. (1951). Regeneration and budding in tunicates. *Biol. Rev.* **26**, 456–475.
BERRILL, N. J. (1961). "Growth, Development and Pattern." Freeman, San Francisco.
FREEMAN, G. (1964). The role of blood cells in the process of asexual reproduction in the tunicate *Perophora viridis*. *J. exp. Zool.* **156**, 157–184.
GOOCH, J. L. (1975). Mechanisms of evolution and population genetics. *In* "Marine Ecology—A comprehensive, integrated Treatise on Life in Oceans and Coastal Waters" (O. Kinne, ed., Vol. 2, pp. 349–409. Springer Verlag.
GOODBODY, I. (1963). The biology of *Ascidia nigra* (Savigny). II. The development and survival of young ascidians. *Biol. Bull.* **124**, 31–44.
MILKMAN, R. (1967). Genetic and developmental studies on *Botryllus schlosseri*. *Biol. Bull.* **132**, 229–243.
MILLAR, R. H. (1966). Evolution in Ascidians. *In* "Some contemporary studies in Marine Science" (H. Barnes, ed.), pp. 519–534. G. Allen and Unwin Ltd, London.
OKA, H. (1970). Colony specificity in compound ascidians. *In* "Profiles of Japanese Science and Scientists 1970" pp. 195–206. Kodansha, Tokyo.
OKA, H. and WATANABE, H. (1957a). Vascular budding, a new type of budding in *Botryllus*. *Biol. Bull.* **112**, 225–240.
OKA, H. and WATANABE, H. (1957b). Colony specificity in compound ascidians as tested by fusion experiments. *Proc. Japan Acad.* **33**, 657–659.
SABBADIN, A. (1958). Analisi sperimentale dello sviluppo delle colonie di *Botryllus schlosseri*. *Arch. It. Anat. Embr.* **63**, 178–221.
SABBADIN, A. (1962). Le basi genetiche della capacità di fusione fra colonie in *Botryllus schlosseri*. *Rend. Accad. Lincei* **32**, 1031–1035.
SABBADIN, A. (1971). Self- and cross-fertilization in the compound ascidian *Botryllus schlosseri*. *Devl Biol.* **24**, 379–391.
SABBADIN, A. and GRAZIANI, G. (1967a). New data on the inheritance of pigments and pigmentation patterns in the colonial ascidian *Botryllus schlosseri*. *Riv. Biol.* **60**, 559–598.

SABBADIN, A. and GRAZIANI, G. (1967b). Microgeographical and ecological distribution of colour morphs of *Botryllus schlosseri*. *Nature, Lond.* **213**, 815–816.

SABBADIN, A., CAVALIERI, G. and VISCOVICH, L. (1970). Transmissione delle cellule germinali per via sanguigna nella colonia di *Botryllus schlosseri*. *Acta Embr. Morph. exp.* **1970**, 198–199.

SABBADIN, A., ZANIOLO, G. and MAJONE, F. (1975). Determination of polarity and bilateral asymmetry in palleal and vascular buds of the ascidian *Botryllus schlosseri*. *Devl Biol.* **46**, 79–87.

WATANABE, H. (1962). Further studies on the regulation in fused colonies in *Botryllus primigenus*. *Sc. Rep. Tokyo Kyoiku Daigaku*, B **10**, 253–284.

25 | Colonial Breeding in Sea-birds

J. C. COULSON AND FIONA DIXON

Department of Zoology, University of Durham, England

Abstract: There has been much confusion in ornithology over the meaning of the word "colony". As a result, other terms, such as "breeding station", have been used to avoid the social and behavioural implications.

Recent studies have given a better insight into the concept of a colony. In the cases examined, a colony is not an inbreeding unit contrary to what has often been implied. It is true in many seabird species that the breeding birds maintain complete fidelity to their colony and often the place of previous breeding, but the young produced in each colony recruit to many colonies over a large area.

Studies on the kittiwake gull *Rissa tridactyla* have indicated that an appropriate model of a colony consists of a series of interacting and interlinked groups; a chain-mail effect.

Evidence exists that birds behave differently at the centre and edge of a colony, for example, the divorce rate is higher at the edge than in the centre. In addition, breeding success and adult survival are greater at the centre.

Four explanations for colonial breeding in birds have been advanced:
1. Defence against predators.
2. Social stimulation.
3. Population regulation.
4. "Information centre" for finding food.

The first two are well established; the last two are merely hypotheses at present.

INTRODUCTION

The word colony is apparently derived from the Roman word *colonia*, meaning the human settlement of a new area. Within zoology the word colony has different interpretations, yet it is frequently used without definition. Amongst the lower invertebrates, the term usually implies a group of physically and physiologically linked individuals whilst in entomology and in the vertebrates,

---

Systematics Association Special Volume No. 11, "Biology and Systematics of Colonial Organisms", edited by G. Larwood and B. R. Rosen, 1979, pp. 445–458. Academic Press, London and New York.

it has retained much of its original use and describes breeding groups of aggregated individuals which have originally occupied a new site.

In the lower invertebrates the physical links have usually been developed by the multiplication of the same parent stock and the individuals are therefore of the same genetic constitution. Although this does not happen to the same extent in vertebrates, it has been suggested that some sea-bird colonies may tend to exhibit inbreeding (Wynne-Edwards, 1962; Coulson and Brazendale, 1968) but this is not the case in all species (Chabrzyk and Coulson, 1976).

Despite the lack of direct physical links, colonial vertebrates show strong relationships with their neighbours. These are established and maintained by subtle behavioural interactions, often involving the senses of sight and hearing.

The ornithological literature refers to three types of distribution of breeding birds; solitary (often territorial), semi-colonial and colonial. In reality there is probably a continuum from one extreme to the other with colonial breeding at one limit. Few attempts have been made to investigate the properties of bird colonies and the manner in which they function. Most frequently, the use of the term colony for species such as rooks, penguins and many gulls merely indicates that the species are intensely aggregated for breeding. The "New Dictionary of Birds" (Thomson, 1964) defines a colony as a number of individuals breeding gregariously, but there is no attempt to indicate the properties of a colony nor to offer a more precise definition which is of practical use.

We have frequently encountered practical difficulty in identifying the limits of a sea-bird colony and particularly the point at which a group breaks up into two distinct colonies. For example, Fig. 1 shows the distribution of kittiwake gulls *Rissa tridactyla* nesting along a sea-cliff. The maps clearly show the practical problems. How large a gap is necessary between nesting birds before one colony becomes two? Further, what influence does the headland have upon colony structure where birds on each side of the headland cannot see each other (D1 and D2)? These problems have been encountered by Fisher (1952) in his study of the spread of the fulmar *Fulmarus glacialis*, where to avoid the unknown limits of a colony and the associated implications, he used the term "breeding station", which he defined as being separate from the next breeding station by a gap of more than one mile between the nearest breeding pairs. This is an entirely arbitrary distinction but it clearly indicates the lack of knowledge concerning colony structure, and the interactions between individuals within a colony.

An aggregation of mobile organisms will disperse unless some active process prevents them from separating. There must be a selective advantage in grouping them together, and there must be a mechanism for maintaining the group and preventing dispersion. We have attempted to identify this mechanism and to obtain an understanding of how a colony functions. Ultimately we hope to be able to present a better insight into the selective advantage of social or colonial breeding which at present is still confused.

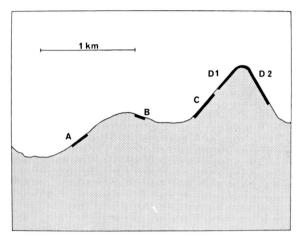

FIG. 1. The position of groups of kittiwakes nesting on a cliff face. This illustrates the difficulties in determining firstly, where one colony stops and another starts, e.g. whether A and B or C and D1 are different colonies, and secondly, what influence the headland has upon colony structure—see D1 and D2.

COLONY STRUCTURE

The colony structure of kittiwake *Rissa tridactyla* colonies has been explored by examining the behavioural responses which involve birds belonging to different pairs. Three activities have been found in which there are clear responses in neighbouring individuals.

(1) Panic flights. These occur early in the breeding season and are most frequent just after the annual re-occupation of the colony. These flights appear spontaneously and without an obvious external stimulus but in all other respects are identical with the fear flights induced by an avian predator or a helicopter. The birds leave the nest with a characteristic plunge from the nest to gain speed and then rapid flight out to sea. The whole activity takes place in complete silence. Such behaviour is triggered off by one individual which is followed, momentarily later, by its neighbours. This chain-reaction gives an indication of the distance over which birds will respond to a single individual initiating the panic flight. Although there is considerable variation in the distance over which birds are influenced, we have found that, on average, birds are affected over a radius of 25 m.

(2) Nest building. Nest building in the kittiwake is often carried out by individual pairs, but on some occasions, many pairs synchronize nest building activities. In such situations, the birds all go to the same source of mud or

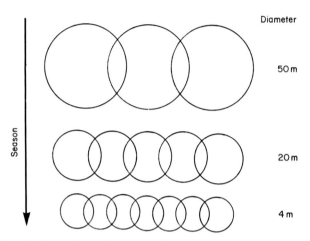

FIG. 2. Model of a kittiwake colony. The "chain-mail" links become progressively smaller as the breeding season approaches.

grass and the activity takes on the appearance of a continuous stream of birds flying to and from the source and the colony. It is relatively easy to identify the areas of the colony connected with such flights and, although extremely variable, these have a mean radius of about 10 m within the colony.

(3) Greeting ceremony. Although the kittiwake has an extensive vocabulary and a large number of sexual behaviour patterns, only one, the pair greeting ceremony, can be seen to have an influence on neighbouring pairs. When a pair becomes reunited at the nest site, it invariably results in a complex display of bowing and calling. The call is the characteristic "kittiwaak" from which the species obtains its vernacular name. This ceremony often produces reaction in neighbouring pairs and a less marked response in birds which are, at the time, on their nest site. The infectious nature of this ceremony has been commented on by many naturalists but little precise observation has been carried out. We find that the infectious nature of this ceremony is usually restricted to a radius of 2 m from the nest where the pair have reunited.

Examination of these behavioural reactions leaves a clear impression that the colony never functions as a whole (unless it is very small) but is constructed of a series of sub-units which interlink and inter-react with neighbouring units.

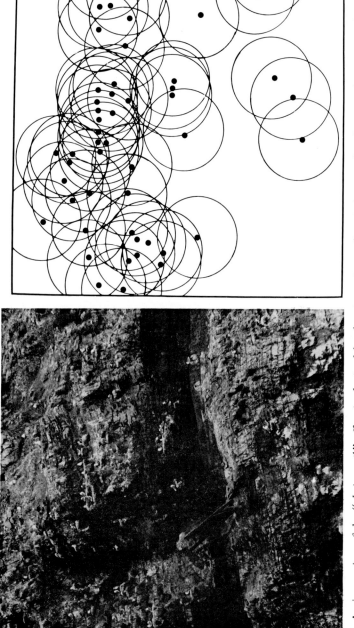

FIG. 3. An impression of the "chain-mail" effect is obtained by drawing a 2 m radius circle around each nest. The way in which neighbouring units interlink is clearly shown.

The size of these units probably changes with season and the behaviour response involved (Fig. 2). We envisage a colony as a structure depending upon a "chain-mail" or "chain-armour" type of interrelationship, with the links only

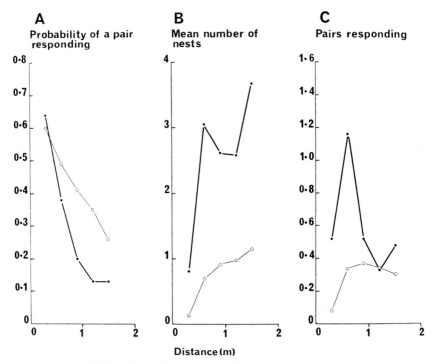

FIG. 4. An examination of the distance over which pairs show a response to a greeting ceremony during the pre-egg laying phase. Two colonies of different density are represented. Solid dots and heavy lines represent the dense colony (mean number of nests within a 1·7 m radius = 12·73): open circles and thin lines represent the sparse colony (mean number of nests within a 1·7 m radius = 3·87).

A, shows the probability of a pair responding to a greeting ceremony at successive distances from it (from 0·3 m to 1·7 m).

B, shows the mean number of nests present in the colony between 0·3 and 1·7 m. These data are obtained from photographs by calculating the mean number of nests at each distance.

C, shows the mean number of pairs responding to a greeting ceremony at successive distances between 0·3 and 1·7 m. This is the product of A and B.

a few metres in diameter and probably becoming progressively smaller as the breeding season approaches.

An impression of the chain-mail effect is shown in Fig. 3, which shows circles of 2 m radii drawn around each nest. Although the density of nests varies considerably, the bonding effect of the neighbours is clearly indicated.

The kittiwake, like many other sea-birds, is an obligatory colonial breeder. Single pairs do not breed in isolation and there is good reason to believe that they are unable to do so. Not only does the female kittiwake require the attentions of the male to bring her into breeding condition, she also needs the stimulation derived from other pairs. It seems most likely that the additional stimulation is obtained from the greeting ceremony as it initiates additional display in neighbouring pairs.

A detailed examination of the distance over which other pairs show a response to a greeting ceremony is shown in Fig. 4. There is clearly a rapid decrease in the proportion of pairs which respond as distance increases, with the most frequent response occurring in birds with nests immediately adjacent (Fig. 4A). In a less dense colony, the response rate is significantly higher at distances greater than 0·5 m. There may well be a compensating effect, with birds nesting at low densities responding more readily to greeting ceremonies which take place at greater distances from them.

The number of nests at successive distances from any individual nest tends to increase with distance (Fig. 4B). This does not follow the expected trend, which is based on the assumption that the nests are randomly distributed because kittiwakes can only nest where there are suitable ledges and these are determined by the geological structure of the cliff face. By multiplying the probability of response by the number of available pairs for each distance, the mean number of pairs influenced at progressive distances is obtained. This also indicates the distance from which each pair receives the maximum stimulation, which, on the data presented, is at about 0·5 m in a dense colony and at 1 m in a sparse group.

PROPERTIES OF A COLONY: CENTRE AND EDGE EFFECT

In the kittiwake and in many other sea-birds, sites in the centre of the colony are more attractive to recruits to the breeding population than sites at the edge. On the other hand, sites are more limited in the centre as many are occupied by established breeders who usually return first in the new breeding season and claim the best sites or the site which they used in the previous year. Unoccupied nesting sites in the centre of the colony are either of poor quality or are sites which have recently been made available by the previous owner having died. The severity of competition for the central sites is considerable. If the breeding population is increasing, this results in more recruits which makes the competition even more intense.

The preference of young birds for central sites is an important requirement in maintaining the colony structure. This is particularly evident if the colony is decreasing; this preference then ensures that the shrinking colony maintains its social grouping and that the peripheral areas are deserted first.

The intense competition for centre sites, which is an important aspect of social organization, results in a segregation of individuals, particularly males. In the kittiwake, once a male has obtained a site and mate, it rarely changes its nest site. When this does occur, the move is usually minimal, to a neighbouring nest site. Thus, birds which nest at the centre or the edge of a colony usually remain in that area for life, an effect which is only modified if the colony is rapidly increasing or decreasing. Several major differences have been found between the individuals which recruit to the centre and edge of a kittiwake colony and these differences lead to the conclusion that the better quality individuals obtain the central sites.

The differences between centre and edge birds are:

(1) The males which obtain central sites have a higher survival rate than those which recruit at the edge. This is despite the greater competition for central sites and the greater amount of time central males have to spend defending their nest sites.

(2) Central males divorce their mates less frequently than edge males.

(3) Central adults rear more offspring in each breeding season than do edge birds.

(4) As a result of (1) and (3) above, centre males produce more young in their lifetime than do edge males.

1. Survival Rates of Kittiwakes at the Edge and Centre of the Colony

The survival rates for male and female kittiwakes have been calculated from the disappearance of individually colour-marked breeding birds. It is known from extensive searches of other colonies that breeding kittiwakes do not move to other colonies. Accordingly, colour-marked birds which disappear from the colony can, with confidence, be assumed to be dead (Coulson and Wooller, 1976). The colony under study (at North Shields, Tyne and Wear, England) went through a period of expansion and then stabilized. During the former period, the survival rate of birds nesting in the centre of the colony was higher than those at the edge (significantly so in the male) (Table I). In the stable phase, the overall survival rate decreased, and the differences between edge and centre birds became less marked and non-significant. Table I also shows the survival rates converted to average expectation of further life once birds had started to breed, that is, the mean number of years that the birds bred. In the expansion phase, the edge males bred on 40% fewer occasions in their lifetime than centre males and the comparable value for the females was 28%. In the stable phase, the difference in life expectancy was much less.

TABLE I. The survival rate and expectatron of further life of male and female kittiwake recruits to a breeding population during (A) the period of expansion and (B) the period of colony stability

	Males			Females		
	No. bird-years	Survival rate	Expectation (years)	No. bird-years	Survival rate	Expectation (years)
A. Expansion (1954–55)						
Centre	404	0·88	8·1	412	0·92	11·5
Edge	307	0·81	4·9	332	0·89	8·3
B. Stable (1967–74)						
Centre	303	0·78	4·0	335	0·84	5·8
Edge	307	0·77	3·8	357	0·82	5·0

2. Divorce Rate in Centre and Edge Birds

Divorce is used to denote the change of mate between successive breeding seasons when the partner of the earlier year is still alive and therefore available to reform the pair. This definition removes any influence of the survival rate between centre and edge birds on the divorce rate, since only surviving birds are considered. Typically, about 66% of the breeding birds retain their mates between successive years. The remaining 34% change mates due either to the partner's death since the previous breeding season, or to divorce.

TABLE II. The divorce rate amongst old and young kittiwakes nesting at the centre and edge of a colony

	Young females		Old females	
	No.	%	No.	%
Centre	127	35	194	23
Edge	112	43	120	38

The divorce rate is shown in Table II sub-divided for older and younger breeding birds. Divorce is less likely amongst older birds and is, age for age, significantly lower in birds nesting in the centre of the colony. A correlation exists between breeding success and the retention of the mate from one breeding season to the next (Coulson, 1972). It seems likely that the effects of both age and nest site on the divorce rate may be explained in terms of differences in breeding success between these groups (see Table III).

3. Breeding Success in Centre and Edge Sites

The breeding success in centre and edge birds is shown in Table III in relation to the breeding experience of the female (although an almost identical relationship exists with male breeding experience) and whether the pair bred together in the previous year. Clearly previous breeding experience (age) and the past status of the pair greatly influence the breeding success, but when these factors are taken into account, it is also evident that the number of young reared per pair is always higher at the centre of the colony.

TABLE III. *The mean number of young fledged per pair of kittiwakes according to pair status, position in the colony and breeding experience (age). These data are based on 610 clutches. The appropriate pairs for comparison of edge and centre are bracketed*

	Breeding experience			
	Breeding for 2nd–4th time		Breeding for 5th–16th time	
	Edge	Centre	Edge	Centre
Same mate as last year	1·37	1·43	1·51	1·62
New mate	1·20	1·31	1·27	1·35

4. Young Produced in an Adult's Lifetime

Using the results presented in Tables I and III, it is possible to calculate the mean number of young fledged in the lifetime of typical edge and centre adults. In the expanding phase of the colony, the divorce rate and the mortality rate are both higher at the edge of the colony and, since both of these factors result in more younger birds which change their mates more frequently, the breeding success is appreciably lower there than at the centre. On average a male at the edge of the colony fledged 6·1 young in its lifetime whereas males at the centre fledged 11·4 young, which is 88% more. In the stable phase of the population, the numbers of young produced by centre males are only 13% greater than those at the edge.

ADVANTAGES OF COLONIAL BREEDING

Although this review is directed towards sea-birds, the functions of colonial breeding, and particularly its selective advantages, are of much wider application, involving not only birds but also most of the vertebrates and some invertebrates. Since colonial breeding occurs in many taxonomic groups and, for example among the birds, in most Orders, it is most likely that it has evolved not just once, but several times independently. It is possible that the selective forces bringing it about have not been the same in each case and that the advantage to one group may be different to that in another. Further, once colonial breeding has become established in a species, other and secondary functions may become attached to it although these do not represent the selective forces involved in its original development.

This is not an appropriate place to critically review the functions of colonial breeding and we present below only a brief summary of the major views of the advantages in colonial breeding.

1. Defence

There is considerable evidence that groups of organisms can deter predators earlier than isolated individuals or pairs, thus improving their chances of escape. Many avian predators are deterred from entering sea-bird colonies and taking young and eggs whereas they are known to predate commonly on solitary species. This is clearly the case with carrion crows predating ground nesting birds. For example tern colonies, although conspicuous, suffer little nest predation from this species or from gulls. Colonial nesting does not, however, deter mammalian predators and thus most colonial bird species nest in localities where they are relatively inaccessible to mammals. Sea birds often nest on small islands and precipitous cliffs where mammalian predators cannot reach the nesting site; similarly, colonial land birds nest in the tops of trees (herons, rooks), on inaccessible ledges, (swallows and house martins) or in holes and cavities (sand martins, starlings and house sparrows).

2. Social Stimulation

It is possible that social breeding arose from advantages in social stimulation resulting in more synchronized and co-ordinated breeding, particularly in harsh environments such as in polar regions. It is more likely that this phenomenon arose out of existing social breeding systems. Fraser Darling (1938) first postulated the concept of social stimulation and his views have been modified from more extensive information. Essentially, birds nesting at similar densities breed at the same time whilst birds nesting at higher densities breed earlier. In the herring gull, this synchronization has been shown to result in more successful breeding owing to a swamping effect of the predators (Parsons, 1975). In contrast, the kittiwake, which nests on precipitous cliffs and is relatively free from predation, has a longer breeding season in colonies with high densities of nests. In this species, the maximum density is determined by the geological structure of the cliff on which the birds nest, and all colonies have low density areas (where breeding is late). Early nesting and hence the spread of breeding, is dependent on the maximum density the birds achieve in any part of the colony (Coulson and White, 1960).

It should be realized that social stimulation has resulted in the breeding birds requiring more stimulation than can be obtained solely from a breeding pair. Stimulation from other pairs is needed and in species where this situation is well established, the species has become obligatorily colonial. Such a system of social stimulation ensures the continuation of colonial breeding and prevents birds dispersing and breeding as isolated pairs. In some species, such as the shag and possibly the herring gull, only the young recruits need stimulation from other pairs to ensure that breeding can take place. Older, experienced pairs appear to be able to breed in isolation.

3. Population Regulation

It is possible that colonial breeding can bring about some constraints on the growth of the breeding population, brought about by the social system. Personal experience of the development of kittiwake colonies shows that they do not grow in the manner expected of an expanding population but that the rate of increase becomes progressively less as the colony increases in size. There is also evidence that the social requirements cause many (usually young) birds to breed less well than they would otherwise do. This is clearly the case in the shag, where, when many of the old birds were removed from the breeding population, the young birds bred much more successfully after taking better nest sites (Coulson, 1971).

Wynne-Edwards (1962) has put forward a concept that animal populations, at least of some species, can regulate their own populations in relation to available food supply. He envisages that social gatherings, including breeding colonies, are one means whereby this self regulation takes place. This concept is, at present, highly controversial and evidence of this type of system, involving self regulation of numbers, is lacking. However, the extensive work reviewed by Wynne-Edwards does draw attention to the vast and complex nature of animal behaviour and how little of this is understood.

4. Information Centre

It is now well established that hive bees communicate the direction and distance of food supplies to other workers in the hive. Ward and Zahavi (1973) have put forward a similar concept for birds, suggesting that the advantage in groups and colonies is that these act as centres where information about food sources can be obtained from other individuals. It is not suggested that a "language" exists between birds but rather that individuals can follow others which, by their behaviour, can be considered to "know" where good food supplies exist.

This concept, like that of Wynne-Edwards, is still in a controversial state: it has not been established that sea-birds obtain any information about feeding sites from other birds at the colony. However, the concept is one which clearly excites interest and will be investigated further in future years.

CONCLUSIONS

Clearly, colonial breeding in sea-birds influences the social status of the birds involved. The presence of so many birds in close proximity results in each individual having the opportunity of testing its "status" against many others in the same colony in only a few minutes. Consequently, it is an ideal situation for animals to segregate themselves and this has been shown to occur in the kittiwake gull, where fitter individuals tend to obtain sites in the central part of the colony.

When animals congregate as in a colony, it is possible to develop complex social structures. Sea-birds are long lived animals, often using the same nest site year after year. In such conditions, it would not be surprising to find that a system of individual recognition exists and that the social structure involves a major individual component.

Much still remains to be explained about the functions and effects of colonial breeding. Nevertheless, this paper draws attention to the complexity of behaviour and organization within a colony, a complexity which is probably much greater than is usually accepted. Further, an attempt has been made to show the nature of coloniality, an approach which strangely has been lacking in studies on birds.

REFERENCES

CHABRZYK, G. and COULSON, J. C. (1976). Survival and recruitment in the herring gull, *Larus argentatus*. *J. anim. Ecol.* **45**, 187–203.

COULSON, J. C. (1971). Competition for breeding sites causing segregation and reduced young production in colonial animals. *Proc. Adv. Study Inst. Dynamics Numbers Popul.* (Oosterbeek), 257–268.

COULSON, J. C. (1972). The significance of the pair-bond in the kittiwake. *Proc. XV Int. Ornithol. Congr.* 424–433.

COULSON, J. C. and BRAZENDALE, M. G. (1968). Movements of cormorants ringed in the British Isles and evidence of colony-specific dispersal. *Brit. Birds* **61**, 1–21.

COULSON, J. C. and WHITE, E. (1960). The effect of age and density of breeding birds on the time of breeding of the kittiwake *Rissa tridactyla*. *Ibis* **102**, 71–86.

COULSON, J. C. and WOOLLER, R. D. (1976). Differential survival rates among breeding kittiwake gulls *Rissa tridactyla* (L.). *J. anim. Ecol.* **45**, 205–213.

DARLING, F. FRASER (1938). "Bird flocks and the Breeding Cycle." Cambridge University Press.

FISHER, J. (1952). "The Fulmar." Collins, London.

PARSONS, J. (1975). Seasonal variation in the breeding success of the herring gull; an experimental approach. *J. anim. Ecol.* **44**, 553–573.

THOMSON, A. L., ed. (1964). "A New Dictionary of Birds." Thomas Nelson & Sons Ltd., London.

WARD, P. and ZAHAVI, A. (1973). The importance of certain assemblages of birds as "information-centres" for food finding. *Ibis* **115**, 517–534.

WYNNE-EDWARDS, V. C. (1962). "Animal dispersion in relation to social behaviour." Oliver and Boyd, Edinburgh, London.

26 | Habitat Selection, Directional Growth and Spatial Refuges: Why Colonial Animals Have More Hiding Places

L. W. BUSS

Department of Earth and Planetary Sciences, The Johns Hopkins University, Baltimore, USA

Abstract: The complex of adaptations evolved in marine invertebrates for the location of spatial refuges is investigated. Field observations and experiments suggest that two classes of mechanisms occur repeatedly: habitat selection and directional growth. It is suggested that habitat selection evolves when the location of a refuge can be sensed by the recruiting larvae; whereas directional growth develops when the position of the refuge is spatially unpredictable. Both solitary and colonial animals employ habitat selection and colonial animals alone employ directional growth. It is argued that the ability of colonial organisms to suffer some degree of zooid/polyp mortality without experiencing whole-colony mortality allows them to exploit a class of spatial refuges unavailable to solitary organisms.

A graphic model is introduced in attempt to outline the relationship between various directional growth morphologies and life-history characteristics. Consideration of the interaction between curves generated for the biomass per unit surface area and cumulative probability of locating a refuge leads to a number of predictions. It is suggested that more runner-like forms should exhibit higher recruitment and growth rates, but lower interference competitive capabilities than more sheet-like forms. These predictions are consistent with the existing data in the literature.

INTRODUCTION

The divergent modes of form and function of solitary and colonial organisms represent one of the earliest recognized patterns in metazoan organization. Interpretation of the evolution of these life modes has classically been (and con-

---

Systematics Association Special Volume No. 11, "Biology and Systematics of Colonial Organisms", edited by G. Larwood and B. R. Rosen, 1979, pp. 459–497. Academic Press, London and New York.

tinues to be) made virtually exclusively on the basis of form (e.g. Boardman *et al.*, 1973a). This purely morphological approach has met with considerable interpretative success, particularly within taxonomic groups, but has suffered from the difficulty that no single set of morphological criteria seem to apply universally to all groups (Hartman and Reiswig, 1973; Simpson, 1973). This problem has led to considerable discussion regarding the necessity of an analytical framework which allows unambiguous definition of colonial traits (Boardman *et al.*, 1973b; Mackie, 1963).

Differences in form are simply reflections of differing solutions to varying problems of adaptation. Organisms, though, are not limited to morphological change in response to such problems, as a host of behavioural and physiological options may exist (Slobodkin, 1970). Given this point, alternative approaches to the solitary–colonial problem do exist (see for example Jackson, 1977). Rather than considering a single adaptive response (morphological) to an unspecified problem of adaptation, one may specify a single problem and investigate the full range (behavioural, physiological, morphological) of responses available to organisms as a solution to this problem. Differential responses to similar problems will allow definition of solitary and colonial traits.

As any individual in a population of sessile marine invertebrates settles and metamorphoses, it gains access to a complex of resources necessary for further growth and development. Upon settlement, the individual also is exposed to a host of processes (e.g. competition, predation and physical disturbance) which may result in its death. Mortality processes do not act uniformly over all substrata (e.g. Connell, 1961; Dayton, 1971, 1974; Hayward and Harvey, 1974b). Thus on any given substrate there will exist spatial positions where the fitness of an individual occupying that position will be higher than for individuals of the same species occupying different spatial positions (Fig. 1). I will, hereafter, refer to these "safe positions" as spatial refuges. Clearly, for any perennial species, an individual occupying a spatial refuge will contribute disproportionately to future generations by virtue of their higher survivorship. Selection should therefore strongly favour evolution of mechanisms for the exploitation of spatial refuges.

In the discussion below I consider the adaptive responses of solitary and colonial sessile organisms inhabiting marine, hard substrata (Table I) in relation to the common problem of adaptation of location and exploitation of spatial refuges. The nature of the adaptive response exhibited by organisms for location of spatial refuges is found to be dependent upon whether or not the recruiting larva is capable of sensing the exact position of the refuge. Solitary organisms are found to employ behavioural responses (habitat selection) colonial organisms

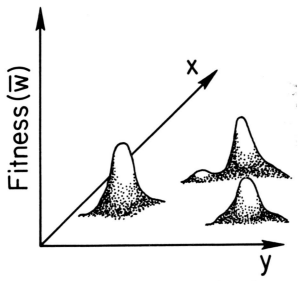

FIG. 1. Fitness as a function of spatial position. x–y co-ordinate plane represents marine hard substratum. Refuges are positions on the x–y plane with high fitness. The discussion will consider organism adaptations evolved to locate such refuges under different conditions of refuge distribution in space and time.

employ both behavioural (habitat selection) and morphological responses (directional growth).

ADAPTIVE RESPONSES

1. Introduction

Spatial refuges, by definition, provide some escape from mortality processes. Since distribution patterns in non-vagile organisms are the product of the interaction of settlement, growth and mortality processes (Connell, 1959), adaptive responses for the location of spatial refuges must be represented in patterns of settlement and growth. Such responses may further be expected to vary as a function of the spatio–temporal distribution of the refuges themselves. Thus consideration of the problem of location of spatial refuges requires analysis of the manner in which the distribution of these refuges, through settlement and growth processes, leads to observable species distribution patterns (see Fig. 1).

Presented below are a series of examples of species distribution patterns, related to spatial refuges, drawn from the literature and my own field research. Consideration of the distribution of refuges in space and time allows identification of the adaptive responses of both solitary and colonial organisms for the location and subsequent exploitation of spatial refuges.

TABLE I. Groups of marine epifaunal Metazoa that attach permanently to hard substrates

Taxon	Solitary	Colonial
Porifera	.	+
Cnidaria		
C. Hydrozoa		
O. Hydroida	+	+
O. Milleporina	.	+
O. Stylasterina	.	+
C. Anthozoa		
S.C. Alcyonaria	.	+
S.C. Zoantharia		
O. Actinaria*	+	.
O. Scleractinia (Madreporaria)	+	+
O. Zoanthidea	+	+
O. Corallimorpharia	+	+
O. Antipatharia	.	+
Entoprocta	.	+
Mollusca		
C. Gastropoda		
S.C. Prosobranchia		
O. Mesogastropoda		
S.F. Cerithiacea		
F. Vermetidae	+	.
F. Turitellidae	+	.
F. Tenagodidae (= Siliquariidae)	+	.
C. Bivalvia		
O. Filibranchia		
S.F. Arcacea	+	.
S.F. Mytilacea	+	.
S.F. Pteriacea	+	.
S.F. Pectinacea	+	.
S.F. Anomiacea	+	.
S.F. Ostreacea	+	.
O. Eulamellibranchia		
S.F. Chamacea	+	.
Annelida		
C. Polychaeta		
F. Sabellidae	+	.
F. Serpulidae	+	.
F. Sabellaridae	+	.
Arthropoda		
C. Crustacea		
S.C. Cirripedia	.	.
O. Thoracica	+	.

TABLE I.—continued

Taxon	Solitary	Colonial
Phoronida	+	.
Ectoprocta	.	+
Brachiopoda		
C. Inarticulata		
O. Neotremata	+	.
C. Articulata	+	.
Echinodermata		
C. Holothuroidea†	+	.
C. Crinoidea		
O. Articulata		
S.O. Isocrinida	+	.
Hemichordata		
C. Pterobranchia	+	+
Chordata		
S.P. Urochordata		
C. Ascidiacea	+	+

(Compiled from Barnes, 1974; Gosner, 1971; and Ricketts et al., 1968.)

* Most Actinaria are vagile.

† Some suspension-feeding holothurians do not move but none are permanently attached.

The terms spatial and temporal predictability will recur throughout this paper. Temporal predictability refers to the season-to-season or year-to-year predictability in the availability of refuges. Spatial predictability refers to predictability in the position of the refuge on a particular substratum. Spatially predictable refuges are those which occur on a substratum when an organism settles and this position remains invariate over the lifespan of the organism. Spatially unpredictable refuges are those which either do not exist on the substratum when the organism settles or those where the exact position of the refuges changes over time.

2. Examples

(a) *Example 1.* A wide variety of sessile marine invertebrates inhabit caves, crevices and undersurface environments. These cryptic environments are frequently free from the effects of predation (Jackson and Buss, 1975), physical disturbance processes (Connell, 1961; Jackson and Buss, 1975) and physico-chemical stress

TABLE II. Larvae which exhibit a photonegative response prior to metamorphosis

Porifera

Halicoma sp. (Bergquist et al., 1970)
Mycale macilenta (Bergquist et al., 1970)
Ophlitaspongia seriata (Fry, 1971)

Cnidaria: Hydroidea

Clava squamata (Williams, 1965b)

Cnidaria: Anthozoa

Fungia actiniformis (Abe, 1937)
Dendrophyllia manni (Edmondson, 1929, 1946)
Pocillopora damicornis var. *cespitosa* (Atoda, 1947; Edmondson, 1929)
Acropora bruggemanni (Atoda, 1951)
Cyphastraea ocellina (Edmondson, 1929, 1946)
Galaxea aspera (Atoda, 1951)

Mollusca: Bivalvia

Ostrea edulis (Cole and Knight-Jones, 1949)
Mytilus edulis (Bayne, 1964)

Annelida: Polychaeta

Spirorbis borealis (Knight-Jones, 1951, 1953b)
Spirorbis tridentatus (de Silva, 1962)
Spirorbis pagenstecheri (Knight-Jones, 1951, 1953b)
Sabellaria alveolata (Wilson, 1968)
Spirorbis rupestris (Gee and Knight-Jones, 1962)
Sabellaria spinulosa (Wilson, 1970)

Arthropoda: Cirripedia

Balanus improvisus (Bousfield, 1955)
Balanus perforatus (Ewald, 1912)
Balanus balanoides (Visscher, 1928)
Balanus eburneus (McDougall, 1943)
Pollicipes spinosus (Batham, 1946)

Ectoprocta

Scrupocellaria reptans (Ryland, 1960)
Bugula simplex (Ryland, 1960)
Scrupocellaria scruposa (Ryland, 1960)
Bugula plumosa (Ryland, 1960)
Bugula flabellata (Grave, 1930; Lynch, 1949, 1955, 1956, 1960)
Bugula fulva (Ryland, 1960)
Celleporella hyalina (Ryland, 1960)
Bugula turrita (Lynch, 1949, 1955, 1956, 1960; Ryland, 1960)
Bowerbankia pustulosa (Hasper, 1913)
Watersipora cucullata (Wisely, 1958b)
Bugula neritina (Lynch, 1947, 1949, 1955)
Cryptosula pallasiana (Ryland, 1962a)

Brachiopoda

Terebratella inconspicua (Percival, 1944)
Frenulina sanguinolenta (Mano, 1960)

Chordata: Ascidacea

Amaroucium constellatum (Mast, 1921)
Symplegma virde (Grave, 1935)
Amaroucium pellucidum (Grave, 1920: Mast, 1921)
Diplosoma listerianum (Crisp and Ghobashy, 1971)
Polyandrocarpa tincta (Grave, 1935)
Ciona intestinalis (Berrill, 1947; Dybern, 1962; Millar, 1953)
Polyandrocarpa gravei (Grave, 1935)
Botryllus schlosseri (Grave and Woodbridge, 1924; Woodbridge 1926)
Ascidia nigra (Goodbody, 1963; Grave, 1935)

gradients (e.g. sedimentation stress, Jackson, 1977). Such environments then can serve as spatial refuges from these mortality sources. They will be distributed predictably in time in that any boulder will always have an undersurface, any cave an interior, etc. The exact spatial position of the refuge may be predictable to the recruiting individual in that a wide variety of environmental cues are available indicating the existence of a cryptic environment.

Perhaps the most obvious of the environmental cues indicative of cryptic environments is the absence, or low level of light. Larvae of sedentary marine invertebrates are known to demonstrate complicated behaviour patterns in response to light (reviews by Meadows and Campbell, 1972; Newell, 1970; Thorson, 1964). Early larval life is often characterized by positive phototactic behaviour. This is, at least partly, a response to dispersal requirements. Many organisms, though, demonstrate a negative phototaxis prior to settlement. Although this behaviour must, in part, reflect the requirement of escaping the pelagic zone and seeking the bottom, it will also tend to favour settlement in cryptic environments. The degree of negative phototrophy and timing of this response will determine the degree to which cryptic environments are favoured.

Thus a wide variety of sessile marine invertebrates possess behavioural responses capable of locating spatial refuges provided by cryptic environments. Listed in Table II is a non-exhaustive survey of organisms known to show negative phototrophic responses prior to metamorphosis. Although not all authors cited specify whether the organism under consideration occurs commonly or exclusively in cryptic environments, such distributional evidence does exist for some species (e.g. Porifera: Bergquist *et al.*, 1970; Bryozoa: Ryland, 1960; serpulids: de Silva, 1962; Ascidea: Crisp and Ghobashy, 1971). It appears that both solitary and colonial organisms employ an identical behavioural response for location of spatial refuges provided in cryptic environments.

(*b*) *Example 2.* Species aggregations, both in uniform and non-uniform local environments, are a common distributional phenomenon. Such aggregations may arise in several ways (Knight-Jones and Moyse, 1961). Community development may reach a stable point, leaving a single species stand. Some organisms multiply by budding or fission to produce clumped progeny, others have crawling young, which move only a short distance from adults. The great majority of sessile marine invertebrates, though, have swimming larvae which are widely dispersed. These larvae may tend to re-aggregate by settling in massive "spatfalls", settling preferentially in positions near adults, or suffering differential mortality.

Positions near adults of the same species will represent spatial positions where fitness is likely to be higher than alternative positions if (a) internal fertilization is required for larval production (i.e. barnacles, but see Barnes and Crisp, 1956),

Table III. Larvae which exhibit a gregarious settling behaviour

---

Porifera

Ophlitaspongia seriata (Fry, 1971)

Cnidaria: Hydroidea

Tubularia larynx (Pyefinch and Downing, 1949)

Cnidaria: Anthozoa

Dendrophyllia manni (Edmondson, 1929)

Mollusca: Bivalvia

Ostrea edulis (Cole and Knight-Jones, 1949)

Annelida: Polychaeta

Merierellia enigmatica (Straughan, 1972)   *Sabellaria alveolata* (Wilson, 1968)
Spirorbis borealis (Knight-Jones, 1951, 1953b)   *Sabellaria spinulosa* (Wilson, 1970)
Spirorbis pagenstecheri (Knight-Jones, 1951, 1953b)   *Hydroides norvegica* (Wisely, 1958a)
Spirorbis rupestris (Gee and Knight-Jones, 1962)

Arthropoda: Cirripedia

Balanus amphitrite (Daniel, 1955)   *Balanus crenatus* (Knight-Jones, 1953a)
Balanus balanoides (Knight-Jones, 1953a)   *Elminius modestus* (Knight-Jones, 1953a; Knight-Jones and Stephenson, 1950)

Ectoprocta

Watersipora cucullata (Wisely, 1958b)

Chordata: Ascidacea

Amaroucium constellatum (Grave, 1941)   *Phallusia nigra* (Grave, 1935)
Polyandrocarpa tincta (Grave, 1935)

(b) if external fertilization of distant organisms is unlikely due to physical conditions such as high turbulence, or (c) sources of post-settlement mortality are acting in the same manner as when the adults present settled. If any of these three conditions is met, then positions near adults may serve as spatial refuges. These refuges will be distributed in a temporally predictable fashion since adults of the same species must always exist and will be distributed in a spatially predictable fashion if the larvae are capable of locating the exact spatial position of the refuge.

Location of positions near adults by free-swimming larvae has been investi-

TABLE IV. Larvae which exhibit preferential settlement on algal substrates

Larval type	Substrate	Source(s)
Cnidaria: Hydroidea		
Coryne ochidai	*Cystoseira barata*	Nishihira, 1967
Sertularella miurensis	*Cystoseira barata*	Nishihira, 1968a, b
Clava squamata	*Ascophyllum nodosum*	Williams, 1965b
Tubularia mesembryanthemum	*Zostera marina*	Nishihira, 1968c
Plumaria undulata	*Phyllospadix iwatensis*	Nishihira, 1968c
Clytia edwardsi	*Zostera marina*	Nishihira, 1968c
Clytia volubilis	*Sargassum fulvellum*	Kato et al., 1961
Obelia dichotoma	*Sargassum fulvellum*	Kato et al., 1961
Mollusca: Bivalvia		
Brachyodantes lineatus	*Cystoseira barata*	Kisseleva, 1966
Annelida: Polychaeta		
Spirorbis borealis	*Fucus serratus*	Gross and Knight-Jones, 1957; Knight-Jones, 1951, 1953b; de Silva, 1962; Wisely, 1960
Spirorbis corallinae	*Corallina ollcinalis*	de Silva, 1962
Spirorbis rupestris	*Lithophyllum polymorphum*	Gee, 1965; Gee and Knight-Jones, 1962; de Silva and Knight-Jones, 1962
Ectoprocta		
Alcyonidium hirsutum	*Fucus serratus*	Ryland, 1959
Alcyonidium polyoum	*Fucus serratus*	Ryland, 1959
Flustrellidra hispida	*Fucus serratus*	Ryland, 1959
Membranipora membranacea	*Laminaria digitata* *Laminaria hyperborea*	Ryland, 1970
Scrupocellaria reptans	*Laminaria digitata* *Laminaria saccharina*	Stebbing, 1972a; Ryland and Stebbing, 1971

gated in some detail (reviews by Meadows and Campbell, 1972; Newell, 1970; Scheltema, 1974). Two mechanisms appear to act generally: (a) direct chemical recognition of adults (see e.g. Bayne, 1969; Crisp, 1965, 1967; Crisp and Meadows, 1962, 1963; Wilson, 1968), (b) direct chemical recognition of particular substrates (e.g. algal substratum preferences) commonly inhabited by adults (see e.g. Crisp and Williams, 1960; Gee, 1965; Williams, 1964). Gregarious settling behaviour by larvae of marine sedentary invertebrates is, therefore, well established. Table III represents a non-exhaustive compilation of organisms in which this settling behaviour is known to occur. Table IV is a compilation of species known to aggregate through settlement on algal substrata. It is clear upon examination of these tables that both solitary and colonial animals exhibit similar behavioural responses which may function for location of spatial refuges provided in positions near adults of the same species.

(c) *Example 3*. A variety of sessile marine invertebrates are found to occur predominantly on one or a few species of alga which cover many rock surfaces in both intertidal and subtidal zones (Ryland, 1962b). These animals commonly exhibit a highly localized distribution on the algal fronds themselves, many being restricted to the midline regions of the fronds (Hayward, 1973; Hayward and Harvey, 1974a; Ryland, 1972, 1973; Wisely, 1960). Individuals inhabiting midlines are likely to experience higher fitness than those individuals occupying other positions on the frond due to:

(1) The increased rate of physical disturbance associated with outer margins of the frond.
(2) The increased rate of desiccation associated with outer margins of the frond.
(3) The periodic defoliation of many such plants. *Fucus serratus* in England, for example, undergoes winter defoliation. Plants lose foliage by a process of basal denudation and by shedding of distal branches bearing fruiting tips. Those epibionts occurring in the middle regions of the frond are most likely to survive winter defoliation of the plant (Hayward and Harvey, 1974a).

TABLE V. Larvae which exhibit preferential settlement on algal midlines

Annelida: Polychaeta	
Spirorbis borealis (Wisely, 1960)	*Spirorbis pagenstecheri* (Stebbing, 1972a)
Spirorbis corallinae (Stebbing, 1972a)	
Ectoprocta	
Alcyonidium hirsutum (Hayward, 1973)	*Celleporella hyalina* (Ryland, 1959)
Alcyonidium polyoum (Hayward, 1973; Ryland, 1959)	*Flustrellidra hispida* (Ryland, 1959)
	Membranipora membranacea (Ryland, 1970)

Positions along the midline of algal fronds, can therefore represent a spatial refuge for epiphytes. Such refuges are distributed in a temporally predictable manner due to the predictable occurrence of algal beds, and will be distributed in a spatially predictable manner if the larvae are capable of locating algal midlines.

Location of algal substrata (for review, see Sheltema, 1974) and, in particular, location of positions along midline regions of the frond by larvae of sedentary marine invertebrates, has been studied in some detail (Hayward, 1973; Hayward

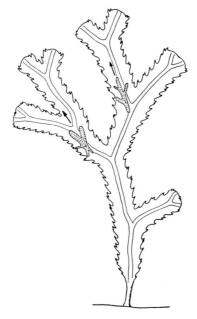

FIG. 2. Distally directed vine-like growth of *Electra pilosa* on the fronds of *Fucus serratus*. (Reproduced from Ryland and Stebbings, 1971.)

and Harvey, 1974a, b; Ryland, 1959; Wisely, 1960). The primary mechanism for algal substratum selection appears to be chemical (Crisp and Williams, 1960; Gee, 1965; Nishihira, 1968b; Williams, 1965a) and for midline selection to be physical (rugophilic behaviour) (Ryland, 1959, Wisely, 1960). Table V presents a list of species for which a preference for algal substratums has been documented, and should be compared with Table IV. It is clear from these tables that solitary and colonial animals employ similar behavioural responses for location of spatial positions along the midline of algal fronds which may serve as spatial refuges.

(d) *Example 4.* The cheilostomatous ectoproct, *Electra pilosa* (Fig. 2), is a widely distributed species which commonly inhabits *Fucus serratus* fronds on sheltered shores (Ryland and Stebbings, 1971). The composition and zonation of the epibiont community on *Fucus* has been described by Boaden et al. (1975). The community is characterized by 11 common space occupiers comprising four ectoprocts (*Membranipora, Flustrellidra, Electra, Alcyonidium*), two tunicates (*Polyclinum, Diademnum*), two sponges (*Grantia, Sycon*), two hydroids (*Dynamena, Gonothyrea*) and one serpulid (*Spirorbis*). The structure of this system appears similar to that described for *Fucus serratus* by Stebbing (1973). Organisms are distributed in a non-random fashion along the length of the frond, many being restricted to a particular section (see Example 3). Of particular interest here is the observation that one of the eleven common species, *Electra pilosa*, is commonly found on the most distal portions of the frond.

Interpretation of any single species zonation pattern on the frond is dependent upon knowledge of patterns of that species settlement and of mortality as a function of position on the frond. Selective settling behaviour of a number of these species for specific frond positions has been demonstrated (Hayward, 1973; Hayward and Harvey, 1974a; Stebbing, 1973; Ryland, 1959). In particular, Ryland and Stebbing (1971) have investigated the pattern of settlement and growth of *Electra pilosa* on *Fucus serratus* fronds. *Electra* ancestrulae were found to be non-randomly oriented on *Fucus* fronds. Of some 589 ancestrulae observed, only 43 were found to be oriented at right angles to the median longitudinal axis of the frond. In the majority (375/589), the median longitudinal axis of the ancestrulae and the frond were exactly parallel. In addition, the overwhelming majority of ancestrulae faced frond apices, not frond bases. *Electra* colonies exhibit a characteristic runner-like or oriented growth pattern (Ryland and Stebbing, 1971). The initial stages of astogeny have been illustrated and described in some detail (Marcus, 1926; Waters, 1924). The apically oriented ancestrula of *Electra* on *Fucus* fronds lead to apically oriented runner-like colonies.

Mortality processes affecting epibionts have also been investigated in detail as a function of frond position (Hayward, 1973; Hayward and Harvey, 1974b; Hayward and Ryland, 1975; Stebbing, 1972b, 1973; Ryland and Stebbing, 1971). The probability of mortality due to direct interspecific interference competition on *Fucus* fronds is higher on basal portions of the fronds than on distal portions (Stebbing, 1972b, 1973; Ryland and Stebbing, 1971). This is largely due to the increased density of competitors on the basal portions of the fronds (Boaden et al., 1975; Stebbing, 1973; Ryland and Stebbing, 1971). Stebbing (1972b, 1973) has investigated mortality from interspecific interference competition among seven common *Fucus* epibionts. In all cases, *Electra pilosa* was consistently overgrown by competing colonial species.

Spatial positions toward the apices of *Fucus* fronds therefore represent a refuge from spatial competition for *Electra pilosa* colonies (Stebbing, 1973; Ryland and Stebbing, 1971). For reasons discussed earlier, this refuge will be temporally predictable. Unlike previous examples, though, the exact spatial position of the refuge will vary in time (as the *Fucus* frond grows) and so will be spatially unpredictable. *Electra pilosa* colonies employ a morphological response, directional growth, to locate spatial refuges provided along the apices of *Fucus* fronds.

(e) *Example 5*. A number of sessile marine species may exhibit both runner-like and sheet-like growth morphologies (see Jackson, this volume). Although the genetic basis of such growth-form plasticity is largely uninvestigated, a number of studies have demonstrated that specific growth morphologies can be induced as a function of nutritional regime (Table VI).

TABLE VI. Species which adopt directional growth patterns as a function of nutritional regime

Ectoprocta	
Forella repens	Jebram, 1970
Electra monostachys	Jebram, 1973
Electra pilosa	Jebram, 1973
Bowerbankia gracilis	Jebram, 1973
Bowerbankia imbricata	Jebram, 1973
Conopeum tenuissimum	Winston, 1976
Cnidaria: Hydroidea	
Campanularia flexuosa	Crowell, 1957

Winston (1976), for example, has investigated the influence of nutrition on colony morphology in the ectoproct, *Conopeum tenuissimum*. Under laboratory conditions, colonies which had been fed on algal cells of varying nutritional value developed varying colony morphologies (Fig. 3). Colonies fed on a good diet (Fig. 3A) developed buds not only along the major growth axis but also distally and laterally to fill the spaces between the branches. As food quality was decreased (Figs 3C, D), growth became restricted to the major growth axis, producing vine-like patterns in growth.

Winston (1976) suggests that development of growth in all directions (vine-like growth) is an adaptation of the colony for location of a more nutritionally favourable microenvironment. Under natural conditions, if colonies are not able to grow, store energy and reproduce within approximately four weeks, colonies will be overgrown by other fouling organisms. Thus an individual which settles in a nutritionally unfavourable microhabitat is likely to have a lower fitness than other individuals occupying nutritionally more favourable positions. Nutritionally favourable positions, then, may serve as spatial refuges for *Conopeum*. The runner-like growth pattern by maximizing the distance

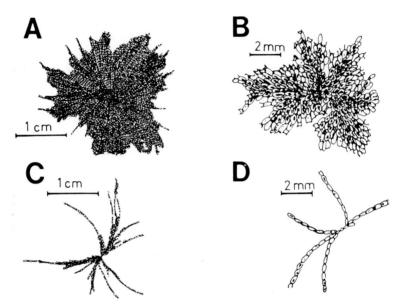

FIG. 3. Examples of variation in growth form among *Conopeum tenuissimum* colonies raised in nutritionally dissimilar microhabitats. (**A**) Example of a colony raised on a good food, the chlorophyte flagellate, *Dunaliella tertiolecta*, after 40 days of growth. (**B**) Morphology of a colony grown under natural conditions, after 20 days of growth. (**C**) Example of a colony raised on a fair food, the chrysophyte flagellate, *Monochrysis lutheri*, after 42 days of growth. (**D**) Example of a colony raised on a poor food, the diatom, *Cyclotella nana*, after 41 days of growth. Note that as the nutritional value of food available is decreased colony morphologies become increasingly runner-like in character. (Reproduced from Winston, 1976.)

reached from the point of settlement per unit energy expended, may represent a morphological response for location of refuges provided in regions with access to necessary resources.

(*f*) *Example 6.* The cyclostomatous ectoproct, *Stomatopora* sp., occurs commonly on the undersurfaces of foliaceous corals, e.g. *Agaricia* and *Montastrea* spp., in Jamaica, West Indies. This species is found almost exclusively occupying the secondary free space (Dayton, 1971) provided by serpulid tubes (e.g. *Pseudovermilus* sp.) and bivalve shells (e.g. *Chama* sp). This distributional pattern is reminiscent of those resulting from specific settling preferences discussed earlier (Examples 2, 3). Analysis of experimental settling panels, though, reveals a fundamentally different origin for this distribution pattern.

TABLE VII. Data from experiment panels on the total percent cover of all organisms, the percent cover of *Stomatopora* on primary free space and on secondary free space over time. Note that as total percent cover increases, *Stomatopora* is found to shift its distribution from occupation largely of primary space to largely of secondary space. Percentages do not total 100, since occupation of serpulid tubes was not considered.

Time (months)	Total cover (%)	Cover secondary space (%)	Cover primary space (%)
7	15	2	98
14	51	31	59
26	96	68	7

In 1975, Jackson (1977) completed an experimental study of recruitment and development of encrusting communities on Transite (asbestos-cement) panels placed in artificial cryptic environments at 40 m depth on a Discovery Bay (Jamaica, West Indies) reef. Replicate panels were submerged for 7, 14 and 26 month periods, collected, preserved, and observed under a stereomicroscope. The distribution of *Stomatopora* on these panels, as a function of time, results from the interacting influences of: (a) patterns of *Stomatopora* recruitment, growth and mortality; (b) patterns in bivalve recruitment; (c) patterns in recruitment and growth of a variety of species capable of overgrowing and killing *Stomatopora* colonies. The sequence of events leading to the observed *Stomatopora* distribution is outlined in Table VII and presented schematically in Fig. 4.

After seven months, *Stomatopora* had successfully recruited at high densities. The species grows in a characteristically runner-like fashion; colonies being linear at first, stellate later (Fig. 4a). At the same time, a number of other ectoprocts, sponges and ascidians capable of overgrowing *Stomatopora* had recruited and overgrown up to 15% of the available free space. Where these colonies contact *Stomatopora*, overgrowth occurs. Bivalves (primarily *Chama* sp.) had only begun to recruit at seven months and were represented by very low population densities (Fig. 4a). These interacting patterns result in *Stomato-*

pora being distributed primarily on primary free space at seven months (Table VII).

After 14 months, *Stomatopora* had criss-crossed the substratum with threads

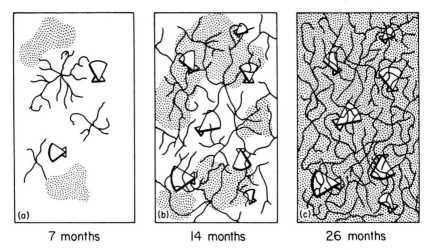

FIG. 4. Pattern in the distribution of *Stomatopora* (lines), bivalves (shells), competitors superior to *Stomatopora* but not to bivalves (shading) over time. Space occupancy of bivalves and *Stomatopora* is grossly exaggerated for purposes of illustration. (a) After 7 months; (b) after 14 months; (c) after 26 months.

of zooids. By this time, many bivalves (primarily *Chama*) had recruited. *Stomatopora* zooids, when contacting bivalve shells, simply overgrew them. Such overgrowth does not appear to affect bivalves adversely. Those species capable of outcompeting *Stomatopora* have, by one year, covered up to 50% of the available free space, overgrowing *Stomatopora* whenever it is encountered. Bivalve species also experience overgrowth by these same species, but at a much slower rate (Fig. 4b). At 14 months, then, *Stomatopora* is found to be distributed on both primary and secondary free space (Table VII).

By 26 months, *Stomatopora* is found to occur almost exclusively on the secondary free space provided by bivalve shells (Table VII). Those species capable of defeating *Stomatopora* in spatial competition occupy nearly 100% of the available free space, leaving *Stomatopora* colonies restricted in distribution to the surfaces of bivalve shells (Fig. 4c). Since bivalves are overgrown more slowly, those *Stomatopora* occupying bivalve shells will experience a higher survivorship than those occupying primary free space.

Bivalve shells therefore represent a refuge from spatial competition for *Stomatopora*, in that colonies occupying bivalves have higher survivorship and, presumably, higher fitness, than colonies occupying other spatial positions. This refuge is temporally predictable in that bivalves occur predictably on all panels (or coral plates) observed (Jackson, unpublished data). The refuge, though, is spatially unpredictable in that bivalves have a lower recruitment rate than *Stomatopora*. The exact spatial position of the refuge is therefore temporally unpredictable to the recruiting *Stomatopora* larvae. The directional growth pattern of *Stomatopora* may represent a morphological response for the location of refuges which are relatively free from interspecific interference competition.

(g) *Example 7*. The experimental panels described earlier exhibit another example of an organism which appears to locate and exploit spatial refuges. This example is slightly more complicated than those preceding and so requires a brief introduction to the nature of competitive interactions in cryptic coral reef environments.

Competition for living space is intense in cryptic coral reef environments (Jackson and Buss, 1975). Although interference competitive mechanisms for living space are only beginning to be understood in any detail (Connell, 1961; Jackson and Buss, 1975; Stebbing 1972b), competitive dominance is commonly represented by an organism's ability to overgrow a potential competitor. Rankings of the interference competitive (overgrowth) ability of several co-existing species may result in a simple hierarchy (Species A > Species B, Species A > Species C, Species B > Species C) or in a competitive network (Species A > Species B, Species B > Species C, but Species C > Species A) (Jackson and Buss, 1975; Gilpin, 1975). Competitive hierarchy has been described in intertidal environments among barnacle species (Connell, 1961) and in open coral reef environments among coral species (Lang, 1973). Cryptic coral reef communities, though, are characterized by competitive networks (Buss, 1976; Buss and Jackson, 1979).

The existence of competitive networks in cryptic coral reef environments is of particular interest to the discussion of spatial refuges. Where competitive networks, as opposed to competitive hierarchies, occur, the exact spatial position an organism occupies relative to potentially competing neighbours may determine the resultant pattern of distribution on the substratum (Buss and Jackson, 1979). This situation is presented schematically in Fig. 5. Consider three substrates, occupied by three species with approximately equivalent growth rates, positioned relative to one another as illustrated. In the hierarchy case (Fig. 5a), over time, the competitive dominant, Species A, will monopolize all the available substratum, independent of the initial positioning of the three species. In the networks case (Fig. 5b), this does not occur. Given identical positioning of the three species as considered in the hierarchy case, the dominant, over time, will depend upon the initial positioning of the three species (Fig. 5b). Thus in environments where competitive networks occur, the exact

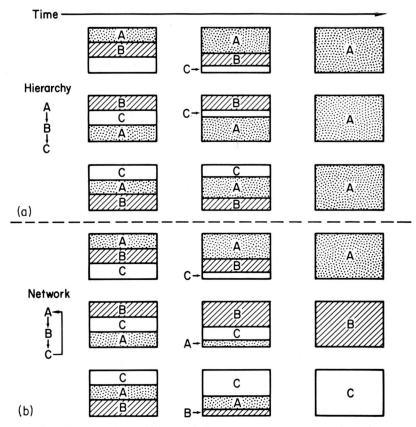

FIG. 5. The influence of spatial position relative to competitors on substratum monopolization; hierarchies (a), compared with networks (b). Letters represent species, arrows point away from the dominant species, squares represent substrata. Given conditions of near equal growth rates, the dominant species over time is independent of spatial position in the hierarchy case (a), but strongly dependent upon spatial positioning in the networks case (b).

spatial position occupied by an organism relative to its neighbours may be an important determinant of species distribution patterns. This result will hold even under conditions of more complex competitive networks and unequal growth rates (Karlson and Jackson unpublished).

An undescribed sponge species, which occurs on the experimental settling panels described earlier, exhibits a distribution pattern which appears to be related to location and subsequent exploitation of these optimal spatial positionings in a competitive network. The distribution of this sponge on the settling

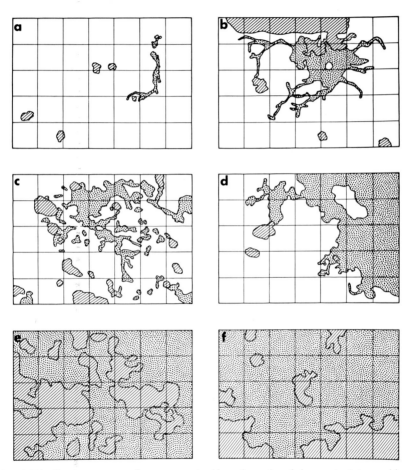

FIG. 6. Distribution pattern of sponge species (dotted areas) and the species it is capable of overgrowing (lined areas) on experimental settling panels over time. (a) 6 months. (b) 6 months. (c) 1 year. (d) 1 year. (e) 2 years. (f) 2 years. Panels are not temporal antecedents of one another, but represent recurrent patterns on separate panels.

panels over time is represented in Fig. 6. Note that these panels are not temporal antecedents of one another, since destructive sampling was required in this experiment. Each illustration is, then, a tracing from photographs of panels submerged for different time periods. At six months, colonies recruit and begin to grow in a runner-like fashion (Fig. 6a, b). By one year, various arms of the sponge have fused forming sheets, although the sponge retains its runner-like character at the colony margin (Fig. 6c, d). By two years, this sponge has formed extensive sheets occupying from 50–100% of the available space on the panel (Fig. 6e, f).

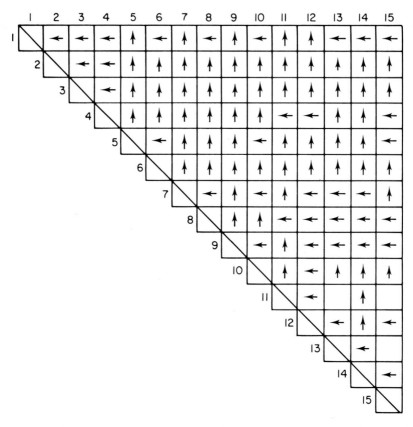

FIG. 7. Data from microscopic examination of competitive interactions (overgrowth phenomenon). Arrows point toward the dominant of a two-species interaction. All organisms are sponges. Numbers refer to different species. For details see Buss (1976).

Since, as described in the previous example, competition on the panels has become intense after one year, one would expect that the near-dominance achieved by this sponge could only be explained by virtue of its being a superior competitor for space. To investigate this hypothesis, rankings of the interference competitive abilities of the nineteen important competitors frequenting these panels were made (for methods, see Buss and Jackson, 1979). Two facts of particular interest emerged from this study: (a) Competitive networks exist on these panels (Fig. 7); (b) The sponge species under consideration was not a superior interference competitor for space. This ranking is illustrated in Table

TABLE VIII. Data on the number of species overgrown by the sponge species in question and by those species which this sponge is capable of overgrowing

Species	No. species-superior competitors	No. species-inferior competitors
S-4	3	11
S-11	9	3
S-12	7	7
S-15	5	7

VIII. The sponge species (S-4) was capable of overgrowing only three of the fifteen species. Examination, though, of Table VIII, shows that the three species it is capable of overgrowing can, in turn, overgrow all but one of the remaining species.

How, then, does this sponge achieve near-dominance on experimental panels? The conclusion appears inescapable that it locates a spatial position near species it can overgrow. These species in turn, overgrow those species which the sponge is unable to overgrow. In this manner, dominance may be achieved.

Clearly, if a competitive network exists on the substratum in question, organisms occupying spatial positions next to species they are capable of overgrowing will experience higher survivorship and thus higher fitness. If competitive networks always exist on substrata frequented by this species, this refuge will be temporally predictable. The refuge, though, will be spatially unpredictable because the exact spatial position will be determined in part by stochastic recruitment effects and will change in space through time due to growth of competing species. The directional growth pattern of the sponge species considered here appears to be a morphological response for location of the spatial refuge provided in an environmental context of competitive networks.

3. Summary

A number of consistent patterns emerge from the above examples. These are: (a) In all cases the refuge occurred predictably in time. Cryptic environments always exist, as do positions near adults, algal midlines, apices of algal fronds, bivalve shells, etc. This is not a particularly surprising result. Any organism which evolved an adaptive response for location of a refuge which occurred

TABLE IX. Review of behavioural responses for the location and exploitation of spatial refuges

Refuge	Mortality source	Behavioural response	Solitary groups	Colonial groups
Cryptic environments	Predation, competition, physical disturbance	Habitat selection	Hydroidea Bivalvia Cirripedia Polychaeta Ascidea	Demospongiae Hydroidea Anthozoa Ectoprocta Ascidea
Position near adults	Competition, predation, physical disturbance	Habitat selection	Hydroidea Bivalvia Polychaeta Cirripedia Ascidea	Demospongiae Hydroidea Anthozoa Ectoprocta Ascidea
Algal midlines	Physical disturbance, habitat alteration	Habitat selection	Polychaeta	Ectoprocta

unpredictably in time, would undergo sudden reductions in population sizes when the refuge was unavailable. Such species would be expected to suffer high extinction probabilities and thus rarely be observed.

(b) The exact spatial position of the refuge relative to the sensory capacities of the recruiting individual is found to be distributed in two ways. In Examples 1, 2, 3, the exact spatial position occurred predictably in space and was invariable over time. However, in Examples 4, 5, 6, 7, the exact spatial position was variable in space through time.

(c) In cases where the refuge was distributed in a spatially predictable fashion (Examples 1, 2, 3), organisms were always found to respond with behavioural mechanisms for refuge location. In each case the response took the same form, that of active "habitat selection". (For review see Table IX.)

(d) In cases where refuges were distributed in spatially unpredictable fashions (Examples 4, 5, 6, 7), organisms were always found to respond with morphological mechanisms for refuge location. In each case the response took the form of runner-like growth patterns. (For review see Table X.)

(e) In cases where the refuge was distributed in a spatially predictable fashion

(Examples 1, 2, 3), both solitary and colonial organisms responded with identical behavioural (habitat selection) mechanisms (Table XI). In cases where the refuge was distributed in a spatially unpredictable fashion (Examples 4, 5, 6, 7),

TABLE X. Review of morphological responses for location and exploitation of spatial refuges

Refuge	Mortality source	Morphological response	Solitary groups	Colonial groups
Apices of algal fronds	Interference, competition	Directional growth form	None	Cheilostomous ectoproct
Favourable micro-habitat for feeding	Physical disturbance, interference, competition	Directional growth form	None	Cheilostomous ectoproct
Bivalve shells	Interference, competition	Directional growth form	None	Cyclostomous ectoproct
Position in competitive network	Interference, competition	Directional growth form	None	Demospongiae

only colonial organisms were found to be capable of locating such refuges. The colonial organisms responded with identical (runner-like growth) responses (Table X). Solitary organisms, apparently have not developed mechanisms for location of refuges distributed in this manner and must rely on higher fecundity (i.e. greater larval production) to locate these refuges at a high rate.

TABLE XI. Review of responses of both solitary and colonial organisms to the adaptive problem of location and exploitation of spatial refuges

Refuge distribution	Adaptive response	
	Solitary	Colonial
Spatially predictable	Behavioural response: habitat selection, examples 1, 2, 3	Behavioural response: habitat selection examples, 1, 2, 3
Spatially unpredictable	Physiological response: increased fecundity	Morphological response: runner-like growth forms, examples 4, 5, 6, 7

(f) When refuges are distributed in a spatially unpredictable fashion, morphological strategies for their location may be exceedingly complex. Such strategies, by definition, require analysis of growth processes and will therefore commonly require a time-series or experimental approach in analysis.

DISCUSSION

Differential suitabilities of spatial positions on marine hard substrata give rise to organism responses for location of positions where fitness is high. Summarized in Table XI are the adaptive responses of both solitary and colonial organisms to the problem of location of spatial refuges. Refuges were found to be distributed in two fundamentally different manners: (a) cases where the exact spatial position of the refuge is predictable to the recruiting individual (Examples 1, 2, 3); (b) cases where it is unpredictable (Examples 4, 5, 6, 7). Location of refuges requires responses in either settlement or growth processes. Growth responses will require a time interval for location of the refuge which will be longer than that required in the case of settlement (behavioural) responses. During this time interval the organism is exposed to a wide variety of potential mortality agents. As a result, whenever possible, one would expect there to be strong selection for behavioural responses (habitat selection) for the location of refuges. Clearly, only two criteria must be met for this response to occur: free-swimming larvae should exist and these should be capable of locating the exact spatial position of the refuge. Indeed, we find that both solitary and colonial organisms meet these criteria and react in this manner if refuges are so distributed. (See Tables IX and XI.)

In cases where the exact spatial position of the refuge is unpredictable to the recruiting larvae, adaptive responses for the location of refuges must take the form of growth responses. In all such examples considered (Examples 4, 5, 6, 7), organisms were found to respond to this problem with the same morphological mechanism, that of runner-like growth patterns (Table X). Such organisms settle and begin growing in a directional fashion. This growth form increases the distance from the initial point of settlement achieved in a given time interval and in so doing increases the probability of locating the refuge. Those portions of the organism occupying the refuge survive while those portions of the organism occupying unfavourable spatial positions suffer mortality. This mechanism will be successful as long as one criterion is met: the organism must be capable of suffering loss of some proportion of the selective unit. Such events are frequently fatal to most solitary organisms, but are often limited to portions of colonial animals and are commonly not fatal to the entire colony (Dayton et al., 1974; Glynn et al., 1972; Randall and Hartman, 1968; Reiswig, 1973; Ryland, 1970; Storr, 1976; Wyer and King, 1973). This criterion can only be met in colonial organisms. The location of "hiding places" in the form of refuges distributed in a spatially unpredictable fashion, therefore appears to be an adaptation restricted to colonial animals.

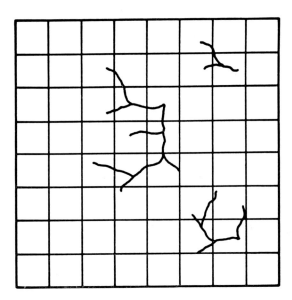

FIG. 8. Derivation of the monotonically increasing relationship between the probability of locating a refuge and the distance from the point of settlement. Consider the grid above to represent a marine hard substratum. The substratum is divided into quadrats the size of a refuge. Since settlement of an organism is random relative refuge position, the problem for the organism is to cross as many quadrat boundaries as possible. As the distance from the point of settlement increases, so does the cumulative probability of locating a refuge.

In the preceding discussion I have used the terms "runner-like" or "directional growth" as if this morphology was invariate. This, of course, is untrue. Runner-like growth-forms are represented in all major colonial groups and the degree of runner-like versus sheet-like behaviour (see Jackson, this volume) varies considerably both within and between groups. Starting from the premise that directional growth serves as an adaptation for location of spatial refuges, we may attempt to define directional growth in terms of a given form's degree of directionality and its life-history characteristics. Presented below is a graphical model designed as a preliminary investigation of these questions. The graphical techniques employed here are patterned after those of Janzen's (1970) work on the analogous problem of the actions of herbivores on recruitment processes in rainforest trees. This model generates a number of predictions, many of which may be partially verified from the literature.

For any colonial organism settling and growing on a marine hard substratum,

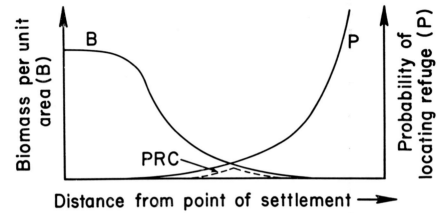

FIG. 9. A model showing the probability of colony survival at a given spatial position as a function of (1) colony morphology, (2) growth rate, (3) distance from point of settlement, and (4) location of spatial refuge. With increasing distance from the point of settlement, the biomass per unit area (B) declines. The nature of this curve will be dependent upon colony growth form. With increasing distance from the point of settlement, the probability (P) that a refuge exists increases. The product of the B and P curves yields a curve for the probability of a colony locating a refuge (PRC), with a peak at the distance from the point of settlement, that maximizes the likelihood of refuge location and, thus, fitness. The area under the PRC curve will represent an approximation of the population survivorship function. The curves in this and in other Figs (9–13) are not precise quantifications of empirical observations or theoretical considerations, but are intended to illustrate general relationships only.

the biomass per unit surface area will decrease with distance from the point of settlement. Although the biomass per unit area versus distance from point of settlement will always be a decreasing function, the nature of this curve will be strongly dependent upon the degree of directionality (sheet-like or runner-like) in growth exhibited by the organism (see Fig. 10). Similarly, for any organism which exploits are fuge distributed in a spatially unpredictable fashion, the probability of locating that refuge will increase with increasing distance from the point of settlement. If settlement of the organism is random relative to refuge distribution, as is the case in examples 5, 6 and 7, the relationship will be exponential. (See Fig. 8 for visualization of this relationship.) A simple graphical model summarizes these two relationships (Fig. 9).

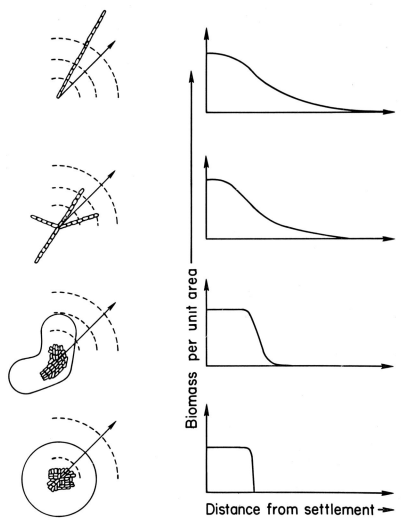

FIG. 10. The influence of colony growth form on the nature of the B curve (see Fig. 9). Schematic diagrams in the column on the left are colonies with differing degrees of directionality in growth. The concentric boundaries are provided to aid visualization of the relationship between the biomass per unit area and the distance from point of settlement (B curve). B curves corresponding to each colony are shown on the right. Note that, over some given time interval, the distance attained from the point of settlement is greatest in forms which are more runner-like.

Consider first the question of the degree of directionality in colony growth forms. Fig. 9 shows the relationship between biomass per unit area and distance from point of settlement for four commonly occurring growth forms which

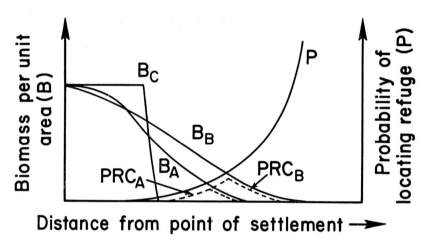

FIG. 11. The influence of colony growth form on the PRC curve. As colonies become more sheet-like (B_A to B_B to B_C), the area under the PRC curve drops drastically. Sheet-like forms would therefore be expected to have much lower population levels occupying refuges than runner-like forms.

form a progressive series from vine-like to sheet-like forms. Fig. 11 defines a curve for the probability of a colony locating a refuge (PRC) under a given condition of refuge distribution with varying growth morphologies. Note that the optimum degree of directionality for refuge location corresponds to the distance along the abscissa from the point of settlement to the point corresponding to the peak of the PRC curve.

Given this basic model we may now investigate the influences of varying life-history parameters on maximization of the area under the PRC curve. Any increase in the area under the PRC curve implies increased fitness and this will be selected for. Although a number of topics may be considered using this technique, I consider here only three: (a) competitive ability; (b) growth rate; (c) recruitment rate.

(a) *Competitive Ability*. Consider again Fig. 11. Two additional relationships are evident: (1) as colonies become more sheet-like (curves A to D), the area under PRC curves drops sharply; (2) the area under the PRC curves decreases as colonies become more sheet-like. This, then, means that as colony form becomes more sheet-like, the likelihood of locating a refuge will decrease. As a result, one would predict that sheet-like forms must be better adapted to confront mortality agents than runner-like forms.

I consider here only defences against mortality from interference competition (see Jackson, 1977 for arguments on the relative importance of competition-predation or physically induced mortality agents in subtidal marine environments dominated by colonial sessile invertebrates). Figure 11 predicts that sheet-like forms would be better spatial competitors than more runner-like forms, since runner-like forms are more likely to locate refuges from mortality. This prediction may be tested to some degree by examination of the literature. The relative competitive ability, as measured by overgrowth ability, of sheet-like and runner-like forms have been documented in at least four environments. The ectoproct *Electra pilosa* (See Example 4) commonly grows in a runner-like fashion when inhabiting the fronds of the intertidally occurring alga *Fucus serratus*. Stebbings (1973) has shown that *Electra* is consistently overgrown by all sheet-like competitors, whereas no single sheet-like competitor is overgrown by all others. Fuller (1946) reports similar findings in his observations of overgrowth phenomena between *Electra* and more sheet-like forms in fouling communities off the northeast coast of the United States. Dudley (1973, 1976) notes that the ectoproct *Conopeum tenuissimum* which may grow in a runner-like pattern (Example 5, Fig. 2) is an inferior spatial competitor relative to forms in epizooic communities inhabiting oyster shells which are more sheet-like. Buss (1976) has documented the interference competitive ability of some 25 species of the colonial organisms inhabiting the undersurfaces of foliaceous corals (e.g. *Montastrea* and *Agaracia* spp.) in Jamaica, West Indies. Table XII shows that sheet-like forms are superior spatial competitors to vine-like forms in this environment as well. The prediction, then, that sheet-like forms should be superior spatial competitors is supported with data from a number of natural situations.

(b) *Growth Rates.* Consider Fig. 12. For any given degree of directionality in growth (i.e. set nature of B curve), differential growth rates will, (B_A, B_B, B_C) over some time interval, t, result in different points of intersection of the B curve with the P curve. The area under the PRC curve thus increases with increasing growth rates. This relationship will hold even with changes in the nature of the B curve. More sheet-like forms, though, will require greater changes in growth rate to realize similar probabilities of touching refuges than in runner-like forms. As a result, runner-like forms should be expected to have higher growth rates than more sheet-like forms. This prediction may also be partially verified by literature reference. *Electra pilosa* has been found to have higher growth rates than that of coexisting sheet-like forms (Fuller, 1946; Ryland, personal communication). Various species of Jamaican cryptofauna

TABLE XII. Data on the average number of species overgrown (a), or the reverse event (b), or either result possible (c), for runner-like versus sheet-like growth forms from cryptic coral reef environments at Rio Bueno, Jamaica, West Indies. Species considered include sponges, ectoprocts, colonial foraminifera, coralline algae, ascidians, and a colonial ahermatypic coral. For details see Buss (1976).

Growth form	No. of species	Competitive ability		
		a	b	c
Runner-like forms	5	3	11	2
Sheet-like forms	13	10	4	2

exhibiting runner-like growth forms also have higher growth rates than coexisting sheet-like forms (Jackson, unpublished data).

(c) Recruitment Rates. On any given substratum, increased recruitment of an organism which exploits spatial refuges will have the effect of increasing the probability of locating a spatial refuge at a given distance from the point of settlement (i.e. change the position of P curves along the abscissa) (Fig. 13). This has the effects of increasing the area under the PRC curve and will do so to a greater extent in more runner-like, than sheet-like forms. As a result, one would predict high recruitment rates in organisms exhibiting runner-like growth. This prediction is borne out in a number of environments (see Dudley, 1976 for *Conopeum* on oyster shells; Jackson, 1977, for the encrusting fauna of Jamaican cryptic coral reefs; Ryland, personal communication for *Electra* on *Fucus* fronds).

In summary, then, directional growth may represent an adaptation for the location and subsequent exploitation of spatial refuges distributed in a spatially unpredictable fashion. The degree of directionality in growth found in any given species is found to be a function of two relationships: that between biomass per unit area and distance from point of settlement, and that between probability of locating a spatial refuge and distance from point of settlement. These functions will also vary with organism growth form, growth rate and recruitment role. Sheet-like growth forms are found to represent a morphological response for a confrontation strategy whereas runner-like growth represents a morphological

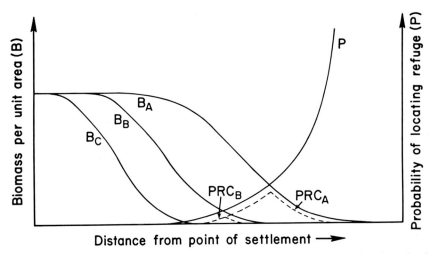

FIG. 12. The influence of colony growth rate on the PRC curve. Note that for a fixed colony growth form, as growth rates (B_C to B_B to B_A) increase, the area under the PRC curves increases.

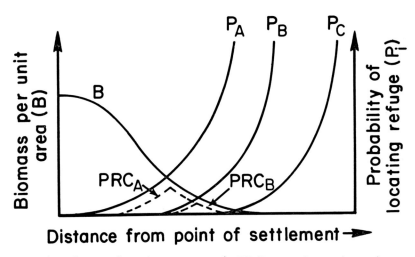

FIG. 13. The influence of recruitment rate on the PRC curve. As recruitment increases, the probability of locating a refuge at a given distance from the point of settlement increases (P_C to P_B to P_A). For a fixed colony growth form and rate (B), the area under the PRC curve increases with increasing recruitment rate.

response for the refuge exploitation strategy. These basic differences lead to the predictions that forms which are more runner-like should be characterized by low interference competitive ability, high growth rates and high recruitment rates relative to forms which are more sheet-like. These predictions are verified by existing data in the literature.

SUMMARY

Jackson's (1977) approach to definition of individual and colonial traits is employed. This encompasses behavioural and physiological adaptations in addition to the traditional morphological considerations. A single problem of adaptation is considered: that of the location of spatial refuges by sessile marine invertebrates inhabiting marine hard substrata.

Refuges are spatial positions on marine hard substrata where the fitness of an individual is high relative to other spatial positions. Such refuges most commonly result from the non-uniform action of mortality agents. A series of examples of organism adaptations for the location of refuges are presented and discussed. The following patterns are reflected in these examples:

(1) Refuges are distributed predictably in time. The exact position of the refuge relative to the sensory capacities of a recruiting individual may be predictably or unpredictably distributed.

(2) If the refuge is distributed predictably in space, organisms respond with behavioural adaptations (habitat selection) for refuge location.

(3) If the refuge is distributed unpredictably in space, organisms respond with morphological adaptations (runner-like growth) for refuge location.

(4) Solitary organisms are capable of exploiting only refuges distributed in a predictable fashion, whereas colonial organisms are capable of exploiting refuges distributed in either fashion. These results serve to further clarify the distinction between solitary and colonial forms as well as underscore the fact that apparently very different adaptations (habitat selection and runner-like growth morphologies) may represent differing solutions to a common problem of adaptation.

Using the premise that directional growth represents an adaptation for location of spatial refuges, a graphic model is introduced. This model attempts to define the relationship between a given form's degree of directionality and its life-history characteristics. Consideration of the interactions between curves generated for the biomass per unit surface area and cumulative probability of locating a refuge leads to a number of predictions. It is suggested that runner-like forms should exhibit higher recruitment rates and growth rates but lower interference competitive abilities than sheet-like forms. These predictions are consistent with existing data in the literature.

ACKNOWLEDGEMENTS

Jeremy B. C. Jackson has contributed fundamentally to the development of the ideas presented here. Without his support and encouragement, this work could never have been attempted. J. P. Gilman, B. D. Keller, I. Lubchenco and K. Rylaarsdam read earlier versions of the manuscript, providing many useful comments.

This work was supported by NSF Grant DES 72-01559-410, Sigma Xi, and a Department of Earth and Planetary Sciences Summer Field Grant.

REFERENCES

ABE, N. (1937). Post-larval development of the coral *Fungia actiniformis* var. *palawensis*. *Palaeo. trop. biol. Stn Stud.* **1**, 73–93.

ATODA, K. (1947). The larval and post-larval development of some reef-building corals. I. *Pocillopora damicornis cespitosa*. *Sci. Rep. Tohoku Univ.* 4 Ser. (Biol.) **18**, 24–47.

ATODA, K. (1951). The larval and post-larval development of some reef-building corals. III. *Acropora bruggemanni*. *J. Morph.* **89**, 1–13.

ATODA, K. (1951). The larval and post-larval development of some reef-building corals. IV. *Galaxea aspera*. *J. Morph.* **89**, 17–35.

BARNES, H. and CRISP, D. J. (1956). Evidence for self-fertilization in certain species of barnacles. *J. mar. biol. Ass., U.K.* **35**, 631–639.

BARNES, R. D. (1974). "Invertebrate Zoology" 3rd Edition. W. B. Saunders, Philadelphia.

BATHAM, E. G. (1946). *Pollicipes spinosus*. II. Embryonic and larval development. *Trans. R. Soc., N.Z.* **75**, 405–418.

BAYNE, B. L. (1964). The response of the larvae of *Mytilus edulis* to light and gravity. *Oikos* **15**, 162–174.

BAYNE, B. L. (1969). The gregarious behaviour of the larvae of *Ostrea edulis* at settlement. *J. mar. biol. Ass., U.K.* **49**, 327–356.

BERGQUIST, P. R., SINCLAIR, M. E. and HOGG, J. J. (1970). Adaptation to intertidal existence: reproductive cycles and larval behaviour in Demospongiae. In "Biology of the Porifera" (W. G. Fry, ed.), pp. 247–271. Symp. zool. soc., London. Academic Press, London.

BERRILL, N. J. (1947). The development and growth of *Ciona*. *J. mar. biol. Ass., U.K.* **26**, 616–625.

BOADEN, P. J. S., O'CONNOR, R. J. and SEED, R. (1975). The composition and zonation of a *Fucus serratus* community in Stratigford Lough, Co. Down. *J. exp. mar. Biol. Ecol.* **17**, 116–136.

BOARDMAN, R. S., CHEETHAM, A. H. and OLIVER, W. A. eds. (1973a). "Animal Colonies: Development and Function Through Time" Dowden, Hutchinson, and Ross, Stroudsburg, Pennsylvania.

BOARDMAN, R. S., CHEETHAM, A. H., OLIVER, W. A., COATES, A. G. and BAYER, F. M. (1973b). Introducing coloniality. In "Animal Colonies: Development and Function Through Time" (R. S. Boardman, A. H. Cheetham, and W. A. Oliver, eds.), pp. v–ix. Dowden, Hutchinson, and Ross, Stroudsburg, Pennsylvania.

BOUSFIELD, E. L. (1955). Ecological control of the occurrence of barnacles in the Miramichi Estuary. *Bull. natn. Mus. Can.* **137**, 1–66.

BUSS, L. W. (1976). Better living through chemistry: the relationship between competitive networks and allelochemical interactions. *In* "Aspects of Sponge Biology" (F. W. Harrison and R. R. Cowden, eds), pp. 315-327. Academic Press, New York.

Buss, L. W. and Jackson, J. B. C. (1979). Habitat structure and community development on marine hard substrata. *Proc. natn. Acad. Sci., USA.* (In Press.)

Buss, L. W. and Jackson, J. B. C. (1979). Competitive networks: non-transitive competitive relationships in cryptic coral reef environments. *Am. Nat.* (In Press.)

Cole, H. A. and Knight-Jones, E. W. (1949). The settling behaviour of larvae of the European flat oyster, *Ostrea edulis*, and its influence on methods of cultivation and spat collection. *Fishery Invest. Lond.* (Ser. 2), **17**, 1-39.

Connell, J. H. (1959). Studies of some factors affecting recruitment and mortality of natural populations of intertidal barnacles. *In* "Marine Boring and Fouling Organisms" (D. L. Ray, ed.), pp. 226-233. Univ. of Washington Press, Seattle.

Connell, J. H. (1961). The effects of competition, predation by *Thais lapillus* and other factors on natural populations of the barnacle *Balanus balanoides*. *Ecol. Monogr.* **31**, 61-106.

Connell, J. H. (1975). Some mechanisms producing structure in natural communities. *In* "Ecology and Evolution of Communities" (M. L. Cody and J. M. Diamond, eds), pp. 460-490. Belknap Press, Cambridge, Mass.

Crisp, D. J. (1965). Surface chemistry, a factor in settlement of marine invertebrate larvae. *Proc. 5th Mar. Biol. Symp. Botanica Gothoburgensis* **3**, 51-65.

Crisp, D. J. (1967). Chemical factors inducing settlement in *Crassostrea virginica*. *J. anim. Ecol.* **36**, 329-335.

Crisp, D. J. and Ghobashy, A. F. A. A. (1971). Responses of the larvae of *Diplosoma listerianum* to light and gravity. *In* 4th Europ. Mar. Biol. Symp. (D. J. Crisp, ed.), pp. 443-465. Cambridge Univ. Press.

Crisp, D. J. and Meadows, P. S. (1962). The chemical basis of gregariousness in cirripedes. *Proc. R. Soc.* B **156**, 500-520.

Crisp, D. J. and Meadows, P. S. (1963). Absorbed layers: the stimulus to settlement in barnacles. *Proc. Roy. Soc.* B **158**, 364-387.

Crisp, D. J. and Williams, G. B. (1960). Effects of extracts from fucoids in promoting settlement in epiphytic Polyzoa. *Nature, Lond.* **188**, 1206-1207.

Crowell, S. (1957). Differential responses of growth zones to nutritive level, age, and temperature in the colonial hydroid, *Campanularia*. *Biol. Bull.* **131**, 213-252.

Daniel, A. (1955). Gregarious attraction as a factor influencing the settlement of barnacle cyprids. *J. Madras. Univ.* B **25**, 97-107.

Dayton, P. K. (1971). Competition, disturbance, and community organization: the provision and subsequent utilization of space in a rocky intertidal community. *Ecol. Monogr.* **41**, 351-389.

Dayton, P. K. (1973). Dispersion, dispersal, and persistence of the annual intertidal alga, *Postelsia palmaeformis*. *Ecology* **54**, 433-438.

Dayton, P. K. (1974). Two cases of resource partitioning in an intertidal community: making the right prediction for the wrong reason. *Am. Natur.* **107**, 662-670.

Dayton, P. K., Rolbillard, G. A., Paine, R. T. and Dayton, L. B. (1974). Biological accommodation in the benthic community at McMurdo Sound, Antarctica. *Ecol. Monogr.* **44**, 105-128.

Dudley, J. E. (1973). Observations on the reproduction, early larval development, and colony astogeny of *Conopeum tenuissimum*. *Chesapeake Sci.* **14**, 270-278.

DUDLEY, J. E. (1976). See J. E. Winston (1976).
DYBERN, B. I. (1962). Biotope choice in *Ciona intestinalis*. The influence of light. *Zool. Bidr. Upps.* **35**, 589–601.
EDMONDSON, C. H. (1929). Growth of Hawaiian corals. *Bull. Bernice P. Bishop Mus.*, **58**, 1–38.
EDMONDSON, C. H. (1946). Behavior of coral planulae under altered saline and thermal conditions. *Occ. Pap. Bernice P. Bishop Mus.* **18**, 283–304.
EWALD, W. F. (1912). On artificial modification of light reactions and the influence of electrolytes on phototaxis. *J. exp. Zool.* **13**, 591–612.
FRY, W. G. (1971). The biology of the larvae of *Ophlitaspongia seriata* from two North Wales populations. In 4th Europ. Mar. Biol. Symp. (D. J. Crisp, ed.), pp. 155–178. Cambridge Univ. Press.
FULLER, J. L. (1946). Season of attachment and growth of sedentary marine organisms at Lamoine, Maine. *Ecology* **27**, 150–157.
GEE, J. M. (1965). Chemical stimulation of settlement in larvae of *Spirorbis rupestris*. *Anim. Behav.* **181**, 1–86.
GEE, J. M. and KNIGHT-JONES, E. W. (1962). The morphology and larval behaviour of a new species of *Spirorbis*. *J. mar. biol. Ass., U.K.* **42**, 641–654.
GILBERT, L. E. and RAVEN, PH. H. (eds) (1975). "Coevolution of Plants and Animals." Proc. Symp. No. 5, Inter. Cong. of System. and Evol. Biol. Univ. of Texas Press, Austin.
GILPIN, M. E. (1975). Limit cycles in competition communities. *Am. Natur.* **109**, 51–60.
GLYNN, P. W., STEWART, R. H. and McCOSKER, J. E. (1972). Pacific coral reefs of Panama: structure, distribution and predators. *Geol. Rundschau* **61**, 483–519.
GOODBODY, I. (1962). The biology of *Ascidea nigra*. II. The development and survival of young ascidians. *Biol. Bull.* **124**, 31–44.
GOSNER, K. L. (1971). "Guide to Identification of Marine and Estuarine Invertebrates." Wiley-Interscience. John Wiley and Sons, New York.
GRAVE, C. (1920). *Amaroucium pellucidum* form *constellatum*. I. The activities and reactions of the tadpole larva. *J. exp. Zool.* **30**, 241–257.
GRAVE, C. (1935). Metamorphosis of ascidian larvae. *Pap. Tortugas Lab.* **29**, 209–292.
GRAVE, C. (1941). The eye shot and light responses of the larvae of *Cynthia partita*. *Biol. Bull.* **81**, 287.
GRAVE, C. and WOODBRIDGE, H. (1924). *Botryllus schlosseri*. The behavior and morphology of the free-swimming larva. *J. Morph.* **29**, 207–247.
GRAVE, B. H. (1930). The natural history of *Bugula flabellata* at Woods Hole, Mass., including the behavior and attachment of the larvae. *J. Morph.* **49**, 355–383.
GROSS, J. and KNIGHT-JONES, E. W. (1957). The settlement of *Spirorbis borealis* on algae. *Proc. Challenger. Soc.* **3** (9), 18
HARTMAN, W. D. and REISWIG, H. M. (1973). The individuality of sponges. In "Animal Colonies: Development and Function through Time" (R. S. Boardman, A. H. Cheetham, W. A. Oliver, Jr., eds), pp. 567–584. Dowden, Hutchinson and Ross, Stroudsburg, Pennsylvania.
HASPER, M. (1913). On a method of rearing larvae of Polyzoa. *J. mar. biol. Ass., U.K.* **9**, 435–436.
HAYWARD, P. J. (1973). Preliminary observations on settlement and growth in populations of *Alcyonidium hirsutum*. In "Living and Fossil Bryozoa" (G. P. Larwood, ed.), pp. 107–113. Academic Press, London.

HAYWARD, P. J. and HARVEY, P. H. (1974a). The distribution of settled larvae of the bryozoans *Alcyonidium hirsutum* and *A. polyoum* on *Fucus serratus*. *J. mar. biol. Ass., U.K.* **54**, 677–684.

HAYWARD, P. J. and HARVEY, P. H. (1974b). Growth and mortality of the bryozoan *Alcyonidium hirsutum* on *Fucus serratus*. *J. mar. biol. Ass., U.K.* **54**, 677–684.

HAYWARD, P. J. and RYLAND, J. S. (1975). Growth, reproduction, and larval dispersal in *Alcyonidium hirsutum* and some other Bryozoa. *Proc. VII Europ. Mar. Biol. Symp. Pubbl. Staz. zool. Napoli* **39** (Suppl.), 226–241.

JACKSON, J. B. C. (1977). Competition on marine, hard substrates: the adaptive significance of solitary and colonial strategies. *Am. Natur.* **110**, 743–767.

JACKSON, J. B. C. and BUSS, L. W. (1975). Allelopathy and spatial competition among coral reef invertebrates. *Proc. natn. Acad. Sci., USA.* **72** (12), 5160–5163.

JANZEN, D. H. (1970). Herbivores and the number of tree species in the tropics. *Am. Natur.* **104**, 501–528.

JEBRAM, D. (1970). Preliminary experiments with Bryozoa in a simple apparatus for producing continuous water currents. *Helgolander wiss. Meeresunters.* **20**, 278–292.

JEBRAM, D. (1973). Preliminary observations on the influences of food and other factors on the growth of Bryozoa with a description of a new apparatus for the cultivation of sessile plankton feeders. *Kieler Meeresforsch.* **29**, 50–57.

KATO, M., NAKAMURA, K., HIRAI, E. and KAKINUMA, Y. (1961). The distribution pattern of Hydrozoa on seaweed with some notes on the so-called coaction among hydrozoan species. *Bull. biol. Stn Asamushi*, **10**, 195–202.

KAUFMAN, K. W. (1973). The effect of colony morphology on life-history parameters of colonial animals. *In* "Animal Colonies: Development and Function Through Time." (R. S. Boardman, A. H. Cheetham, and W. A. Oliver, eds), pp. 221–222. Dowden Hutchinson, and Ross, Stroudsburg, Pennsylvania.

KISSELEVA, G. A. (1966). Factors stimulating larval metamorphosis of the lamellibranch, *Brachyodontes lineatus*. *Zool. Zh.* **45**, 1571–1573.

KNIGHT-JONES, E. W. (1951). Gregariousness and some aspects of the settling behaviour of *Spirorbis*. *J. mar. biol. Ass., U.K.* **30**, 201–222.

KNIGHT-JONES, E. W. (1953a). Laboratory experiments on gregariousness during settlement in *Balanus balanoides* and other barnacles. *J. exp. Biol.* **30**, 584–598.

KNIGHT-JONES, E. W. (1953b). Decreased discrimination during settlement after prolonged planktonic life in larvae of *Spirorbis borealis*. *J. mar. biol. Ass., U.K.* **32**, 337–345.

KNIGHT-JONES, E. W. and MOYSE, J. (1961). Intraspecific competition in sedentary marine animals. *Symp. Soc. exp. Biol.* **15**, 72–95.

KNIGHT-JONES, E. W. and STEVENSON, J. P. (1950). Gregariousness during settlement in the barnacle *Elminius modestus*. *J. mar. biol. Ass., U.K.* **29**, 281–287.

LANG, J. C. (1973). Interspecific aggression by scleractinian corals. 2. Why the race is not only to the swift. *Bull. mar. Sci.* **23**, 260–279.

LYNCH, W. F. (1947). The behavior and metamorphosis of the larvae of *Bugula neritina*: experimental modification of the length of the free-swimming period and the responses of the larvae to light and gravity. *Biol. Bull.* **92**, 115–150.

LYNCH, W. F. (1949). Modification of the responses of two species of *Bugula* larvae from Woods Hole to light and gravity: ecological aspects of the behavior of *Bugula* larvae. *Biol. Bull.* **97**, 302–310.

LYNCH, W. F. (1955). Synergism and anatgonism in the induction of metamorphosis of *Bugula* larvae by neutral red dye. *Biol. Bull.* **109**, 82–98.
LYNCH, W. F. (1956). Experimental modification of the rate of metamorphosis of *Bugula* larvae. *J. exp. Zool.* **133**, 589–612.
LYNCH, W. F. (1960). Problems of the mechanisms involved in the metamorphosis of *Bugula* and *Amaroecium* larvae. *Proc. Iowa Acad. Sci.* **67**, 552–531.
MACKIE, G. O. (1963). Siphonophores, bud colonies, and superorganisms. *In* "The Lower Metazoa: Comparative Biology and Physiology" (E. C. Dougherty, ed.), pp. 329–337. University of California Press, Berkley, California.
MANO, R. (1960). On the metamorphosis of a brachiopod, *Frenulina sanquinolenta*. *Bull. biol. Stn. Asamushi* **10**, 171–175.
MARCUS, E. (1926). Beobachtungen und Versuche und lebendem Meeres-bryozoen. *Zool. Jb. (Syst.)* **52**, 1–102.
MAST, S. O. (1921) Reaction to light in the larvae of the ascidians, *Amaroucium constellatum* and *Amaroucium pellucidum* with special reference to photic orientation. *J. exp. Zool.* **34**, 149–187.
MCDOUGALL, K. D. (1943). Sessile marine invertebrates at Beaufort, North Carolina. *Ecol. Monogr.* **13**, 321-374.
MEADOWS, P. S. and CAMPBELL, J. I. (1972). Habitat selection by aquatic invertebrates. *Adv. mar. Biol.* **10**, 271-382.
MILLAR, R. H. (1953). *Ciona*. *L.M.B.C. Mem.* **35**, 1–123.
NEWELL, R. C. (1970). "Biology of Intertidal Animals." Logos Press, London.
NISHIHIRA, M. (1967). Observations on the selection of algal substrata by hydrozoan larvae, *Sertularella miurensis* in nature. *Bull. biol. Stn Asamushi* **13**, 35–48.
NISHIHIRA, M. (1968a). Experiments on the algal selection by the larvae of *Coryne uchidai*. *Bull. biol. Stn Asamushi* **13**, 83–89.
NISHIHIRA, M. (1968b). Brief experiments on the effect of algal extracts in promoting the settlement of the larvae of *Coryne uchidai*. *Bull. biol. Stn Asamushi* **13**, 91–101.
NISHIHIRA, M. (1968c). Distribution patterns of Hydrozoa on the broad-leaved eelgrass and narrow-leaved eelgrass. *Bull. biol. Stn Asamushi* **13**, 125–138.
PAINE, R. T. (1971). A short-term experimental investigation of resource partitioning in a New Zealand rocky intertidal habitat. *Ecology* **52**, 1096–1106.
PERCIVAL, E. (1944). A contribution to the life history of the brachiopod, *Terebratella inconspicua*. *Trans. R. Soc., N.Z.* **74**, 1–23.
PYEFINCH, K. A. and DOWNING, F. S. (1949). Notes on the general biology of *Tubularia larynx*. *J. mar. biol. Ass., U.K.* **28**, 21–43.
RANDALL, J. E. and HARTMAN, W. D. (1968). Sponge-feeding fishes of the West Indies. *Mar. Biol.* **1**, 216–225.
REISWIG, H. M. (1973). Population dynamics of three Jamaican Demospongiae. *Bull. mar. Sci.* **23**, 191–226.
RICKETTS, E. W. and CALVIN, J. (1968). "Between Pacific Tides" 4th edition, rev. by J. W. Hedgpeth. Stanford University Press, Stanford, California.
RYLAND, J. S. (1959). Experiments on the selection of algal substrates by polyzoan larvae. *J. exp. Biol.* **36**, 613-631.
RYLAND, J. S. (1960). Experiments on the influence of light on the behaviour of polyzoan larvae. *J. exp. Biol.* **37**, 783–800.

Ryland, J. S. (1962a). The effect of temperature on the photic responses of polyzoan larvae. *Sarsia* **6**, 41–48.
Ryland, J. S. (1962b). The association between Polyzoa and algal substrata. *J. anim. Ecol.* **31**, 331–338.
Ryland, J. S. (1970). *Bryozoans*. Hutchinson University Library, London.
Ryland, J. S. (1972). The analysis of pattern in communities of Bryozoa. I. Discrete sampling methods. *J. exp. mar. Biol. Ecol.* **8**, 277–297.
Ryland, J. S. (1973). The analysis of spatial distribution patterns. *In* "Living and Fossil Bryozoa" (G. P. Larwood, ed.), pp. 165–172. Academic Press, London.
Ryland, J. S. and Stebbing, A. R. D. (1971). Settlement and oriented growth in epiphytic and epizoic bryozoans. *In* "Proc. 4th Europ. mar. Biol. Symp." (D. J. Crisp, ed.), pp. 283–300. Cambridge, Cambridge University Press.
Scheltema, R. S. (1974). Biological interactions determining larval settlement of marine invertebrates. *Thalassia Jugoslavica* **10**, 263–296.
Silva, P. H. D. H. de (1962). Experiments on choice of substrate by *Spirorbis* larvae. *J. exp. Biol.* **39**, 483–490.
Silva, P. H. D. H. de and Knight-Jones, E. W. (1962) *Spirorbis corallinae* and some other Spirorbinae common to British shores. *J. mar. biol. Ass., U.K.* **42**, 601–608.
Simpson, T. L. (1973). Coloniality among the Porifera. *In* "Animal Colonies: Development and Function Through Time" (R. S. Boardman, A. H. Cheetham and W. A. Oliver, eds), pp. 549–565. Dowden, Hutchinson and Ross, Stroudsburg, Pennsylvania.
Slobodkin, L. B. (1970). Toward a predictive theory of evolution. *In* "Population Biology and Evolution" (R. C. Lewontin, ed.), pp. 187–205. Syracuse Univ. Press, Syracuse.
Stebbing, A. R. D. (1972a). Preferential settlement of a bryozoan and serpulid larvae on the younger parts of *Laminaria* fronds. *J. mar. biol. Ass., U.K.* **52**, 756–772.
Stebbing, A. R. D. (1972b). Some observations of colony overgrowth and spatial competition. *In* "Living and Fossil Bryozoa" (G. P. Larwood, ed.), pp. 173–183. Academic Press, London.
Stebbing, A. R. D. (1973). Competition for space between the ephiphytes of *Fucus serratus*. *J. mar. biol. Ass., U.K.* **53**, 247–261.
Storr, J. F. (1976). Field observations of sponge reactions as related to their ecology. *In* "Aspects of Sponge Biology" (F. W. Harrison and R. C. Cowden, eds), pp. 261–276. Academic Press, New York.
Straughan, D. (1972). Ecological studies of *Mercierella enigmatica* in the Brisbane River. *J. anim. Ecol.* **41**, 93–136.
Thorson, G. (1964). Light as an ecological factor in the dispersal and settlement of marine bottom invertebrates. *Ophelia*, **1**, 167–208.
Visscher, J. P. (1928). Reactions of the cyprid larvae of barnacles at the time of attachment. *Biol. Bull.* **54**, 327–335.
Waters, A. W. (1924). The ancestrula of *Membranipora pilosa* and cheilostomatous Bryozoa. *Ann. Mag. nat. Hist.* (ser. 9), **14**, 594–612.
Williams, G. B. (1964). The effect of *Fucus serratus* in promoting settlement of *Spirorbis borealis*. *J. mar. biol. Ass., U.K.* **44**, 397–414.
Williams, G. B. (1965a). Settlement-inducing factors in fucoids. *Rep. Challenger Soc.* **3** (17), 31.

WILLIAMS, G. B. (1965b). Observations on the behaviour of the planulae larvae of *Clava squamata*. *J. mar. biol. Ass., U.K.* **45**, 257–273.

WILSON, D. P. (1968). The settlement behaviour of the larvae of *Sabellaria alveolata* (L.). *J. mar. biol. Ass., U.K.* **48**, 387–435.

WILSON, D. P. (1970). The larvae of *Sabellaria spinulosa* and their settlement behaviour. *J. mar. biol. Ass., U.K.* **50**, 33–52.

WINSTON, J. E. (1973). See J. E. Dudley (1973).

WINSTON, J. E. (1976). Experimental culture of the estuarine ectoproct *Conopeum tenuissimum* from Chesapeake Bay. *Biol. Bull.* **150**, 318–335.

WISELY, B. (1958a). The development and settling of a serpulid worm, *Hydroides norvegica*. *Aust. J. mar. Freshwat. Res.* **9**, 351–361.

WISELY, B. (1958b). The settling and experimental reactions of a Bryozoan larva, *Watersipora cucullata*. *Aust. J. mar. Freshwat. Res.* **9**, 362–371.

WISELY, B. (1960). Observations on the settling behavior of larvae of the tubeworm *Spirorbis borealis*. *Aust. J. mar. Freshwat. Res.* **11**, 55–72.

WOODBRIDGE, H. (1924). *Botryllus schlosseri*. The behaviour of the larva with special reference to the habitat. *Biol. Bull.* **47**, 223–230.

WYER, D. W. and KING, P. E. (1973). Relationships between some British littoral and sublittoral pycnogonids. *In* "Living and Fossil Bryozoa" (G. P. Larwood, ed.), pp. 199–208. Academic Press, London.

27 | Morphological Strategies of Sessile Animals

J.B.C. JACKSON

Department of Earth and Planetary Sciences, The Johns Hopkins University, Baltimore, U.S.A.

Abstract: Sponges and colonial animals inhabiting marine hard substrata exhibit a great array of forms. Yet, in spite of this variability, six basic shapes are commonly repeated among many taxonomically distinct groups. These are runners (linear or branching encrustations), sheets (two-dimensional encrustations), mounds (massive three-dimensional encrustations), plates (foliose projections from a restricted zone of substratum attachment), vines (linear or branching, semi-erect forms with restricted zones of substratum attachment) and trees (erect, usually branching projections with a restricted zone of substratum attachment).

A simple model is presented for interpretation of the adaptive significance of these recurrent forms. Each shape is characterized by a series of geometric (size and shape) parameters whose potential survival importance is deduced relative to processes known to influence distribution and abundance patterns of sessile animals. Regardless of animal shape these parameters cannot vary entirely independently of one another. Thus variations in parameter values for different animal shapes can be interpreted as a series of trade-offs (potential costs versus potential benefits) which define the morphological survival strategies of these organisms. In this manner runners and vines are interpreted as being entirely committed to a fugitive (refuge-oriented) strategy. All other growth forms represent increasing commitments (trees $\geqslant$ plates > mounds > sheets) to survival within their areas of settlement and to maintenance and defence of the integrity of colony surfaces. Thus, to some degree, they all exhibit a confrontation strategy.

Deduction of the presumed ecological significance of these six morphological strategies leads to a series of testable hypotheses regarding the importance of sessile animal shape to variations in their:
1. Life-history attributes;
2. Zooid morphology, budding and growth processes and levels of colony integration;
3. Distributions in space and time.

Preliminary data are in agreement with many of the predictions and thus lend support to the assumptions of the model.

Systematics Association Special Volume No. 11, "Biology and Systematics of Colonial Organisms", edited by G. Larwood and B. R. Rosen, 1979, pp. 499–555. Academic Press, London and New York.

INTRODUCTION

Sponges and colonial animals living attached to the sea bottom exhibit a bewildering array of forms, sizes, internal organizations and growth processes. Yet in spite of this variability, a limited number of basic morphological patterns are repeated among taxonomically distinct groups. These animal forms are here termed runners, sheets, mounds, plates, vines and trees. Other forms can be recognized but these are largely derivatives of the above types. Definition of each of the basic growth-forms and representative groups from Caribbean open reef (exposed reef surfaces and vertical walls) and cryptic reef (caves, crevices) environments are given in Table I.

The repetition of similar forms among such widely different organisms as sponges, cnidarians and ascidians must represent convergent adaptations to meet similar selective processes. For Recent organisms, these selective processes should be discernible through conventional ecological methods. Processes which regulate distribution and abundance patterns of sessile animals and thus their survival and reproduction, must also be the agents of morphological selection.

The potential physiological significance of animal shape (e.g. surface/volume ratios) has been extensively studied (Prosser, 1973), as it has also been for interactions of organisms with fluid mechanical properties of their environment (Wainwright et al., 1976). Largely ignored by animal ecologists, however, is the potential significance of variations in shape parameters to interactions among sessile organisms or in relation to their predators. In this paper I describe known mortality agents for sessile animals and relate these to a series of morphological parameters (dimensions and dimensional ratios). Variations in these parameters for a series of geometrically simple sessile organisms as a function of growth-form are interpreted as adaptations to particular categories of mortality processes (selective forces). Results of these analyses suggest which morphological parameters most clearly separate the different shape categories of organisms from one another, and thus which parameters selection presumably acts on most strongly. The procedure is similar to that which I have applied elsewhere to analysis of ecological differences between solitary and colonial sessile animals (Jackson, 1977a).

MORTALITY PROCESSES

The purpose of this section is twofold: first to review the wide variety of processes which may influence the distribution and abundance patterns of sessile animals; secondly to emphasize the largely correlative (i.e. experimentally untested) nature of the evidence for the importance of these different processes in most natural situations.

TABLE I. Six basic growth forms of sessile foraminifera, sponges, and colonial animals

Growth form	Definition	Characteristic groups in Caribbean reef environments	
		Open reef	Cryptic reef
Runners	Linear or branching forms lying parallel to the substratum; more or less continuously encrusting	Demospongiae	Demospongiae Hydroida Ctenostomata Cheilostomata Cyclostomata Ascidiacea
Sheets	Two-dimensional encrustations more or less completely attached to the substratum	Demospongiae Milleporina Scleractinia Zoanthidea Scleraxonia (Gorgonacea) Cheilostomata Ascidiacea	Foraminiferida Demospongiae Sclerospongiae Scleractinia Cheilostomata Cyclostomata Ascidiacea
Mounds	Regular or irregular massive encrustations with vertical as well as lateral growth; usually attached to substratum along most of basal area	Demospongiae Sclerospongiae Milleporina Scleractinia Cheilostomata	Calcarea Demospongiae Sclerospongiae Scleractinia Cheilostomata Ascidiacea
Plates	Flattened, foliose forms more or less parallel to the substratum and projecting into the water column from a limited zone of basal attachment	Demospongiae Milleporina Scleractinia	Sclerospongiae Cheilostomata
Vines	Linear or irregularly branching erect, semi-erect, or climbing forms, with one or more restricted zones of attachment to the substratum	Demospongiae Scleraxonia (Gorgonacea) Scleractinia	Demospongiae Hydroida Cheilostomata Cyclostomata
Trees	Erect, usually regularly branching forms, with a restricted zone of basal attachment to the substratum	Demospongiae Hydroida Milleporina Scleractinia Scleraxonia (Gorgonacea) Holaxonia (Gorgonacea) Antipatharia	Foraminiferida Demospongiae Sclerospongiae Hydroida Stylasterina Scleractinia Holaxonia (Gorgonacea) Antipatharia Cheilostomata Cyclostomata

Consider some area of hard substratum on the sea bottom and the larvae of some sessile species which might settle and grow there. Assume that the prevailing climatic conditions (temperature, salinity, etc.) and potentially available

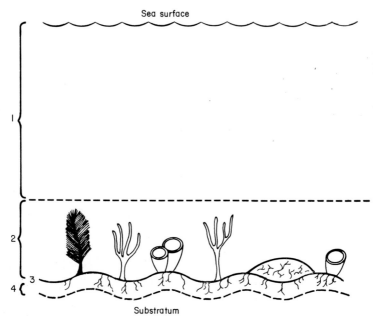

FIG. 1. Vertical zonation of water column and substratum. Zone 1, sea surface to uppermost limit of the tallest sessile organism; zone 2, base of zone 1 to just above substratum surface; zone 3, the substratum surface; zone 4, substratum beneath surface penetrated by boring organisms.

resources are suitable for survival. Then the successful recruitment, growth and survival to maturity of these organisms will be a function of the abundance and activities of other space occupiers (sessile animals and plants) and the nature and intensity of physical and biological disturbance processes (e.g. sedimentation, predation). The importance for any organism of such potential mortality processes may vary considerably with ontogeny (life-history stage) and growth-form (e.g. Reiswig, 1973). For example, predators may be particularly abundant on the substratum surface but absent in the immediately overlying water. In this case, predation would probably be a more serious threat to entirely encrusting animals than to erect forms. Erect juveniles, however, might also be subject to the higher substratum-associated predation. Strong water movements could have an opposite effect.

Enumerated below are categories of mortality processes which will be considered in the interpretation of animal growth form. Processes are considered first for recruitment and secondly for growth and reproductive stages. In each case a distinction is made between those processes occurring in the water just above the substratum surface (zone 2 in Fig. 1) and those acting on or within the substratum (zones 3 and 4 in Fig. 1). Zone 2 is often a region of strong vertical gradients in both concentration of materials (suspended sediments, food, etc.) and the intensity of physical processes (e.g. currents, turbulence) (Neushul, 1972; Riedl, 1971). Such variations may be largely due to the presence of the sessile organisms (Reiswig, 1974). Zone 2 is thus a strongly heterogeneous environment in a vertical as well as a horizontal direction. Zone 3 is a two-dimensional surface; all heterogeneity is thus horizontal. Processes such as primary production or larval predation occurring in the open water (zone 1) can only indirectly affect the survival of different growth forms and are not considered.

1. Recruitment

All sessile, epibenthic marine animals have some form of free-living, usually sexually produced phase generally termed larvae. Recruitment includes all events in the life of an organism from the time the larva first approaches the substratum until the completion of attachment and metamorphosis. Processes which may influence survival during this period are summarized in Table II.

Who gets there first is of primary importance to the successful recruitment of larvae onto most marine hard substrata (Jackson, 1977a). Only organisms

TABLE II. Processes potentially affecting recruitment of sessile organisms onto hard substrata. Climate and substratum composition are assumed to be suitable in all cases. Zones as in Fig. 1

Process	Zone 2	Zones 3 and 4
Activities of other sessile organisms	Prevent larvae from reaching bottom (predation, feeding currents, defence organs)	Occupy space first; prevent recruitment onto adjacent substrata
Activities of vagile organisms	Predation on larvae by swimming predators	Selective or indiscriminate predation on larvae by benthic predators
Water movements and sedimentation	Inhibit settlement by habitat selection	Inhibit settlement; kill new recruits

with exposed external skeletons are susceptible to epizoic larval recruitment. Other sessile forms (the great majority of colonial animals and sponges) are far less fouled, presumably because of their feeding (predation on larvae) or various forms of defensive behaviour. Such processes might also affect recruitment onto adjacent areas of bare substratum, the intensity of the effect falling off with distance from the animal. Settlement experiments (Goodbody, 1961) and observations on sponge feeding (Reiswig, 1971a, b, 1973, 1974) suggest that intensive feeding by sessile animals or allelochemicals could prevent larvae from ever reaching the bottom. Such an effect may be of frequent occurrence but has not to my knowledge been adequately tested. Woodin (1976) has postulated that similar adult–larval interactions are important determinants of distribution and abundance patterns of infauna in sedimentary environments.

Larvae approaching the bottom may also be consumed by vagile predators before, during, or after settlement. Predation may be "deliberate" (plankton-feeding fishes) or accidental (sea urchins grazing on the bottom). Sammarco (1975) has shown that coral recruitment is adversely affected by sea urchins. Numerous other cases remain to be tested but such predation is doubtless of as great importance for recruitment onto marine hard substrata as for seed survival in tropical forests (Connell, 1971; Janzen, 1970).

Sedimentation, sediment, scour, or excessive turbulence may inhibit settlement via habitat selection (Meadows and Campbell, 1972; Newell, 1970), carry larvae away, or kill newly settled forms.

In summary, larval recruitment is often a risky way to maintain or increase a local epibenthic population. Many processes of little consequence to established adults may be fatal to recruiting larvae of the same species. Asexual propagation (budding, "vegetative" growth) may be far more successful on a local scale (Jackson, 1977a; Williams, 1975). As will be demonstrated below, the potential for unlimited asexual propagation is necessarily strongly tied to growth-form. Of course, sexual reproduction and recruitment are necessary for long-distance dispersal and colonization of new and/or physically isolated habitats.

2. *Growth and Sexual Reproduction*

Growth is here defined as any increase in overall animal dimension. The growth phase of sessile animals may be defined as beginning after the completion of larval metamorphosis on the substratum. There is usually some period of growth before sexual reproduction commences (e.g. Crowell, 1957; Sugimoto and Nakauchi, 1974). Growth requires access to space and energy sources (food or light levels) in quantities greater than those necessary for maintenance functions. Reproduction may commonly require access to even higher energy levels (Crowell, 1957; Winston, 1976; Jebram, 1973). Processes which may

TABLE III. Processes potentially affecting growth and reproduction of sessile animals. The general climate and substratum composition are assumed to be suitable in all cases. Zones as in Fig. 1

Process	Zone 2	Zones 3 and 4
Activities of other sessile organisms	Shading/filtering of resources	Occupy space first and defend against encroachment; kill or interfere with activities of neighbours; deplete local food; destabilize (bore) substratum
Activities of vagile organisms	Cropping of canopy; may prevent canopy formation	Create patches of bare substratum; slow growth and reproductive rates (time lost feeding or repair of tissues)
Water movements, sedimentation, and substratum stability	Damage to fragile erect forms by strong movements	Create patches of bare substratum or slow growth and reproductive rates

influence survival during these phases are summarized in Table III. Each of these may be directly responsible for mortality or, by limiting the degree of access of an organism to resources, may indirectly decrease its growth rate or fecundity.

To the best of my knowledge no-one has experimentally demonstrated competition for food or light among sessile marine animals in their natural environment. Circumstantial evidence abounds for local competition for light in coral reef environments (Connell, 1973; Lang, 1970). Evidence for competition for food is much more elusive. All we know is that laboratory growth rates and sometimes growth-form are to some degree food dependent (Crowell, 1957; Jebram, 1973; Winston, 1976) and that many sessile organisms are probably capable of tolerating considerable local reductions in food abundance (Reiswig, 1974). Whether such behaviour ever significantly affects the growth and reproduction of other sessile animals is unknown. The frequent clarity of water just above the surface of coral reefs, or in submarine caves, is certainly suggestive of the potential for food limitation. In the absence of contradictory evidence I will here assume that overtopping strategies (c.f. Horn, 1971; Porter, 1976) are potentially significant in at least some epibenthic environments. There is an obvious urgent need for experimental studies.

In contrast, competition for space among sessile animals has been demonstrated experimentally in a number of hard-substratum marine environments (Connell, 1961, 1975; Dayton, 1971; Lang, 1970, 1973; Paine, 1974). Who gets there first is often as important for sessile animal growth as recruitment (Jackson, 1977b). Growth onto occupied substrata, or areas close to other sessile organisms, often subjects an animal to varying forms of interactions which may result in death, reduced feeding abilities, or changes in the behaviour or morphology of the animals involved (Connell, 1973, 1975; Jackson, 1977b; Lang, 1970, 1973; Chiba and Kato, 1966; Kato et al., 1963, 1967, for experimental data; and Bryan, 1973; Jackson, 1977a; Jackson and Buss, 1975; Stebbing, 1971b, 1973a for field observations or circumstantial evidence). The effects on growth rates may be considerable. Often growth simply stops in the region of contact (Lang, 1970; Karlson, 1978). Growth may then begin, or rates be increased, in other directions. Initial growth rates over unoccupied substrata in settling experiments are commonly much more rapid than growth rates of surviving animals after most of the space has been overgrown (Jackson, unpublished data; Karlson, 1978).

Predators often control the distribution and abundance patterns of sessile marine animals (reviewed in Connell, 1975). Resistance to predation may involve structural, chemical, or behavioural mechanisms (Bakus, 1973; Bakus and Green, 1974; Barnes et al., 1970; Burkholder, 1973; Dayton, 1971; Randall and Hartman, 1968). Predation events are often limited to portions of sponges and colonial animals and may not be fatal to the entire colony (Dayton et al., 1974; Glynn et al., 1972; Reiswig, 1973; Ryland, 1970). Survival of portions of colonies cabable of regeneration and further asexual proliferation provides the potential for rapid recovery of colonial animal populations after episodes of intense predation. In this regard, the foraging patterns and feeding mechanisms of predators are of considerable potential importance to sessile animal morphology (Glynn, 1973; Barnes et al., 1970; Marsden, 1962). Swimming predators may be able to attack their prey from a variety of directions whereas exclusively benthic (e.g. crawling) predators such as snails must approach their prey along the substratum surface.

Most attempts to relate sessile animal distribution patterns, growth rates or reproductive condition to physical environmental processes are based on circumstantial evidence (e,g. Harmelin, 1975; Kinzie, 1973; Lang, 1974; Riedl, 1966, 1971). Yet water movements, sedimentation and substratum stability, as well as most other environmental processes are strongly co-variant (Buzas, 1969; Jackson, 1972) and thus cannot properly be treated as single factors in distributional analyses (Buzas, 1969; Seal, 1964). Certainly delicate animals are not found in excessively turbulent environments and many gorgonians, for example, are torn from the bottom (substratum fails) by storms (Kinzie, 1973). Excessive sedimentation may smother, or at least inhibit feeding by many epibenthic suspension feeders (Jorgensen, 1966; Reiswig, 1973, 1974). Substratum collapse is responsible for the death of many reef organisms (Goreau and Hartman, 1963; Reiswig, 1973) but how often, and to what degree less extreme variations in physical processes may directly or indirectly alter sessile animal growth or survival is very incompletely understood (Brakel, 1976; Connell, 1973, 1975). Laboratory experiments and behavioural observations (Chamberlin and Graus, 1975; Leversee, 1976; Lewis and Price, 1976; Hubbard, 1973) are valuable but cannot confirm the relative importance of different processes in nature. Manipulative experiments are badly needed.

The importance of boring organisms (sponges, algae, etc.) to substratum stability in coral reef environments deserves special attention. Most borers are limited to a few cm of the surface of substrata (Neuman, 1966). Animals whose skeletons or areas of substratum attachment are heavily bored are subject to greater probabilities of detachment than less bored forms (Bertram, 1937; Goreau and Hartman, 1963).

In summary, sufficient circumstantial evidence exists to lend credit to the assumption that most of the above processes sometimes influence growth, reproduction and survival of sessile animals. Demonstration of their action and relative importance, awaits experimental confirmation in most cases. In the following section I explore the potential relationship of these different mortality processes to sessile animal shape.

MORPHOLOGICAL PARAMETERS

1. Size

Nine dimensional parameters of sessile animals are listed in Table IV and illustrated for hypothetical hemispherical and plate-like animals in Fig. 2. It is especially important that, although the dimensional parameters tend to be correlated with one another, the extent or even the sign of such relationships is very strongly shape dependent. The nine size parameters are:

(a) *Tissue surface area* (A_t). Tissue surface area is here defined as the total area of non-skeletal surface tissues of a sessile animal in direct contact with the water column. Excluded are tissues in contact with the substratum, the linings of the canals of sponges, the coelenteron of cnidarians and the gut surfaces of other sessile animals.

The vast majority of sessile marine invertebrates are suspension feeders dependent upon the availability of plant or animal prey in their immediate surroundings. The numbers of autozooids (feeding zooids or feeding polyps)

TABLE IV. Size parameters of sessile animals discussed in text

Parameter	Definition	Formula for hemisphere in fig. 2
1. A_t	Tissue surface area	$2\pi r^2$
2. V_t	Tissue volume	$\dfrac{2\pi r^3 - V_{sk}}{3}$
3. V_{sk}	Skeletal volume (coral)	$\tfrac{2}{3}\pi(r-1)^3$
	Skeletal volume (sponge)	Volume of spicules
4. S_e	Encrusted substratum area	πr^2
5. S_s	Shaded substratum area	—
6. C_s	Length of periphery (circumference) in contact with the substratum	$2\pi r$
7. D_s	Maximum linear dimension along substratum	$2r$
8. H_t	Height above substratum	r
9. A_{sk}	Exposed skeletal surface area	—

of colonial animals and the numbers of ostia (inhalent pores for feeding currents) of sponges are proportional to the tissue surface area of these animals (Hyman, 1940; Reiswig, 1975; Ryland, 1970). Furthermore, symbiotic algae tend to be concentrated in the surface tissues of their hosts (Muscatine, 1974). Thus, barring possible interference effects (e.g. Bishop and Bahr, 1973), the potential of these animals to capture or garden food should increase as some positive function of their tissue surface area.

The same surface area relationships should apply to other categories of interactions between a sessile animal and the water column (Dahl, 1973). Contact with food particles, physical disturbances, and interactions with other organisms are all events which should have some probability of occurring per unit area of surface tissues. So, as A_t increases, so also on average will the number of such interactions per colony.

Most such disturbance interactions can be treated as discrete events with some fixed (or predictably variable) radius of effect on the organism (e.g. Dayton, 1971 for logs crashing onto a rocky shore). Thus for any given probability of disturbance events, not only does the probability that a disturbance will affect

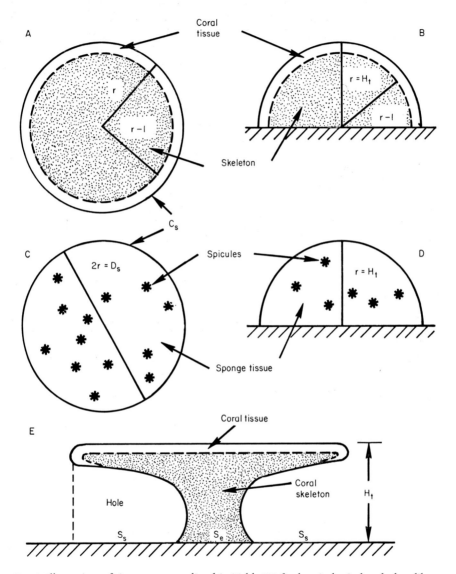

FIG. 2. Illustration of size parameters listed in Table IV for hemispherical and plate-like sessile animals. Top view of hemispherical coral colony (A); side view of same (B); top view of hemispherical sponge (C); side view of same (D); side view of plate-like coral colony (E).

an individual colony increase with its surface area, but so also does the probability that some area of the colony will not be affected. Most sponges or colonial animals can survive some degree of localized damage (mortality) without the necessary death of the entire organism (Dayton et al., 1974; Glynn et al., 1972; Randall and Hartman, 1968; Reiswig, 1973; Storr, 1976). Thus for any given probability of local disturbance events, the greater the surface area of tissues (A_t), the higher the probability that some proportion of these tissues will survive over any interval of time.

The sexual reproductive potential of many colonial animals, as measured by the numbers of reproductive organs (gonads, ovicells, etc.), is commonly proportional to the number of zooids or polyps, and thus the tissue surface area of the colony (Hayward, 1973; Hayward and Ryland, 1973; Sugimoto and Nakauchi, 1974). Some groups, however, have not been examined. For example, the reproductive condition of scleractinian corals as a function of colony surface area is unknown (Connell, 1973). In the absence of contradictory data, I take as a working hypothesis that sexual reproductive potential increases with A_t in colonial animals. More work is needed before this idea can be fully accepted.

(b) *Tissue volume or biomass* (V_t). Tissue volume is here defined as the total volume of non-skeletal tissues.

The numbers of flagellated chambers in sponges, thus their pumping and filtering capacity, increases proportionally to their tissue volume or biomass (Reiswig, 1975). The same is apparently true of sponge sexual reproductive potential (Brien, 1973; Reiswig, 1973). Similar relationships should hold for many alcyonaceans, ascidians and other colonial animals whose zooids are not restricted to the external surface of the colony (Bayer, 1973; Millar, 1971).

(c) *Skeletal volume* (V_{sk}). Skeletal volume is here defined as the total volume of any tissues (fibres, mineralized structures) which serve primarily to strengthen or support an animal. Skeletal tissues may also provide some degree of protection from predation or physical disturbances (reviewed in Jackson, 1977a).

Synthesis of organic skeletons (chitin, spongin, gorgonin, etc.) requires obvious metabolic investment by an animal. The importance of symbiotic zooxanthellae to coral calcification (Goreau and Goreau, 1959, 1960a, b; Pearse and Muscatine, 1971) suggests a significant metabolic investment for such mineralization processes as well, but this has not been demonstrated. Certainly inorganic skeletons require, at the very least, synthesis and maintenance of secretory cells and other template tissues. Thus the value of V_{sk} represents a measure of the cost, integrated over its lifetime, of support and passive protection of an animal.

(d) *Encrusted substratum area* (S_e). This parameter is defined as the surface area of substratum encrusted by an animal. Not included are bottom areas overgrown or "shaded" by tissues elevated above the substratum.

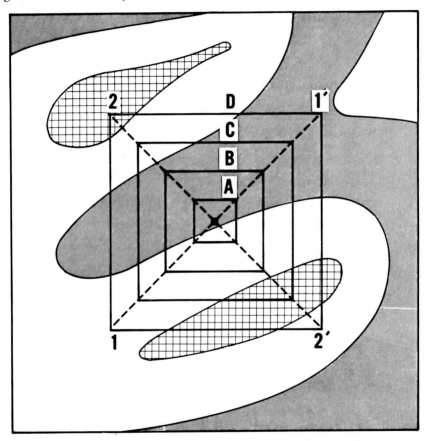

FIG. 3. Map of four, square, sheet-like animals (A, B, C, D) encrusting a heterogeneous substratum. Three substratum conditions shown by stipple and line patterns. The larger the substratum area encrusted the greater the variety of substratum conditions to which the organism is likely to be subjected. For animal dimensions see Table V.

Data are rare, but marine hard substrata, like terrestrial environments (Geiger, 1965), are commonly subject to widely varying environmental conditions (food availability, water movements, sedimentation rates, etc.) over small distances (e.g. Pequegnat, 1964; Riedl, 1971; Wolcott, 1973). Thus the larger the substratum area encrusted by a sessile animal, the greater the variety of substratum conditions to which the organism is likely, on the average,

TABLE V. *Parameters for four, square, sheet-like animals illustrated in Fig. 3. Dimensions in arbitrary units*

	A	B	C	D
Dimensions L × W × H)	6×6×1	14×14×1	22×22×1	30×30×1
Tissue surface area (A_t)	50	252	572	1020
Tissue volume (V_t)	36	196	484	900
Encrusted substratum area (S_e)	36	196	484	900
Number of kinds of environments encrusted	2	2	3	3
Number of environmental patches encrusted	2	2	4	6
Circumference in contact with the substratum (C_s)	24	56	88	120
Number of kinds of environments intersected by C_s	2	2	3	3
Number of environmental patches intersected by C_s	2	2	6	10
Distance along substratum (D_s) (1–1' or 2–2')	9	20	31	42
Number of kinds of environments intersected by D_s (1–1')	2	2	2	2
Number of kinds of environments intersected by D_s (2–2')	2	2	3	3
Number of environmental patches intersected by D_s (1–1')	2	2	2	2
Number of environmental patches intersected by D_s (2–2')	2	2	4	6

to be subjected (Fig. 3, Table V). The probability of many kinds of discrete disturbance events (e.g. contact by a swimming predator, local substratum collapse) will also be proportional to the area of substratum involved.

From the point of view of other sessile organisms, the encrusted substratum equals that area of substratum under the direct influence of a sessile animal.

(*e*) *Shaded substratum area* (S_s). This parameter is defined as the surface area of substratum overgrown or "shaded" by parts of the animal which are elevated above the substratum. Not included are substrata directly encrusted by the animal (S_e).

The shaded area is a measure of the unencrusted substratum area under some influence of a sessile animal.

(f) Length of periphery (circumference) of an animal in direct contact with the substratum (C_s). The line C_s is a sessile animal's border along the substratum. Many processes potentially important to sessile animal survival (e.g. contact with other sessile organisms or crawling predators; shifting of sediments) occur entirely along the substratum surface and thus must occur across C_s. Thus, as the length of C_s increases, so also does the probability that an animal will be exposed to a greater variety of peripheral disturbances (Fig. 3, Table V). Both the probability of the number of kinds of interaction and the probability of any particular kind of interaction, will increase proportionately to C_s.

(g) Maximum linear dimension along the substratum (D_s). The parameter D_s is the longest line which can be drawn between points of substratum attachment of a sessile animal. Thus D_s is a measure of the maximum substratum distance occupied by an animal. As D_s increases, so also, on average, will the variety of substratum conditions to which it is exposed (Fig. 3, Table V).

(h) Height above substratum (H_t). This parameter is defined as the maximum height above the substratum of a sessile animal's physiologically active (non-skeletal) tissues.

As H_t increases, so also, on average, will the numbers and kinds of interactions of an animal with resources and processes active in the water just above the substratum (zone 2 of Fig. 1).

(i) Surface area of exposed skeletal or cuticular tissues susceptible to fouling (A_{sk}). This parameter is defined as the total area of skeletal or cuticular tissues readily susceptible to larval recruitment by other sessile organisms. Values of A_{sk} are widely variable among the different taxa due to basic biological differences between these groups.

Spicular or fibrous skeletons of sponges, alcyonaceans and ascidians are almost entirely internal (i.e. surrounded by live tissues) (Berrill, 1950; Hyman, 1940). When these animals die, their skeletons rapidly disintegrate. Such skeletons contribute little to the direct accumulation of hard substrata (Bathurst, 1971; Scoffin and Garrett, 1973) and usually provide little in the way of attachment surfaces for recruitment or overgrowth by other sessile organisms.

Many other colonial animals (e.g. milleporids, gorgonians, scleractinian corals, many cheilostomes) produce more extensive, usually mineralized skeletons (Bathurst, 1971; Hyman, 1940; Ryland, 1970). These may be morphologically internal, external, or both, but even external skeletons are commonly isolated from the water column by overlying living tissues (e.g. many corals). As such animals grow they may contribute significantly to the mass of hard substratum in their environment, and when portions of colonies die they provide fresh hard substratum for recruitment, overgrowth, or boring by other sessile organisms (Bathurst, 1971; Scoffin and Garrett, 1974; Goreau and Hartman, 1963).

Whereas exposed skeletal surfaces are always subject to rapid fouling, the susceptibility of non-skeletal tissues to biological invasion may be strongly dependent upon the structure and behaviour of the tissues involved. Exposed epidermal layers of most sponges, cnidarians and many colonial ascidians are physiologically active, flexible layers, armed, in the case of cnidarians, with nematocysts, and perhaps in many cases with toxins that may inhibit larval settlement or overgrowth (Hyman, 1940; Sarà and Vacelet, 1973; Berrill, 1950). In cnidarians, these properties characterize both zooidal (polyp surface) and interzooidal (e.g. coenosteal layer) tissues.

Among the gymnolaematous ectoprocts, however, the external cuticle is often fouled as are the "cuticular" surfaces (especially stolons) of many hydroids and colonial ascidians (Gordon, 1972; Stebbing, 1971b, 1973a). In such cases, the spacing and activity of zooids may be important in determining the susceptibility of different portions of an animal to fouling and overgrowth. Thus cuticular surfaces of cheilostome zooids almost certainly receive protection from fouling by the feeding or other activities of their lophophores, vibraculae and avicularia. The same should be true for surfaces of hydroids adjacent to polyps. In contrast, stolons, spines, or other non-motile protective structures, and the basal surfaces of unilaminate cheilostomes do not receive such protection and should be particularly vulnerable to larval invasion. Some sponges characterized by dense dermal structure are also commonly fouled and overgrown, apparently with little or no detrimental effect on the overgrown sponge (Rützler, 1970; Sarà, 1970).

In contrast to such potentially deleterious effects, tough external surfaces may provide increased protection of tissues from various kinds of physical disturbance, especially sediment-scour and predators. External skeletons of many solitary animals appear to be an important factor enabling them to inhabit intertidal substrata (Jackson, 1977a).

2. Shape

Shape parameters are defined as ratios of the nine size parameters. They are presented here from two different perspectives. These are:

(1) An animal's functional (i.e. feeding or reproductive) capacities. This is

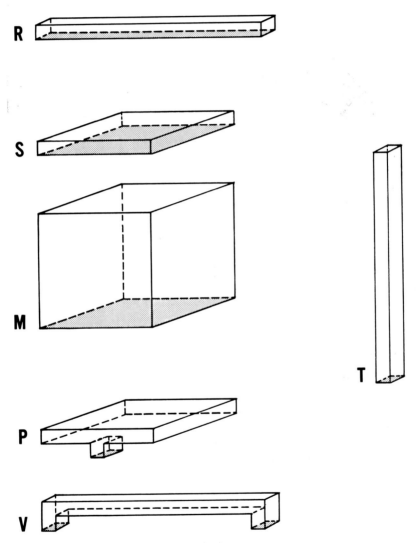

FIG. 4. Rectangular approximations of sessile animal growth-forms used in calculation of shape parameters. R = runner, S = sheet, M = mound, P = plate, V = vine, T = tree. Stippled areas indicate attachment to substratum.

defined as the amount of tissue surface area (A_t) or tissue volume (V_t) relative to any other dimensional parameters,

(2) Substratum utilization, i.e. how size parameters vary as a function of the amount of substratum utilized (S_e). This approach has the advantage of defining shape in terms of the potential utilization of resources (food, space) well known or reasonably assumed to be limiting sessile animal distributions in many natural situations.

TABLE VI. Twenty shape parameters discussed in text. Ratios calculated for a series of simple rectangular solids for all parameters indicated by an asterisk (★)

Parameter	Definition
★V_t/A_t	Tissue volume/tissue surface area
V_{sk}/A_t, V_{sk}/V_t	Skeletal volume/tissue area or volume
★S_e/A_t, ★S_e/V_t	Encrusted substratum area/tissue area or volume
S_s/A_t, S_s/V_t	Shaded substratum area/tissue area or volume
S_s/S_e	Shaded substratum area/encrusted substratum area
★C_s/A_t, ★C_s/V_t	Circumference along substratum/tissue area or volume
★C_s/S_e	Circumference along substratum/encrusted substratum area
★D_s/A_t, ★D_s/V_t	Distance along substratum/tissue area or volume
★D_s/S_e	Distance along substratum/encrusted substratum area
★H_t/A_t, ★H_t/V_t	Tissue height/tissue area or volume
★H_t/S_e	Tissue height/encrusted substratum area
A_{sk}/A_t, A_{sk}/V_t	Exposed skeletal area/tissue area or volume
Holes	Spaces incompletely surrounded by animal projections above the substratum (not a ratio of size parameters)

The 20 shape parameters discussed below are listed in Table VI.

For purposes of comparison, many of the ratio parameters (★ in Table IX) have been calculated for a series of simple rectangular solids which roughly approximate the shape of the six basic growth forms (Fig. 4). Consider a simple functional unit to be a unit cell (cube with n = 1). Then a runner can be considered a uniserial string of cells n units long and a sheet a two-dimensional array of unit cells n × n units in dimension. A mound is approximated by a cube n × n × n units in dimension and a plate by a two-dimensional n × n array of unit cells (a sheet) raised one unit above the substratum by a single unit cell. A vine is represented by a uniserial string of cells n units long (a runner) raised one unit above the substratum by two terminal unit cells. Finally, a tree is approximated by a unit column n units in height.

Of course sponges and colonial animals are not rectangular solids. Runners, vine and trees are usually branched, sheets commonly have a roughly circular outline, plates are rarely flat and usually increase in height with growth; mounds are commonly more hemispherical or cylindrical than cubic. Nevertheless, more accurate approximations of these shapes only change the absolute values of the parameter ratios, but do not alter the relationships of the ratios to one another.

Results of the calculations are introduced as log–log plots under the appropriate shape parameter sections below. Parameter values for a unit cell are indicated in each graph by a point enclosed in a circle. Only the curves for a mound are drawn back to this origin because ratio values for some of the other shapes may be strongly biased by the rectangular approximations at very small sizes.

Patterns of variability of the 20 shape parameters for the six basic growth forms of sponges and colonial animals are described below.

(a) *Tissue surface area and volume* (V_t/A_t). Colonial animals and sponges fall into two groups on the basis of tissue surface area–volume relationships (Fig. 2). In the first and largest group, the living zooids or polyps are restricted to their outer surfaces. Thus, regardless of overall form, these animals (e.g. hydroids, scleractinian corals, cheilostomes) comprise varyingly complex sheets, usually only a single live zooid layer in thickness, with similar low V_t/A_t ratios. Thus the relationships illustrated in Fig. 5a will not apply. Such animals should not be subject to growth-form related surface volume constraints.

In contrast, the live tissues of the second group (sponges, many alcyonaceans and ascidians) are not restricted to their outer surfaces. For these animals surface/volume relationships vary significantly with form (Fig. 5a). Values of V_t/A_t for mound-like forms increase markedly with increase in size whereas ratios rapidly approach constant values (all $\leqslant 1$) for the five other forms (Fig. 5b). Thus surface/volume constraints (e.g. feeding, respiration) may be expected to influence growth patterns of massive, mound-like sponges, alcyonaceans and ascidians. Surface/volume constraints should not affect other growth forms of these animals.

This discussion has assumed no skeletal volume for the animals. This is of no significance for animals of the first group, but is clearly wrong for the second group. As discussed in Section (b) below, V_{sk}/V_t ratios should be highest for erect forms, lowest for entirely encrusting forms. Thus, consideration of true V_t values rather than total animal volume in Fig. 5a would have the effect of decreasing the separation between mounds and sheets and increasing the spread between runners, vines and trees. Nevertheless, the overall relationships (ranks) of the different growth-forms would be the same.

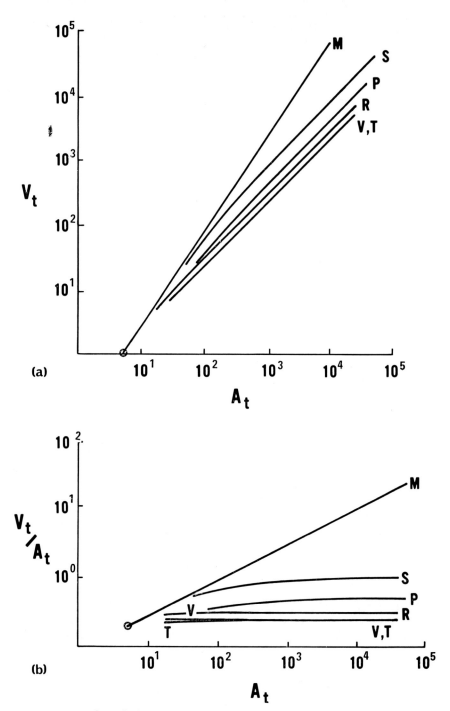

FIG. 5. Tissue volume (V_t)/tissue surface area (A_t) relationships for rectangular approximations of sessile animal forms. Calculations assume that V_t = total animal volume (i.e., that $V_{sk} = 0$). (a) V_t/A_t; (b) V_t/A_t with variation in A_t. R = runner, S = sheet, M = mound, P = plate, V = vine, T = tree.

(b) *Skeletal volume* (V_{sk}/A_t; V_{sk}/V_t). The ratios V_{sk}/A_t and V_{sk}/V_t are measures of the investment of an animal over its lifetime in supportive or protective tissues relative to tissues capable of feeding, maintenance and reproduction. Variations

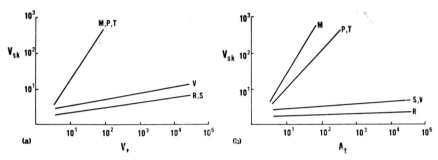

FIG. 6. Skeletal volume (V_{sk})/tissue volume (V_t) and tissue area (A_t) relationships for rectangular approximations of sessile animal forms. (a) V_{sk}/V_t; (b) V_{sk}/A_t R = runner, S = sheet, M = mound, P = plate, V = vine, T = tree.

in these ratios may reflect either of these functions as well as basic organizational features of different animal groups. Thus they may range widely among animals of similar growth form. For this reason no attempt was made to assign skeletal volumes to the geometrical approximations of animal form used for the previous shape parameters.

In spite of this variability, erect forms should on average require considerably more skeletal material for support of their live tissues and these amounts should increase relative to live tissues with increasing animal size. Runners and sheets, being entirely encrusting, should require little skeletal material for support. Vines, with their multiple attachment sites, should require intermediate skeletal volumes. These assumed V_{sk}/V_t relationships are illustrated in Fig. 6a.

Supposed V_{sk}/A_t ratios are shown in Fig. 6b. Mounds will have maximum ratios because of their low surface to volume ratios. Runners have the lowest V_{sk}/A_t ratios because of lack of support problems and high overall surface/volume relationships. Sheets have lower surface/volume values than vines which, however, require more skeleton than sheets or runners for support. Thus sheets and vines cannot be separated without more specific information. Trees and plates should also have approximately similar (high) ratios.

(c) *Encrusted substratum area* (S_e/A_t; S_e/V_t). The parameters S_e/A_t and S_e/V_t are measures of the area of substratum encrusted by a sessile animal relative to its tissue surface area or volume. Low values of these ratios imply high food and light utilization and high sexual reproductive potential per unit substratum area. The percentage of zooids or surface tissues directly susceptible to potentially deleterious substratum-associated processes is also low. The potential costs are twofold. First, low ratio values imply a decrease in the area of sub-

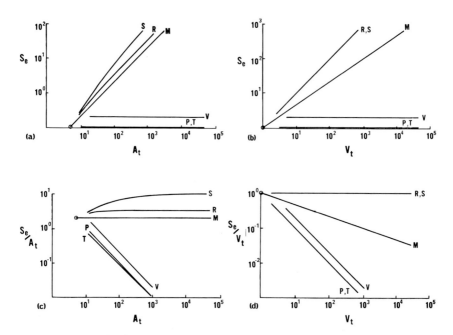

FIG. 7. Encrusted substratum area (S_e)/tissue area (A_t) and volume (V_t) relationships for rectangular approximations of sessile animal forms. (a) S_e/A_t; (b) S_e/V_t; (c) S_e/A_t with variation in A_t; (d) S_e/V_t with variation in V_t. R = runner, S = sheet, M = mound, P = plate, V = vine, T = tree.

stratum such an animal can occupy and protect from occupation by other sessile organisms (except by shading). Secondly, tissues or zooids not directly encrusting the substratum are dependent for mechanical support on encrusting tissues. Mortality or fragmentation of supporting tissues may therefore result in mortality of large numbers of zooids or even the entire colony.

Variations in S_e/A_t and S_e/V_t are illustrated in Fig. 7. The patterns are quite similar for both parameters. Runners and sheets exhibit the highest ratios and plates, vines and trees, the lowest. Mounds are intermediate, especially for S_e/V_t. Both S_e/A_t and S_e/V_t ratios for runners and sheets vary little with animal size, whereas ratio values for plates, vines and trees decrease markedly with increasing size. Mounds show constant S_e/A_t ratios but decreasing S_e/V_t ratios with increasing animal size.

The curves show that, regardless of animal size, runners and sheets exhibit quite low feeding and sexual reproductive potential per unit substratum area as compared with plates, vines and trees. Naturally, the potential to use up available substrata and to encounter a greater number and variety of substratum conditions bears the opposite relationship. Mounds are particularly interesting. Fig. 7c shows that for Group 1 mound-like animals such as corals, the efficiency of substratum utilization cannot increase with animal size. But Group 2 animals such as sponges of the same shape may with growth increase both their feeding and sexual reproductive potentials per unit encrusted substratum area.

(d) *Shaded substratum area* $(S_s/A_t;\ S_s/V_t;\ S_s/S_e)$. The parameters S_s/A_t and S_s/V_t are measures of the substratum shading by a sessile animal per unit area (or volume) of its tissues. The ratio S_s/S_e is the amount of substratum shaded, relative to the amount actually encrusted. High ratio values of all three of these parameters imply relatively high levels of indirect (shading) versus direct (encrusting) influence of an animal over its adjacent substrata. The potential cost, however, lies in the dependence of raised tissues for support by adjacent encrusting tissues.

Except for the zone of influence of individual zooids or polyps, runners and sheets cannot shade substrata adjacent to their areas of attachment. Because of their height, mounds should have some shading effect. Depending on branching patterns, vines and trees may shade widely varying areas of substratum which usually considerably exceed their area of attachment. Plates may completely shade relatively large areas. Thus plates, vines and trees may potentially inhibit local environmental conditions and the activities of other sessile organisms for some distance beyond their usually restricted areas of attachment. Of course this may increase their susceptibility to local disturbance processes within their attachment areas.

(e) *Circumference in contact with the substratum* $(C_s/A_t;\ C_s/V_t;\ C_s/S_e)$. C_s/A_t and C_s/V_t are measures of the length of the periphery of a sessile animal in contact with the substratum (its circumference) relative to its tissue surface area or volume. As the values of these ratios decrease, so also does the percentage of an animal's tissues which lie in direct contact with its substratum periphery. This means that an increasing percentage of tissues are relatively more isolated from peripheral substratum-associated interactions (e.g. competitors, crawling predators, shifting sediments). Furthermore, the relative number of different kinds of substratum conditions encountered at the animal's periphery, and the number of times any one condition is encountered, are also reduced (Fig. 8, Table VII).

Variations in C_s/A_t and C_s/V_t with animal shape are illustrated in Fig. 9.

The patterns for area and volume are similar. Runners exhibit the highest ratios, and these are invariant with animal size. Thus the percentage of tissues directly exposed to interactions with the substratum is also always high. Plates,

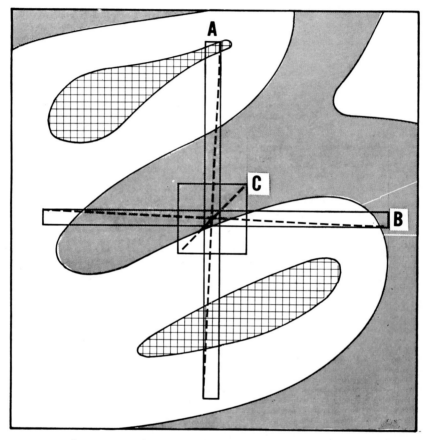

Fig. 8. Map of two, rectangular, runner-like animals and one, square, sheet-like animal encrusting heterogeneous substratum similar to that of Fig. 4. Three substratum conditions shown by stipple and line patterns. D_s indicated by dashed lines. As C_s/S_e decreases, so also does the percentage of encrusted substratum directly exposed to peripheral substratum-associated interactions and the variety of those interactions. For animal dimensions see Table VII.

vines and trees show low ratio values which diminish further with increasing animal size. Thus a low percentage of these forms' tissues are subject to substratum-associated interactions. Sheets and mounds exhibit intermediate ratio values which decrease with increasing animal size.

C_s/S_e is a measure of an animal's circumference relative to the area of substratum which it encrusts. As C_s/S_e decreases, so also does the percentage of encrusted substratum directly exposed to peripheral substratum-associated interactions and the variety of those interactions (Fig. 8, Table VII).

TABLE VII. Parameters for two, rectangular, runner-like animals and one, square, sheet-like animal illustrated in Fig. 8. Dimensions in arbitrary units

	Runner (A)	Runner (B)	Sheet (C)
Dimensions (L × W × H)	50 × 2 × 1	50 × 2 × 1	10 × 10 × 1
Tissue surface area (A_t)	204	204	140
Tissue volume (V_t)	100	100	100
Encrusted substratum area (S_e)	100	100	100
Number of kinds of environments encrusted	3	2	2
Number of environmental patches encrusted	7	4	2
Circumference in contact with the substratum (C_s)	104	104	40
Number of kinds of environments intersected by C_s	3	2	2
Number of environmental patches intersected by C_s	12	6	2
Distance along substratum (D_s)	50	50	14
Number of kinds of environments intersected by D_s	3	2	2
Number of environmental patches intersected by D_s	7	4	2

Variations in C_s/S_e with animal shape are illustrated in Fig. 10a. Runners exhibit high ratios because all the substratum they encrust lies in a peripheral position. Plates, vines and trees show invariant values. This is because their support areas do not vary with size for the geometric approximations used. Real animals of these forms, like sheets and mounds, should have low ratio values.

(*f*) *Distance along substratum* (D_s/A_t; D_s/V_t; D_s/S_e). D_s/A_t and D_s/V_t are measures of the maximum substratum distance occupied by a sessile animal relative to its tissue surface area or volume. Increasing values of these ratios imply increased investment in directional growth along the substratum and decreasing commitment to the point of settlement. For any heterogeneous substratum, this will result in a greater probability of locating larger numbers of different

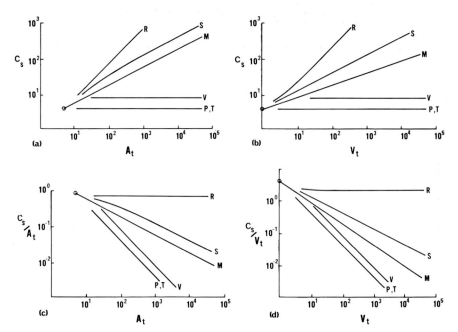

Fig. 9. Circumference (C_s)/tissue area (A_t) and volume (V_t) relationships for rectangular approximations of sessile animal forms. (a) C_s/A_t; (b) C_s/V_t; (c) C_s/A_t with variation in A_t; (d) C_s/V_t with variation in V_t. R = runner, S = sheet, M = mound, P = plate, V = vine, T = tree.

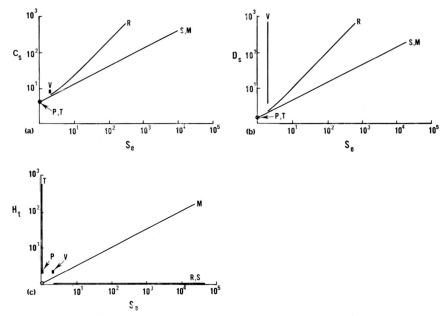

Fig. 10. Circumference (C_s), distance along substratum (D_s), and tissue height (H_t) relationships to encrusted substratum area (S_t) for rectangular approximations of sessile animal forms. (a) C_s/S_e; (b) D_s/S_e; (c) H_t/S_e. R = runner, S = sheet, M = mound, P = plate, V = vine, T = tree.

kinds of substratum patches (Fig. 8, Table VII). The potential costs of such a strategy are decreased potential exploitation of local, relatively favourable environments, and high ratio values of C_s/A_t, C_s/V_t, S_e/A_t and S_e/V_t.

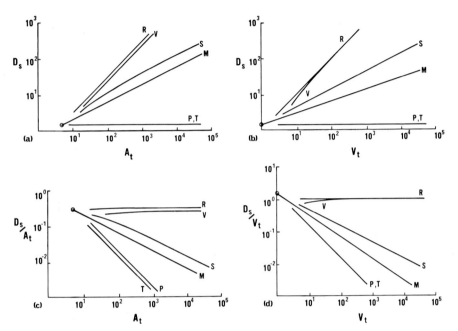

FIG. 11. Distance along substratum (D_s)/tissue area (A_t) and volume (V_t) relationships for rectangular approximations of sessile animal forms. (a) D_s/A_t; (b) D_s/V_t; (c) D_s/A_t with variation in A_t; (d) D_s/V_t with variation in V_t. R = runner, S = sheet, M = mound, P = plate, V = vine, T = tree.

Variations in D_s/A_t and D_s/V_t with animal shape are shown in Fig. 11. Runners and vines exhibit strongly directional growth. They have the highest ratios and these are invariant with animal size. In contrast, plates and trees are entirely committed to their point of settlement and its immediate surroundings. They exhibit very low ratio values which decrease further with increasing animal size. As for C_s ratios, sheets and mounds have intermediate ratio values which decrease with increasing animal size. Note that the positive correlations of D_s/A_t and D_s/V_t with the different C_s and S_e shape parameters are strongly dependent on animal shape (as in Fig. 8, Table VII for runners and sheets). These different parameters are not redundant, although they may appear to be in many cases.

D_s/S_e is a measure of the distance a sessile animal grows along the substratum (or between its points of attachment) relative to the area of substratum which it encrusts. Variations in D_s/S_e with animal size are illustrated in Fig. 10b. Again, plates and trees show invariant values due to the nature of the geometric approximations of these forms. Vines exhibit high ratio values which increase greatly with animal size. (Real vines would require intermediate support points; thus ratios would lie somewhere between those illustrated for runners and vines in Fig. 9b). Runners also exhibit high D_s/S_e values. Thus these two forms achieve considerable length without requiring encrustation of large substratum areas (Fig. 7, Table VII). Sheets and mounds show lower ratios which do not vary with increasing animal size.

(g) *Height above substratum* $(H_t/A_t;\ H_t/V_t;\ H_t/S_e)$. H_t/A_t and H_t/V_t are measures of the maximum height of a sessile animal above the substratum relative to its tissue surface area or volume. Increasing values of these ratios imply increased

TABLE VIII. Parameters for the four rectangular animals illustrated in Fig. 12. Dimensions in arbitrary units

	Runner	Sheet	Mound	Tree
Dimensions (L × W × H)	50 × 2 × 1	10 × 10 × 1	10 × 10 × 10	2 × 2 × 50
Tissue surface area (A_t)	204	140	500	404
Tissue volume (V_t)	100	100	1000	200
Encrusted substratum area (S_e)	100	100	100	4
Height above substratum (H_t)	1	1	10	50
Number of environments encountered vertically	1	1	3	5

accessibility of surface tissues to the water column above the substratum. Recall that this water (zone 2 of Fig. 1) may show marked heterogeneity in concentrations of materials or in the intensity of different physical processes. Assume that such heterogeneities are vertically zoned, either in the simple geometric manner illustrated in Fig. 12, or in some more complex pattern. Then increasing H_t/A_t or H_t/V_t ratios imply a greater probability that an animal will encounter an increased number of zone 2 water conditions. (Fig. 12, Table VIII). The potential costs are decreased potential exploitation of substratum area and an increase in the amount of live tissues dependent on the small encrusted area for support (both consequences of low S_e/A_t and S_e/V_t).

In such an analysis of vertical growth it is especially important that the scale of the environment and attached organisms be clearly defined. This is readily apparent in comparison of cryptic (under coral) and open (exposed) reef surface

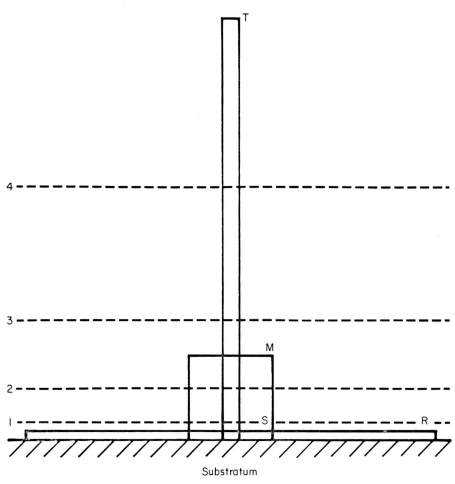

FIG. 12. Comparative tissue heights (H_t) of rectangular approximations of sessile animal forms. For animal dimensions see Table VIII. Vertical heterogeneity of water column indicated by horizontal dashed lines (levels 1–5). R = runner, S = sheet, M = mound, P = plate, V = vine, T = tree.

environments. Encrusting colonial animals and algae covering coral undersurfaces are usually less than 1 cm thick and many competitive dominants are sheets less than 1 or 2 mm in thickness. Thus vertical growth of 1 or 2 cm

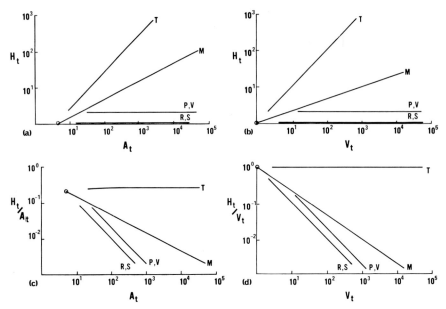

FIG. 13. Tissue height (H_t)/tissue area (A_t) and volume (V_t) relationships for rectangular approximations of sessile animal forms. (a) H_t/A_t; (b) H_t/V_t; (c) H_t/A_t with variation in A_t; (d) H_t/V_t with variation in V_t. R = runner, S = sheet, M = mound, P = plate, V = vine, T = tree.

should be adequate to guarantee access to resources and safety from overgrowth. The system is essentially two-dimensional. In contrast, encrusting and massive colonial animals on open reef surfaces are commonly 10 to 50 cm in thickness. Thus vertical growth of a metre or more may be necessary for free access to the water column and safety from overgrowth.

Variations in H_t/A_t and H_t/V_t with sessile animal shape are illustrated in Fig. 13. Runners, sheets, plates and vines have low ratio values which decrease with increasing animal size. Trees exhibit the highest ratios which do not change with size for the simple columnar solids considered. Branching of trees increases their H_t/A_t and H_t/V_t ratios even more and in this case these ratios increase with animal size. Mounds exhibit intermediate ratio values which, as for all forms except trees, decrease with increasing animal size.

H_t/S_e is a measure of the height achieved by a sessile animal's tissues relative to its basal area of support (encrusted substratum area). Trees show high ratio values which increase markedly with increasing size (Fig. 10c). Runners and sheets have very low H_t/S_e ratios which decrease with increasing animal size. Mounds show intermediate ratio values which are invariant with size. Plates and vines exhibit invariant values for the approximations used; real animals of these forms may increase their areas of basal attachment with increasing size. Vines do not grow high above their substratum. Thus their curves should resemble those for runners and sheets. Real plate-like animals, however, often do grow well above the substratum. In such cases their H_t/S_e curves may more nearly resemble those for trees or mounds.

(h) *Exposed skeletal area* $(A_{sk}/A_t; A_{sk}/V_t)$. A_{sk}/A_t and A_{sk}/V_t are measures of the area of exposed skeletal tissues of a sessile animal relative to the amount of its non-skeletal (and physiologically active) tissues. Increasing values of these ratios imply potentially increased susceptibility of an animal to fouling, overgrowth and boring processes. The potential benefits of such a strategy are decreased susceptibility to potentially deleterious physical processes (e.g. sedimentation or scour) and predation (Jackson, 1977a).

Although much variation in A_{sk} shape parameters may be taxonomic, certain correlations with shape are likely. Closely spaced branches of tree-like and irregular mound-like animals may baffle water currents in their basal regions. Thus local food availability may be reduced and sedimentation rates increased in these areas (e.g. Chamberlain and Graus, 1975). Similarly, the undersurfaces of plate-like animals may also be exposed to restricted seawater circulation. These growth-forms will also be subject to some self-shading. The probability of tissue death for surface areas of animals subject to such conditions is likely to be high, and may result in local exposure of skeletal surfaces. Thus, on the average, large and complex erect animal forms may have relatively more exposed skeletal surface than runners, sheets, regular mounds or vines.

(i) *Spaces incompletely surrounded by animal projections above the substratum (holes).* Erect or irregular growth patterns may result in formation of spaces incompletely bounded by the branches, lobes, ridges, or other projections of an animal. Such spaces on coral reefs provide shelter for a diverse assemblage of free-living animals which are among the most abundant on the reef. Well known examples include damselfish, gobiosomids, moray eels, sea urchins, shrimps, crabs, brittle stars and anemones (e.g. Bruce, 1976; Clark, 1976; Goldman and Talbot, 1976; Hobson, 1976; Patton, 1976). Kelps, sponges and other sessile organisms provide homes for innumerable similar animals in temperate seas.

Many of these animals have been the focus of considerable research by ecologists and behaviourists because of their characteristically territorial behaviour which greatly increases their potential effect on the wellbeing of their habitat "hosts" or other neighbouring sessile organisms. For example, Vine (1974) and Kaufman (1977) have argued that territorial reef damselfish may significantly alter both the species composition and relative abundance of plants and sessile animals within their territories. The importance of such animals as determinants of plant distributions in many terrestrial systems is well known (e.g. Bartholomew, 1970; Gilbert and Raven, 1975; Janzen, 1970) but their potential significance to marine sessile animal distribution patterns remains largely unexplored.

Whatever their activities, the availability of spaces or holes should be strongly dependent on animal growth-form. By definition, runners, sheets and regular mound-like animals lack well developed projections, and thus lack hole habitats. In contrast, complex-branching trees and plates should provide abundant holes for animals. Vines and irregular mounds should be of intermediate value.

3. Morphological Strategies

There are 36 possible sessile animal shape parameters (plus 36 reciprocals thereof) definable in terms of the nine size parameters. Of these ratios, only eight are required to fully describe the shape characteristics of an animal. Additional ratios, although valuable for perspective, are definable in terms of the first eight and so add no new information (Seal, 1964). Eight shape parameters for the six growth-forms are presented in Table IX. All ratios include A_t. Size parameter/V_t ratios showed similar patterns and ratios involving S_e were often impractical for the geometric approximations used (see previous section). Ratio values in Table IX are comparative ranks for the animal forms considered. Rank values range from 1 (lowest ratio value for any shape parameter) to 6 (highest ratio value for the parameter). Ties (similar ratios) were ranked according to standard procedures outlined by Campbell (1975). No attempt was made to estimate ranks for A_{sk}/A_t due to the strong taxonomic dependence of this parameter. Availability of holes is the eighth shape parameter in Table IX.

Kendall rank order correlations (τ) of shape parameter values for the six sessile animal forms are given in Table X. Only three of the possible 15 pairwise correlations are significantly different from zero at $P \leq 0.05$, $\tau = \pm 0.60$ (5 of 15 at $P \leq 0.11$). All of these are negative associations. Clearly the animal forms exhibit considerable differences in the shape parameters considered.

Runners and vines are entirely committed to a fugitive (escape or refuge-oriented) morphological strategy. All other growth-forms represent increasing commitments (trees ≥ plates > mounds > sheets) to survival within and around their areas of settlement and to maintenance of the integrity of their colony surfaces. Thus, to some degree, they may be interpreted to exhibit a confrontation strategy. The basis for these conclusions is summarized below.

TABLE IX. Rank values of eight shape parameters used to characterize morphological strategies. Rank values from 1 (lowest) to 6 (highest)

Form	V_t/A_t	V_{sk}/A_t	S_e/A_t	S_s/A_t	C_s/A_t	D_s/A_t	H_t/A_t	Holes
Runner	3	1	5	1·5	6	6	1·5	1·5
Sheet	5	2·5	6	1·5	5	4	1·5	1·5
Mound	6	6	4	3	4	3	5	3
Plate	4	4·5	1·5	6	1·5	1·5	3·5	4·5
Vine	1·5	2·5	3	4·5	3	5	3·5	4·5
Tree	1·5	4·5	1·5	4·5	1·5	1·5	6	4·5

(a) *Runners.* Very high C_s/A_t values distinguish runners from all other sessile animal forms. High D_s/A_t ratios separate runners from all forms except vines, and low H_t/A_t ratios from all forms except sheets. High D_s/A_t ratios should maximize the ability of runner-like animals to locate favourable microenvironments (habitat patches) (Fig. 8, Table VII). The potential costs of such a strategy, however, are enormous. High S_e/A_t and C_s/A_t ratios greatly increase the susceptibility of runners to virtually all potentially deleterious substratum-associated processes (Fig. 8, Table VII). Minimal H_t/A_t ratios restrict runners to only those resources available at the substratum–water interface (Fig. 12, Table VIII).

In summary, the probability that any one zooid or unit area of tissue of a runner-like animal will die over any interval of time is high, but the probability that at least one zooid somewhere in the colony will survive long enough for sexual reproduction to occur should also be very high. Buss (this volume) has analysed the morphology of runners differently and comes to similar conclusions.

(b) *Sheets.* Sheets and runners are morphologically quite similar ($\tau = +0.39$ in Table X). The primary difference between them is in their C_s/A_t values. Much lower values of these ratios for sheets decrease the percentage of their tissues exposed to potentially deleterious, peripheral, substratum-associated processes. The potential cost is decreased directional growth (lower D_s/A_t), a matter of considerable importance if the point of recruitment is less favourable than neighbouring substratum areas (see Buss, this volume).

Sheets, like runners, differ from all other sessile animal growth-forms in their restriction to the substratum surface (maximum S_e/A_t and minimum H_t/A_t ratios) which may restrict their access to resources in the water column

TABLE X. Kendall rank order correlations of rank values of eight shape parameters for six animal forms. Ranks data from Table IX. $P = 0.05$ for $\tau = \pm 0.60$; $P = 0.11$ for $\tau = \pm 0.50$. R = runner, S = sheet, M = mound, P = plate, V = vine, T = tree.

S	M	P	V	T	
+0·39	−0·21	−0·61	+0·11	−0·54	R
	+0·21	−0·50	−0·32	−0·61	S
		−0·07	−0·68	0·00	M
			+0·04	+0·39	P
				+0·11	V

and leaves all their tissues exposed to substratum-associated disturbances. Their high S_e/A_t ratios do, however, guarantee that sheets cover the substratum with maximum efficiency and so exclude larvae of other sessile organisms.

(c) *Mounds.* High V_t/A_t ratios distinguish mounds from all other sessile animal forms (for animals whose live tissues are not restricted to their outer surfaces). V_{sk}/A_t ratios are also high. Both of these relationships point towards determinant growth patterns of Group 2 mound-like animals such as sponges, alcyonaceans and ascidians (see p. 517). Otherwise, the mound-like growth form represents a series of adaptive compromises for almost all the shape parameters considered. Mounds achieve greater access to the water column (high H_t/A_t) and relative isolation of surface tissues from potentially deleterious substratum-associated processes (lower S_e/A_t and C_s/A_t) than do sheets, at a potential cost of increased commitment to survival at or near their point of settlement (low D_s/A_t). They are surpassed in these characteristics by plates and trees (and vines except for D_s/A_t). Maintenance of relatively large areas of attachment (higher S_e/A_t), however, should make mounds more resistant to strong water movements, boring organisms, or bottom instability than are vines, plates, or trees.

(d) *Plates.* Like trees, plates differ from all other sessile animal forms in their high degree of isolation of surface tissues from substratum-associated processes (low S_e/A_t and C_s/A_t) at a potential cost of maximum commitment for attachment and support to the animal's area of settlement (low D_s/A_t), increased requirements for support structures and determinant growth (high V_{sk}/A_t). Plates cannot usually grow as high as trees and thus may have less access to the water column. Depending on their orientation, however, their large flattened surfaces may be more effective than the branches of trees in obtaining suspended resources in unidirectional water regimes. A potential cost is greater vulnerability to strong water movements. Plates are potentially superior to trees in their ability to alter bottom conditions adjacent to their area of attachment (high S_s/A_t).

(e) *Vines.* Vines are elevated runners. They achieve the advantages of low commitment to their point of settlement (high D_s/A_t) without many of the potential disadvantages of runners (i.e. high S_e/A_t and C_s/A_t). Minimal V_t/A_t ratios distinguish vines from all other sessile animal forms except trees. Unlike trees they potentially benefit from many areas of substratum support but at the possible cost of decreased accessibility to resources in the water column (low H_t/A_t).

(f) *Trees.* High access to resources in the water column (maximal H_t/A_t) distinguishes trees from all other sessile animal forms. Trees differ from plates in their much lower V_t/A_t ratios; otherwise the two forms are quite similar ($\tau = +0.39$ in Table X) in their considerable isolation of tissues from substratum-associated processes and strong commitment to their limited areas of attachment (minimal S_e/A_t, C_s/A_t and D_s/A_t ratios). The potential cost is complete dependence of distal tissues on the structural integrity of their areas of basal attachment.

PREDICTIONS

Deduction of morphological strategies is of value only to the extent that one can predict properties of real sessile animals in their natural environments as a function of growth form. For example, if the conclusion is valid that runners are escapists incapable of confrontation, then one might predict that the extension rates of runner-like animals along the substratum would, on the average, be more rapid than those of sheets or mounds. This is a testable hypothesis and there are at least three general categories of such predictions of animal traits which fall out from the model. These are their:
1. Life-history parameters;
2. Internal morphology and colony integration;
3. Distributions in space and time.

Here I will detail some of the more straightforward predictions in each category and briefly indicate what I believe are some relevant data. Numerous supporting examples for these ideas exist. Nevertheless, the following discussion by no means constitutes a systematic effort to disprove my contentions and thus fails as a serious test of their generality.

The problem of proof lies in the high degree of significant correlations which exists among natural processes. Suppose, for simplicity, the unlikely proposition that the size or shape of an animal structure is a response (developmental or evolutionary) to a single environmental process such as sedimentation rate. The amount of sedimentation at any locality will almost certainly be significantly correlated with many other processes such as current velocity, wave action, etc. Identification of sedimentation stress (and not, for example, wave action) as the process important to the development of the structure must therefore involve systematic comparisons of the intensity of these processes over a wide range of natural environments and reasonable demonstration (ideally experimental) of the mechanism(s) of cause and effect. The failure of biologists to recognize this problem has greatly obscured our understanding of sessile animal form.

1. Life-History Parameters

Here I consider variations in age-specific growth rate, age of first sexual reproduction, and fecundity as a function of growth form. In each case, predictions are based on the assumption of the principle of allocation, i.e. the assumption that there is only a limited amount of energy available to any organism. Thus increased allocation of that energy to growth, for example, must result in a decreased allocation to some other process such as sexual reproduction. Support for this assumption is provided by various theoretical constructs (Charnov and Schaffer, 1973; Schaffer and Gadgil, 1975; Williams, 1975) and numerous studies of growth, competition and reproduction in plants (Abrahamson and Gadgil, 1973; Gadgil and Solbrig, 1972; Solbrig, 1972). Life-history predictions for the different growth forms are summarized in Table XI and discussed below.

(a) *Variations in potential growth rates.* Problems associated with high V_t/A_t and/or high V_{sk}/A_t ratios should result in determinant growth patterns for plates, trees, and Group 2 mounds (animals whose live tissues are not restricted to their colony surfaces). Initial growth rates of these animals should be high in order to raise animal tissues above the substratum before overgrowth by other forms. Numbers of zooids or A_t values should thus initially show exponential increase and then level off at near constant values. Runners, sheets, vines and

Group 1 mounds are not subject to such restrictions and should exhibit indeterminant growth. Numbers of zooids or A_t values of these forms should increase exponentially (or linearly for non-branching runners and vines) throughout life.

TABLE XI. Predicted life-history parameters for different animal growth forms

Growth Form	Change in growth rate with age	Age of first reproduction	Fecundity
Runner	None (indeterminant growth)	Early	High
Sheet	None (indeterminant growth)	Early to Middle	Medium to high
Mound			
(Group 1)	None (indeterminant growth)	Early to Middle	Medium to high
(Group 2)	Decreases (determinant growth)	Middle to Late	Medium to low
Plate	Decreases (determinant growth)	Late	Low
Vine	None (indeterminant growth)	Early	High
Tree	Decreases (determinant growth)	Late	Low

Growth data for sponges and colonial animals are poor, with entire groups (e.g. encrusting sheet-like sponges) unstudied. What we know, however, generally supports the above predictions. Thus, well documented cases of determinant growth are of tree-like animals such as gorgonians (Grigg, 1974) and closely branched corals (Buddemeier and Kinzie, 1976), and of Group 2 mounds such as massive sponges (Reiswig, 1973). Otherwise, regardless of form, most colonial animals exhibit indeterminant growth (Braverman, 1963; Buddemeier and Kinzie, 1976; Davis, 1971; Elvin, 1976; Hayward, 1973; Hayward and Ryland, 1973; Stebbing, 1971a). Ascidian growth patterns are exceedingly complex (Millar, 1971), but often involve alternating periods of asexual exponential growth and low growth (episodes of sexual reproduction) (Haven, 1971; Sugimoto and Nakauchi, 1974).

(b) *Age of first sexual reproduction.* Age of first sexual reproduction should be inversely correlated with growth rate for those animals whose growth rates vary with age. Thus runners, sheets, vines and many Group 1 mounds may be expected to commence sexual reproduction early in colony development whereas trees, plates and Group 2 mounds should delay sexual reproduction until after their initial rapid growth phase.

What data there are tend to support these predictions. Thus sexual development of erect (mound- or tree-like) octocorals tends to be delayed relative to other growth forms of these animals (Bayer, 1973; Grigg, 1970). Similarly, hydroid sexual reproductive polyps and ovicells of cheilostome ectoprocts often appear later in the development of erect colonies than in encrusting, especially runner-like forms (Boardman and Cheetham, 1973; Hyman, 1940). Sexual reproduction in mound-like sponges does not occur until after an initial phase (perhaps several years) of relatively rapid growth (Reiswig, 1973) whereas encrusting forms may show earlier sexual development (Fell, 1976). Colonial ascidian life-histories are too complex to generalize (Millar, 1971) and we know almost nothing about the age of first reproduction of scleractinian corals (Connell, 1973).

(c) *Fecundity*. To the extent that they may require specialized maintenance, support, or defence structures (see next section), trees, plates and Group 2 mounds should exhibit lower fecundities (rates and total average fecundity per zooid) than runners and vines. Group 1 mounds and sheets with defence structures should show intermediate fecundity.

Direct data on fecundity comparable to seed production data for plants are virtually non-existent for sponges and colonial animals. Recruitment data, however, provide indirect evidence for fecundity which support the model's prediction. Recruitment studies in cryptic reef environments (Jackson, unpublished data) and for Massachusetts subtidal hard substrata (Osman, 1977) show that runners, vines and sheets apparently have higher fecundities than do trees, plates and mounds.

2. Internal Morphology and Colony Integration

Features included here are zooid shape, frequency of zooid inter-connections and colony integration (co-ordination of functions), and the development and intracolony distribution of zooid polymorphism or specialized colony "organs". Predictions for each growth-form are summarized in Table XII. Sponges are not discussed because of the uncertain status of their functional units (Simpson, 1973; Hartman and Reiswig, 1973).

(a) *Zooid shape*. Animal forms characterized by strongly directional growth (runners, vines and trees) should have relatively more elongate and/or more widely spaced zooids than other taxonomically related sheet-, mound-, or plate-like animals.

Persual of almost any monograph of colonial animal morphology lends strong support to this prediction. Zooids of runner-, vine- and tree-like hydroids, octocorals, ctenostome ectoprocts and colonial ascidians are commonly more widely spaced than zooids of related sheet- or mound-like forms (e.g. illustrations and descriptions in Bayer, 1973, Beklemishev, 1969; Berrill, 1950; Hyman, 1940, 1959). Polyps of ahermatypic scleractinian coral runners, vines and trees

TABLE XII. Predicted zooid features and levels of colony integration for different animal growth forms

Growth form	Zooid shape	Zooid polymorphism	Intracolony distribution of polymorphs	Other special characteristics	Degree of integration and regeneration
Runner	Elongate	±	Uniform	—	Low
Sheet	Equidimensional	+	Clumped	Smooth periphery	Medium to high
Mound	Equidimensional	+	Clumped	Smooth periphery	Medium to high
Plate	Equidimensional	+	Clumped	"Roots" and flexible skeletons	High
Vine	Elongate	±	Uniform	Attachment	Low
Tree	Elongate	+	Clumped	"Roots" and flexible skeletons	High

are more widely spaced than other forms of the same genera or species (Wells, 1956). Among the cheilostomes, zooids of runners like *Pyripora* and *Hippothoa* are more elongate than those of sheets like *Wilbertopora* and other *Hippothoa* (Cheetham, 1974; Cheetham and Lorenz, 1976; Jackson, unpublished data). Similarly, zooids of erect unilaminar fronds of *Membranipora arborescens* are more elongate than zooids in sheet-like portions of the same colonies (Cook, 1968). However, zooid outlines of tree-like cheilostomes vary considerably as a function of their mode of growth and thus may be elongate or equidimensional (Cheetham, 1971; Cheetham and Lorenz, 1976).

(b) *Zooid polymorphism.* Zooids (or tissue areas) specialized for sexual reproduction or defence should be more varied and abundant among animal forms committed to survival within and around their areas of attachment (trees ⩾ plates > mounds ⩾ sheets) than among those with refuge-oriented growth forms (runners, vines). In the latter case, the probability of zooid mortality is high, and sexual reproduction may often depend on the activities of a few isolated zooids.

Thus polymorphism would decrease the probability that a surviving zooid would be capable of sexual reproduction.

The predicted patterns seem well substantiated. Stoloniferous growth forms (runners, vines) of octocorals are monomorphic whereas sheets, mounds and trees of these animals are commonly polymorphic (Bayer, 1973). Cheilostomes are characteristically polymorphic except for some runner-like forms such as *Pyripora* (Cheetham, 1974). Polymorphism is also more highly developed among sheet- and mound-like colonial ascidians than among stoloniferous forms (Berrill, 1950). True polymorphism is apparently absent among scleractinian corals (Coates and Oliver, 1973) but this subject requires histological investigation. Apical polyps of *Acropora* are larger and more exsert than other polyps of these animals (Pearse and Muscatine, 1971), and interestingly in this context, the remarkably diverse *Acropora* are predominantly tree- and plate-like forms.

(c) *Intracolony distribution of polymorphic zooids.* Animals with forms varyingly adapted for confrontation (sheets, mounds, plates and trees) should exhibit non-uniform intracolony distributions of polymorphic zooids. Defence structures (spines, nematocysts, avicularia) or allelochemicals might tend to be concentrated along animal boundaries (C_s) where probabilities of deleterious substratum-associated interactions are greatest. Reproductive zooids should show opposite distributions. In contrast, any polymorphs which might occur among runners or vines should be uniformly distributed throughout such colonies.

Numerous data tend to support these predictions for cheilostomes (Al-ogily and Knight-Jones, 1977; Boardman and Cheetham, 1973; Cheetham, 1974; Stebbing, 1973a, b). Ascidians, hydroids and octocorals show intracolonial separation of reproductive, pumping or feeding, and defensive polymorphs (Bayer, 1973; Berrill, 1950; Braverman, 1963; Hyman, 1940; Sugimoto and Nakauchi, 1974).

(d) *Other intracolony features.* Sheets and mounds may be expected to adopt a smooth, circular outline which decreases the probability per zooid of adverse substratum-associated peripheral interactions (minimizes C_s/A_t). Frontal budding, or other specialized vertical growth processes, and elevated growth margins of sheets and mounds would also reduce the frequency of potentially deleterious peripheral interactions. Such features are characteristic of cheilostome and cyclostome ectoprocts (Boardman and Cheetham, 1973; Jackson and Buss, 1975; Stebbing, 1973b).

Support problems for plates and trees are potentially severe (depending on animal size and environmental conditions); much energy may be invested in attachment structures ("roots"), protection of the basal attachment area (exclusion of boring organisms from substratum) and specialized, possibly flexible skeletal materials. Flexible skeletons must require production of energetically expensive organic materials such as gorgonin or spongin fibres. These features are well developed among erect forms of hydroids, octocorals and cheilostomes (Bayer, 1973; Cheetham, 1971; Hyman, 1940).

Runners should lack intracolony variations; vines should be the same except for attachment structures. This is indeed the case for many hydroids, octocorals, cheilostomes and ascidians (Bayer, 1973; Berrill, 1950; Boardman and Cheetham, 1973; Cheetham, 1974; Hyman, 1940).

(e) *Colony integration and regeneration.* Unlike runners, mortality of small areas of a sheet-like animal significantly increases its periphery in contact with the substratum (greater C_s/A_t). Thus the probability of mortality of adjacent zooids or tissues also increases (due to overgrowth by organisms which recruit onto a newly killed area of the animal). The same will be true for other Group 1 animals, i.e. death of small areas of surface tissues opens up sites for larval recruitment. For mounds, plates and trees the potential cost is severe due to local concentrations of zooids and high (all or nothing) dependence on areas of attachment. Like runners, this is not a problem for vine-like forms. Thus the physiological integration of colonies (flow of food resources, co-ordination of behaviour) and the ability for rapid regeneration of surface tissues should tend to be higher for sheets, mounds, plates and trees than for runners and vines. Such increased integration presumably requires abundant interzooidial connections.

Boardman and Cheetham's (1973) analysis of morphological integration (and thus inferred physiological integration) among cheilostomes strongly supports these predictions. So also do Bayer's (1973) review of octocorals and Berrill's (1950) of ascidians. Much, however, is based on inference; systematic comparative studies of actual functional differences are rare.

3. *Distribution in Space and Time*

Processes which affect local distributions are no different from those which affect geographic range, or temporal distributions in ecological time (community progression or "succession") and evolutionary time (Jackson, 1974, 1977a; MacArthur, 1972). Predictions for each of these distributional categories as a function of growth form are summarized in Table XIII and presented below.

(*a*) *Between habitat distributions.* Here I discuss variations in animal shape distributions within the same local geographic area. Discussion of tests of predictions is limited to Jamaican reef environments. Consider four general situations:
1. Environments in which physical disturbance processes or predators prevent utilization of all available substrata or food.
2. Environments with low disturbance, all space occupied, but unlimited availability of food or light.
3. Environments with low disturbance, all space occupied, and limited availability of food or light.
4. Epibiotic habitats.

TABLE XIII. Predicted environmental distributions for different animal growth forms

Type of environment	Ranks of predicted success
High disturbance	
Strong water movements	R, S, M > V > P, T
High sedimentation	T > P, M, V > S, R
High substratum instability	R, S, V > M > P, T
Abundant predators	R, S, V > M > P, T
Low disturbance	
Unlimited food/light	S, M ⩾ V ⩾ P, T > R
Limited food/light	T, P > M, V > S > R
Epibiosis	V, R > S > M, P, T

Most attention to sessile animal growth forms has been focused on the first category. In such disturbance environments, any growth-forms which can survive a particular disturbance may be present. Strong water movements or substratum instability should be more deleterious to plates and trees (especially those with inflexible skeletons) than to other forms. In contrast, high sedimentation rates should be most harmful to animals lying on or near the substratum (runners, sheets, low mounds, or vines). Excluding probable differences in palatability, predators should have the least effect on low lying, cryptic, or rapidly growing (regenerating) forms.

There are a vast amount of descriptive data relevant to these predictions, little of which leads to any rigorous conclusions. Here I briefly mention two Jamaican reef habitats. The environments and associated growth forms discussed are illustrated in the references cited. The *Acropora palmata* zone is a highly disturbed environment characterized by very strong water movements, sediment scour and intense predation/grazing by the sea urchin *Diadema antillarum* (Goreau, 1959; Goreau and Goreau, 1973; Kinzie, 1973). Growth-forms here include tall, tree-like *Acropora palmata* (almost no small colonies are present); mound-like corals, and low-lying, sheet-like scleraxonid gorgonians and zoanthids. There is a great deal of bare and/or closely cropped substratum.

Just seaward of this region is the mixed zone, an area also exposed to strong water and sediment movements, and intense predation (Kinzie, 1973). As the name of the zone implies, no small group of species or growth-forms predominates. Tall, tree-like gorgonians and corals; mound- and sheet-like sponges and corals; sheet-like zoanthids and scleraxonids; and numerous vine-like animals (sponges, scleraxonids, corals) are all in evidence within small areas of the reef. Considerable local variations in the relative abundance of different growth forms occurs along the coast within the mixed zone, which may even be replaced by a buttress zone in many areas (Goreau, 1959; Goreau and Goreau, 1973; Kinzie, 1973). These variations presumably reflect differences in the relative intensity of different disturbance processes.

If disturbance levels are relatively low, space or food/light may limit sessile animal distributions. If food is super-abundant, animals able to cover the entire substratum surface (sheets, mounds) should be able to exclude, or maintain at low levels, animals of more erect form (low S_e/A_t) such as trees and plates. In contrast, at high food levels trees and plates should tend to have little effect on the sessile animals beneath them. Dense growths of vines which trap sediments may also cover large areas. Otherwise vines should be easily excluded or limited to epibiotic habits. Runners, except epibionts, will be easily excluded in such competition environments. If food levels are potentially limiting (either reduced overall supply or locally decreased food levels) then the advantage should swing towards more erect growth-forms, especially trees and plates, with increased access to the water column (upper levels of zone 2, Fig. 1). In either case (food abundant or limiting), increased competition should result in dominance by fewer growth-forms than are prominent in high-disturbance environments.

Reef environments showing both high percentage cover ($> 90\%$) by organisms and abundant signs of competitive interactions include the walls of buttresses, tops of pinnacle reefs, vertical walls not subjected to much sediment downpour, undersurfaces of corals, and the sides and ceilings of reef caves (Goreau, 1959; Hartman and Goreau, 1970; Jackson, 1977a; Jackson and Buss, 1975; Kinzie, 1973; Lang, 1970, 1973, 1974). Many of these environments are physically and biologically heterogeneous (often zoned).

Areas of similar physical conditions, however, are commonly dominated by only one or two growth forms. Examples include coral plates on tops of pinnacles; mound- and sheet-like corals, sponges, and ectoprocts on vertical walls; and sheet sponges, cheilostomes and ascidians on the distal undersurfaces of foliaceous corals (see above references).

One strategy of avoiding competition, or even of taking advantage of other larger organisms, is to grow as epibionts on surface tissues or skeletons of large, well established plants and animals. This may be entirely a refuge-oriented strategy ideal for poor competitors like runners (see Buss, this volume), or vines, which involves no special relationship with the substrate organism. Or, an epibiont might have to pay for its otherwise free substratum (no cost of skeleton production, etc.) by manufacture of defensive chemicals or structures which reduce predation on the two organisms.

Vines and runners (also solitary animals) should obscure comparatively little of their host organism's surface tissues and thus are more likely to be tolerated than sheets or mound-like animals. They also have low mass. Unless much smaller than the host, erect animals are unlikely epibionts for stability reasons. The problem of scale is of fundamental importance to interpretation of epibiotic interactions (Jackson, 1977a).

Epibionts on dead skeletal (e.g. coral) surfaces include all the growth-forms on the reef. Epibionts on living surfaces or small dead areas or lesions of large reef animals (gorgonians, sponges) are predominantly solitary animals (bivalves, worms), runners and vines (parazoanthids, other sponges), and sheets (sponges, milleporids, cheilostomes) (Jackson, 1977a; Kinzie, 1973; Sarà and Vacelet, 1973). Epibionts of other growth forms are uncommon on these animals.

(b) *Geographic range.* Other factors being equal, there should be a positive correlation between the range of environmental conditions an animal can survive, its fecundity, and its geographic range (Jackson 1974, 1977a). Thus if predictions presented in Tables XI and XIII are sustained, species of runners and vines should tend to be more widely distributed than species of mounds and sheets; and species of plates and trees should have the most limited geographic range ($R \geqslant V \geqslant S, M > P, T$) (Hansen, 1978; Jackson 1974, 1977a; Koch, 1977).

Alternatively, the more growth-forms that a species assumes, the greater the variety of environments it may be able to survive in (see Discussion). Thus polymorphic species (in the overall shape sense) should be more widely distributed than monomorphic species. Adequate data are not yet available to evaluate these alternatives.

(c) *"Succession".* In high disturbance environments, changes in physical conditions or predator abundances will be the primary cause of any temporal changes in relative growth form abundance. Newly available substrata (e.g. disturbance patches of Dayton (1971) or Levin and Paine (1974)) should be colonized first by growth forms with highest fecundity (runners, vines > sheets ≥ mounds > plates, trees). If disturbance levels are low, these areas should be overgrown by sheets and mounds. If food/light levels stay high, sheets and mounds may continue to dominate the substratum; if not, then plates and trees may slowly tend to dominate and the frequency of epibionts should also increase. Subsequent disturbance (patch formation) may lead to local repetition of the entire sequence. An important point here is that sheets, mounds, or vines may grow into patches from adjacent areas and thus overgrow new recruits. In contrast, trees and plates, like solitary animals (Jackson, 1977a) can usually only re-establish control of such bared areas through the drawn out process of sexual reproduction, recruitment and growth.

Succession experiments in artificial cryptic reef environments (Jackson, 1977a, b, unpublished data; John Sutherland, unpublished data) reveal a consistent progression from early dominance (relative percent cover) of substrata by runners, vines and solitary animals to coverage of almost the entire substratum by sheet- and mound-like sponges, ascidians and cheilostomes. There is some indication that comparatively enclosed substrata (placed within boxes designed to simulate reef crevices and caves) suffer lower food availability than in more open cryptic environments. Tree-like forms appeared commonly within these boxes after two years, but were absent on the undersurfaces of the boxes which were exposed to more open water conditions.

The temporal patterns observed in these experiments are not unlike those reported from other environments (e.g. algae and invertebrates on kelp forests (Foster, 1975a, b). Whether they result from differences in recruitment and growth rates of organisms with different growth forms (i.e. simple community progression) or are truly successional (i.e. presence of one or more organisms required for colonization by other organisms) is unknown (Connell, 1972; Connell and Slayter, 1977).

(d) *Evolution*. The early evolution of any new animal lineage will be strongly influenced by competition with previously established groups. Thus first representatives of a group may be expected to exhibit a fugitive, refuge-oriented strategy. In this case, the earliest species of new adaptive breakthroughs (e.g. the first scleractinian corals or cheilostomes) should be either solitary animals or morphologically simple, small, runners, sheets, or mounds. Attachment requirements or other intracolony specializations make vines, plates and trees unlikely ancestors.

The early evolutionary history of most sessile animal groups is either missing from the fossil record or so poorly preserved as to preclude meaningful interpretation. One prominent exception is the cheilostome ectoprocts, the record of which is in accordance with the predictions of the morphological strategy model. The first cheilostomes are simple, occasionally branching, runner-like fossils apparently without zooid polymorphism or astogenetic change (Cheetham 1974; Larwood, 1974). These forms (*Pyriporopsis*) persist alone from the Upper Jurassic to Lower Cretaceous when a number of morphologically slightly more complex genera appear (sheet-like colonies with varying degrees of zooid polymorphism and astogenetic change such as *Rhammatopora*, *Wawalia*, and *Wilbertopora*) (Cheetham, 1974; Cheetham and Cook, 1978). Only in the Upper Cretaceous (some 40–50 million years after the first appearance of *Pyriporopsis*) does the great adaptive radiation of the cheilostomes occur (Boardman and Cheetham, 1973; Cheetham and Cook, 1978).

The first scleractinian corals were apparently also small morphologically simple, sheet- to mound-like forms (A. Coates, personal communication).

For sessile animals, the probability of allopatric speciation or chronospeciation (sympatric temporal morphological change) should be inversely proportional to a species' geographic range and extent of larval dispersal (Jackson, 1974; Scheltema, 1977; Hansen, 1978; Koch, 1977). Early "successional" or fugitive species should have a lower probability of speciation than confrontation species of those specialized for survival under exceptional physical environmental conditions. Thus the probability of evolutionary change should be highest for species of plates and trees; intermediate for sheets, mounds and vines; lowest for runners (P, T > S, M ⩾ V ⩾ R). Data are not yet adequate to test these predictions. It is certainly suggestive, however, that apparent *Pyriporopsis* still survive in Recent seas (Cheetham, 1974).

COMPARISONS WITH PLANTS

The study of plant life-forms has advanced considerably beyond that of sessile animals, and much of what is presented here owes a strong intellectual debt to the classical works of Schimper (1930), Raunkiaer (1934), Clements *et al.* (1929), Clausen *et al.* (1948) and the more recent experimental studies of Harper and co-workers in Wales (e.g. Harper and White, 1974; Harper *et al.*, 1970). As the names runner, vine and tree make obvious, comparison of hydroid, ectoproct, or gorgonian forms with those of plants is inevitable. However, although often helpful, such comparisons can also be misleading and I have purposely avoided reference to plants in the development of the model. For example, attachment and support are not the only functions of the root systems of trees and, unlike leaves, photosynthesis may not be the primary function of distal polyps of many tree-like corals. Scaling problems are also vastly different due to differences in the fluid properties of air and water. This problem does not apply to aquatic plants such as giant kelps, but it is a fact that sessile animals fail by an order of magnitude to attain the heights reached by aquatic plants. Thus it seemed preferable to try to build independently a model of sessile animal form, then, if the predictions of the model are sustained, to make comparisons with plants at that time. Well demonstrated similarities of form and function would of course be of tremendous evolutionary interest. The work of Steneck and Adey (1976), Neushul (1972) and Foster (1975a, b) demonstrates the promise of such comparisons.

DISCUSSION: THE IMPORTANCE OF SPECIES OF VARIABLE GROWTH-FORM

Some sessile animal species assume more than one growth-form; others do not. Here I discuss why I believe this variety exists in relation to the ideas of this paper and Braverman's (1969) theory of colonial animal development. I then outline the importance of such variable species for testing the assumptions and predictions of the morphological strategy model.

The advantages of morphological plasticity seem obvious: sponges or colonial animals of a species able to adopt more than one growth-form should be able to survive to reproduce in a greater variety of environments than do monomorphic species. Indeed, although data are sketchy, there appears to be a positive correlation between the number of growth-forms exhibited by hermatypic coral species on the north coast of Jamaica and the number of reef zones in which they occur (Jackson, unpublished data). The same is true in Jamaican cryptic environments where the highly plastic *Madracis pharensis pharensis* (grows in runners, sheets, mounds and vines) is more ubiquitous than any other solitary or colonial cryptic coral species (Jackson, in preparation).

The question, then, is why do not all sponges and colonial animal species grow in a diversity of forms? Different growth-forms within a species may be genetic polymorphs or environmentally induced phenotypes of more or less similar genotypes. In the second case, organisms have the advantage of flexibility of response to changing, post recruitment, environmental conditions but the potential disadvantage of being "duped" into growing in a form of short-term benefit but long-term disadvantage. For example, assume the model prediction that it is energetically more expensive to grow as a Group 2 mound than as a sheet (requires more skeletal material and perhaps special pumping structures). Consider an animal which can grow in either form that settles when substratum surface conditions are favourable. The animal then grows as a sheet but, some time later the substratum is inundated by sediments and the animal dies. In such an environment, pre-programmed, rapid, vertical growth, although initially less advantageous, might have the longer term benefit of more frequent survival to sexual maturity. These arguments apply even more strongly to trees and plates since the advantages of these growth forms can only become apparent after the animals have achieved some significant vertical stature.

Given the potential advantages and disadvantages of such variability, it is of interest to ask which combinations of growth-forms are more or less likely to occur within a single species. This question can be approached in a descriptive (iterative) or generative (recursive) fashion (based on Braverman, 1969; Braverman and Schrandt, 1964). In the first case, one may ask how similar two different growth forms are to one another in terms of their shape parameters (i.e. their values of τ in Table X). Examination of Table X suggests that runner–sheet, plate–tree and sheet–mound combinations are most likely, and that runner–plate, runner–tree, sheet–plate, sheet–tree and mound–vine combinations are less likely. Indeed, this is often the case. Runner–sheet transitions occur frequently among cheilostomes and hydroids; sheet–mound transitions are frequent among cheilostomes with frontal budding and many scleractinian corals; plate–tree transitions occur commonly among gorgonians and cheilostomes (e.g. Barnes, 1973; Boardman and Cheetham, 1973; Brakel, 1976; Wainwright *et al.*, 1976). Strongly negatively correlated form pairs seem less common but there are prominent exceptions such as the sheet- and plate-like forms of the abundant Caribbean coral *Montastrea annularis* (Barnes, 1973; Dustan, 1975) and the cheilostome *Steginoporella magnilabris* (Jackson, unpublished data; Jackson and Buss, 1975) and the common occurrence of sheet–tree transitions among cyclostomes (Harmelin, 1975).

The generative approach requires identification of the sets of operatives (rules) of development involved in the growth of any form (Braverman, 1969). For example, the growth of a runner-like hydroid, *Podocoryne carnea*, can be described by three parameters: (i) stolon extension, (ii) stolon branching, and (iii) new hydranth formation (Braverman, 1969). Of interest here is that change in only one simple rule, the frequency of branching, could alter the form of a loosely-branching runner to a densely-branching sheet, or vice versa. This is also true for the other positively correlated forms: plate to tree (again one branching rule) and sheet to mound (amount of horizontal versus vertical zooid budding or tissue growth). In contrast, transition from one to another significantly negatively correlated form in Table X may require changes in more than one rule (except for the sheet–plate transition which may explain the more common occurrence of such forms). Braverman's ideas provide an important framework for interpretation of the interrelationship between developmental and ecological aspects of morphological strategies.

Multi-form species are ideal tools for testing the assumptions and predictions of the morphological strategy model. One can ask what is the advantage of some form relative to another in terms of a specific process, or what is the importance of different processes in determining expression of form (e.g. Winston's (1976) experimental demonstration of the importance of food quality and quantity to the transition from runner to sheet in the cheilostome *Conopeum tenuissimum* and Crowell's (1957) similar study of the influence of nutrition on the growth form of the hydroid *Campanularia flexuosa*). The advantage of such intraspecific comparisons is manifest in the opportunity for laboratory demonstration of cause and effect. Although such experiments can never demonstrate that other processes might not have a similar effect, they can demonstrate the potential importance of the mechanisms examined. Field experiments may then be used to establish whether the particular mechanism is in fact operating in a particular environment. In contrast, interspecific comparisons tend of necessity to be descriptive. For example, congeneric species of different growth forms are often observed to occur in different environments (e.g. Harmelin, 1975; Riedl, 1966, 1971). As we have seen, however, interpretations of such patterns as due to variations in any particular process are unfortunately always open to serious question.

ACKNOWLEDGEMENTS

Leo W. Buss stimulated me to explore beyond comparisons of solitary and colonial animal strategies to consider variations in colonial animal forms. He, A. H. Cheetham, and J. C. Lang contributed greatly to the development of the ideas presented here. B. D. Keller, N. A. Knowlton, L. S. Land, B. A. Menge, G. J. Vermeij, F. Vosburgh, C. M. Wahle and S. A. Woodin provided additional useful criticism. This paper was prepared while I was in residence as a Research Associate of the Department of Paleobiology, National Museum of Natural History, Smithsonian Institution. Additional support was provided by NSF Grants GA-35443 and DES 72-01559. To all I am grateful.

REFERENCES

ABRAHAMSON, W. G. and GADGIL, M. (1973). Growth form and reproductive effort in goldenrods (*Solidago*, Compositate). *Am. Natur.* **107**, 651–661.
AL-OGILY, S. M. and KNIGHT-JONES, E. W. (1977). Anti-fouling role of antibiotics produced by marine algae and bryozoans. *Nature, Lond.* **265**, 728–729.

BAKUS, G. J. (1973). The biology and ecology of tropical holothurians. *In* "The biology and geology of coral reefs II: Biology 1" (O. A. Jones and R. Endean, eds), pp. 326–368. Academic Press, New York.

BAKUS, G. J. and GREEN, G. (1974). Toxicity in sponges and holothurians: a geographic pattern. *Science* **185**, 951–953.

BARNES, D. J. (1973). Growth in colonial scleractinians. *Bull. mar. Sci.* **23**, 280–298.

BARNES, D. J., BRAUER, R. W. and JORDAN, M. R. (1970). Locomotory responses of *Acanthaster planci* to various species of corals. *Nature, Lond.* **228**, 342–344.

BARTHOLOMEW, B. (1970). Bare zone between California shrub and grassland communities: the role of animals. *Science* **170**, 1210–1212.

BATHURST, R. G. C. (1971). Carbonate sediments and their diagenesis. "Developments in sedimentology 12" Elsevier, Amsterdam.

BAYER, F. M. (1973). Colonial organization in octocorals. *In* "Animal Colonies, Development and Function through Time" (R. S. Boardman, A. H. Cheetham and W. A. Oliver, Jr., eds.), pp. 69–73. Dowden, Hutchinson, and Ross, Stroudsburg, Pennsylvania.

BEKLEMISHEV, W. N. (1969). "Principles of comparative anatomy of invertebrates. Vol. 1. Promorphology." Translated by J. M. MacLennan. University of Chicago Press, Chicago.

BERRILL, N. J. (1950). "The Tunicata, with an account of the British species." Ray. Society, London.

BERTRAM, G. C. L. (1937). Some aspects of the breakdown of coral at Ghardaqa, Red Sea. *Proc. zool. Soc. Lond.* **106**, 1011–1026.

BISHOP, J. W. and BAHR, L. M. (1973). Effects of colony size on feeding by *Lophopodella carteri* (Hyatt). *In* "Animal Colonies, Development and Function through Time" (R. S. Boardman, A. H. Cheetham and W. A. Oliver, Jr., eds), pp. 433–438. Dowden, Hutchinson and Ross, Stroudsburg, Pennsylvania.

BOARDMAN, R. S. and CHEETHAM, A. H. (1973). Degrees of colony dominance in stenolaemate and gymnolaemate Bryozoa. *In* "Animal Colonies, Development and Function through Time" (R. S. Boardman, A. H. Cheetham and W. A. Oliver, Jr., eds), pp. 121–220. Dowden, Hutchinson and Ross, Stroudsburg, Pennsylvania.

BRAKEL, W. H. (1976). The ecology of coral shape: microhabitat variation in the colony form and corallite structure of *Porites* on a Jamaican reef. Ph.D. diss., Yale University, New Haven, Connecticut.

BRAVERMAN, M. H. (1963). Studies on hydroid differentiation. II. Colony growth and the initiation of sexuality. *J. Embryol. exp. Morph.* **11**, 239–253.

BRAVERMAN, M. H. (1969). Towards a general theory of development. Pittsburg Section I.E.E.E. Modeling and Simulation Conference Proceedings, 213–218.

BRAVERMAN, M. H. and SCHRANDT, R. G. (1964). Colony development of a polymorphic hydroid as a problem in pattern formation. *Symp. Zool. Soc. Lond.* **16**, 169–198.

BRIEN, P. (1973). Les démosponges, morphologie et reproduction. *In* "Traité de Zoologie 3 (1). Spongiaires" pp. 133–461. Masson et Cie Éditeurs, Libraires de l'Académie de Médecine, Paris.

BRUCE, A. J. (1976). Shrimps and prawns of coral reefs, with special reference to commensalism. *In* "Biology and Geology of Coral Reefs III: Biology 2" (O. A. Jones and R. Endean, eds), pp. 38–94. Academic Press, New York.

BRYAN, P. G. (1973). Growth rate, toxicity and distribution of the encrusting sponge *Terpios* sp. (Hadromerida: Suberitidae) in Guam, Mariana Islands. *Micronesica* **9**, 237–242.

BUDDEMEIER, R. W. and KINZIE III, R. A. (1976). Coral growth. *Oceanogr. marine Biol. ann. Rev.* **14**, 183–225.

BURKHOLDER, P. R. (1973). The ecology of marine antibiotics and coral reefs. In "Biology and Geology of Coral Reefs" Vol. 2 (O. A. Jones and R. Endean, eds), pp. 117–182. Academic Press, New York.

BUZAS, M. A. (1969). Foraminiferal species densities and environmental variables in an estuary. *Limnol. Oceanogr.* **14**, 411–422.

CAMPBELL, R. C. (1975). "Statistics for biologists" 2nd edition. Cambridge University Press.

CHAMBERLAIN, J. A., Jr. and GRAUS, R. R. (1975). Water flow and hydromechanical adaptations of branched reef corals. *Bull. marine Sci.* **25**, 112–125.

CHARNOV, E. L. and SCHAFFER, W. M. (1973). Life history consequences of natural selection: Cole's result revisited. *Am. Natur* **107**, 791–793.

CHEETHAM, A. H. (1971). Functional morphology and biofacies distribution of cheilostome Bryozoa in the Danian Stage (Paleocene) of southern Scandinavia. *Smithsonian Contrib. Paleobiol.* **6**, 1–87.

CHEETHAM, A. H. (1974). Taxonomic significance of autozooid size and shape in some early multiserial cheilostomes from the Gulf Coast of the U.S.A. *Bryozoa 1974. Docum. Lab. Géol. Fac. Sci. Lyon.* Hors series 3 (fasc. 2). 547–564.

CHEETHAM, A. H. and COOK, P. L. (1978). "General features of the class Gymnolaemata. Treatise on Invert. Paleontol. Bryozoa." Part G, revised. (In press.)

CHEETHAM, A. H. and LORENZ, D. M. (1976). A vector approach to size and shape comparisons among zooids in cheilostome bryozoans. *Smithsonian Contrib. Paleobiol.* **29**, 1–55.

CHIBA, Y. and KATÔ, M. (1966). Interspecific relation in the colony formation among *Bougainvillia* sp. and *Cladonema radiatum* (Hydrozoa, Coelenterata). *Sci. Rep. Tôhoku Univ.* (Ser. IV) **32**, 201–206.

CLARK, A. M. (1976). Echinoderms of coral reefs. In "Biology and Geology of Coral Reefs III: Biology 2" (O. A. Jones and R. Endean, eds), pp. 95–124. Academic Press, New York.

CLAUSEN, J., KECK, D. D. and HIESEY, W. M. (1948). Experimental studies on the nature of species. III. Environmental responses of climatic races of *Achillea. Carnegie Inst. Washington Publ.* No. 581, 1–129.

CLEMENTS, F. E., WEAVER, J. E. and HANSON, H. C. (1929). Plant competition, an analysis of community functions. *Carnegie Inst. Washington Publ.* No. 398, 1–340.

COATES, A. G. and OLIVER, W. A., Jr. (1973). Coloniality in zoantharian corals. In "Animal Colonies, Development and Function through Time" (R. S. Boardman, A. H. Cheetham and W. A. Oliver, Jr., eds), pp. 3–27. Dowden, Hutchinson, and Ross, Stroudsburg, Pennsylvania.

CONNELL, J. H. (1961). The influence of interspecific competition and other factors on the distribution of the barnacle *Chthamalus stellatus. Ecology* **42**, 710–723.

CONNELL, J. H. (1971). On the role of natural enemies in preventing competitive exclusion in some marine animals and in rain forest trees. In "Proc. Advan. Study Inst.

Dynamics of Populations, Oosterbeek" (P. J. den Boer and G. Gradwell, eds), pp. 298–312. Centre for Agricultural Publishing and Documentation, Wageningen, The Netherlands.

CONNELL, J. H. (1972). Community interactions on marine rocky intertidal shores. *An. Rev. Ecol. Systematics* **3**, 169–192.

CONNELL, J. H. (1973). Population ecology of reef-building corals. In "Biology and Geology of Coral Reefs" (O. A. Jones and R. Endean, eds) Vol. 2, pp. 205–245. Academic Press, New York.

CONNELL, J. H. (1975). Some mechanisms producing structure in natural communities: a model and evidence from field experiments. In "Ecology and Evolution of Communities (M. L. Cody and J. M. Diamond, eds) pp. 460–490. Belknap Press, Cambridge, Mass.

CONNELL, J. H. and SLAYTER, R. O. (1977). Mechanisms of succession in natural communities and their role in community stability and organization. *Am. Natur.* **111**, 1119–1144.

COOK, P. L. (1968). Polyzoa from West Africa, the Malacostega, Part I. *Bull. Brit. Mus. (Nat. Hist.) Zool.* **16**, 115–160.

CROWELL, S. (1957). Differential responses of growth zones to nutritive level, age, and temperature in the colonial hydroid *Campanularia*. *J. exp. Zool.* **134**, 63–90.

DAHL, A. L. (1973). Surface area in ecological analysis: quantification of benthic coral-reef algae. *Marine Biol.* **23**, 239–249.

DAVIES, L. V. (1971). Growth and development of colonial hydroids. In "Experimental Coelenterate Biology" (H. M. Lenhoff, L. Muscatine, and L. V. Davies, eds), pp. 16–36. Univ. Hawaii Press, Honolulu.

DAYTON, P. K. (1971). Competition, disturbance, and community organization: the provision and subsequent utilization of space in a rocky intertidal community. *Ecol. Monogr.* **41**, 351–389.

DAYTON, P. K., ROBILLIARD, G. A., PAINE, R. T. and DAYTON, L. B. (1974). Biological accommodation in the benthic community at McMurdo Sound, Antarctica. *Ecol. Monogr.* **44**, 105–128.

DUSTAN, P. (1975). Growth and form in the reef-building coral *Montastrea annularis*. *Marine Biol.* **33**, 101–107.

ELVIN, D. W. (1976). Seasonal growth and reproduction of an intertidal sponge, *Haliclona permollis* (Bowerbank). *Biol. Bull.* **151**, 108–125.

FELL, P. E. (1976). Analysis of reproduction in sponge populations: an overview with specific information on the reproduction of *Haliclona loosanoffi*. In "Aspects of Sponge Biology" (F. W. Harrison and R. R. Cowden, eds), pp. 51–68. Academic Press, New York.

FOSTER, M. S. (1975a). Algal succession in a *Macrocystis pyrifera* forest. *Marine Biol.* **32**, 313–329.

FOSTER, M. S. (1965b). Regulation of algal community development in a *Macrocystis pyrifera* forest. *Marine Biol.* **32**, 331–342.

GADGIL, M. and SOLBRIG, O. T. (1972). The concept of "r" and "K" selection: evidence from wildflowers and some theoretical considerations. *Am. Natur.* **106**, 14–31.

GEIGER, R. (1965). "The Climate Near the Ground" (transl. 4th German edition "Das Klima der bodennahen Luftschicht", 1961). Harvard University Press, Cambridge, Mass.

GILBERT, L. E. and RAVEN, P. H. eds (1975). "Coevolution of Animals and Plants" University of Texas Press, Austin.

GLYNN, P. W. (1973). Aspects of the ecology of coral reefs in the western Atlantic region. In "Biology and Geology of Coral Reefs" (O. A. Jones and R. Endean, eds), Vol. 2, pp. 271–324. Academic Press, New York.

GLYNN, P. W. STEWART, R. H. and McCOSKER, J. E. (1972). Pacific coral reefs of Panama: structure, distribution and predators. Geol. Rundschau 61, 483–519.

GOLDMAN, B. and TALBOT, F. H. (1976). Aspects of the ecology of coral reef fishes. In "Biology and Geology of Coral Reefs III: Biology 2" (O. A. Jones and R. Endean, eds), pp. 125–154. Academic Press, New York.

GOODBODY, I. (1961). Inhibition of the development of a marine sessile community. Nature, Lond. 190, 282–283.

GORDON, D. P. (1972). Biological relationships of an intertidal bryozoan population. J. nat. Hist. 6, 503–514.

GOREAU, T. F. (1959). The ecology of Jamaican coral reefs. I. Species composition and zonation. Ecology 40, 67–90.

GOREAU, T. F. and GOREAU, N. I. (1959). The physiology of skeleton formation in corals. 2. Calcium deposition by hermatypic corals under various conditions in the reef. Biol. Bull. 117, 239–250.

GOREAU, T. F. and GOREAU, N. I. (1960a). The physiology of skeleton formation in corals. III. Calcification as a function of colony weight and total nitrogen content in the reef coral Manicina areolata (Linnaeus). Biol. Bull. 118, 419–429.

GOREAU, T. F. and GOREAU, N. I. (1960b). The physiology of skeleton formation in corals. IV. On isotopic equilibrium exchanges of calcium between corallum and environment in living and dead reef-building corals. Biol. Bull. 119, 416–427.

GOREAU, T. F. and GOREAU, N. I. (1973). The ecology of Jamaican coral reefs. II. Geomorphology, zonation, and sedimentary phases. Bull. mar. Sci. 23, 299–464.

GOREAU, T. F. and HARTMAN, W. D. (1963). Boring sponges as controlling factors in the formation and maintenance of coral reefs. In "Mechanisms of Hard Tissue Destruction" (R. F. Sognnaes, ed.), pp. 25–54. Am. Assoc. Adv. Sci. Publ. 75, Washington, D.C.

GRIGG, R. W. (1970). Ecology and population dynamics of the gorgonians. Muricea californica and Muricea fruticosa—Coelenterata: Anthozoa. Ph.D. diss., University of California, San Diego.

GRIGG, R. W. (1974). Growth rings: annual periodicity in two gorgonian corals. Ecology 55, 876–882.

HANSEN, T. A. (1978). Larval dispersal and species longevity in Lower Tertiary gastropods. Science 199, 885–887.

HARMELIN, J. G. (1975). Relations entre la forme zoariale et l'habitat chez les bryozoaires cyclostomes, conséquences taxonomiques. Bryozoa 1974. Docums Lab. Géol. Fac. Sci. Lyon. (Hors Series 3) (Fasc. 2), 329–384.

HARPER, J. L. and WHITE, J. (1974). The demography of plants. A. Rev. Ecol. Systematics 5, 419–464.

HARPER, J. L., LOVELL, P. H. and MOORE, K. G. (1970). The shapes and sizes of seeds. A. Rev. Ecol. Systematics 1, 327–356.

HARTMAN, W. D. and GOREAU, T. F. (1970). Jamaican coralline sponges: their morphology, ecology, and fossil relatives. Proc. zool. Soc. Lond. 25, 205–243.

HARTMAN, W. D. and REISWIG, H. M. (1973). The individuality of sponges. *In* "Animal Colonies, Development and Function through Time" (R. S. Boardman, A. H. Cheetham and W. A. Oliver, Jr., eds), pp. 567–584. Dowden, Hutchinson, and Ross, Stroudsburg, Pennsylvania.

HAVEN, N. D. (1971). Temporal patterns of sexual and asexual reproduction in the colonial ascidian *Metandrocarpa taylori* Huntsman. *Biol. Bull.* **140**, 400–415.

HAYWARD, P. J. (1973). Preliminary observations on settlement and growth in populations of *Alcyonidium hirsutum* Fleming). *In* "Living and fossil Bryozoa" (G. P. Larwood, ed.), pp. 107–114. Academic Press, London.

HAYWARD, P. J. and RYLAND, J. S. (1973). Growth, reproduction and larval dispersal in *Alcyonidium hirsutum* (Fleming) and some other Bryozoa. *Pubbl. Staz. Napoli* **39** (suppl.), 226–241.

HOBSON, E. S. (1976). Feeding patterns among tropical reef fishes. *Am. Scientist* **63**, 382–392.

HORN, H. S. (1971). "The adaptive geometry of trees". Princeton University Press, Princeton, New Jersey.

HUBBARD, J. A. E. B. (1973). Sediment-shifting experiments: a guide to functional behavior in colonial corals. *In* "Animal Colonies, Development and Function through Time" (R. S. Boardman, A. H. Cheetham and A. H. Oliver, Jr., ed.), pp. 31–42. Dowden, Hutchinson and Ross, Stroudsburg, Pennsylvania.

HYMAN, L. H. (1940). "The invertebrates: Protozoa through Ctenophora". McGraw-Hill, New York.

HYMAN, L. H. (1959). "The invertebrates, Vol. 5, smaller coelomate groups." McGraw-Hill, New York.

JACKSON, J. B. C. (1972). The ecology of the molluscs of *Thalassia* communities, Jamaica, West Indies. II. Molluscan population variability along an environmental stress gradient. *Marine Biol.* **14**, 304–337.

JACKSON, J. B. C. (1974). Biogeographic consequences of eurytopy and stenotopy among marine bivalves and their evolutionary significance. *Am. Natur.* **108**, 541–560.

JACKSON, J. B. C. (1977a). Competition on marine hard substrata: the adaptive significance of solitary and colonial strategies. *Am. Natur* **111**, 743–767.

JACKSON, J. B. C. (1977b). "Habitat area, colonization, and development of epibenthic community structure." *In* "Biology of Benthic Organisms" (B. F. Keegan, P. O. Ceidigh and P. J. S. Boaden, eds), pp. 349–358. Pergamon Press, Oxford.

JACKSON, J. B. C. and BUSS, L. W. (1975). Allelopathy and spatial competition among coral reef invertebrates. *Proc. natn. Acad. Sci., U.S.A.* **72**, 5160–5163.

JANZEN, D. H. (1970). Herbivores and the number of tree species in tropical forests. *Am. Natur.* **104**, 501–528.

JEBRAM, D. (1973). Preliminary observations on the influences of food and other factors on the growth of Bryozoa with a description of a new apparatus for the cultivations of sessile plankton feeders. *Kieler Meeresforsch.* **29**, 50–57.

JØRGENSEN, C. B. (1966). "Biology of Suspension Feeding." Pergamon, Oxford, England.

KARLSON, R. (1978). Predation and space utilization patterns in a marine epifaunal community. *J. exp. mar. Biol. Ecol.* **31**, 225–239.

KATÔ, M., HIRAI, E. and KAKINUMA, Y. (1963). Further experiments on the interspecific relation in the colony formation among some hydrozoan species. *Sci. Rep. Tôhoku. Univ.* (Ser. IV) **29**, 317–325.

Katô, M., Hirai, E. and Kakinuma, Y. (1967). Experiments on the coaction among hydrozoan species in the colony formation. *Sci. Rep. Tôhoku Univ.* (Ser. IV) **33**, 359–373.

Kaufman, L. (1977). The threespot damselfish: effects on benthic biota of Caribbean coral reefs. *Proc. Third Int. Coral Reef. Symp.* **1**, 559–564.

Kinzie, R. A. III. (1973). The zonation of West Indian gorgonians. *Bull. Mar. Sci.* **23**, 93–155.

Kinzie, R. A. (1974). *Plexaura homomalla*: the biology and ecology of a harvestable marine resource. *Studies in Tropical Oceanogr.* **12**, 22–38.

Koch, C. F. (1977). Evolutionary and ecological patterns of Upper Cenomanian (Cretaceous) mollusk distribution in the Western Interior of North America. Ph.D. diss., The George Washington University, Washington, D.C.

Lang, J. C. (1970). Inter-specific aggression within the scleractinian reef corals. Ph.D. diss., Yale University, New Haven, Connecticut.

Lang, J. C. (1973). Interspecific aggression by scleractinian corals. 2. Why the race is not only to the swift. *Bull. mar. Sci.* **23**, 260–279.

Lang, J. C. (1974). Biological zonation at the base of a reef. *Am. Sci.* **62**, 272–281.

Larwood, G. P. (1974). Preliminary report of early (Pre-Cenomanian) cheilostome Bryozoa. *Bryozoa 1974. Docums Lab. Géol. Fac. Sci. Lyon.* (Hors séries 3 (fasc. 2)), 539–545.

Leversee, G. J. (1976). Flow and feeding in fan-shaped colonies of the gorgonian coral, *Leptogorgia*. *Biol. Bull.* **151**, 344–356.

Levin, S. A. and Paine, R. T. (1974). Disturbance, patch formation, and community structure. *Proc. natn. Acad. Sci. USA* **71**, 2744–2747.

Lewis, J. B. and Price, W. S. (1976). Patterns of ciliary currents in Atlantic reef corals and their functional significance. *J. Zool. Lond.* **178**, 77–89.

MacArthur, R. H. (1972). "Geographical Ecology." Harper and Row, New York.

Marsden, J. R. (1962). A coral-eating polychaete. *Nature, Lond.* **193**, 598.

Meadows, P. S. and Campbell, J. I. (1972). Habitat selection by aquatic invertebrates. *Adv. mar. Biol.* **10**, 271–382.

Millar, R. H. (1971). The biology of ascidians. *Adv. mar. Biol.* **9**, 1–100.

Muscatine, L. (1974). Endosymbiosis of cnidarians and algae. In "Coelenterate Biology" (L. Muscatine and H. M. Lenhoff, eds), pp. 313–358. Academic Press, New York.

Neumann, A. C. (1966). Observations on coastal erosion in Bermuda and measurements on the boring rate of the boring sponge, *Cliona lampa*. *Limnol. Oceanogr.* **11**, 92–108.

Neushul, M. (1972). Functional interpretation of benthic algal morphology. In "Contributions to the Systematics of Benthic Marine Algae of the North Pacific" (I. A. Abbott and M. Kurogi, eds), pp. 47–73. Japanese Society of Phycology, Kobe.

Newell, R. C. (1970). "Biology of Intertidal Animals." Logos Press, London.

Osman, R. W. (1977). The establishment and development of a marine epifaunal community. *Ecol. Monogr.* **47**, 37–63.

Paine, R. T. (1974). Intertidal community structure. Experimental studies on the relationship between a dominant competitor and its principal predator. *Oecologia* (Berl.) **15**, 93–120.

Patton, W. K. (1976). Animal associates of living reef corals. In "Biology and Geology of Coral Reefs III: Biology 2" (O. A. Jones and R. Endean, eds), pp. 1–37. Academic Press, New York.

PEARSE, V. B. and MUSCATINE, L. (1971). Role of symbiotic algae (zooxanthellae) in coral calcification. *Biol. Bull.* **141**, 350–363.

PEQUEGNAT, W. E. (1964). The epifauna of a California siltstone reef. *Ecology* **45**, 272–283.

PORTER, J. W. (1976). Autotrophy, heterotrophy, and resource partitioning in Caribbean reef-building corals. *Am. Natur.* **110**, 731–742.

PROSSER, C. L., ed. (1973). "Comparative Animal Physiology. Vol. 1. Environmental Physiology." W. B. Saunders, Philadelphia.

RANDALL, J. E. and HARTMAN, W. D. (1968). Sponge-feeding fishes of the West Indies. *Mar. Biol.* **1**, 216–225.

RAUNKIAER, C. (1934). "The Life-forms of Plants and Statistical Plant Geography." Oxford University Press, Oxford.

REISWIG, H. M. (1971a). *In situ* pumping activities of tropical Demospongiae. *Mar. Biol.* **9**, 38–50.

REISWIG, H. M. (1971b). Particle feeding in natural populations of three marine demosponges. *Biol. Bull.* **141**, 568–591.

REISWIG, H. M. (1973). Population dynamics of three Jamaican Demospongiae. *Bull. mar. Sci.* **23**, 191–226.

REISWIG, H. M. (1974). Water transport, respiration and energetics of three tropical marine sponges. *J. exp. mar. Biol. Ecol.* **14**, 231–249.

REISWIG, H. M. (1975). The aquiferous systems of three marine Demospogiae. *J. Morph.* **145**, 493–502.

RIEDL, R. (1966). "Biologie der Meeresholen." Paul Parey, Hamburg.

RIEDL, R. (1971). Water movement: animals. *In* "Marine ecology 1, part 2" (O. Kinne, ed.), pp. 1123–1156. Wiley-Interscience, New York.

RÜTZLER, K. (1970). Spatial competition among Porifera: solution by epizooism. *Oecologia* **5**, 85–95.

RYLAND, J. S. (1970). "Bryozoans." Hutchinson University Library, London.

SAMMARCO, P. W. (1975). Grazing by *Diadema antillarum* Philippi (Echinodermata: Echinoidea): density-dependent effects on coral and algal community structure. Abstr., Assoc. Intern. Marine Laboratories Carib., St. Croix, Virgin Islands.

SARÀ, M. (1970). Competition and cooperation in sponge populations. *Symp. Zool. Soc. Lond.* **25**, 273–284.

SARÀ, M. and VACELET, J. (1973). Écologie des démosponges. *In* "Traité de Zoologie 3(1). Spongiaires" (P.-P. Grassé, ed.), pp. 462–576. Masson et Cie Editeurs, Libraries de l'Académie de Médicine, Paris.

SCHAFFER, W. M. and GADGIL, M. D. (1975). Selection for optimal life histories in plants. *In* "Ecology and Evolution of Communities" (M. L. Cody and J. M. Diamond, eds), pp. 142–157. Belknap Press, Cambridge, Ma.

SCHELTEMA, R. S. (1977). Dispersal of marine invertebrate organisms: paleobiogeographic and biostratigraphic implications. *In* "Concepts and Methods of Biostratigraphy" (E. G. Kauffman and J. E. Hazel, eds), pp. 73–108. Dowden, Hutchinson and Ross, Stroudsburg, Pennsylvania.

SCHIMPER, A. F. W. (1903). "Plant Geography upon a Physiological Basis." (Transl. by W. R. Fisher. P. Groom and I. B. Balfour, eds), Clarendon Press, Oxford.

SCOFFIN, T. P. and GARRETT, P. (1974). Processes in the formation and preservation of internal structure in Bermuda patch reefs. *Proc. Second Intern. Coral Reef Symp.* **2**, 429–448.

SEAL, H. L. (1964). "Multivariate Statistical Analysis for Biologists." Methuen, London.

SIMPSON, T. L. (1973). Coloniality among the Porifera. *In* "Animal Colonies: Development and Function through Time" (R. S. Boardman, A. H. Cheetham and W. A. Oliver, Jr., eds), pp. 549–565. Dowden, Hutchinson and Ross, Stroudsburg, Pennsylvania.

SOLBRIG, O. T. (1972). The population biology of dandelions. *Am. Scientist* **59**, 686–694.

STEBBING, A. R. D. (1971a). Growth of *Flustra foliacea* (Bryozoa). *Mar. Biol.* **9**, 267–273.

STEBBING, A. R. D. (1971b). The epizoic fauna of *Flustra foliacea* (Bryozoa). *J. mar. biol. Ass. U.K.* **51**, 283–300.

STEBBING, A. R. D. (1973a). Competition for space between the epiphytes of *Fucus serratus* L. *J. mar. biol. Ass. U.K.* **53**, 247–261.

STEBBING, A. R. D. (1973b). Observations on colony overgrowth and spatial competion. *In* "Living and Fossil Bryozoa" (G. P. Larwood, ed.), pp. 173–183. Academic Press, London.

STENECK, R. S. and ADEY, W. H. (1976). The role of environment in control of morphology in *Lithophyllum congestum*, a Caribbean algal ridge builder. *Botanica Marina* **19**, 197–215.

STORR, J. F. (1976). Ecological factors controlling sponge distribution in the Gulf of Mexico and the resulting zonation. *In* "Aspects of Sponge Biology" (F. W. Harrison, and R. R. Cowden, eds), pp. 261–276. Academic Press, New York.

SUGIMOTO, K. and NAKUACHI, M. (1974). Budding, sexual reproduction, and regeneration in the colonial ascidian, *Symplegama reptans*. *Biol. Bull.* **147**, 213–226.

VINE, P. J. (1974). Effects of algal grazing and aggressive behavior of the fishes *Pomacentrus lividus* and *Acanthurus sohal* on coral-reef ecology. *Mar. Biol.* **24**, 131–136.

WAINWRIGHT, S. A. and DILLON, J. R. (1969). On the orientation of sea fans (genus *Gorgonia*). *Biol. Bull.* **136**, 130–139.

WAINWRIGHT, S. A., BIGGS, W. D., CURREY, J. D. and GOSLINE, J. M. (1976). "Mechanical Design in Organisms." Edward Arnold, London.

WELLS, J. W. (1956). Scleractinia. *In* "Treatise on Invertebrate Paleontology" Part F (R. C. Moore, ed.). pp. 328–444. University of Kansas Press, Lawrence, Kansas.

WILLIAMS, G. C. (1975). "Sex and Evolution." Princeton University Press, Princeton, New Jersey.

WINSTON, J. E. (1976). Experimental culture of the estuarine ectoproct *Conopeum tenuissimum* from Chesapeake Bay. *Biol. Bull.* **150**, 318–335.

WOLCOTT, T. G. (1973). Physiological ecology and intertidal zonation in limpets (*Acmaea*): a critical look at "limiting factors." *Biol. Bull.* **145**, 389–422.

WOODIN, S. A. (1976). Adult-larval interactions in dense infaunal assemblages: patterns of abundance. *J. mar. Res.* **34**, 25–41.

Author Index

The numbers in *italic* indicate the pages on which names are mentioned in the reference lists

A

Abe, N., *491*
Abeloos, M., 227, *236*
Abrahamson, W. G., 534, *547*
Absolon, K., 287, *293*
Adelberg, E. A., 5, 15, *26*
Adey, W. H., 282, *293*, 545, *555*
Adler, J., 10, *25*
Agassiz, A., 282, *293*
Ahmadjian, V., 22, *25*
Alexopoulos, C. J., 21, *26*
Allain, J-Y., 379–380, *394*
Al-Ogily, S. M., *296*, 538, *547*
Amante, E., *374*
Anderson, D. T., *293*
Anderson, P. A. V., 145, 147, *151*, *154*
Andreski, S., xiii, *xxx*
Andrews, J. C., 288, *293*
Andrews, P. B., 282, 289, *293*
Ashworth, J. M., 12, 13, 14, 16, 20, 21, *25*
Atoda, K., *491*
Augener, H., 291, *293*
Ayala, F. J., 163, *170*

B

Bagby, R. M., 57, *74*
Bahr, L. M., 235–6, *237*, 508, *548*
Bailey, J. H., *296*, *326*
Baker, R., 366, 370, 371, *372*, *374*
Bakus, G. J., 72, *74*, 506, *547–8*
Balavoine, P., 227, *236*
Band, R. N., 32, *37*
Banta, W. C., 195, 200, 208, *209*, 213, 214, 219, 222, 224, *236–7*
Barnes, D. J., 506, 546, *548*
Barnes, H., 289, *293*, 320, 324, *325*, 466, *491*

Barnes, R. D., 463, *491*
Barrat, R. J., *396*
Barrington, E. J. W., 258, *278*
Bartholomew, B., 530, *548*
Bartnicki-Garcia, S., 16
Basalingappa, S., *347*
Bassindale, R., 313, *317*
Batham, E. G., *491*
Bathurst, R. G. C., 267, *278*, 513–4, *548*
Bayer, F. M., *xxx*, *491*, 510, 535, 537, 538, 539, *548*
Bayne, B. L., 467, *491*
Beach, D. H., 389, *394*
Behrens, E. W., 289, *293*
Bein, A., 259–60, 270, 273, *278*, *278*
Beklemishev, W. N., xv, xx, xxi, xxii, xxvi, *xxx*, 54, *74*, 188, 191, *192*, 212, 215, 235, *237*, 404, 408, *409*, 537, *548*
Bellan, G., 292, *298*
Benham, W. B., 283, *294*
Bennet, I., *294*
Bergquist, P. R., 54, 71, *74*, 465, *491*
Berrill, N. J., 434–5, *443*, *491*, 513–14, 537, 538, *548*
Berry, W. B. N., 421, *431*
Bertram, G. C. L., 507, *548*
Beug, H., *26*
Bidder, G. P., 55, 57, 61, *74*
Bierl, E. D., *373*
Biggs, W. D., *396*, *555*
Birkeland, C., *396*
Bishop, J. W., 235–6, *237*, 508, *548*
Blaich, R., 19, 24, *25*
Blum, M. S., 365–6, *371*, *373*, *374*
Blumer, M., *369*
Boaden, P. J. S., 469, *491*
Boardman, R. S., xi, xv, xxi, xxii, *xxx*, *101*, 174, 178, *192*, 194, 209, *209*, 215,

Boardman, R. S.—*continued*
  217, 218, 219, 226, *237*, 460, *491*, 536, 538, 544, 546, *548*
Bobin, G., 213, 215, 224–6, 227, 231, 234, *237*
Bohlin, B., 397, 404, 409, *409*
Bonner, J. T., 20, *25*
Borg, F., 213, 219, 224, 227, 232, 235, *237*
Bornhold, R. D., 305, *317*
Borojevič, R., 47, *47*, 51, 68, 70, 71, *74–5*
Bosence, D. W. J., xiv, 282, 292, *294*, 300, 306, *317*
Bossert, W. H., 365, *371*
Bouillon, J., 122, 123, 126, 128–30, 133, 138–9
Boury Esnault, N., 60, *75*
Bousfield, E. L., 317, *491*
Bowerbank, J. S., *75*
Boyer, H. W., 24, *25*
Bradley, S., 14
Bradshaw, J. W., 366, *371*
Braem, F., 226, 227, *237*
Brahm, C., 95, *101*
Brakel, W. H., 507, 546, *548*
Brauer, R. W., 389–90, *395*, *548*
Braverman, M. H., 535, 538, 545–7, *548*
Brazendale, M. G., 446, *458*
Brien, P., 54, 58, *75*, 224, *237*, 510, *548*
Briner, P. H., *372*
Brinkman-Voss, A., 107, 120, 136, *139*
Broch, H., 107, *139*
Brock, H., 320, *325*
Bromley, G. R., 54, *75*
Bronstein, G., 221, 227–8, *237–8*
Brood, K., 218, *238*
Broom, D. M., 379–80, *395*
Brossut, R., *373*
Brown, W. L., 353, *371*
Brownlee, R. G., *374*
Bruce, A. J., 529, *548*
Brulé, G., *373*
Brun, E., 379, 388, *395*
Bryan, P. G., 506, *548*

Buchanan, J. B., 382, *395*
Buddemeier, R. W., 535, *549*
Buisonjé, P. H. de, 282, *294*
Buisson, B., 145, *152*
Bull, A. T., 16, 18, *25*
Buller, A. R. H., 17, 18, 22, *25*
Bullivant, J. S., 235–6, *238*
Bullock, A. L. C., xiii, *xxx*
Bulman, O. M. B., 397, 398, 399, 404, *409*, 411–12, 414, 417, 420, 422, 424, 426, *431*
Bulvanker, E. Z., 158, *170*
Burchette, T. P., 282, *294*
Burkhardt, D., *373*
Burkholder, P. R., 506, *549*
Burnet, F. M., 24, *25*, 53, 75
Burnett, J. H., 16, 19, *25*
Burton, M., 54, 68, 71, 73, *75*
Bushnell, J. H., 235, *238*
Buss, L. W., xi, xiv, xxviii, 463, 475, 479, 487, *491–2*, *494*, 506, 531, 538, 541–2, 546, *552*
Butler, C. G., 364, *371*
Butler, G. M., 18, *25*
Buzas, M. A., 506, *549*

C

Cabot, E. C., 282, *294*
Calvet, L., 214, 224, *238*
Calvin, J., *495*
Campbell, A. C., *396*
Campbell, J. I., 465, 467, *495*, 504, *553*
Campbell, R. C., 530, *549*
Carlile, M. J., xi, xiv, xv, xx, xxi, xxii, 10, 15, 18, 21, 22, 24, *25*, 30
Carr, N. G., 14, *25*
Carriker, M. R., 289, *298*
Carter, H. J., 51, *75–6*, 223, *238*
Case, J. F., 145, 147, *151*
Castilla, J. C., 390, *395*
Caten, C. E., 19, *25*
Cavill, G. W. K., 368, *371*
Chabrzyk, G., 446, *458*
Chamberlin, J. A. Jr., 507, 529, *549*
Chambers, D. M., 279

Chapman, D. M., 83, 87, 97, *101*
Charnov, E. L., 534, *549*
Chater, K. F., 12, 13
Chaudhury, M. F. B., *373*
Cheetham, A. H., xxx, *101*, 174, *192*, 194, 209, *209*, 215, 217, 218, 219, 235, *237*, *238*, *491*, 536, 537, 538–9, 544, 546, *548–9*
Chesher, R. H., 389, *395*
Chiba, Y., 506, *549*
Clark, A. M., 529, *549*
Clark, D. V., 368, *371*
Clausade, M., 286, *294*
Clausen, J., 544, *549*
Clements, F. E., 544, *549*
Coates, A. G., xxii, xxviii, *xxx*, 180–3, 184, 186, 191, *192*, *491*, 583, *549*
Coaton, W. G. H., 369, *371*
Cocke, J. M., 157, *170*
Cole, H. A., 321, *325*, *492*
Collins, A. R. S., 389, *395*
Connell, J. H., 460, 461, 463, 475, *492*, 504, 505–7, 510, 536, 543, *549–50*
Cook, P. L., xvi, xix, 195, 196, 197, 198, 200, 208, *209–10*, 213, 214, 217, 222, 223, *235*, *237*, *238*, *241*, 537, 544, *549*, *550*
Cooke, I. M., 149, *153*
Cooper, A. L., 8–10, *25*
Cooper, R. A., 421, *431*
Cornelius, P. F. S., xxv
Corrêa, D. D., 229–31, *238*
Coulson, J. C., x, xiv, xxii, 446, 452, 454, 456, *458*
Cowden, R. R., 73, *76*
Cresp, J., 283, *294*
Crewe, R. M., 365–6, *371*
Crisp, D. J., xi, xiv, xxi, 250, *253*, 320, 321–5, *325–6*, 390, *395*, 465, 466–7, 469, *491*, *492*
Crisp, M., 321, *326*
Crowell, S., *492*, 504–5, 547, *550*
Crowther, P. R., xxv
Curds, C. R., xi, xxi, 30, *37*
Currey, J. D., *396*, *555*
Curtis, A. S. G., xiv, xv, xix, xx, xxi, xxii, xxv, 40, *42*, 46, *47*, 54, *76*

D

Dagg, A. I., 353, *372*
Dahl, A. L., 508, *550*
Dakin, W. J., 290, *294*
Dales, R. P., 283, *294*
Dana, T. E., 389, 390, *395*
Daniel, A., *492*
Darling, F. Frazer, 456, *458*
Darwin, C., 222, *238*, 329
Davenport, E., *317*
Davis, L. V., 535, *550*
Day, P. R., 19, *25*
Dayton, L. B., *492*, *550*
Dayton, P. K., *396*, 460, 472, 482, *492*, 506, 508, 510, 543, *550*
Dean, A. C. R., *25*
Dee, J., 20, 21, *25*
Denby, A., 64, *76*, 78
Desor, M. E., 282, *294*
Dew, B., 286, 290, *294*
D'Hondt, J. L., 236, *238*
Dicquemarke, Abbé, 144, *152*
Dillon, J. R., *555*
Dixon, F., x, xiv, xxii
Doty, M. S., 289, *294*
Douvillé, H., 273, *279*
Downing, F. S., *495*
Dubois, P., *373*
Dudley, J. E., 487, 488, *492–3*
Dujardin, F., *76*
Durston, A. J., 21, *25*
Dunstan, P., 546, *550*
Dworkin, M., 13
Dybern, B. I., 290, 292, *294*, 314, 317, *493*

E

Eberhard, M. J. W., 348, 358, *372*
Ebling, F. J., *317*, 386, *395*
Edmondson, C. H., *493*
Efremova, S. M., 51, *76*
Eggleston, D., 227, 229, *239*
Eisele, C. R., 156, 157, *170*
Eisner, T., 369, *371*
Ekman, S., 282, *294*

Elder, R. H., 10, 11, *26*
Elles, G. L., 418, 426
Ellington, C. P. Jr., *396*
Ellis, J., *76*
Elvin, D. W., 535, *550*
Emerson, A. E., 329, *343*, 345, 363, *372*
Endean, R., 291, *294*
Esser, K., 19, 24, *25*
Evans, D. A., 370, *372*
Evans, H. E., 348, 352, 354, 358, *372*
Evans, S. L., *374*
Ewald, W. F., *493*

F

Fager, E. W., *395*
Fagerstroem, J. A., 156, 157, *170*
Fairey, E. M., 364, *371*
Fales, H. M., *373*
Faulkner, G. H., 283, *294*
Fauré-Fremiet, E., 33-5, *37*
Fauvel, P., *294*, 305-6, *317*
Favre, J., 269, *279*
Fay, P., *26*
Fedorowski, J., x, xxvii, 155, 156, 157, 158, 160, *170*
Fell, P. E., 536, *550*
Fields, W. G., 148-9, *152*
Fincham, J. R. S., 19, *25*
Finks, R. M., xviii-xix, *xxx*
Fischer-Piette, E., 291, *294*
Fish, J. D., 385, *395*
Fisher, D. W., 53, *76*
Fisher, J., 446, *458*
Fletcher, D. J. C., 349, *372*
Fogg, G. E., 14, *26*
Fontaine, A. R., 379, 382. *395*
Foster, B. A., 322, *327*
Foster, J. B., 353, *372*
Foster, M. S., 545, *550*
Fowler, G. H., 191, *192*
Fox, H. M., 385, *395*
Francis, L., 144, *152*, 235, *239*, 324, 326
Franzén, Å., 212, *239*
Fraser, C. McLean, 120, *139*
Freeman, G., 434, *443*

Frisch, K. von, 371, *372*
Frost, T. M., *76*
Fry, W. G., xi, xiv, xv, xvii, xviii-xix, xxiv, *xxxi*, 51, 53, 54, 58, 59, *75*, *76*, 324, *493*
Fuller, J. L., 487, 488, *493*
Fulton, C., 149, *152*
Furssenko, A., 34, *37*

G

Gabbott, P. A., 322, *326*, *327*
Gadgil, M., 534, *547*, *550*, 554
Gaertner, H. R. von, 282, *294*
Garrett, P., 513-14, *555*
Garrone, R., *78*
Garstang, W., 110, *139*
Garwood, E. J., 282, *294*
Gause, G. F., 10, *26*
Gee, J. M., 288, *295*, 324, *326*, 467, 469, *493*
Geiger, S. R., 95, *101*
Geiger, R., 511, *550*
George, C. J., *374*
George, J. D., 286, *295*
Gerisch, G., *26*
Gerwerzhagen, A., 221, *239*
Ghobashy, A. F. A. A., 465, *492*
Gilbert, L. E., *493*, 530, *551*
Gillete, M. U., *253*
Gilpin, M. E., 475, *493*
Glaessner, M. F., 83, *102*
Glynne, P. W., 482, *493*, 506, 510, *551*
Goette, A., 108-9, *139*
Goldacre, R. J., 7, *26*
Goldman, B., 529, *551*
Golubić, S., 14
Gooch, J. L., 163, *170*, 441, *443*
Good, R. A., 53, *76*
Goodbody, I., 387, *395*, 440, *443*, *493*, 504, *551*
Gordon, D. P., 197, *210*, 215, 220, 223, 224-5, 227-9, *239*, 514, *551*
Goreau, N. I., 510, 541, *551*
Goreau, T. R., 58, *76-7*, 507, 510, 514, 541, *551*

Author Index

Gorianov, V. B., *170*
Gorzula, S. J., 383, *395*
Gosline, J. M., *396*, *555*
Gosner, K. L., *493*
Grant, R. E., 59, *76*
Grassé, P.-P., 357, 362, *372*
Graus, R. R., 507, 529, *549*
Grave, B. H., *493*
Grave, C., *493*
Gravier, C., 283, *295*
Gray, J. S., 322, *326*
Gray, W. D., 21, *26*
Graziani, G., 441, *443–4*
Green, G., 506, 548
Gregory, R. L., x, *xxxi*
Grigg, G. C., 359, *372*
Grigg, R. W., 535, *551*
Grime, J. P., xxvii, *xxxi*
Gross, J., *493*
Gutmann, W. F., 320, *356*

H

Hadfield, M. G., 244, 250, 251, *253*
Hadži, J., 50, *76*
Haeckel, E., 53, *76*
Håkansson, E., 194, 209, *210*, 239
Haldane, J. S., 330, 331
Hale, M. E., 22, *25*, 26
Hamilton, W. D., xx, 330–1, 340, *343*, 346, *372*
Hanscomb, N. J., *394*
Hansen, T. A., 542, *551*
Hanson, H. C., *549*
Hanson, J., 283, *295*
Hare, H., 330–1, *343*, 347, *374*
Hargitt, C. W., 282, *295*
Harlett, J. C., *279*
Harmelin, J-G., 203, *210*, 213, 214, 218, *239*, *242*, 506, 546–7, *551*
Harmer, S. F., 197, 208, *210*, 213, 214, 216, 218, 227, *239*
Harper, J. L., xv, xviii–xix, xx, xxi, xxii, xxiii, xxiv, xxv, xxvi, xxvii–xxviii, *xxxi*, 544, *551*
Harrington, H. J., 84, *102*

Harris, H., 163, *170*
Harris, W. V., 361, *372*
Harrison, F. W., 73, *76*
Hartlaub, C., 124, *139*
Hartman, W. D., 50, 58, 63, 68, *74*, 76–7, 460, 482, *493*, *495*, 507, 510, 514, 536, *551*, *554*
Hartmann-Schröder, G., 289, 291, *295*
Harvey, P. H., 460, 468–9, 470, *493–4*
Haskins, C. P., 348–9, *372*
Hasper, M., *493*
Hastings, A. B., 215, 218, 227, *239*
Hauenschild, C., 46, *47*
Haven, N. D., 535, *552*
Hayward, P. J., 216, *241*, 460, 468–9, 470, *493–4*, 510, 535, *552*
Head, S. M., *396*
Hedgecok, D., *170*
Hedgepeth, J. W., 282, *295*
Hedley, R. H., 32, *37*, 305, *318*
Heeb, M. A., 390, *395*
Heilprin, A., 282, *295*
Heinrich, B., 358, *372*
Heisey, W. M., *549*
Heldt, J. H., 282, 292, *295*
Hermans, C. O., 283, 290, 291, *297*
Hibberd, D. J., 51, *77*
Hill, M. B., 288, 292, *295*
Hiller, S., 220, 221, *240*
Himmer, A., 257–8, *372*
Hincks, T., 216, 222, *240*
Hinshelwood, C., *25*
Hintenberger, H., 368, *371*
Hirai, E., *494*, *552*, *553*
Hobson, E. S., 529, *552*
Hoffman, H., 6, *26*
Hogg, J. J., *74*, *491*
Holland. C. H., 408, *410*
Holldöbler, B., 364–5, *372*, *374*
Holme, N. A., *396*
Hopwood, D. A., 12, 13
Horn, H. S., 505, *552*
Horridge, G. A., 144, 145, *152*
House, M. J., 429
Hove, T. ten, xi, xiv, 287, 291, *295*, 314, 315–16, *318*

Howse, P. E., xi, xiii, xxii, 348, 366, 368, 371, *372*, *374*
Hrabe, S., 287, *293*
Hubbard, J. A. E. B., 144, *152*, 188, *192*, 507, *552*
Hughes, R. G., 394, *395*
Hughes, R. N., xiv, xxi, 244–6, 249, 251–2, 253
Humphries, E., 218, *240*
Humphreys, T., 54, *77*
Hungate, R. E., 8, *26*
Hutt, J. E., 417, 418, 421, *431*
Huxley, J., *344*
Huysmans, G., 224, *237*
Hyman, L. H., 212, 213, *240*, 508, 513–14, 536, 537, 538, 539, *552*

I

Ingraham, J. L., 5, 15, *26*
Isaac, M. J., *296*, *326*
Ishay, J., 358, 361, *372*

J

Jackson, J. C. B., xi, xiv, xxi, xxvii, 184, 289, 460, 463, 465, 471, 473, 475–6, 479, 483, 487, 488, 490, *492*, *494*, 500, 503–4, 506, 510, 514, 529, 538–9, 541–3, 544, 546, *552*
Jaeger, H., 429, *431*
Jaisson, P., 367, *372*
James Clark, H., 51, *77*
Jander, R., *373*
Janet, C., 36, *37*
Janzen, D. H., 353, *372*, 483, *494*, 504, 530, *552*
Jeanne, R. L., 354, 356, 361, *373*
Jebram, D., 213, 232, *240*, *494*, 505, *552*
Jinks, J. L., 24, *26*
Johnson, M. G., *26*
Johnston, G., *77*
Jones, N. S., 385, *395*
Jones, W. E., 324, *326*
Jordan, M. R., 389–90, *395*, *548*
Jørgensen, C. B., 507, *522*

Josephson, R. K., *152*
Jull, R. K., 155, 158, 160, *170*

K

Kain, J. M., 385, *395*
Kakinuma, Y., *494*, *552*, *553*
Karakashian, S., 218, *240*
Karlson, R., 476, 506, *552*
Katô, M., 156, *170*, 184, *192*, *494*, 506, 549, *552*, *553*
Kauffman, E. G., 259–60, 267, 270–3, 278, 279
Kaufman, L., 530, *553*
Kaufmann, K. W., 200, *210*, *494*
Kawahara, T., 288, 293, *295*, *296*
Kay, E. A., *253*
Keck, D. D., *549*
Keegan, B. F., 316, *318*, 379, 383, 385, 386, *395*
Kenny, R., *294*
Ketchum, B. H. 316, 317, *318*
Kiderlen, H., 83, *102*
Kilgore, D. L. Jr., *396*
Kilian, E. F., 51, *77*
King, P. E., 482, *497*
Kinzie, R. A., III, 506–7, 535, 541–2, *549*, *553*
Kirk, N. H., xi, xvi, xxv, 397–8, 404, 408, *410*, 412, 414, 415, 427, *431*
Kiseleva, G. A., *494*
Kitching, J. A., *317*, 386, *395*
Klöckner, K., 288, 289, 290, *295*
Kloft, W., 367, *373*
Knight, J. B., 83, *102*
Knight-Jones, E. W., 151, *152*, 218, *240*, 286, 288, *295–6*, 320, 321–3, *325*, *326*, 465, *492*, 493, *494*, *496*, 538, *547*
Knight-Jones, P., 286, 288, *296*
Koch, C. F., 542, *553*
Koch, G. V., 155, *170*
Kogane, F., 18, *27*
Koltun, V. M., 61, 64, 65, 67, *77*
Komai, T., 84, 96, *103*
Konijn, T. N., 21
Konnecker, G., 385, *395*

Author Index

Korringa, P., 316, *318*
Kozlowski, R., 83, *101*, 397, 404, *410*, 412, 414, *431*
Kramp, P. L., 84, *102*, 107, 108, 110, *139*
Kruije, H. A. M. de, 149, *152*, *154*
Krumbein, W. E., 385, *395*
Kühlmann, D. H. H., 282, *296*
Kühn, A., 110, *139*
Kulm, L. D., 265, *279*

L

Labedz, L., xxv, *xxxi*
Lang, J. C., 144, *152*, 475, *495*, 505–6, 541, *553*
Larman, V. N., 322, *326*, *327*
Larwood, G. P., xii, *xxxi*, 214, 215, 216, *240*, 544, *553*
Laverack, M. S., *153*, *242*
Lawn, I. D., 146, *152*
Layman, M., 266, *279*
Lazaroff, N., 14
Leadbeater, B. S. C., 51, *77*
Lecompte, M., 164, *170*
Leeder, M. R., 282, 289, *296*
Leloup, E., 84, 85, *102*
Lentz, T. L., 53, *77*
Leopold, L. B., 18, *26*
Leversee, G. J., 507, *553*
Lévi, C., 51, 71, *75*, *77*
Levin, S. A., 543, *553*
Levinsen, G. M. R., 202, *210*, 213, 214, 215, 217, 227, *240*
Lewis, A. H., 244–6, 249, *253*
Lewis, J. B., 144, *152*, 385, *396*, 507, *553*
Lewis, J. R., 289, *296*, 313, *318*
Lewontin, R. C., 163, *170*
Lieberkéhn, N., *77*
Lin, N., 347, *373*
Lihdauer, M., 358–9, *373*
Lloyd, M. C., *253*
Loomis, W. F., 20, *26*, 30, 32, *37*
Looper, J. B., 33, *37*
Lovell, P. H., *551*
Lorenz, D. M., 537, *549*
Ludwig, F.-D., 100, *102*

Luscher, M., xviii, 350, 357, 359–60, *373*
Lutaud, G., 213, 220, 221, 224, *240*
Lynch, W. F., *494–5*

M

MacArthur, R. H., 539, *553*
McCosker, J. E., *493*, *551*
McDougall, K. D., *495*
McDowell, P. G., *372*
McFarlane, I. D., 146, 147, 149, *152*, *153*
MacGillavry, H. J., *279*
MacGintie, G. E., 283, 286, *296*
MacGintie, N., 283, 286, *296*
Machado, A. de B., 363, *373*
McIntosh, W. C., 305, *318*
Mackie, G. O., ix, xv–xvi, xviii, xxvii, *xxxi*, 50, 77, 121, *139*, 141–2, 148–50, 151, *152*, *152–3*, *192*, 221, 222, *240*, 460, *495*
McKinney, F. K., *209*, *237*
McLaurin, A. N., 421, *431*
MacLennan, A. P., 54, *78*
Majone, F., *444*
Manning, J. T., 343, *344*
Mano, R., *495*
Manton, S. M., 117, *139*
Marcus, E., 194, 208, *210*, 220, 221, 227, 235, *240–1*, 470, *495*
Marsden, J. R., 506, *553*
Maschwitz, U., 365, 367, 370, *373*
Mast, S. O., *495*
Matricon, I., 226, *241*
Mauzey, K. P., 390, *396*
Maynard-Smith, J., 330, *343*
Meadows, P. S., 322, *326*, 465, 467, *492*, *495*, 504, *553*
Meainwald, J., *374*
Mergner, H., 51, *78*
Meyer, D., 350, *373*
Meyer, D. L., 378–9, 394, *396*
Michener, C. D., 332, *344*, 347, *373*
Miles, A., 3, *27*
Milkman, R., 218, *240*, 436, 438, *443*
Millar, R. H., 434, *443*, *495*, 510, 535, *536*, *553*

Millard, N. A. H., 107, *139*
Milliman, J. D., 305, *317*
Minato, M., 156, *170–1*
Minchin, E. A., 63, *78*
Mohammad, M-B. M., 290, *296*
Mohrlok, S. H., 32, *37*
Montagner, H., 348, *373*
Moore, B. P., 369, 370, *373*
Moore, D. R., 95, *102*
Moore, K. G., *551*
Moore, R. C., 84, *102*
Moore, R. J., *396*
Mörch, O. A. L., 282, *296*
Morin, J. G., 149, *153*
Morton, C., 51, *77*
Moscona, A. A., 46, *47*, 54, *78*
Moser, J. C., *374*
Moser, S. L., *374*
Moyano, G. H. I., 218, *241*
Moyse, J., 151, *152*, 218, *140*, 288, *296*, 323, 465, *494*
Müller, A. H., 84, *102*
Muller, E., *26*
Muscatine, L., 508, 510, 538, *553*, *554*
Murray, J., 282, *296*
Murray, R. G. E., 10, 11, *26*

N

Nagai, T., *78*, 114
Nakamura, K., *494*
Nakauchi, M., 510, 535, 538, *555*
Nakazawa, K., *171*
Nardo, G. D., 282, *296*
Naumov, D. V., 84, *102*
Neff, J., 305, *318*
Nelson-Smith, A., 305–6, *318*
Neudek, R. H., *279*
Neumann, A. C., 507, *553*
Neushul, M., 503, 545, *553*
Newell, R. C., 465, 467, *495*, 504, *553*
Newman, H. N., 8, *26*
Newman, W. A., *395*
Nicol, J. A. C., 145, *153*
Nieh, M., *374*
Nielson, C., 224, *241*

Nishihira, M., 469, *495*
Noirot, C., 357, 362, *374*
Norman, A. M., 216, *241*
Nott, J. A., 322, *327*
Nudds, J. R., xv, xxi
Numakunai, T., 227, *241*
Nutting, W. L., 368, *373*

O

O'Connor, R. J., *491*
O'Donoghue, C. H., 196, 202, *210*
O'Donoghue, E., 202, *210*
Odum, E. R., *279*
Ogden, C. G., 32, *37*
Oka, H., 46, *47*, 434, 439, *443*
Olive, L. S., 20, 21, *26*, 30, *37*
Oliver, W. A. Jr., xxii, xxviii, *xxx*, *101*, 155, 162, *171*, 174, 180–3, 184, 186, 186, 191, *192*, *491*, 538, *549*
Orlove, M. J., x, xxii, 331, 340, 343, *344*
Ormond, R. F. G., 389, *394*, *396*
Osmond, R. W., 536, *553*
Ott, J. A., 212, *241*

P

Packard, A. S. Jr., 282, *296*
Paine, R. T., *492*, *495*, 506, *550*, *553*
Palumbo, S. A., 8, *26*
Parenzan, P., 292, *296*
Paris, J., 46, *47*, *79*
Parker, G. H., 53, 58, *78*, 145, *153*
Parsons, J., 456, *458*
Patton, W. K., 529, *553*
Paul, C. R. C., 262, *279*
Pavans de Ceccaty, M., 48, 53, *78*, *79*
Pearse, V. B., 510, 538, *554*
Pearson, T., 316, *318*
Pequegnat, W. E., 512, *554*
Percival, E., *495*
Pérès, J-M., 282, *296*
Pers, J. N. C., 385, *395*
Perty, J. A. M., 51, *78*
Peterson, K. W., xi
Philip, J., 259–60, 270, 273–6, *279*

Author Index

Pichon, M., xxxi
Pirt, J. S., 8–10, *26*
Pocock, M. A., 36, *37*
Pohowsky, R. A., 215, 219, *241*
Pontecorvo, G. C., 19, *26*
Poole, D. F. G., 8, *26*
Pope, E., *294*
Porter, J. W., 505, *554*
Powell, H. T., 320, *325*
Powell, N. A., 217, *241*
Prenant, M., 266, *237*
Prestwich, G. D., 369, 370, *373*
Price, W. S., 144, *152*, 507, *553*
Prosser, C. L., 53, *78*, 500, *554*
Pyefinch, K. A., 291, *296*, *495*

Q

Quennedey, A., 368, *373*
Quinlan, M. S., 13, *26*

R

Rainbow, P. R., *396*
Randall, J. E., 482, *495*, 510, *554*
Ranzoli, F., *298*
Raper, J., 20, *26*
Raper, K. B., 13, *26*
Rasmont, R., *75*
Raunkiaer, C., 544, *554*
Raven, P. H., *493*, 530, *551*
Reed, C. T., 282, *296*
Rees, W. J., 107, 110, 117, 119, 127, *139*
Reese, E. S., 376–7, 391, *396*
Regenhardt, H., 282, *297*
Regnier, F. E., 365, *374*
Reimer, A. A., 291, *297*
Reish, D. J., 291, 292, *297*
Reiswig, H. N., 50, 53, 55, 68, 69, 73, *77*, *78*, 460, 482, *493*, *495*, 502–3, 504, 505–7, 508, 510, 535, 536, *551*, *554*
Relini, G., 290, *297*
Remane, A., 212, 235, *241*
Remy, P., 287, *297*
Renard, A. F., 282, *296*
Rensch, B., 100, *102*

Rhoads, D. F., 265, *279*
Richard, A., 269, *279*
Richards, M. J., 347, 353, *374*
Richards, O. W., 347, 348–9, 353, 354, 356, 361, *374*
Rickards, R. B., xxv, 397–8, 399, 400, 404, 409, *410*, 424, 427, *413*
Ricketts, E. W., 463, *495*
Riding, R., 282, *294*
Ridley, H. N., 353, *374*
Ridley, S. O., 64, *78*
Rieck, V. T., *26*
Riedal, V., 21, *26*, 512
Riedl, R., 503, 506, 547, *554*
Rigaud, R., *373*
Riva, J., *431*
Roberts, J. L., 10, *26*
Robinson, S. W., 364, *377*
Roitt, I., 53, *79*
Rolbillard, G. A., *492*, *550*
Rosen, B. R., xii, *xxxi*
Roush, H. J., *279*
Różkowska, M., 155, 164, *171*
Ruedemann, R., 417, 418, 426, *431*
Ruelle, J. E., 360, *374*
Rullier, F., 287, *297*
Russell, F. S., 107, 124, 133, *139*
Ruth, J. M., *374*
Rützler, K., 73, *79*, 514, *554*
Ryder, T. A., 155, 156, *171*
Ryland, J. S., xi, xv, xvi, xix, xxi, 194, 195, 202, 206, 212, 213, 216, 220, 221, 224, 227, 232, 234, *239*, *241*, 465, 468–70, 482, 488, *494*, *495–6*, 506, 508, 510, 514, 535, *552*, *554*

S

Sabbadin, A., xx, 54, 324, *327*, 435, 436–41, *443–4*
Sadeh, D., 361, *372*
Safriel, U., 247, *253*
Saint-Joseph, A. de, 283, 291, *297*
Salsman, G. G., 387, 393, *396*
Sammarco, P. W., 504, *554*
Sands, W. A., 369, *374*

Sannasi, A., 350, *374*
Sarà, M., 51, *75*, *79*, 290, *297*, 514, 542, *554*
Sasaki, N., 71, *79*
Saunders, A. P., *396*
Saville Kent, W., 51, *79*
Schäfer, W., 258, 277, *279*
Schaffer, W. M., 534, *549*, *554*
Scheltema, R. S., 467, 469, *496*, 544, *554*
Schimper, A. F. W., 544, *554*
Schmidt, O., 54, *79*
Schmidt, R. S., 362, 363, *374*
Schmidt, W. J., 282, *297*
Schneirla, T. C., 356, 357, *374*
Schnopf, T. J. M., *xxxi*, 163, *170*, 213, *241*
Schrandt, R. G., 546, *548*
Schroeder, P. C., 283, 290, 291, *297*
Schulze, E., 63, *79*, 118
Schuhmacher, H., 385, *396*
Scoffin, T. P., 513–14, *555*
Seal, H. L., 506, 530, *555*
Seed, R., *396*, *491*
Segrove, F., 288, *297*
Sen-Sarma, P. K., *374*
Sentz-Braconnot, E., 288, 290, *297*, 314, *318*
Shelton, G. A. B., 145, 147, *153*, *242*
Sieburth, J. M., 8, *26*
Silén, L., 197, 200. *210*, 212, 213, 214, 215, 216, 217, 218, 219, 222, 229–31, *241–2*
Silva, P. H. D. H. de, 314, *317*, 465, *496*
Silverstein, R. M., *374*
Simma-Krieg, B., 227, *242*
Simon-Papyn, L., 283, *297*
Simons, J. R., *79*
Simpson, T. L., 54, 79, 460, *496*, 536, *555*
Sinclair, M. E., 71, 74, *491*
Skelton, P. W., xiv, xxi, 259, 267, 270–3, 277, *279*
Skevington, D., 420, 421, *431*
Slama, K., 350, *374*
Slayter, R. O., 543, *550*
Sleigh, M. A., *317*
Sloan, J. F., *317*
Slobodkin, L. B., 460, *496*

Smith, D. G., 10, *26*
Smith, F., x, *xxxi*
Smith, G. M., 36, *37*
Smith, J. E., 12, 13, 14, 16, *25*
Smith, S., 155, 156, *171*
Smitt, F. A., 216, 222, *242*
Sohl, N. F., 259–60, 267, 270–3, *278*, *279*
Solbrig, O. T., 534, *550*, *555*
Soldatova, I. M., 292, *297*
Sollas, W. J., 62, *79*
Sorauf, J. E., 162, *171*
Soulier, A., 282, *297*
Sousa, M. A. B. de, 40, 46, *47*
Spassky, N. Ya., *171*
Spencer, A. N., 149, *153*
Srinivasagam, R. T., 288, *297*
Stach, L. W., 229, *242*
Stammer, H. J., 287, *297*
Stanier, R. Y., 5, 15, *26*
Stebbing, A. R. D., 218, *242*, 469–70, 475, 487, *496*, 506, 514, 535, 538, *555*
Stempien, M. F. Jr., *79*
Steneck, R. S., 545, *555*
Stephenson, A., 247–8, *253*
Stephenson, T. A., 247–8, *253*
Stephenson, W., *294*
Stevenson, J. P., *494*
Stewart, R. H., *493*, *551*
Stewart, W. D. P., *26*
Stokes, D. R., 149, *153*
Størmer, L., 399, 410, *431*
Storr, J. F., *79*, 482, *496*, 510, *555*
Straughan, D., 288, 289, 290, 291, 292, *297*, 314, *318*, *496*
Sugimoto, K., 504, 510, 535, 538, *555*
Sweeney, B. M., 15, *26*
Summers, F. M., 34–5, *37*

T

Talbot, F. H., 529, *551*
Tardy, J., 322, *327*
Tavener-Smith, R., 232, *242*
Taylor, J. D., 266, *279*
Taylor, P. D., 208, *210*
Teitelbaum, M., 322, *327*

Tendal, O. S., 54, *75*
Theodor, J. L., 218, *242*
Thiel, H., 83, *102*
Thiney, Y., *78*
Tho, Y. P., *373*
Thomas, H. D., 215
Thompson, T. E., 322, *327*
Thomson, A. L., 446, *458*
Thomson, C. W., 282, *297*
Thorpe, J. P., 150, *153*, 217, 220, 221, *242*
Thorson, G., 283, 286, 287, 289, 291, *298*, 320, *327*, 465, *496*
Thorsteinsson, R., 427, *431*
Tokuoka, Y., 96, *102*
Tolbert, W. H., 387, 393, *396*
Tomlinson, J., 324, *327*
Topsent, E., *79*
Torgerson, R. L., *374*
Towe, K. M., 404, *410*, 412, 415, *432*
Trinci, A. P. J., 16, 18, *25*, 27
Travers, R. L., xx, 330-1, *344*, 347, *374*
Truckenbrodt, W., *374*
Tumlinson, J. H., 364, *374*
Turpaeva, E. P., 292, *297*
Tuzet, O., *79*

U

Uchida, T., 83, *102*, 114
Uhr, J. W., 53, *80*
Ultina, L. M., 155, 160, *171*
Urbanek, A., xix, *xxxi*, 397, 404, 408-9, *410*, 412, 415, 417, 423, 425, *426*, *427*, *432*

V

Vacelet, J., 70, *79*, 514, 542, *554*
Valentine, J. W., *170*
Vandyke, J. M., 30, *37*
Vannini, E., *298*
Vaughan, T. W., 186, *192*
Veron, J. E. N., *xxxi*
Verrill, A. E., 282, *298*
Vervoort, W., 126, *139*
Vevers, H. G., 383, *396*
Vine, P. J., *296*, 530, *555*

Viscovich. L., *444*
Visscher, J. P., 496
Vitteta, E. S., 53, *80*
Vogel, S., 73, *80*, 387, *396*
Vosmaer, G. C. J., 53, 62, 64, 72, *88*
Vuillemin, S., 282, *298*
Van de Vyver, G., 39-40, *47*, 53, 54, *80*

W

Wadhams, L. J., 369, *374*
Wainwright, S. A., 394, *396*, 500, 546, *555*
Wallis, D. L., 367, *374*
Walne, P. R., 317, *318*
Walsby, A. E., *26*
Ward, P., 457, *458*
Warner, G. F., xi, xiv, xxii, xxviii, 379-80, 382, 393, 394, *396*
Warren, P. T., *410*
Watanabe, H., 434, *443*, *444*
Waters, A. W., 470, *496*
Watteville, D. de, 196, 202, *210*
Weaver, J. E., *549*
Webster, A. P. J., 16, *27*
Weerdenburg, J. C. A., 287, *295*
Weir, J. S., 359, 361, *374*
Wells, G. P., *343*
Wells, H. G., 329, *343*
Wells, J. W., xv, *xxxi*, 141, *153*, 162, *171*, 174, 179, 186, *192*, 286, *298*, 537, *555*
Werner, B., 83-99, *101*, 102-3, 106, 107, 137, *139*
West-Eberhard, M. J., 330, *343*
Wheeler, D. M., 329, *343*
Wheeler, J. W., *374*
Whittle, K. J., *396*
Whitton, B. A., 14, *25*
White, E., 456, *458*
White, J., *551*
Wijsman-Best, M. B., xxvii, *xxxi*
Wilcox, M., 14, *27*
Wilkins, N. P., 325, *327*
Williams, C. M., 350, *374*
Williams, G. B., 288, *298*, 326, 467, 469, *492*, *496*

Williams, G. C., 324, 325, *327*, 504, 534, *555*
Willmer, E. N., 51, *80*
Wilson, D. P., 290, *298*, 321–2, *327*, 467, *497*
Wilson, E. O., xiii–xiv, xviii, xx, xxii, xxiv, xxix, *xxxi*, 329, *344*, 347, 364, 365, *371*, *372*, *374*
Wilson, G. S., 8, *27*
Wilson, J. B., 382, *396*
Winfree, A. T., 15, *27*
Winston, J. E., 471–2, *497*, 505, 547, *555*
Wintermann, G., *80*
Wiseley, B., 323, *327*, 468–9, *497*
Witter, L. D., *26*
Wobber, D. R., 391, *396*
Wolcott, T. G., 512, *555*
Wood, C. L., 343, *344*
Wood, E. J. F., 290, *294*
Wood, E. M. R., 418, 426
Wood, W. F., 370, *374*
Woodbridge, H., *493*, *497*
Woodin, S. A., 504, *555*
Woodley, J. D., 379–80, 382, 394, *396*
Woolacott, R. M., 202, *210*, 214, 231, *242*
Wooller, R. D., 452, *458*

Wyer, D. W., 482, *497*
Wynne-Edwards, V. C., 446, 457, *458*

Y

Yakovlev, N. N., 158, 160, *171*
Yanagita, R., 18, *27*
Yonge, C. M., 244, *253*, 269, *279*
Young, J. Z., x, xxvi, *xxxi*
Young, K., 259–60, *279*

Z

Zafiriov, O, 390, *396*
Zahavi, A., 457, *458*
Zaneveld, J. S., 282, *298*
Zaniolo, G., *444*
Zibrowius, H., 283, 286, 292, *298*, 305, *318*
Zimmer, R. L., 202, *209*, *210*, 214, 231, *237*, *242*
Zirpollo, G., 227–8, *242*
Zonneveld, J. I. S., 282, *294*
Zottoli, R. A., 289, *298*
Zumwalt, G. S., *170*

Index of Genera and Species

A

Abiesograptus, 429
Acacia drepanolobium, 353
Acanthaster, 375, 390–3
  planci, 388–90
Acartia clausi, 235
Acetabularia, 291
Acharadia larynx, 121
Acromyrmex, 365–6
  octospinosus, 364, 366
Acropora, 538
  bruggemanni, 464
  palmata, 540–1
Acrothoracica, 320
Actinomyces, 12
Actinophrys, 33
  sol, 33
Adamsia palliata, 146
Adelograptus lapworthi, 418
Adeonella, 197, 208
Aerobacter, 9
Aetea, 194, 202–3
  sica, 227
Agaricia, 472, 487
Aglauropsis, 109
Alaesma, 321
Alcyonidium, 196, 218, 470
  gelatinosum, 217
  hirsutum, 217–18, 224, 467–8
  nodosum, 196
  polyoum, 218, 226, 467–8
Alcyonium digitatum, 145
Alysidium parasiticum, 214
Amaroucium constellatum, 464, 466
  pellucidum, 464
Amblyopone, 349
Amiternes, 359, 368

  evuncifer, 357, 369–70
  herbertensis, 370
  laurensis, 359, 370
  lonnbergianus, 369–70
  messiane, 370
  unidentatus, 369–70
  vitiosus, 370
Amphiura chiajei, 382
  filiformis, 376, 382
Amplexograptus, 422
Analcidometra caribbea, 379
Ancistrotermes cavithorax, 369, 370
Antedon, 378–9, 383
  bifida, 376, 378
Anthopleura elegantissima, 144, 235
Aphroceras, 58
Apicotermes, 363
  desneuxi, 362–3
  gurgulifex, 362–3
  lamani, 362–3
Apiograptus, 421
Apis, 356
  mellifica, 358
Aplousobranchiata, 433
Aplysilla rosacea, 60
Apoica, 354
  pallida, 354
Archotermopsis, 350
Artemia salina, 235
Arthrobacter, 7, 12
Ascidia nigra, 441, 464
Ascophyllum nodosum, 313, 467
Aspercoccus, 313
Aspidostoma, 204
  giganteum, 204, 206
Astalotermes quietus, 369
Asterias, 375, 390–2

Asterias—continued
   rubens, 310, 376, 381, 388–90
Asyncoryne, 128
   ryniensis, 127
Atta, 365–6
   cephalutes, 364
   sexden, 364
   texana, 364
Augochloropsis, 347
Averianowograptus, 426–7
Azteca, 353, 361

B

Bacillus, 11
   cereus var. *mycoides*, 10
   circulans, 11
   mycoides, 7, 9–10
Balanus, 290–1, 317
   amphitrite, 466
   balanoides, 321–3, 464, 466
   crenatus, 466
   eburneus, 464
   hameri, 320
   improvisus, 464
   perforatus, 464
   psittacus, 320
Balella, 133
Barrettia, 272
   multivata, 260
Beania discordermae, 218
Belonogaster, 348, 354
Bembix, 352
Berenicea patina, 310
Bicellariella, 216
   ciliata, 229
Bicorona, 123–4
   elegans, 123
Bicosoeca, 33
Bimeria, 134
Biradiolites, 278
Blastrotrochus nutrix, 158
Bombus vosnesenskii, 358
Bothrophyllum dobroljubovae, 158, 160
Botrylloides, 435
Botryllus, 433, 435, 438–9, 441–2
   schlosseri, 218, 435–6, 438–43, 464

   primigenus, 439
Botrytis cinerea, 19
Bowerbankia, 224, 227
   candata, 226
   gracilis, 226, 471
   imbricata, 224–5, 471
   pustulosa, 464
Brachionus, 235
Brachyodontus lineatus, 467
Buccinum undatum, 310
Bugula, 200, 214, 216, 224, 227, 231
   calathus, 198
   flabellata, 229–31, 464
   fulva, 464
   neritina, 202–3, 214, 227, 464
   plumosa, 464
   simplex, 464
   turbinata, 217
   turrita, 464

C

Caberea, 222
Calliactis parasitica, 144, 146
Callopora, 213–14, 216, 231
   dumerilii, 219, 230–1, 310
   lineata, 216
   rylandi, 216
Callyspongia vaginalis, 64
Campanularia flexuosa, 471, 547
Camponotus truncatus, 352
Comptoplites, 218
   atlanticus, 218
Cancer pagurus, 310
Candelabrum, 121
Carbasea indivisa, 229
Carchesium, 33
   limneticum, 33
Cassiopea andromeda, 100
Catenaria parasitica, 214
Catenicula, 202–3
Celleporella hyalina, 464, 467
Cephalodiscus, 409
Ceratophyllum eifelense, 158
Ceratoporella nicolsoni, 60
Chaetopterus variopedatus, 235
Chama, 472–4

Chartergus, 354–6
　chartarius, 353
Chelonibia patula, 321
Chitinopoma surrula, 291
Chlamys varia, 310
Chorda filum, 313
Chorizopora, 217
Chrondoclafia gigantea, 60–1
Ciona intestinalis, 464
Cladocoryne floccosa, 128
　littoralis, 128
Clathrina, 58
　blanca, 63
　coriacea, 63
Clava, 131–2
　squamata, 464, 467
Clavelina epadiformis, 310
Clavopsella, 133–4
Climacograptus, 422
　innutatus, 408
　inuiti, 408
　typicalis, 402, 406, 409
Cliona, 61, 69
　celata, 310
Clisaxophyllum ava atetsuense, 158
Clonograptus tenellus, 417–18, 421
Clytia edwardsi, 467
　volubilis, 467
Cnidotiara, 133
Codium tomentosum, 312
Codosiga botrytis, 51
Comminella papyracea, 196
Canopea, 320–1
Conopeum, 213, 472, 488
　tenuissimum, 471–2, 487, 547
Conularia, 98
Corallina ollcinalis, 467
Cordylophora, 132, 149
Coronula, 321
Corydendrium, 132
Coryne, 124–6
　ochidai, 467
Corynoides, 420
Crambe crambe, 53
Craspedacusta, 109
Crassimarginatella, 202–3, 208
　similis, 198, 206, 208

Crassostrea, 319, 324
　virginica, 320–2
Craterophyllum, 157
　verticillatum, 157
Crematogaster, 353
Crenilbrus melops, 310
Cribrilaria, 216
Cribrilina, 216
Crisia, 227
Cristatella, 234, 236
　mucedo, 220, 235
Cryptocerus punctulatus, 350
Cryptosula pallasiana, 217, 464
Cryptotermes, 352
Ctenolabrus rupestris, 310
Cubitermes, 357, 368
　intercalatus, 357
Cucullograptus, 425
Cucumaria elongata, 385
Cunina, 108
Cupuladria, 198, 222, 235–6
　biporosa, 235
Cyclotella nana, 472
Cyphastraea ocellina, 464
Cyphomyrmex, 366
Cyrtograptus, 402, 408, 426–7
　linarssoni, 427
Cystoseira barata, 467

D

Daphnia pulex, 235
Dendrograptus, 408
Dendrophyllia manni, 464, 466
Dendropoma, 243, 247–9, 251–2
　corallinaceum, 248–9, 251
　irregulare, 247
　maximum, 246, 249–50
　petraeum, 247
Dendya, 58
Desmarella moniliformis, 51
Diadema antillarum, 540
Diademnum, 470
Dicaulograptus hystrix, 424
Dicellograptus, 420
　caduceus, 420
Diceras, 268

Dichograptus, 419
  octobranchiatus, 418
Dictyonema, 399–400, 408, 417
  flabelliforme, 399–400, 417, 419
Dictyostelium discoideum, 20, 30
Dicyclocoryne, 123–4, 137
  filamenta, 123
Didymograptus, 404, 406, 408
Dimorphograptus, 397
Diplosolen obelia, 213
Diplosoma listerianum, 464
Dipurena, 110, 124–6
Discoporella, 236
  umbellata, 235
Disporella, 235
Disyringa, 61
  dissimilis, 62
Dolichoderus scabridus, 368
Doryporella alcicornis, 202, 204
Drepanotermes rupiceps, 370
Dunaliella tertiolecta, 472
Dynamena, 470

E

Echinus, 385
  esculentus, 310, 385
Eciton, 356
  hamatum, 357
Ectopleura, 106, 120–1, 136
  crocea, 121
  dumortieri, 121
  larynx, 121
  ochracea, 121
  wrighti, 121, 128, 135–6
Electra, 213, 470, 487
  monostachys, 471
  pilosa, 213, 219–20, 227–8, 269–71, 487–8
  verticellata, 213, 215
Elminus modestus, 466
Entelophyllum articulatum, 160
Eodiceras perversum, 260
Eoradiolites, 278
Ephydatia fluviatilis, 39, 53
Epistylis, 33
Escharella immisca, 310

Escharina porosa, 198
Escharoides coccineus, 310
Escherichia coli, 3, 7–8, 10, 18, 20
Esperiopsis digitata, 64, 66
Eudendrium, 134
Eudorina elegans, 31
Euglypha rotunda, 32
Eunicella stricta, 218
Euphysa, 106
  aurata, 110
  farcta, 135
Eurystomella, 228
  foraminigera, 227–9
Eusmilia, 147
  fastigiata, 146
Euthyroides episcopalis, 214
Exchonella, 217

F

Ficopomatus, 287, 292
  enigmaticus, 285, 287, 291–2
  macrodon, 285
  miamiensis, 285
  uschakovi, 285, 288–9, 292
Filograna, 281, 283, 286–7, 319
Filogranula, 281
  gracilis, 286
Flabellum rubrum, 157
Flustrellidra, 213, 470
  hispida, 218, 467–8
Forella repans, 471
Formica 367
  polyctena, 367
Fredericella, 235
Frenulina sanguinolenta, 464
Fucus, 469–71
  serratus, 467, 469–70, 487
Fulmarus glacialis, 446
Fungia actiniformis, 464
Fusarium oxysporum, 19

G

Galathea squamifera, 310
Galaxea aspera, 464
Galeolaria, 291

Index of Genera and Species

Galeolaria—continued
  caespitosa, 284, 288
  hystrix, 284
Geodia muelleri, 72
  phleraei, 65, 71
Globitermes, 346
  sulphureus, 369
Gonionemus, 109
Gonium pectorale, 31
  sociale, 31
Gonothyrea, 470
Gorgonia, 321
Gothograptus, 424
  nassa, 424
Gracilaria, 324
Grantia, 470
Guynia annulata, 286

H

Halichondria, 310
  panicea, 40, 59, 235
Haliclona, 71
  oculata, 66–7
  permollis, 64
  ventilabra, 66–7
Halicoma, 464
Halitiara, 133
Halmomises, 111
Halocoryne epizoica, 137
Heliophyllum halli, 158, 160, 164
Heritschioides, 155, 157–8, 160, 163–4, 166, 168
Heteropora, 218
Hippopodinella, 197
  adpressa, 223
Hippoporidra, 196–7, 208, 223
  senegambiensis, 223
Hippothoa, 219
Hippuritella toucasi, 260, 274, 276
Hippurites bioculatus, 274
  socialis, 260, 274
Holmograptus lentus, 420–1
Holoretiolites, 402, 424
  balticus, 424
Hornera, 218
Hyalonema, 64

  hazawai, 60–1
Hybocodon, 120–1
Hydra, 148
Hydractinia, 126, 149
  echinata, 322
Hydroides, 290, 292
  dianthus, 285, 289
  dirampha, 285
  elegans, 285, 287–8, 291–3
  norvegica, 287–8, 291–2, 466
  pacificus, 292
  sanctaecrucis, 285
  spongicola, 287
  uncinata, 292
Hymeniacidon, 41, 43, 46, 62
  perleve, 40, 60, 71
Hypolytus, 118
Hypotermes obscuriceps, 370

I

Ianthella, 66
Iovaphyllum, 164, 169
Iridomyrmex humilis, 348
  rufonigra, 368
Isograptus gibberulus, 420

J

Josephella, 281, 286
  marenzelleri, 286

K

Kalotermes, 350
Kinetoskias, 22, 236
Klebsiella aerogenes, 8–10
Kodonophyllum truncatum, 160
Koellikerina, 134

L

Labioporella, 198
  dipla, 198
Laminaria digitata, 467
  hyperborea, 467
  saccharina, 467

Lampropedia, 6
Lasiograptus harknessi, 424
Leptocheilopora tenuilabrosa, 214
Leptogenys, 349
Leptomonas, 32
Leuconia, 58
Leucosolenia, 62
  blanca, 60, 63
  botryoides, 63
  cavata, 63
  complicata, 55
  cordata, 62
  coriacea, 58
  osculum, 53
Lichenopora, 235
  hispida, 310
Limnocnida, 109
Limulus, 163
Linograptus, 429
  posthumus, 426–8
Linuche, 93
  unguiculata, 85, 89–95, 99
Linvillea, 116, 137
  agassizi, 115
Lithophyllum polymorphum, 467
Lithostrotion, 160, 183, 192
  aramea, 180
  decipiens, 180–2, 184, 187
  martini, 178, 180
  pauciradiale, 178, 180–1
Lithothamnion, 248–9, 282
Littorina, 320
Lobocoryne travancorensis, 128
Lophopodella carteri, 235
Lophopus cristallinus, 235

M

Macaranga, 353
Macrotermes, 356–7, 369
  bellicosus, 357, 359–60, 362, 370
  carbonarius, 370
  falciger, 357
  gilvus, 370
  natalensis, 359
  subhyalinus, 350, 356, 359, 361, 370
Madrepora durvillei, 191

Margelopsis, 120–1
Marifugia, 287
  cavatica, 285, 287
Mayorella palestinensis, 32
Megalodon, 268
Megaponera foetens, 349, 357, 368
Mellita quinquiesperforata, 376, 386–8
  sexiesperforata, 387
Melonanchora kobjakovae, 66–7
Membranipora, 150–1, 215, 221, 224, 270
  hyadesi, 218
  membranacea, 150, 213, 217–18, 220–2, 224, 227–8, 467–8
Membraniporella nitida, 216
Mercierella enigmatica, 288, 291–2, 317, 466
Merona, 132
Metalcyonidium gantieri, 236
Metridium, 144, 310
Microecia suborbicularis, 218
Microporella ciliata, 310
Microtermes globicola, 370
Millepora, 130, 149
Miroserpula inflata, 291
Mischocyttarus, 354
  drewseni, 354
Modiolus, 255
  modiolus, 255
Moerisia, 111, 115
  hori, 114, 134
  inkermanica, 114, 134
  lyonsi, 110–11, 114–15, 121, 128, 134–6
  pallasi, 120
Monia patelliformis, 310
Monobrachium, 109
Monobryozoon, 236
  ambulans, 212
  bulbosum, 212
  limicola, 212
Monochrysis lutheri, 472
Monocoryne gigantea, 119
Monograptus, 400, 408, 426
  flemingii, 402
  limatulus, 425–6
  priodon, 402
  sedgwickii, 402
  turriculatus, 426
Monopleura, 268–9

Monopleura—continued
  marcida, 270
  urgonensis, 270
Montastrea, 472, 487
  annularis, 546
Mycale macilenta, 464
  richardsoni, 72
Mycobacterium, 12
Myrionema, 134
Myrmecia, 349
Mytilus edulis, 235, 464
Myxococcus xanthus, 13

N

Nannocystis, 13
Nanomia, 150
Nasutitermes, 353
  ephratae, 361
Nausithoë, 85–7, 91–2, 94, 96, 98
  punctata, 84–5, 88, 90–5, 99
Nemagraptus gracilis, 420
Nemaster grandis, 378–9
  rubiginosa, 378
Nemertesia antennina, 394
Neofibularia mordens, 63
Neokoninckophyllum kansasense, 157
Nicholsonograptus fasciculatus, 420–1
Niobia, 133
Nocardia, 12

O

Obelia, 149
  dichotoma, 467
Octolasmis, 321
Odessia, 111
  maeotica, 114, 134
Odontotermes, 369
  badius, 370
  horni, 370
  praevalens, 370
  redemanni, 370
  stercorivorus, 370
Oecophylla, 353
  longinoda, 366–7
Olindias, 109

Omphalopoma gracilis, 286
Ophiocomina, 382–3
  nigra, 379, 382
Ophiothrix, 382
  fragilis, 376, 379–83, 392–4
Ophlitaspongia seriata, 59, 69, 72, 464, 466
Orionastraea, 188
  edmondsi, 186–7
  ensifer, 182, 184
  indivisa, 186–7
  phillipsii, 182, 184
  tuberosa, 184, 186–7
Orthograptus gracilis, 408
Ostrea edulis, 235, 464, 466
Ostroumovia, 111

P

Pachymatisma johnstoni, 40
  johnstonia, 63–4
Paludicella articulata, 227
Paracentrotus lividus, 316, 376, 386, 392
Paracoryne, 122
  huvei, 122–3
Paramecium caudatum, 30
Paramurrayona corticata, 70
Parasmittina trispinosa, 218
Paratiara, 133
Paravespula germanica, 365
Pelagia noctiluca, 109
Pelagohydra, 120–1
Pennatula, 145
  phosphorea, 143, 145, 147
Pentapora foliacea, 217, 227
Perophora, 434
Petaloconchus montereyensis, 244, 251
Petalograptus, 422–3
Phakellia arctica, 64
Phallusia mamillata, 235
  nigra, 466
Phyllospadix iwatensis, 467
Physalia, xxv, 149, 150
Platylepas, 321
Pleodorina, 36
  californica, 31
Pleurocystis, 29

Plumaria undulata, 467
Plumatella fungosa, 222
Pocillopora damicornis, 464
Podangium lichenorum, 13
Podocoryne carnea, 546
Pogonomyrmex, 364–5
   barbatus, 365
   rugosus, 365
Polistes, 332, 347–8, 354, 356–7
Pollicipes spinosus, 464
Polyandrocarpa gravei, 464
   tincta, 464, 466
Polybia, 353
   dimidiata, 354, 356
   scutellaris, 358
Polyclinum, 470
Polydora ciliata, 385
Polykrikos, 30
   kofoidi, 31
Polymastia, 60, 69
   bursa, 59
Polyrhachis, 353
Pomatoceros lamarckii, 284, 291
   triqueter, 235, 284, 288–92, 306, 310, 314
Pomatoleios, 290, 319, 324
   krausii, 284, 320–1
Porites, 144, 150
Porpita, 236
Poterion neptuni, 64
Proboscidactyla, 149
Procubitermes, 357
Proterospongia haeckeli, 51
Proteus, 10
   mirabilis, 11
Protiara, 133
Protopolybia, 353–4
Psammechinus miliaris, 310
Pseudalcyonidium bobinae, 236
Pseudocucumis mixta, 376, 385
Pseudovermilia multispinosa, 287
Pseudovermilus, 472
Pteroclava, 128
Pyrgoma, 321
Pyripora, 215, 538
Pyriporopsis, 215, 219, 544

R

Radiolites, 272
Radiophrya, 30
   hoplites, 31
Rastrites, 426–7
Renilla köllikeri, 145
Reptadeonella, 202–3
Requinia ammonia, 270
Reteporellina evelinae, 214
Retiolites, 400
Rhammatopora, 215, 544
Rhizaxinella burtoni, 60–1
Rhizogeton, 131–2
Rhizopus stolonifer, 18
Rissa tridactyla, 445–7
Ropalidia, 354
Rosalinda, 106, 130
   incrustans, 130
   williami, 130

S

Sabellaria, 319, 322
   alveolata, 320–1, 464, 466
   spinulosa, 464, 466
Sacculina carcini, 321
Salmacina, 283
Sarcina, 6–7
Sargassum fulvellum, 467
Sarsia, 124–6
   eximia, 124
   gemmifera, 125
   princeps, 125
   prolifera, 124
   tubulosa, 124–5
Sauvageria, 259, 268
Scalpellum, 320
Schedorhinotermes, 368
Scruparia, 213
Scrupocellaria, 198, 200, 202–3, 214, 217
   reptans, 464, 467
   scabra, 214
   scruposa, 464
Scypha, 58
   ciliata, 58
   okadai, 71
Semperella schulzei, 63

Index of Genera and Species

Serpula massiliensis, 285
 vasifera, 291
 vermicularis, 285, 299, 305–6, 310–11, 314–17
Serpulorbis, 246, 251
 aureus, 246–7
 navalensis, 245–6, 249
 squamigerus, 250
Sertularella miurensis, 467
Silhouetta, 133–4
Sinodiversograptus, 426–7
Siphonodictyon, 61, 69
Skatitermes, 369
Sphaerocoryne, 110, 115–16, 137
 bedoti, 115
 multitentaculata, 115–16
Sphex, 352
Spinograptus, 402
Spirobranchus, 129
 cariniferus, 284
 giganteus, 287
 paumotanus, 284
 polycerus, 284
 polytrema, 284
 tetraceros, 129, 287
Spirophyllum geminum, 158
Spirorbis, 323, 470
 borealis, 464, 466–8
 corallinae, 467–8
 pagenstecheri, 464, 466, 468
 rupestris, 464, 466–7
 tridentatus, 314, 464
Spisula, 381, 389
Stauridiosarsia, 124
 producta, 124
Straurocoryne, 124
Steganoporella, 200, 208
 magnilabris, 198, 546
Stelopolybia, 361
 angulata, 354, 361
 areata, 361
 flavipennis, 361
 multipicta, 361
 myrmecophila, 361
 testacea, 354, 361
Stephanoscyphus, 83–4, 94, 98, 100, 108
 allmani, 85

 bianconis, 84
 cornifurmis, 84
 komaii, 85
 mirabilis, 84
 racemosus, 85, 90–6, 99–100
 sibogae, 84
 simplex, 84
 striatus, 84
Stomatopora, 472–5
Streptococcus, 6–7
 faecalis, 8
Streptomyces, 12
 coelicolor, 13
Stylocordyla, 64
 borealis, 60–1
Stylopoma, 198
Stylotella heliophyla, 59
Sycon, 470
 defendens, 70
Symplegma virde, 464
Synocoryne, 124
Synoeca, 354, 356
 surinama, 354
Synops anceps, 63–4

T

Tabulophyllum rotundum, 158
 schlueteri, 158
Tedania actiniformis, 63–4
Tegella verrucosa, 196
Tendra, 202–3
Teissiera, 106, 128, 130
 australe, 128
 medusifera, 128
 milleporoides, 128–9
Tenuirostritermes tenuirostris, 368
Terebratella inconspicua, 464
Terebripora comma, 226
Termitomyces, 356
Termitopone, 349, 368
Tessaradoma boreale, 235
Tethya lyncurium, 46
Tetraclita stalactifera panamensis, 291
Tetracoccus, 6–7
Tetragraptus phyllograptoides, 420
Thalamoporella, 202–3

Thoracotermes, 357
Tiaricodon, 111
Timania, 157
  rainbowensis, 157
Titanosarcolites, 270
Toucasia texana, 270
Trachymyrmex, 365
  septentrionalis, 366
  seminole, 366
  urichi, 366
Tremogasterina, 202–3, 217
Tricellaria ziczac, 213
Tricyclusa, 117
  singularis, 117–18
Tripneustes esculentus, 385
Triporula, 217
Tropidozoum, 208
  cellariiforme, 200
Tropiometra carinata, 378
Trypanosoma, 32
Trypostega, 208
  claviculata, 197, 202, 204
  venusta, 202, 208
Tubularia, 106, 120–1, 135, 148–9, 320
  ceratogyne, 120–1
  cornucopia, 120
  crocea, 120
  indivisa, 120–1
  larynx, 120, 466
  mesembryanthemum, 467
  prolifera, 120
  regalis, 120–1
  unica, 120
Tubulipora phalangea, 310
Turritella, 251–2

U

Uscia mexicana, 200, 208

V

Vaccinites, 272–3, 277
  gasaviensis, 270
  giganteus, 274
Vaginicola, 33
Vallentina, 109
Vannuccia, 106
Velella, 148–9, 236
Veretillum cynomorium, 145
Vermetes, 321
Vermetus allii, 244
Vermicularia, 252
Vermiliopsis, 285, 287
Verongia, 66
Vespa orientalis, 361
Vespula, 348
  vulgaris, 365
Victorella pavida, 227
Volvox, 36
Vorticella, 32
  convallaria, 32

W

Wafkeri uva, 227
Watersipora cucullata, 464, 466
  nigra, 224
Wawalia, 544
Wilbertopora, 215, 544

X

Xestospongia muta, 64, 66

Z

Zanclea, 110, 128
Zoobotryon, 227
  verticellatum, 227, 235
Zootermopsis, 350
Zoothamnium, 34–5
  alternans, 34
  arbuscula, 34
Zostera, 441–2
  marina, 467
Zygocystis, 29
Zyzzyzus solitarius, 121

Subject Index

The numbers in *italic* refer to figures

A

actinula larva, 109, 121–2
adaptive responses, 461
age-correlated zones, 229–31
ageing, 226, 227, 236
aggregation, xxvi, xxvii, 446, 465
  barnacles, 320–3, 325
  disadvantages, 393
  echinoderms, 375–94
  molluscs, 243, 244, 247, 249, 251, 252, 257–9
  myxobacteria, 13
  protozoans, 30, 32
  serpulids, 283, 287–91, 299, 305–9, 315–17
  slime moulds, 20–1
  social insects, 347
aggression
  intercolonial in Cnidaria, 144
  interspecific in Cnidaria, 144
  social insects, 354–5
alarm pheromones, 363, 365–7
Albian, 270
allelo chemicals, 504, 538
allografts, 42, 44, 45, 218
alternation of generations, 81, 94, 100
altruism, 330–42, 346
amoeboid cells, 49, 51, 52
ancestrula, 212, 216, 217, 227
antibodies, 46
aquiferous system, 49–51, 53–69, 71–3
Arenig, 400
Arundian, 173, 178
Asbian, 178, 180
associative settlement, 321–2
astogeny, see colony formation
astogenetic repetition, see cormidial hierarchy
atrichous isorhiza, 131, 134

autografts, 42, 44
autosyndrome, 218, 219
autotheca, 412, 415–17, 419
autotomy, 227
avicularia, 193, 198, 199, 200, 206–8, 213, 216, 217, 514, 538

B

b-zooids, 208
baffle effect, 375, 378, 379, 380, 382, 383, 385, 392
bandages, 397, 402–9
basal disc, 114, 117, 119–20, 122, 135, 176–7, *114*, 182, 184
behaviour
  barnacle larvae settlement, 323
  colonial in Cnidaria, 144, 145, 149
bilateral symmetry, 435
biochemical halo, 267
biochemical tissue integrity, 51
biodeposition, 255
biohermal distribution, 259
biomass of barnacle aggregates, 323
biomicrudite, 259
biopelmicrite, 259
biosparite, 259
biostromal distribution, 259
biota of serpulid reefs, 310–11
biotic factors, in serpulid distribution, 290–1
bitheca, 414, 416, 417, 419, 420
blastogeny, 162
blastogenic generations, 435–8
blastostyle, 118, 127
body chamber, rudist, 263, 264
body wall composition in bryozoans, 216, 220, 227

boring sponges, 61, 62
breeding
  population, 451
  success in, 454
Brigantian, 178, 184, 186
brood chambers, 193, 199–207, 212, 213, 229–31
brooding
  bryozoans, 193, 196, 197, 199–208
  serpulids, 283, 291, 316, 317
brown body, 226, 227, 229
budding, 465, 499, 504
  ascidians, 434–8, 442
  bryozoans, 194–8, 206, 212, 215, 217, 219, 223, 225, 229, 232
  cnidarians, 178, 179, 189
  graptolites, 411, 412, 414, 416, 421, 423, 425, 427, 430
  hydrozoans, 108, 111, 117
  monopodial, 412
  protozoans, 30
  rugose corals, 156–8, 162–8, 178, 179, 189
  scyphozoans, 93, 96, 97, 100
  serpulids, 282, 283, 293
  sponges, 58
  sympodial, 412
buoyancy, 398, 399, 400, 417, 423, 427
burrowing, in Cnidaria, 145

C

calcification
  blue-green algae, 14
  bryozoans, 194, 200, 233
calice, 176, 183, 184, 190
cnidarians, 142, 143, 176, 184, 185
calicoblast layer, 176
Cambrian, 399, 402, 412
camouflage, 266, 267
Campanian, 260
Caradoc, 402
carbonate sediments, 1
Carboniferous, 173, 399, 412
castes, xxxiv–xxvi
  social insects, 329–43
cauda, 418, 421

cell adhesion, 40, 42, 44, 45, 46
cell communication, sponges, 48
cell differentiation, 51, 52
cell lineages, 51, 52
centre effect, 451–5
chemoresponse, 390, 392, 393, 467, 469, 514
chemosensory organs, 322
chemotaxis
  Eukaryotes, 15, 20, 21, 23
  myxobacteria, 13
  Prokaryotes, 10
  serpulids, 288
  slime moulds, 21
chemotropism, in Eukaryotes, 15
chimneys, 58, 60–4, 69, 195–7, 206, 222–3, 263–5, 277
choanocytes, 52
choanocyte chambers, 49, 53, 55, 58, 69, 70–2, 510
cilia, 194
ciliary currents, 414
cincture cell, 224
circadian rhythms, 15, 19
cirri, 378
cladia, 402, 426–8
cleaning, 193, 195, 198, 200, 213, 258, 262, 263, 265, 266, 416
clearance rate, 235, 236
cloaca, 68
clone, xvii–xxx
  eukaryotes, 24
  social insects, 330, 343
clumping, 32, 33, 465
clustering, 273–8, 375–83, 385, 388
cnidophores, 127–8
coefficient of relationship, 343, 346–8
coelenteron, 86
coelome, 55, 194, 202–4, 206, 211, 214, 218, 222, 228, 230, 232
  colony-wide, 194, 202, 203, 206, 211, 234
coelomic hydrostatic pressure, 222, 228
coenenchyme, 176
coenoecium, 412
coenosarc, 122, *123*, *129*, 176, 178, 179, 186–90, 192

coenosteum, 176, 178, 188–90
colonialism, xvii, xxvii
coloniality
  advantages of, 14, 33, 151, 435, 438, 440–2, 455–7
  development in Hydrozoa, 134–7
  functional significance, iv–v
colonization, xviii
colony boundaries, 68
colony control, 53, 180, 193–209, 219–22, 414–16, 427, 536, 537
colony formation, xix
  ascidians, 435–8
  bryozoans, 193, 195, 204, 212, 215–18, 225, 544
  corals, 162–4, 168
  graptolites, 398, 399
  hydrozoans, 134–7
  molluscs, 244, 246–8, 251
  protozoans, 30–6
  rugose corals, 176–81, 184–9
  sponges, 51–5, 57, 58, 60–4, 66, 68–74, 478
colony individuality, 39–48, 217
colony integration, 33–6, 173, 178, 180, 183, 184, 188, 189, 191, 193–209, 499, 533, 536, 537, 539
colony movement, 221, 222
colony polarity, 229
colony traits and definition, xvi–xviii, xxiii–xxx, 4, 31, 36, 174, 243, 393, 394, 460, 468, 469, 481, 482, 490, 506
colony water currents, 42, 193, 195–8, 208, 375, 378–80, 382, 387, 388, 391, 392, 414, 417, 420, 503, 538, 545; see also baffle effect
column wall, 176, 177, 184
commensalism, 223, 529, 530
common canal, 413
communal feeding, 249, 250, 252
commune, xviii, xxvi–xxix
communication, 13, 15, 193–5, 200, 211, 234
communication pores, 194, 202, 206
competition, 23, 244, 247, 251, 252, 255, 290–2, 313, 314, 316, 321, 323, 391, 436–41, 451–2, 470, 480, 486, 487, 505
  spatial, 473, 474, 475, 479, 505, 506, 520, 521, 539, 541, 542, 543
competitive hierarchy, 475
competitive networks, 475, 476, 479
complemental males, 320
confrontation strategy, 499
coniconchines, 429
contact recognition, 321
conus, 417, 418
convergence, 500
co-ordination
  in defining colonies, xxv–xxx
  inter zooidal, 220
  mechanical, 221, 222, 394
  nervous, 220, 221, 222
  protozoans, 33
  sponges, 48, 51
coral reef, 378, 388, 389, 390, 391, 475, 488, 500, 501, 505, 507, 529, 530, 540–2, 545
coral reef spaces, 529, 530
cormidia, 206, 215, 216, 217, 236
cormidial hierarchy, xix
cormus, 50
cortex, 399, 404, 406, 408, 409, 412–15, 424
cortical skeleton, 71, 72
cortical zooids, 197
costae, 177, 178, 189, 190, 216
Cretaceous, 209, 215, 216, 219, 258, 268, 269, 544
critical spaces, 258
cross-fertilization, 320
cryptic environments, 463, 465, 473, 475, 479, 480, 483, 488, 500, 501, 526, 536, 543, 545
cryptocyst, 202, 204, 213
culturing, 3, 42
current effects, 244, 247, 249–52, 259, 260, 262, 264–7, 377, 378, 380, 382, 386, 387, 391–4, 502–7, 511, 534, 540, 541
cuticle, 194, 202, 514
cyclic A.M.P., 20, 21
cytoplasmic connections, 32, 34, 36

D

dactylozoids, 122, *123*, 126, 128, *129*, 130
de-colonization, 430
defence strategies, 198, 200, 244, 250, 255, 265–8, 277, 352–7, 355, 380, 455, 459, 463, 465, 504, 514, 537, 538
defensive secretions, 367–71
degeneration, 195, 226–31
denticles, 213
desmoneme, 117, 122, 127, 131
Devonian, 268, 402, 428, 429, 430
dietellae, 214, 215, 219, 220
differentiation
  blue-green algae, 14
  mycelial colonies, 13, 23
  protozoa, 33, 34, 36
digestion, 176
dimorphism, 82
direct development, 108
directional growth, 459, 461, 471, 472, 475, 482–90, 499, 506, 523, 525, 526, 528, 531, 533–6, 538, 546
dispersal, 306, 309, 465
dissepiments, 173, 182, 186–90, 400, 416, 417, 419
division of labour, 329
divorce rate, 453–4

E

ecology, 310, 311
ectooecium, 214
edge effect, 451–5
edge zone, 176–9, 181, 189
egg, 109, 131
electrophoresis, 217
electrophysiology, 146–51
embryos, 52, 202, 226, 229, 230, 434
Emsian, 429, 430
endodermal canals, 122, *123*, 128
entooecium, 214
environmental effects, 59, 68, 71–3, 180, 181, 196, 197, 244, 250, 257–60, 262–7, 375, 377–9, 385, 386, 390, 391, 394, 429, 463, 479–81, 500, 502–4 511, 534, 540, 542, 545

ephyrae, see medusa larva
epitheca, 176, 177, 179, 181, 182, 183, 185, 191
epithelia, 82, 218, 412, 413, 415, 427
erosion, 248
eusocial insect colonies, 347, 348
evolution of coloniality, 23, 24
  insects, 351
evolution of colony form, 543, 544
excretion, 55, 176, 188, 262–4, 272, 277
exopinacoderm, 57
extinction, 429, 430

F

fecundity, 454, 455, 534, 535, 536, 542, 543
feeding, 504, 506, 508, 515, 517, 538
  bryozoans, 193–5, 198, 235, 236
  cnidarians, 144, 145, 147
  currents, 244, 262–4, 266, 270, 272, 273, 276, 277, 415, 416, 419, 423, 425, 503
  echinoderms, 375–83, 385–94
  gastropods, 244, 245, 249, 252
  graptolites, 419–23, 425–8
  rudists, 257, 259, 262, 270, 274, 276, 277
  rugose corals, 180–3
  synchronized, 249–50, 252
female zooids, 192, 202
fertilization, 218, 244, 251, 252, 253
filter feeding, xxxii; see also feeding
filtration fan, 378, 379
fitness to environment, 460, 461, 466, 468, 472, 475, 479, 482, 484, 486–90
flagellate cells, 49, 51, 52, 53
follicle cells, 232
foraging, 352
fossilization, 429, 430
fouling, 513, 514, 529; see also sedimentation
fugitive strategy, 499
functional design, 267–73
funiculus, 211, 218, 219, 223, 225, 226, 234
fusella, 404, 406, 409, 412–15
fusion, 32, 33, 39–48, 53, 54, 60, 62, 173, 176–84, 189, 192, 217, 218, 258, 439–44, 478

Subject Index 583

G

gametes, 331, 333
gasterozooid, 122, 126, 128, 130, 132
gastral cavity, see paragaster
gastrovascular system, 82, 173, 176, 182–6, 188, 191, 192
gene frequency, 163, 334, 338
genet, xxii, xxvii
genetic composition, 394, 545, 547
  ascidians, 438–44
  barnacles, 324, 325
  bryozoans, 195, 211, 217, 218
  graptolites, 423, 427
  rugose corals, 180
  sponges, 47, 48, 54, 71, 73, 74
genetic drift, 334
gonads, 108, 109, 115, 116, 124, 125, 132, 133
gonophores, 117, 120, 121, *123*, 124, 125, 132, 133
gonozooid, 122, *123*, *129*, 132, 213, 234
graft
  acceptance, 40, 42, 43, *43*, 44
  hierarchy, 46, 47
  rejection, 39–48, *43*
greeting ceremony, 448
gregariousness, iv, xvi
  barnacles, 321, 323
  molluscs, 243, 251, 253, 257, 258
  protozoans, 31, 32
  scyphozoans, 82
  serpulids, 288–90, 293, 316, 317
grex, see plasmodium
group phenomena, 29–33
growing edge, 228, 229, 404
growth, 459–61, 470, 473, 476, 482, 484, 487, 499, 503, 504, 507, 510, 525, 526, 528, 533, 534, 536, 538, 541, 546
  fungal, 18
  graptolite, 409, 413–16, 418
  molluscan, 248, 257, 258, 263, 268, 269
  protozoan, 33, 34
  scyphozoan, 83, 91–3
  sponge, 52, 68, 73
growth form, 459, 460, 480, 485, 486, 488–90, 499–503, 509, 515–39, 541–6; see also morphological strategies

aphroid, 173, 175, *175*, 186–9, 192
ascon, 58, 62
astraeoid, 173, 175, *175*, 182, 184, 185, 187, 189, 192
bryozoans, 209
cerioid, 173–5, *174*, *175*, 180–4, 187, 189–92, *190*
dendroid, 173, 174, 175, *175*, 178–80, 399, 402
elevator, 264, 265, 267, 272, 273, 276, 277
encruster, 264, 265, 267
factors affecting, 53, 68, 471, 472, 475, 479, 545–7; see also morphogens
fasciculate, 173–5, 177–80, *174*, 184, 189–92, *190*
graptolites, 411, 415–18, 420, 421, 424, 430
indivisoid, 173, 175, *175*, 186–8, 192
leucon, 58
life-history attributes, 499, 533
massive, 174, *174*, 175
mounds, 499–501, 515–22, 524–32, 534–9, 541–6
pedunculate, 61–8
phaceloid, 173, 175, *175*, 178–80
plates, 499–501, 515–22, 524–39, 541–4, 546
plocoid, 189, *190*
ramose, *190*
recumbent, 264, 265, 267, 269, 278
rugose corals, 156, 162
runner, 478, 480–3, 487, 488, 490, 499–501, 515–32, 534–9, 541–4, 546
scandent, 421–3
sheet, 478, 483, 486–90, 499–501, 515–32, 534–9, 541–6
sycon, 58
tabulate, 190, *190*, 191
thamnastraeoid, 173, 175, *175*, 184–7, 192
tree, 499, 500–1, 515–22, 524–39, 541–4, 546
vines, 471, 481, 485–8, 499–501, 515–22, 524–39, 541–4, 546
growth lines, 413, 414, 415
growth rate, 8, 22, 283
gymnocyst, 213–15

H

habitat
  eubacteria, 8
  rudists, 260
  scyphozoans, 95, 96, 100
  selection of, 459–91
  serpulids, 287
haplo-diploid incompatibility, 439–40
Hardy-Weinberg principle, 325
hemichordate affinities, 402, 408
hermatypic colonies, 143
heterogenous colonies, 330
heteromorphy, 193–209
hibernacula, 227
holdfast, 399, 415, 416, 417
Holkerian, 178, 180
holotheca, 185
homeostasis, 357
homosyndrome, 218
hydranth, 115–23, *118*, 128, 130
hydrocaulus, 118, 123, 137
hydrodynamics, 53, 55–62, 68, 74; see also colony water currents
hydroid, 111, *114*, *115*, 117–20, 123, 125–7, 130
hydrology, 301, 302, 303, 304
hydrorhiza, 131
hyphal network, 19
hypostome, 131, 134

I

inbreeding, 324
incipient colonies, 158–63
inclusive fitness, 330
individual, definition and terminology, xiii, xviii–xxiii, xxvi
infertility, in serpulid aggregates, 292
information centre, 457
injury, 227, 228, 459, 482, 506, 507, 510
integration gradient, 193–209, *199*
interaction modulation factors, see morphogens
intercommunication, 194, 204, 206
interdigitating cells, 194
internal morphology, 533, 536, 537
interstitial system, 71
irritability, 198, 220, 221

J

jaws, 245
Jurassic, 219, 215, 258, 268, 544
juvenile hormone, 350

K

K-selection, 330–1
kenozooids, 196, 202, 203, 204–5, 212, 213, 215
Kimmeridgian, 269
kin selection, stimulation, 330–43

L

lamella, 412, 413, 414
larva, 108, 196, 263, 503, 504, 514
larval attachment, 263
larval development in Hydrozoa, 121
larval dispersal, 324–5
larval ecology, 287–90
larval fusion, 53
larval gregarity, 317
larval retention, see brooding
learning mechanism, 367
life cycle, of *Zoothamnium*, 34, *35*, 36
ligament in rudists, 267, 268
limiting cells, 224, 225, 226
lipids, 226
Llandovery, 400, 402, 424, 425
lophophore, see tentacle crown
Ludlow, 399, 402, 424, 428

M

macrobasic eurytele, 131, 134
macrobasic mastigophore, 130
macrocolonies, 8–12, 23
macrozooids, 34–6
male zooids, 193, 197, 206, 208, 223
mammillae, 223
mandibles, 198, 199, 200
mantle, 411, 413, 415, 416, 418, 419, 421, 427
mantle cavity, 266
mantle edge, 245
manubrium, 115, 124–5
maternal zooid, 202, 203, 204–5, 206, 207

maturity, relationship to budding, 168-9
medullary skeleton, 71, 72
medusa, 83, 86-7, 107-10, 115-16, 120-7, 130-3
medusa bud, 111, *116*, *118*, 125, 127, *129*, 133
medusa larva, 86
mesenterial connections, 144
mesenteries, 142-3, 176, 177, 178, 183, 184
mesogloea, 84, 97, 123, 142, 176
mesolamella, 119
Mesozoic, 1, 257
metabolite
 accumulation in Scyphozoa, 93-4, 96
 distribution, 55-7, 183, 184, 188, 191, 194, 195, 211, 212, 225-9, 232, 234-6, 539
metamorphosis
 of graptolites, 416
 of hydrozoan polyp, 108-9
 of sponges, 52, 53
metasicula, 415, 416, 418, 421
microbasic euryteles, 131, 134
microcolonies, 14
microscleres, 69
microvilli, 225
microzooid, of *Zoothamnium*, 34-5
modular society, xxiv-xxx; see also colony
monomorphy, 194
morphogens, 40, 42, 44, 45, 46; (see also growth form, factors affecting)
morphological parameters, 507
 circumference in contact with substrate, 508, 513, 516, 521, 522, 524, 525, 531, 532, 533, 538, 539
 encrusted substrate area, 508, 511, 512, 516, 519, 520, 521, 523, 524, 526, 528, 529, 530, 531, 532, 533
 exposed skeletal surface area, 508, 513, 516, 529, 530
 height above substrate, 508, 513, 516, 524, 526, 527, 528, 531, 532, 533
 maximum linear dimension, 508, 512, 513, 516, 523, 524, 525, 526, 531, 532, 533

shaded substrate area, 508, 512, 513, 516, 521
skeletal volume, 508, 510, 511, 516, 517, 519, 533, 534
tissue surface area, 507, 508, 510, 512, 516, 517, 518, 519, 520, 521, 523, 524, 525, 526, 528-34, 538, 539
tissue volume, 508, 510, 512, 516, 517, 518, 519, 520, 521, 523, 524, 525, 526, 528, 529, 530, 532, 533, 534
morphological strategies, 480, 481, 499-547
 comparison with plants, 545
 of cnidarian colonies, 142-4, *143*
 of microbial colonies, 10, 12, 15-23, 30-6
 of scyphozoa, 85-91
 of serpulid aggregates, 289
 of serpulid tubes, 305-6
mortality processes, 251, 460, 461, 470, 473, 480, 481, 482, 487, 490, 500, 502, 503, 505, 506, 507, 510, 520, 529, 537, 538, 539, 546
movement
 of bacterial colonies, 11
 of eukaryotes, 23
mucous net, 244, 245, 249, 250, 251, 252
mutualism, 249

N

Namurian, 173
nanozooids, 213
nema, 397, 398, 399, 400, 409, 417, 418, 419, 421, 422, 425, 427, 428, 430
nemal vanes, 400
nematocyst, 110, 114-17, 122, 126-8, 130, 133, 514, 538
nematophores, 132
neoteny, 430
nerve impulse, 221
nerves
 cystidial, 220, 221
 cerebral ganglion, 220
 motor, 220, 221
 pericystidial ring, 220, 221
 peripharyngeal ring, 220
 sensory, 220, 221
 tentacle sheath, 220

nervous communication, 194, 200
nervous organization, 206
nervous system, 53, 146, 147, 151, 211, 213, 219, 220, 221, 222
nests
  thermoregulation, 357–60
  insect defence, 352–7, *355*
  insect taxonomy, 360–3
nest building, in sea-birds, 444
nesting density, 451, 456
non-coalescence, 40, 42, 43, *43*, 44, 45, 46
nutrient storage, 208, 229
nutrition, 55, 225, 226, 229, 235, 236, 471, 472, 505, 511, 519, 529, 541, 543, 546
nutritional regime, 471

O

obligatory colonial breeding, 451
obligatory cross-fertilization, 320
ocellus, 115, 124, 128, 130, 132
offsetting, *see* budding
ontogeny, 51, 71, 73, 74, 82, 85, 97, 163–8, 195, 197, 202, 204, 216, 502
oocyte, 52, 229, 230, 231
ooecium, 213, 214, 215, 216, 219
open reef environment, 500, 501, 526
operculum, 194, 198, 208, 220
oral disc, 176, 177
Ordovician, 232, 414, 423
orientation, 244, 246, 247, 249, 250, 259, 268, 269, 270, 378, 380, 387, 397, 398, 400, 402, 414, 421, 422, 424, 470, 471
orifice, 194, 197, 198, 200, 201, 208
oscula, 49, 54, 55, 57, 58, 59, 60, 62, 63, 64, 66, 67, 68, 69, 70, 71
oscula grouping, 60, 62, 63, 64, 65, 66, 67, 68, 73
ostia, 55, 59, 62, 63, 66, 68, 69, 70, 71, 72, 508
overgrowth, 46, 47, 471, 473, 474, 475, 477, 479, 487, 513, 514, 528, 529, 534, 538, 539, 543
ovicells, 213, 215, 231
ovum, 200, 202, 229
ovum storage, 229, 231
Oxfordian, 258, 260, 267

P

palaeobiology, 1
Palaeocene, 258
Palaeozoic, 1
panic flights, 447
papillae, 59, 60, 61, 62, 63, 69
paragaster, 54
parenchyma, 119, 122, *123*
pedal disc, see basal disc
pedal gland, 244, 245, 252
pelagic organisms, xxxii, 120
periderm, 84–5, 97, 402, 404, 406, 408, 411, 412, 414, 421, 423, 424, 430
periostracum, 267
perisarc, 116–17, *119*, 120–1, 130, 132–4
perithecal membrane, 404, 408
Permian, 155–64
pharynx, 220
pheromones, 352, 358, 363–71
phosphorescence, 145, 147
photic zone, 427
photoresponse, 21, 36, 221, 222, 289–90, 383, 464, 465
phragmosis, 352
phylogeny
  bryozoans, 216, 217, 232, 233
  cnidarians, 107, 108
  graptolites, 397–409
  hydrozoans, 108, 109, 112, 113
  insects, 365
  rudist, 267–9
  rugose corals, 173, 178, 180–2, 184, 186, 191, 192
  scyphozoans, 97–100
  stromatoporoids, 1
physical selection, 469
physical stability, 263, 264, 265, 268, 273, 274, 379, 380, 386, 388
physiological gradients, 211, 226–9
physiology, 211, 226
pigmentation, 42, 223, 267
pinacoderm, 42, 47, 51
pinnules, 378
placentation, 229, 230
plagues, in starfish, 389–93
planktonic habit, 417–20, 421, 423–8
planktonic larva, 82, 286–8

planula larva, 82, 85, 97
plasmodium, 21
podocyst, 114, *114, 115*, 135
poison, 266
polarity changes, 225, 226, 227, 228, 229
polyembryony, 218
polymorphism, xix–xx, xxvi–xxxii, 82, 130–2, 137, 143, 188, 191–3, 209, 211–17, 350, 414, 415, 419, 441–2, 536–8, 542-5
polyp, 82, 84–100, 107–9, 114–15, 123, 131, 134–5, 137, 157, 160, 163, 167–8
  bud, 111, *118*
  enteron, 54
  fusion, 176–81, 183, 184, 189, 192
polypide, 220, 221, 226, 227, 228, 229
population
  definition in micro-organisms, 3–5
  density, 376, 378, 382, 383, 385–7, 389–93, 473
  dynamics, 440–2
  regulation, 456–7
  stability, 251
population biology, of plants, xxi–xxii, xxxi
pore, 190, 191, 194, 213, 214, 215, 218, 219, 224, 225, 226, 232, 233
  chamber, 214, 215
  plates, 218, 219
pore-canal system, 259, 260, 261, 270, 272, 274, 275
porus, 415
predation, 255, 266, 277, 290, 292, 310, 375, 389, 390, 391, 394, 429, 455, 480, 487, 500, 502–6, 512–14, 529, 540-3
primary free space, 473, 474
procladia, 427
prosicula, 415-19, 421, 425
protandry, 438–40
protective behaviour, 144–5, 150
protein, 228
  polymorphism, 163
proto-co-operation, 257
protocorallites, 156–8, 164, 166–7
protoplasmic streaming, 14, 16, 22
protopolyp, 157, 164–8, 180

pseudocolumella, 164
pseudohormones, 53
pseudo-stolons, 276

Q

quasi-coloniality, 156–7, 163

R

r-selection, 330–1
radial canal, 133
radial ribbing, 273, 276
radial septa, 176, 182, 183, 187
radial symmetry, 259, 260, 261, 262, 268
radula, 244, 245, 247, 249
ramet, xxvii, xxx
recruitment, 488, 502–4
  dance, 359
reef, 259
  distribution, 311–16
  formation, 247, 248, 282, 290, 305–9, *308*
regeneration, xxix, 48, 52, 91–3, *92*, 195, 217, 226–31, 537, 539
regimes in sponges, 51
rejuvenescence, 158, 163
reproduction, 52, 226, 229–31, 423, 481, 500, 503, 504–7, 515, 535; see also budding
  asexual, xxvii, xxx, 6, 13, 20, 34, 108, 109, 114, 143, 144, 156–8, 160, 162, 164–8, 178, 179, 189, 229, 231, 283, 286, 412, 430, 436, 504
  sexual, 34–6, 93–4, 230, 244, 251, 252, 253, 283, 324, 375–8, 383, 385, 388–91, 428, 438, 466, 504, 510, 531, 534–8, 543
reproductive rate, 286–7
respiration, 517
reticulum zooid, 213
rhabdosome, 397–400, 402, 408, 409, 411, 417, 418, 420, 422–7, 430
rhizocaulus, 132, 136
rhizoids, 212, 216, 218, 222, 227, 228, 229
rugophilic behaviour, 469

S

salinity, effect on serpulids, 292, 302–4
Santonian, 260, 274
scissiparity, see budding
scutum, 213, 216
scyphoriza, see stolonal plate
secondary free space, 472, 473, 474
sedimentation, 502, 503, 504, 506, 511, 514, 529, 534, 541
    avoidance, 246, 247, 265, 266, 268, 269, 274, 416
    removal, 188
selection, stabilising, 331
semi-social insects, 332, 347, 348
Senonian, 270, 274
septal carinae, 162
septotheca, 176, 177
septulum, 202, 206, 207, 214, 215, 219, 224
septum, 164, 182, 184, 186, 188, 189, 190, 226
    interzooidal, 212, 219, 224, 225, 232, 233
sessile habit, xxxiii, 82, 91
seta, 200, 213, 216, 222, 514
settlement, 255, 259, 260, 263, 288–91, 306, 309, 313–16, 321, 322–4, 376, 380, 428, 439, 459–61, 465–73, 477, 480–6, 488, 490, 503, 504, 506, 514
    patterns, 246, 248, 249, 250, 251, 252
    behaviour of serpulid larvae, 290-1
    stimulus, 322
settling plates, 313–14
shales, 429
shelly marls, 259
sicula, 400, 419, 420, 421, 422, 424, 425, 426, 427, 430
sicular aperture, 397
Silurian, 402, 425
siphonal bands, 273
size, xxxii, 85, 89, 98, 287
skeletal fusion, 179
skeletal structure, 49, 50, 69, 70, 71, 72, 73
skeleton, 508, 509, 510, 511, 519, 545
    in graptolites, 402, 404, 406, 408, 409, 411, 412, 423, 425
    in rugose corals, 176, 178, 179, 180, 184, 185, 187, 188, 192

social behaviour, 375, 376, 379, 380, 383, 386, 391, 392
social bivalves, 258
social breeding, obligatory, 456
social stimulation, 456
sociality, xvii
somatic cycle, 229–31
spatfall, 288–9, 465; see also settlement
spatial
    distribution, 461, 499, 533, 539, 542, 544
    predictability, 463, 465–8, 471, 475, 479–84, 488, 490
    refuges, 459–61, 463, 465–70, 472, 475, 480–4, 488, 490
special cells, 224, 225, 226
speciation, 544
sperm, 52, 197, 229, 243, 244, 252
spicules, 69, 72, 73, 143, 509
spines, 213, 214, 215, 216, 217, 514, 538
spirogyrate juvenile stage, 267, 268, 269, 273
sponge factors, 42, 44
spongin, 69, 72
spongozoa, 51
sporosac, 134
statoblasts, 227
statocyst, 109
stenotele, 116, 122, 127
stings, in social insect evolution, 352
stolon, 82, 115–16, *116*, 119, 121, 123, 128, 131–3, 136–7, 212, 225–8, 411, 412, 413, 415, 418, 426, 427, 428, 514, 546
stolonal plate, 89–90, 94, 98–9, 128, 130, *89*
stomach, 110, 119, 122, 124, 132–3
stomodaeum, 176, 177
stridulation, 352
strobilation, xxxii, 86–7, 91–4, 96, 99–100, 108
stromatolite, 1
structure
    bryozoan, 211–36
    graptolite, 412, 413
substrate
    stability, 506, 507, 512, 532, 540
    utilization, 304–5, 516

Subject Index

sub-social insects, 347, 348
supercolony, 348
survival, 459, 460, 474, 479, 482, 484, 499, 500, 502, 503, 506, 507, 510, 513, 542
    rates in sea-birds, 452–3
suspension feeding, 306–9
swarming, 13, 34
symbiosis, xxxii, 22, 95, 259, 356–7, 361, 508, 510
synchrony of reproduction, 283, 376, 385
synergists, 364, 366–7
synoecium, 202, 203
synrhabdosome, 427, 428

T

tabulae, 173, 186, 190
tata, 216
temperature, effect on aggregates, 292
temporal distribution, 461, 499, 533, 543, 544
temporal predictability, 463, 465, 466, 468, 470, 471, 475, 479, 480, 490
temporary aggregations, 32–3, 292–3
tentacle, 110–11, 114–20, *116*, 122–34, *127*, 142, 176, 177, 193, 195, 196, 197, 200, 220, 223
    crowns, 193, 195, 197, 208, 220, 221, 222, 223, 226, 236, 413, 414, 415, 416, 418, 421, 425, 514
    heteromorphy, 195
    number, 197
    pedal, 244, 245, 252
tentacular bulb, 111, 116, 124, 127, 132
territoriality, 323
Tethys, 259
theca, 397, 402, 403, 412, 413, 414, 415, 416, 418, 419, 420, 421, 422, 423, 424, 425, 426, 428, 430
thecal gradient, 419, 425
thermoregulation, 350, 357–60
thermosiphon effect, 359–60, *360*

toxins, 514
Tremadoc, 399, 400, 402, 416
trophic group, 310
tubercles, 213

U

ultrastructure, of graptolites, 397, 406, 408, 412, 415, 418
upright rhabdosome, 397, 398
Urgonian, 269, 270

V

Valanginian, 269
vascular systems, 435–6, 439–40
vegetative growth, 87, 94, 100
vertical distribution, 95–6, 313, 502, 503, 513, 526
vibracula, see seta
virgella, 421, 426, 427
virgula, see nema
vitello genesis, 226, 231

W

wall structure, 176, 194, 200, 202, 203, 204, 211, 219, 224
wave action, 247, 248, 252, 378, 386
web structures, 400, 402

X

xenomorphic growth, 263

Z

zigzag suture, 415, 416, 418
zooid, 33–6, 144–57, 193–209, 282, 435–8
    bilateral symmetry of, 435
    density, 536, 539
    shape, 233
zooxanthella, xxxii, 95
zygote, 331-3

The Systematics Association Publications

10. MODERN APPROACHES TO THE TAXONOMY OF RED AND BROWN ALGAE (1978)★
    Edited by D. E. G. IRVINE and J. H. PRICE
11. BIOLOGY AND SYSTEMATICS OF COLONIAL ORGANISMS (1979)★
    Edited by G. LARWOOD and B. R. ROSEN

★Published by Academic Press for the Systematics Association

The Systematics Association Publications

2. FUNCTION AND TAXONOMIC IMPORTANCE (1959)
   Edited by A. J. CAIN
3. THE SPECIES CONCEPT IN PALAEONTOLOGY (1956)
   Edited by P. C. SYLVESTER-BRADLEY
4. TAXONOMY AND GEOGRAPHY (1962)
   Edited by D. NICHOLS
5. SPECIATION IN THE SEA (1963)
   Edited by J. P. HARDING and N. TEBBLE
6. PHENETIC AND PHYLOGENETIC CLASSIFICATION (1964)
   Edited by V. H. HEYWOOD and J. MCNEILL
7. ASPECTS OF TETHYAN BIOGEOGRAPHY (1967)
   Edited by C. G. ADAMS and D. V. AGER
8. THE SOIL ECOSYSTEM (1969)
   Edited by J. SHEALS
9. ORGANISMS AND CONTINENTS THROUGH TIME (1973)†
   Edited by N. F. HUGHES

LONDON. Published by the Association

Systematics Association Special Volumes

1. THE NEW SYSTEMATICS (1940)
   Edited by JULIAN HUXLEY (Reprinted 1971)
2. CHEMOTAXONOMY AND SEROTAXONOMY (1968)*
   Edited by J. G. HAWKES
3. DATA PROCESSING IN BIOLOGY AND GEOLOGY (1971)*
   Edited by J. L. CUTBILL
4. SCANNING ELECTRON MICROSCOPY (1971)*
   Edited by V. H. HEYWOOD
5. TAXONOMY AND ECOLOGY (1973)*
   Edited by V. H. HEYWOOD
6. THE CHANGING FLORA AND FAUNA OF BRITAIN (1974)*
   Edited by D. L. HAWKSWORTH
7. BIOLOGICAL IDENTIFICATION WITH COMPUTERS (1975)*
   Edited by R. J. PANKHURST
8. LICHENOLOGY: PROGRESS AND PROBLEMS (1977)*
   Edited by D. H. BROWN, D. L. HAWKSWORTH and R. H. BAILEY
9. KEY WORKS (1978)*
   Edited by G. J. KERRICH, D. L. HAWKSWORTH and R.W. SIMS

*Published by Academic Press for the Systematics Association
†Published by the Palaeontological Association in conjunction with the Systematics Association